Abel Symposia

Volume 16

Series Editor

Norwegian Mathematical Society, Trondheim, Norway

More information about this series at https://link.springer.com/bookseries/7462

Sigbjørn Hervik • Boris Kruglikov • Irina Markina •
Dennis The
Editors

Geometry, Lie Theory and Applications

The Abel Symposium 2019

Springer

Editors
Sigbjørn Hervik
University of Stavanger
Stavanger, Norway

Boris Kruglikov
UiT - The Arctic University of Norway
Tromsø, Norway

Irina Markina
University of Bergen
Bergen, Norway

Dennis The
UiT - The Arctic University of Norway
Tromsø, Norway

ISSN 2193-2808 ISSN 2197-8549 (electronic)
Abel Symposia
ISBN 978-3-030-81295-9 ISBN 978-3-030-81296-6 (eBook)
https://doi.org/10.1007/978-3-030-81296-6

Mathematics Subject Classification: 53Bxx, 53C15, 53C30, 53A55, 53Z05, 22E46, 22E65, 22E70, 83C20, 58H05, 83E50

This Springer imprint is published by the registered company Springer Nature Switzerland AG.
The registered company address is: Gewerbestrasse 11, 6330 Cham, Switzerland

Participants of the Abel Symposium 2019

Preface

The ideas of symmetry and invariance, forming Lie theory, are central in many applications, in particular in differential geometry and mathematical physics. The Abel Symposium 2019 brought together leading experts from all around the world. They presented the state-of-the-art, new trends, and advances in their areas of research. Among them we mention:

- The equivalence problems of geometric structures
- Classification of invariant differential operators and structures with abundant symmetries in parabolic geometries
- Description of low-dimensional orbits of the action of Lie groups in tensor spaces and their projectivizations
- Finite generation of classical and differential invariants of infinite dimensional transformation groups
- Invariance methods in the integration of equations of mathematical physics
- Classification problems in general relativity
- Application of representation theory to the construction of compatibility complexes of overdetermined systems
- Prolongation of graded nilpotent Lie algebras with application to optimal control theory
- Integrability techniques in classical mechanics

The Abel Symposium took place on 24–28 June 2019 in Ålesund, which is often called the Norwegian Venice. The main organisers were:

- José Figueroa-O'Farrill, University of Edinburgh
- Sigbjørn Hervik, University of Stavanger
- Boris Kruglikov, UiT - The Arctic University of Norway
- Irina Markina, University of Bergen
- Jan Slovák, Masaryk University
- Dennis The, UiT - The Arctic University of Norway
- Bent Ørsted, Aarhus University

Leading researchers from 17 different countries working in these areas were invited to participate. The symposium hosted 47 participants in total, including six postdoctoral researchers. The daily program consisted of lectures and discussions held in an open and encouraging atmosphere.

The following 28 plenary lectures were given:

1. Ilka Agricola (Marburg): *Generalizations of 3-Sasakian manifolds and skew torsion*
2. Dmitri Alekseevsky (Moscow): *Homogeneous sub-Riemannian manifolds*
3. David Calderbank (Bath): *GL_2 geometry and quaternionic manifolds with symmetry*
4. Andreas Čap (Vienna): *The bundle of Weyl structures*
5. Michael Cowling (Sydney): *Lie groups as metric spaces*
6. Boris Doubrov (Minsk): *Extrinsic geometries and linear PDEs on filtered manifolds*
7. Maciej Dunajski (Cambridge): *Conformally isometric embeddings*
8. Michael Eastwood (Adelaide): *The parabolic geometry of a flying saucer*
9. Eugene Ferapontov (Loughborough): *On integrability of dispersionless Hirota type equations*
10. Anna Fino (Torino): *Closed G_2 structures and Laplacian flow*
11. Eduardo Garcia-Rio (Santiago de Compostela): *Conformally Einstein metrics in dimension four*
12. Peter Gilkey (Oregon): *Geodesic completeness of homogeneous affine surfaces*
13. Rod Gover (Auckland): *Distinguished curves and integrability in Riemannian, conformal, and projective geometry*
14. Erlend Grong (Paris-Sud & Bergen): *Curvature for sub-Riemannian manifolds*
15. Jun-Muk Hwang (Seoul): *Recognizing symplectic Grassmannians by their varieties of minimal rational tangents*
16. Andreas Juhl (Berlin): *On the structure of Branson–Gover and Q-curvature operators on differential forms*
17. Jorge Lauret (Cordoba): *Soliton geometric structures on Lie groups*
18. Thomas Leistner (Adelaide): *Semi-Riemannian cones and their holonomy*
19. Valentin Lychagin (Tromsø): *On equivalence of linear differential operators*
20. Karin Melnick (Maryland): *A D'Ambra Theorem in conformal Lorentzian geometry*
21. Hans Munthe-Kaas (Bergen): *A survey of Lie–Butcher theory*
22. Katharina Neusser (Brno): *Projective geometry of Sasaki–Einstein structures and their compactification*
23. Pawel Nurowski (Warsaw): *The parabolic geometry of a car*
24. Andrea Santi (Bologna): *Symmetry superalgebras in geometry and physics*
25. Martin Schlichenmaier (Luxembourg): *N-point Virsasoro algebras are multi-point Krichever–Novikov type algebras*
26. Vera Serganova (Berkeley): *Homological tensor functors for representations of supergroups and superalgebras*

27. Michael Singer (London): *Boundary value problems for Einstein metrics*
28. Lode Wylleman (Stavanger): *Invariantly-defined curvature properties and integrability in general relativity*

In addition to the above speakers (and organisers), the following invited guests also attended the symposium:

29. Matthias Hammerl (Vienna)
30. Denson Hill (Stony Brook)
31. Omid Makhmali (Warsaw)
32. David McNutt (Stavanger)
33. Paul de Medeiros (Stavanger)
34. Katja Sagerschnig (Warsaw)
35. Eivind Schneider (Tromsø)
36. Eldar Straume (Trondheim)
37. Cynthia Will (Cordoba)
38. Travis Willse (Vienna)
39. Henrik Winther (Brno)
40. Ilya Zakharevich (Berkeley)

We are grateful to all the participants for their valuable contributions and for making the symposium a successful event. We would also like to express our gratitude to the Abel Foundation, Trond Mohn Research Foundation, Tromsø Research Foundation, and Norwegian Mathematical Society for giving us the opportunity to arrange the Abel Symposium 2019. Moreover, we would like to thank the Scandic Parken Hotel in Ålesund for providing us practical assistance related to the hosting of the symposium.

Stavanger, Norway — Sigbjørn Hervik
Tromsø, Norway — Boris Kruglikov
Bergen, Norway — Irina Markina
Tromsø, Norway — Dennis The

30 June 2020

Contents

Four-Dimensional Homogeneous Generalizations of Einstein Metrics 1
E. Calviño-Louzao, E. García-Río, I. Gutiérrez-Rodríguez, and R. Vázquez-Lorenzo

Conformal and Isometric Embeddings of Gravitational Instantons 21
Maciej Dunajski and Paul Tod

Recent Results on Closed G_2-Structures 49
Anna Fino and Alberto Raffero

Almost Zoll Affine Surfaces 69
Peter B. Gilkey

Distinguished Curves and First Integrals on Poincaré–Einstein and Other Conformally Singular Geometries 79
A. Rod Gover and Daniel Snell

A Car as Parabolic Geometry 93
C. Denson Hill and Paweł Nurowski

Legendrian Cone Structures and Contact Prolongations 131
Jun-Muk Hwang

The Search for Solitons on Homogeneous Spaces 147
Jorge Lauret

On Ricci Negative Lie Groups 171
Jorge Lauret and Cynthia E. Will

Semi-Riemannian Cones 193
Thomas Leistner

Building New Einstein Spaces by Deforming Symmetric Einstein Spaces 223
David McNutt, Sigbjørn Hervik, and Eivind E. Dahl

Remarks on Highly Supersymmetric Backgrounds of 11-Dimensional Supergravity ... 253
Andrea Santi

Krichever–Novikov Type Algebras. A General Review and the Genus Zero Case ... 279
Martin Schlichenmaier

Four-Dimensional Homogeneous Generalizations of Einstein Metrics

E. Calviño-Louzao, E. García-Río, I. Gutiérrez-Rodríguez, and R. Vázquez-Lorenzo

Abstract We review some generalizations of Einstein metrics and focus on the four-dimensional homogeneous case to exhibit classification results and new examples.

1 Introduction

A Riemannian manifold is said to be Einstein if its Ricci curvature is proportional to the metric tensor. Einstein metrics, being critical for the Hilbert-Einstein functional, have had a privileged role not only in Differential Geometry but also in Mathematical Physics. Various generalizations of Einstein metrics are important. Since the Einstein condition is linear on the Ricci tensor one may consider at a first glance some different linear generalizations [24]. Manifolds with cyclic-parallel Ricci tensor as well as manifolds with vanishing Cotton tensor have been extensively investigated. Higher dimensional versions of the Gauss-Bonnet Theorem give rise to universal curvature identities, which lead to different weakly-Einstein conditions generalizing the Einstein one [17]. The existence of Einstein representatives of a given conformal class of a metric gives rise to conformal generalizations of the

E. Calviño-Louzao
Consellería de Cultura, Educación e, Universidade, Xunta de Galicia, Santiago de Compostela, Spain
e-mail: estebcl@edu.xunta.gal

E. García-Río (✉)
Faculty of Mathematics, University of Santiago de Compostela, Santiago de Compostela, Spain
e-mail: eduardo.garcia.rio@usc.es

I. Gutiérrez-Rodríguez
Department of Mathematics, University of Vigo, Ourense, Spain
e-mail: ixchel.dzohara.gutierrez.rodriguez@uvigo.es

R. Vázquez-Lorenzo
Department of Mathematics, IES de Ribadeo Dionisio Gamallo, Ribadeo, Spain
e-mail: ravazlor@edu.xunta.gal

S. Hervik et al. (eds.), *Geometry, Lie Theory and Applications*, Abel Symposia 16,
https://doi.org/10.1007/978-3-030-81296-6_1

Einstein condition [23]. Some analytical generalizations are provided by Ricci solitons, which naturally appear in the study of the Ricci flow, as well as some more general structures like the quasi-Einstein metrics.

Our purpose in this paper is to review some classification results for some kinds of generalizations of Einstein metrics in the homogeneous setting. Bérard-Bergery showed that complete and simply connected four-dimensional homogeneous Riemannian manifolds are either symmetric or isometric to a Lie group with a left-invariant metric [7]. Hence the classification of the different homogeneous generalizations of Einstein metrics naturally splits into two settings: the symmetric and the Lie group ones. Since any symmetric manifold splits as a product of Einstein ones, this case is usually straightforward. In opposition, the existence of generalized Einstein left-invariant metrics on Lie groups usually leads to systems of polynomial equations that are rather involved. The theory of Gröbner basis (see, for example [15]) provides a useful tool to solve the corresponding polynomial systems. We recall some known results and give explicit realizations of homogeneous conformally Einstein metrics and conformal C-spaces in dimension four. This provides non-trivial examples of homogeneous conformally Einstein manifolds and conformal C-spaces.

1.1 Linear Generalizations of Einstein Metrics

Let (M, g) be a Riemannian manifold and let ρ denote the corresponding Ricci tensor. Since the Ricci tensor of any Einstein metric is parallel, different linear generalizations of Einstein metrics have been considered in the literature after the seminal work of Gray [24]. A Riemannian manifold is said to be an *$\mathcal{A}$-space* (resp., a *$\mathcal{B}$-space*) if the Ricci tensor is cyclic-parallel (resp., the Ricci tensor is Codazzi). Type $\mathcal{A}$ manifolds naturally appear in the study of D'Atri spaces (i.e., Riemannian manifolds with volume-preserving geodesic symmetries), since the cyclic parallelizability condition is the first of the infinite sequence of curvature identities characterizing D'Atri spaces (see, for example [28]). Naturally reductive homogeneous spaces have cyclic-parallel Ricci tensor. The converse is true in dimension three due to the work of Pedersen and Tod [34] (see also [27]). The situation is different in dimension four, where homogeneous $\mathcal{A}$-spaces have been classified by Arias-Marco and Kowalski [4], who showed that if (M, g) is not locally symmetric, then it is locally isometric either to a Riemannian product $\mathbb{R} \times N$, where N is a naturally reductive homogeneous 3-dimensional space with only two distinct Ricci curvatures, or to a four-dimensional Lie group G with a left-invariant metric, whose Lie algebra $\mathfrak{g}$ is explicitly given. Four-dimensional Type $\mathcal{B}$ Riemannian manifolds (or equivalently Riemannian manifolds with harmonic curvature) are known to be symmetric in the homogeneous case [36].

Let $\mathcal{S} = \frac{1}{n-2}(\rho - \frac{\tau}{2(n-1)}g)$ be the Schouten tensor of an n-dimensional Riemannian manifold, where τ denotes the scalar curvature. The Schouten tensor is cyclic-parallel if and only if so is the Ricci tensor. Riemannian manifolds whose

Schouten tensor is Codazzi are, however, a larger class than that of $\mathcal{B}$-spaces. The *Cotton tensor* $C(X, Y, Z) = (n-2)\{(\nabla_X \mathcal{S})(Y, Z) - (\nabla_Y \mathcal{S})(X, Z)\}$, as a measure of the failure of the Schouten tensor to be Codazzi, satisfies $\operatorname{div} W = -\frac{n-3}{n-2}C$, where W denotes the Weyl conformal curvature tensor. Hence the Schouten tensor of a Riemannian manifold is Codazzi if and only if it is Cotton-flat (equivalently, the Weyl tensor is harmonic). Clearly the Schouten tensor of any Einstein metric is parallel and thus any Einstein metric is Cotton-flat. Four-dimensional homogeneous Cotton-flat metrics are symmetric [36] (see [26] for the special case of homogeneous Einstein metrics).

1.2 Weakly-Einstein Conditions

Let M be a compact surface. The Gauss-Bonnet Theorem shows that the functional $g \mapsto \int_M \tau\, vol_g$ is constant on the space of metrics and thus the Ricci tensor of any two-dimensional Riemannian manifold (M, g) satisfies the universal curvature identity $\rho = \frac{1}{2}\tau g$. Generalizations of the Gauss-Bonnet Theorem to higher dimensions provide new universal curvature identities. Indeed, the functional defined by the Gauss-Bonnet integrand $g \mapsto \int_M \{\|R\|^2 - 4\|\rho\|^2 + \tau^2\}\, vol_g$ is constant in dimension four. Hence any compact four-dimensional Riemannian manifold is critical and Berger [8] showed that one has $\check{R} - 2\check{\rho} - 2R[\rho] + \tau\rho = \frac{1}{4}\left(\|R\|^2 - 4\|\rho\|^2 + \tau^2\right) g$. Previous identity can be equivalently written as

$$\left(\check{R} - \frac{\|R\|^2}{4} g\right) + \tau\left(\rho - \frac{\tau}{4} g\right) - 2\left(\check{\rho} - \frac{\|\rho\|^2}{4} g\right) - 2\left(R[\rho] - \frac{\|\rho\|^2}{4} g\right) = 0, \tag{1}$$

where $\check{R}$, $\check{\rho}$ and $R[\rho]$ are the symmetric $(0,2)$-tensor fields given by $\check{R}_{ij} = R_{iabc} R_j{}^{abc}$ $\check{\rho}_{ij} = \rho_{ia}\rho^a{}_j$ and $R[\rho]_{ij} = R_{iabj}\rho^{ab}$. The tensor fields $\check{R}$, $\check{\rho}$ and $R[\rho]$ provide natural Riemannian invariants, algebraically the simplest ones after the Ricci tensor. However it does not seem to have received much attention in the literature.

If (M, g) is Einstein, then all terms in (1) vanish. It is thus a natural question to investigate the tensor fields $\check{R}$, $\check{\rho}$ and $R[\rho]$ separately, in particular whether they are a multiple of the metric tensor. A Riemannian manifold (M, g) is said to be *$\check{R}$-Einstein* if $\check{R} = \frac{1}{4}\|R\|^2 g$. (M, g) is said to be *$\check{\rho}$-Einstein* (resp., *$R[\rho]$-Einstein*) if $\check{\rho} = \frac{1}{4}\|\rho\|^2 g$ (resp., $R[\rho] = \frac{1}{4}\|\rho\|^2 g$). Note that, in contrast with the Einstein condition, any of the conditions above do not imply that $\|R\|^2$ (or $\|\rho\|^2$) be constant. Generically, we say that a four-dimensional Riemannian manifold (M, g) is *weakly-Einstein* if it satisfies any of the previous conditions but (M, g) is not Einstein.

1.3 Conformal Generalizations of Einstein Metrics

From a conformal point of view, one says that a Riemannian manifold (M, g) is *conformally Einstein* if there is an Einstein representative of the conformal class $[g]$. Conformally Einstein metrics are conformally Cotton, but the converse is not true.

The necessary and sufficient conditions for a metric to be conformally Einstein were established by Brinkmann in terms of the existence of positive solutions to the differential equation

$$(n-2)\operatorname{Hes}_\varphi + \varphi\,\rho - \frac{1}{n}\{(n-2)\Delta\varphi + \varphi\,\tau\}g = 0\,, \tag{2}$$

where $\operatorname{Hes}_\varphi = \nabla d\varphi$ is the Hessian tensor and $\Delta\varphi = \operatorname{tr}_g \operatorname{Hes}_\varphi$ denotes the Laplacian. Besides its apparent simplicity, the integration of Eq. (2) is surprisingly difficult. It was shown in [30] that any four-dimensional conformally Einstein manifold satisfies

$$(i)\ \ \mathfrak{B} = 0, \qquad\qquad (ii)\ \ C(\cdot,\cdot,\cdot) + W(\cdot,\cdot,\cdot,\nabla\sigma) = 0, \tag{3}$$

where the conformal Einstein metric is given by $\widetilde{g} = e^{2\sigma} g$, $\sigma = -2\log\varphi$, and $\mathfrak{B} = \operatorname{div}_1 \operatorname{div}_4 W + \frac{n-3}{n-2} W[\rho]$ is the Bach tensor. Conditions in Eq. (3) are also sufficient to be conformally Einstein if (M, g) is *weakly-generic* (i.e., the Weyl tensor viewed as a map $TM \to \bigotimes^3 TM$ is injective).

A Riemannian manifold (M, g) is said to be *conformally Cotton* if there is a Cotton-flat representative of the conformal class $[g]$. This is achieved if there is a smooth function such that Eq. (3)-(ii) is satisfied. More generally, (M, g) is said to be a *conformal C-space* if there is a (not necessarily gradient) vector field ξ on M so that $C + i_\xi W = 0$ (see [23]). It was shown in [22] that a Riemannian four-manifold with non-vanishing Weyl tensor is a conformal C-space if and only if

$$\|W\|^2 C_{kji} - 4W^{dabc} C_{cba} W_{dijk} = 0.$$

It was shown in [23] that any compact conformal C-space is locally conformally Cotton-flat and moreover any compact conformal C-space is conformally Einstein if and only if it is Bach-flat, independently of any weakly-generic assumption.

1.4 Ricci Solitons

A Riemannian manifold (M, g) is called a *Ricci soliton* if $\mathcal{L}_X g + \rho = \lambda g$ for some $\lambda \in \mathbb{R}$ and some vector field X on M. These metrics are natural generalizations of Einstein metrics and correspond to self-similar solutions of the Ricci flow $\frac{\partial}{\partial t} g = -2\rho$. A Ricci soliton (M, g, X) is called *shrinking*, *steady* or *expanding* if $\lambda > 0$, $\lambda = 0$ or $\lambda < 0$, respectively. Moreover, a Ricci soliton is called *trivial* if (M, g) is Einstein.

Homogeneous gradient Ricci solitons (i.e., the vector field X is the gradient of a potential function) are products $\mathbb{R}^k \times N$, where N is an Einstein manifold and the potential function is given by the projection on the Euclidean factor [35].

Ricci solitons are self-similar solutions of the Ricci flow, i.e., they are fixed points of the flow up to diffeomorphisms and rescaling. On a Lie group one may consider a stronger condition and search for fixed points of the flow up to automorphisms of the Lie group instead of diffeomorphisms. This observation led Lauret [32] to introduce algebraic Ricci solitons. Let G be a Lie group with Lie algebra $\mathfrak{g}$. A left-invariant metric $\langle \cdot, \cdot \rangle$ on G is called an *algebraic Ricci soliton* if

$$\mathfrak{D} = \mathrm{Ric} - \lambda \,\mathrm{Id} \tag{4}$$

is a derivation of the Lie algebra, where Ric denotes the Ricci operator $\langle \mathrm{Ric}\, X, Y\rangle = \rho(X, Y)$ and $\lambda \in \mathbb{R}$.

Let $\Delta_X u = \Delta u - g(X, \nabla u)$ be the X-Laplacian on a Ricci soliton structure (M, g, X) (see, for example [14]). Then $\frac{1}{2}\Delta_X \tau = \lambda\tau - \|\rho\|^2$, which shows that a steady Ricci soliton with constant scalar curvature is Ricci-flat, and hence flat in the homogeneous setting [2]. Furthermore, four-dimensional homogeneous shrinking Ricci solitons have bounded curvature and thus they are gradient [33] (see also [5]). Non-gradient homogeneous expanding Ricci solitons are isometric to a solvsoliton in the four-dimensional case [5, 25].

2 Homogeneous Four-Manifolds with Cyclic-Parallel Ricci Tensor

(M, g) has cyclic-parallel Ricci tensor if $\mathfrak{C}(X, Y, Z) = (\nabla_X \rho)(Y, Z) + (\nabla_Y \rho)(Z, X) + (\nabla_Z \rho)(X, Y) = 0$ for all vector fields X, Y, Z on M. The following result has been essentially proven as a sequence of works by Podestà and Spiro [36] and Arias-Marco and Kowalski [4], providing a complete description in the four-dimensional homogenous setting. We provide a slightly different proof which also corrects some inaccuracies in [4]. Further note that since the cyclic parallelizability of the Ricci tensor is invariant by homotheties, we work at the homothetic level to simplify the exposition.

Theorem 1 *Let (M, g) be a complete and simply connected non-symmetric four-dimensional homogeneous manifold with cyclic-parallel Ricci tensor. Then one of the following holds:*

(a) (M, g) is a four-dimensional naturally reductive homogeneous space.
(b) (M, g) is homothetic to the Lie group $\mathbb{R} \ltimes H^3$ with left-invariant metric determined by the Lie algebra

$$[e_1, e_2] = e_3, \quad [e_1, e_4] = -\frac{1}{2}e_1, \quad [e_2, e_4] = \frac{1}{2}e_2,$$

where $\{e_1, \dots, e_4\}$ is an orthonormal basis.

(c) (M,g) *is homothetic to the Lie group* $\mathbb{R} \ltimes \mathbb{R}^3$ *with left-invariant metric determined by the Lie algebra*

$$[e_1,e_4] = -4(e_2+e_3), \quad [e_2,e_4] = 4e_1+3e_2-e_3, \quad [e_3,e_4] = 4e_1+e_2-3e_3,$$

where $\{e_1,\dots,e_4\}$ *is an orthonormal basis.*

(d) (M,g) *is homothetic to the Lie group* $\widetilde{SL(2,\mathbb{R})}\times\mathbb{R}$ *or to the Lie group* $SU(2)\times\mathbb{R}$ *with left-invariant metric determined by the Lie algebra*

$$\begin{array}{lll} [e_1,e_2] = \frac{2(\alpha^2-\alpha+1)}{\alpha+1}e_3, & [e_2,e_3] = e_1, & [e_3,e_1] = \alpha e_2, \\ [e_1,e_4] = \alpha\sqrt{\frac{(\alpha-2)(2\alpha-1)}{(\alpha+1)^2}}e_2, & [e_2,e_4] = -\sqrt{\frac{(\alpha-2)(2\alpha-1)}{(\alpha+1)^2}}e_1, & \end{array}$$

where $\{e_1,\dots,e_4\}$ *is an orthonormal basis and* $\alpha \in (-\infty,\frac{1}{2})\cup(2,+\infty)$, $\alpha \neq -1, 0$.

Remark 1 Left-invariant metrics corresponding to Theorem 1-(b) have exactly two-distinct Ricci curvatures $\varrho_1 = \varrho_2 = \varrho_3 = -\frac{1}{2}$, $\varrho_4 = \frac{1}{2}$. Metrics in Theorem 1-(c) have exactly four-distinct Ricci curvatures given by $\varrho_1 = 3(1+\sqrt{33})$, $\varrho_2 = 3(1-\sqrt{33})$, $\varrho_3 = -18$ and $\varrho_4 = -6$. On the other hand metrics corresponding to case (d) have exactly four-distinct Ricci curvatures unless $\alpha = 3$ or $\alpha = \frac{1}{3}$ in which cases there are three-distinct Ricci curvatures. Furthermore the scalar curvature is positive in the interval $(3-2\sqrt{2}, 3+2\sqrt{2})$, it vanishes at the boundary and is negative otherwise.

Proof By a result of Bérard-Bergery [7], a complete and simply connected homogeneous four-manifold is either symmetric or a Lie group. In particular, either (M,g) is isometric to one of the groups $\widetilde{SL(2,\mathbb{R})}\times\mathbb{R}$, $SU(2)\times\mathbb{R}$ or it is a solvable Lie group whose Lie algebra is a one-dimensional extension of the three-dimensional unimodular Lie algebras: the abelian Lie algebra $\mathfrak{r}^3$, the Heisenberg algebra $\mathfrak{h}^3$, the Poincaré algebra $\mathfrak{e}(1,1)$ and the Euclidean algebra $\mathfrak{e}(2)$. Note that, instead of varying the metric we fix an orthonormal basis $\{e_1,\dots,e_4\}$ and vary the brackets to express the corresponding left-invariant metrics. Next we analyze each case separately.

- **Left-invariant metrics on** $\mathbb{R}\ltimes H^3$. Following [11], there exists an orthonormal basis $\{e_1,\dots,e_4\}$ of $\mathbb{R}\oplus\mathfrak{h}^3$ such that

$$\begin{array}{ll} [e_1,e_2] = \gamma e_3, & [e_4,e_1] = \frac{1}{R}\{ae_1 - ce_2 + (h+k_2\gamma)e_3\}, \\ [e_4,e_3] = \frac{1}{R}(a+d)e_3, & [e_4,e_2] = \frac{1}{R}\{ce_1 + de_2 + (f-k_1\gamma)e_3\}, \end{array} \tag{5}$$

where $R>0$ and $\gamma\neq 0$. In order to simplify the expressions we use the notation $F = f - k_1\gamma$ and $H = h + k_2\gamma$. A long but straightforward calculation shows that the cyclic-parallel condition, considering the structure constants in Eq. (5), is equivalent to the vanishing of the polynomials $\widetilde{\mathfrak{C}}_{ijk}$ given by

$$\begin{aligned}
&\widetilde{\mathfrak{C}}_{111} = R^2\mathfrak{C}_{111} = 3aF\gamma, \quad \widetilde{\mathfrak{C}}_{112} = R^2\mathfrak{C}_{112} = -aH\gamma,\\
&\widetilde{\mathfrak{C}}_{113} = R^2\mathfrak{C}_{113} = 2(d-a)c\gamma,\\
&\widetilde{\mathfrak{C}}_{222} = R^2\mathfrak{C}_{222} = -3dH\gamma, \quad \widetilde{\mathfrak{C}}_{212} = R^2\mathfrak{C}_{212} = dF\gamma,\\
&\widetilde{\mathfrak{C}}_{213} = R^2\mathfrak{C}_{213} = 2(a^2-d^2)\gamma,\\
&\widetilde{\mathfrak{C}}_{313} = R^2\mathfrak{C}_{313} = -((2a+d)F+cH)\gamma,\\
&\widetilde{\mathfrak{C}}_{323} = R^2\mathfrak{C}_{323} = -(cF-(a+2d)H)\gamma,\\
&\widetilde{\mathfrak{C}}_{114} = R^3\mathfrak{C}_{114} = aF^2 - cFH - (2c^2-4d^2+R^2\gamma^2)a + 2c^2d,\\
&\widetilde{\mathfrak{C}}_{214} = 2R^3\mathfrak{C}_{214} = -c(F^2-H^2) - (a+d)FH + 2(a^2-d^2)c,\\
&\widetilde{\mathfrak{C}}_{224} = R^3\mathfrak{C}_{224} = dH^2 + cFH + 2ac^2 + 4a^2d - (2c^2+R^2\gamma^2)d,\\
&\widetilde{\mathfrak{C}}_{314} = 2R^3\mathfrak{C}_{314} = -(3a-d)cF - (4a^2+c^2-d^2+8ad)H,\\
&\widetilde{\mathfrak{C}}_{324} = 2R^3\mathfrak{C}_{324} = (a^2-c^2-4d^2-8ad)F - (a-3d)cH,
\end{aligned}$$

where $\mathfrak{C}_{ijk} = \mathfrak{C}(e_i, e_j, e_k)$. Note that since the structure constant of $\mathfrak{h}^3$ satisfies $\gamma \neq 0$, one may work with a homothetic basis $\widetilde{e}_k = \frac{1}{\gamma}e_k$ so that we may assume $\gamma = 1$ in the rest of the proof. We analyze by separate the cases $a = 0$ and $a \neq 0$.

Case (1): $a = 0$ In this case, $\widetilde{\mathfrak{C}}_{213} = -2d^2$ implies $d = 0$. Now, $\widetilde{\mathfrak{C}}_{313} = -cH$ and $\widetilde{\mathfrak{C}}_{323} = -cF$. Hence, either $c = 0$ or $c \neq 0$ and $F = H = 0$, and in both cases we get a cyclic-parallel manifold. If $c = 0$, then the Ricci operator has eigenvalues $\{0, \varrho_1 = \varrho_2 = -\frac{1}{2R^2}(F^2 + H^2 + R^2), \varrho_3 = -\varrho_1\}$, while if $c \neq 0$ and $F = H = 0$ the eigenvalues of the Ricci operator are given by $\{0, \varrho_1 = \varrho_2 = -\frac{1}{2}, \varrho_3 = -\varrho_1\}$. Furthermore since the eigenspace corresponding to the kernel of the Ricci operator is parallel, we conclude that, in both cases, the manifold splits as a product of a real line and a naturally reductive homogeneous three-dimensional manifold. Hence (M, g) is naturally reductive (see [29]).

Case (2): $a \neq 0$ The expressions for $\widetilde{\mathfrak{C}}_{111}$, $\widetilde{\mathfrak{C}}_{112}$ and $\widetilde{\mathfrak{C}}_{113}$ imply that $F = H = 0$ and either $d = a$ or $d \neq a$ and $c = 0$.

If $F = H = 0$ and $d = a$, then the cyclic-parallel condition $\{\widetilde{\mathfrak{C}}_{ijk} = 0\}$ reduces to $4a^3 - aR^2 = 0$. Since $a \neq 0$, we get $a = \frac{\varepsilon}{2}R$, where $\varepsilon^2 = 1$. In this case, a straightforward calculation shows that the manifold is locally symmetric.

If $F = H = c = 0$ and $d \neq a$, then $\{\widetilde{\mathfrak{C}}_{ijk} = 0\}$ reduces to

$$\widetilde{\mathfrak{C}}_{114} = (4d^2 - R^2)a, \quad \widetilde{\mathfrak{C}}_{213} = 2(a^2 - d^2), \quad \widetilde{\mathfrak{C}}_{224} = (4a^2 - R^2)d\,.$$

Since $a \neq 0$ and $d \neq a$, it follows that the manifold is cyclic-parallel for $a = \frac{\varepsilon}{2}R$ and $d = -a$, where $\varepsilon^2 = 1$. Finally, observe that a substitution of e_4 by $-e_4$ is an isometry which interchanges $\varepsilon = 1$ by $\varepsilon = -1$. It corresponds to Theorem 1-(b).

- **Left-invariant metrics on $\mathbb{R} \ltimes \mathbb{R}^3$.** All left-invariant metrics on $\mathbb{R} \ltimes \mathbb{R}^3$ are described by the Lie brackets (see [11])

$$[e_4, e_1] = \tfrac{1}{R}\{ae_1 + be_2 + ce_3\}, \qquad [e_4, e_2] = \tfrac{1}{R}\{-be_1 + fe_2 + he_3\},$$
$$[e_4, e_3] = \tfrac{1}{R}\{-ce_1 - he_2 + pe_3\}, \tag{6}$$

where $R > 0$, and $\{e_1, \ldots, e_4\}$ is an orthonormal basis of $\mathbb{R} \oplus \mathfrak{r}^3$. A long but straightforward calculation shows that the cyclic-parallel condition, considering the structure constants in Eq. (6), is equivalent to the vanishing of the polynomials $\widetilde{\mathfrak{C}}_{ijk}$ given by

$$\begin{aligned}
\widetilde{\mathfrak{C}}_{114} &= -2(f+p)a^2 - 2(b^2+c^2-f^2-p^2)a + 2(b^2f + c^2p),\\
\widetilde{\mathfrak{C}}_{214} &= -(bp+ch)a + bfp - (f-2p)ch,\\
\widetilde{\mathfrak{C}}_{224} &= 2fa^2 + 2(b^2-f^2)a - 2b^2f - 2(f-p)(h^2+fp),\\
\widetilde{\mathfrak{C}}_{314} &= (bh-cf)a - (2f-p)bh + cfp,\\
\widetilde{\mathfrak{C}}_{324} &= (2bc-(f-p)h)a - (f+p)bc,
\end{aligned}$$

where $\widetilde{\mathfrak{C}}_{ijk} = R^3\mathfrak{C}(e_i, e_j, e_k)$. We consider the diagonal matrix diag$[a, f, p]$ in the decomposition of elements of der$(\mathfrak{r}^3)$ and we analyze by separate the cases of null and non-null determinant.

Case (1): $afp \neq 0$ Without loss of generality, one may work with a homothetic basis $\widetilde{e}_k = \frac{1}{a}e_k$ so that we may assume $a = 1$. Let $\mathcal{I} \subset \mathbb{R}[c, b, f, h, p]$ be the ideal generated by the polynomials $\widetilde{\mathfrak{C}}_{ijk}$. We compute a Gröbner basis $\mathcal{G}$ of $\mathcal{I}$ with respect to the lexicographical order and a detailed analysis of the Gröbner basis shows that the polynomials

$$\mathbf{g} = fp(p-1)(p^2+p+1) \quad \text{and} \quad \mathbf{g}' = fp(f-p^2)$$

belong to $\mathcal{G}$. Since $fp \neq 0$, it follows that necessarily $f = p = 1$ and, in such a case, the manifold is cyclic-parallel and locally symmetric.

Case (2): $afp = 0$ In this case at least one of a, f and p must be zero. Thus, without loss of generality, we may assume $a = 0$. Considering $\widetilde{\mathfrak{C}}_{324} = -(f+p)bc$ we are led to the following cases:

$$(i)\ a = b = 0, \qquad (ii)\ a = 0,\ b \neq 0,\ c = 0, \qquad (iii)\ a = 0,\ bc \neq 0,\ p = -f\,.$$

Case (2.i): $a = b = 0$ In this case, $\widetilde{\mathfrak{C}}_{114} = 2c^2p$. Hence, either $c = 0$ or $c \neq 0$ and $p = 0$. Suppose first that $c = 0$. Then the cyclic-parallel condition $\{\widetilde{\mathfrak{C}}_{ijk} = 0\}$ reduces to $\widetilde{\mathfrak{C}}_{224} = -2(f-p)(h^2+fp) = 0$. If $p = f$ or $fp = h = 0$ then the manifold is locally symmetric. The last possibility corresponds to $fp \neq 0$ and $f = -\frac{h^2}{p}$. A straightforward calculation shows that, in this latter case, the Ricci operator has eigenvalues $\{0, \varrho_1 = \varrho_2 = -\frac{h^4+p^4}{p^2R^2}, \varrho_3 = \frac{2h^2}{R^2}\}$ and hence the manifold is locally isometric to a naturally reductive homogeneous space (Theorem 1-(a)) since the kernel of the Ricci operator is a parallel line field.

If $c \neq 0$ and $p = 0$, then the condition $\{\widetilde{\mathfrak{C}}_{ijk} = 0\}$ reduces to $\widetilde{\mathfrak{C}}_{214} = -cfh = 0$ and $\widetilde{\mathfrak{C}}_{224} = -2fh^2 = 0$. Hence $fh = 0$ and the manifold is locally symmetric.

Case (2.ii): $a = 0$, $b \neq 0$, $c = 0$ In this case, $\widetilde{\mathfrak{C}}_{114} = 2b^2 f$ and therefore $f = 0$. Now, $\{\widetilde{\mathfrak{C}}_{ijk} = 0\}$ reduces to $\widetilde{\mathfrak{C}}_{224} = 2h^2 p = 0$ and $\widetilde{\mathfrak{C}}_{314} = bhp = 0$. Therefore $hp = 0$ and, in any case, the manifold is locally symmetric.

Case (2.iii): $a = 0$, $bc \neq 0$, $p = -f$ In this case, $\widetilde{\mathfrak{C}}_{114} = 2(b^2 - c^2)f$. If $f = 0$ then the manifold is locally symmetric. If $f \neq 0$ and $c = \varepsilon_1 b$, with $\varepsilon_1^2 = 1$, then the cyclic-parallel condition reduces to $\widetilde{\mathfrak{C}}_{214} = \widetilde{\mathfrak{C}}_{224} = \widetilde{\mathfrak{C}}_{314} = 0$, where

$$\widetilde{\mathfrak{C}}_{214} = -(f + 3\varepsilon_1 h)bf, \quad \widetilde{\mathfrak{C}}_{224} = -2(b^2 - 2f^2 + 2h^2)f, \quad \widetilde{\mathfrak{C}}_{314} = -(3h + \varepsilon_1 f)bf.$$

Thus, we get $f = -3\varepsilon_1 h$ and $b = 4\varepsilon_2 h$, where $\varepsilon_2^2 = 1$. Hence the left-invariant metric is isometric to one of the following

$$[e_1, e_4] = -\tfrac{4\varepsilon_2 h}{R}(e_2 + \varepsilon_1 e_3), \qquad [e_2, e_4] = \tfrac{h}{R}(4\varepsilon_2 e_1 + 3\varepsilon_1 e_2 - e_3),$$
$$[e_3, e_4] = \tfrac{\varepsilon_1 h}{R}(4\varepsilon_2 e_1 + \varepsilon_1 e_2 - 3e_3)\,.$$

Now one has that the left-invariant metrics given by $(\varepsilon_1, \varepsilon_2) = (1, 1)$ and $(\varepsilon_1, \varepsilon_2) = (1, -1)$ are isometric considering the isometry obtained by replacing e_1 by $-e_1$. In an analogous way, $e_2 \mapsto -e_3$ and $e_3 \mapsto e_2$ defines an isometry between the metrics corresponding to $(\varepsilon_1, \varepsilon_2) = (1, 1)$ and $(\varepsilon_1, \varepsilon_2) = (-1, -1)$. Furthermore, the replacement $e_2 \mapsto e_3$ and $e_3 \mapsto -e_2$ defines an isometry between the cases $(\varepsilon_1, \varepsilon_2) = (1, 1)$ and $(\varepsilon_1, \varepsilon_2) = (-1, 1)$. Hence all left-invariant metrics above are homothetic to the one in Theorem 1-(c).

- **Left-invariant metrics on $\mathbb{R} \ltimes E(1, 1)$ and $\mathbb{R} \ltimes E(2)$.**

  It follows from the work in [11] that there exists an orthonormal basis $\{e_1, \ldots, e_4\}$ of $\mathbb{R} \oplus \mathfrak{e}(1, 1)$ or $\mathbb{R} \oplus \mathfrak{e}(2)$ such that

$$\begin{aligned} &[e_2, e_3] = \lambda_1 e_1, && [e_3, e_1] = \lambda_2 e_2, \\ &[e_4, e_1] = \tfrac{1}{R}\{b e_1 - \lambda_2(\tfrac{a}{\lambda_1} + k_3)e_2\}, && [e_4, e_2] = \tfrac{1}{R}\{(a + k_3\lambda_1)e_1 + b e_2\}, \\ &[e_4, e_3] = \tfrac{1}{R}\{(c - k_2\lambda_1)e_1 + (d + k_1\lambda_2)e_2\}, && \end{aligned} \tag{7}$$

  where $R > 0$ and $\lambda_1\lambda_2 \neq 0$. Moreover, it corresponds to $\mathbb{R} \oplus \mathfrak{e}(2)$ (resp., $\mathbb{R} \oplus \mathfrak{e}(1, 1)$) if $\lambda_1\lambda_2 > 0$ (resp., $\lambda_1\lambda_2 < 0$). In order to simplify the expressions, we use the notation $A = \frac{a}{\lambda_1} + k_3$, $C = c - k_2\lambda_1$ and $D = d + k_1\lambda_2$. A long but straightforward calculation shows that the cyclic-parallel condition, considering the structure constants in Eq. (7), is equivalent to the vanishing of the polynomials $\widetilde{\mathfrak{C}}_{ijk}$ given by

$$\begin{aligned}
\widetilde{\mathfrak{C}}_{111} &= R^2\mathfrak{C}_{111} = -3bD\lambda_2, \qquad \widetilde{\mathfrak{C}}_{112} = R^2\mathfrak{C}_{112} = -bC(2\lambda_1 - 3\lambda_2),\\
\widetilde{\mathfrak{C}}_{222} &= R^2\mathfrak{C}_{222} = 3bC\lambda_1, \qquad \widetilde{\mathfrak{C}}_{212} = R^2\mathfrak{C}_{212} = -bD(3\lambda_1 - 2\lambda_2),\\
\widetilde{\mathfrak{C}}_{113} &= R^2\mathfrak{C}_{113} = -Ab(\lambda_2^2 - \lambda_1^2) - CD(\lambda_1 + \lambda_2),\\
\widetilde{\mathfrak{C}}_{114} &= R^3\mathfrak{C}_{114} = bR^2(\lambda_1^2 - \lambda_2^2) + ACD(\lambda_1 + \lambda_2) - b(C^2 - D^2),\\
\widetilde{\mathfrak{C}}_{213} &= R^2\mathfrak{C}_{213} = 2(A^2 + R^2)\lambda_1^2\lambda_2 - (2b^2 - C^2)\lambda_2\\
&\qquad - (2(A^2 + R^2)\lambda_2^2 - 2b^2 + D^2)\lambda_1,\\
\widetilde{\mathfrak{C}}_{214} &= R^3\mathfrak{C}_{214} = -2A(A^2 + R^2)\lambda_1^2\lambda_2 + A(2b^2 - C^2)\lambda_2\\
&\qquad + (2A(A^2 + R^2)\lambda_2^2 - A(2b^2 - D^2))\lambda_1 - 2bCD,\\
\widetilde{\mathfrak{C}}_{313} &= R^2\mathfrak{C}_{313} = -AC\lambda_2^2 + (2AC\lambda_2 + 3bD)\lambda_1,\\
\widetilde{\mathfrak{C}}_{314} &= 2R^3\mathfrak{C}_{314} = C(A^2 - R^2)\lambda_2^2 + AbD\lambda_2\\
&\qquad - (2(A^2 - R^2)C\lambda_2 + 3AbD)\lambda_1 + b^2C,\\
\widetilde{\mathfrak{C}}_{323} &= R^2\mathfrak{C}_{323} = -AD(\lambda_1^2 - 2\lambda_1\lambda_2) - 3bC\lambda_2,\\
\widetilde{\mathfrak{C}}_{324} &= 2R^3\mathfrak{C}_{324} = (A^2 - R^2)D\lambda_1^2 - (2(A^2 - R^2)D\lambda_2 + AbC)\lambda_1\\
&\qquad + 3AbC\lambda_2 + b^2D,
\end{aligned}$$

where $\mathfrak{C}_{ijk} = \mathfrak{C}(e_i, e_j, e_k)$. Note that since the structure constants of $\mathfrak{e}(1, 1)$ or $\mathfrak{e}(2)$ satisfy $\lambda_1\lambda_2 \neq 0$, one may work with a homothetic basis $\widetilde{e}_k = \frac{1}{\lambda_1}e_k$ so that we may assume $\lambda_1 = 1$ in the rest of the proof. Considering the expressions of $\widetilde{\mathfrak{C}}_{111}$ and $\widetilde{\mathfrak{C}}_{222}$, we analyze the following cases by separate:

$$(1)\ b = 0, \qquad (2)\ b \neq 0, C = D = 0\,.$$

Case (1): $b = 0$ In this case, $\widetilde{\mathfrak{C}}_{113} = -CD(\lambda_2 + 1)$, so we are led to the cases:

$$(i)\ b = 0,\ C = 0 \qquad (ii)\ b = 0,\ C \neq 0,\ D = 0, \qquad (iii)\ b = 0,\ CD \neq 0,\ \lambda_2 = -1\,.$$

Case (1.i): $b = 0$, $C = 0$ Note that, in this case, $\widetilde{\mathfrak{C}}_{323} = AD(2\lambda_2 - 1)$ and $\widetilde{\mathfrak{C}}_{324} = -(A^2 - R^2)D(2\lambda_2 - 1)$. Hence, either $D = 0$ or $\lambda_2 = \frac{1}{2}$. If $D = 0$ then the cyclic-parallel condition $\{\tilde{\mathfrak{C}}_{ijk} = 0\}$ reduces to $\lambda_2 = 1$ and the manifold is locally symmetric. If $D \neq 0$ and $\lambda_2 = \frac{1}{2}$ then the condition $\{\tilde{\mathfrak{C}}_{ijk} = 0\}$ reduces to $A^2 - 2D^2 + R^2 = 0$. Hence, $D = \frac{\varepsilon}{\sqrt{2}}\sqrt{A^2 + R^2}$, with $\varepsilon^2 = 1$. The replacement defined by $e_1 \mapsto -e_1$ and $e_2 \mapsto -e_2$ defines an isometry between the two cases corresponding to $\varepsilon = \pm 1$. Hence the corresponding left-invariant metric is isometric to one of the following

$$\begin{aligned}
&[e_2, e_3] = e_1, && [e_3, e_1] = \tfrac{1}{2}e_2, && \alpha \in \mathbb{R},\\
&[e_1, e_4] = \tfrac{1}{2}\alpha e_2, && [e_2, e_4] = -\alpha e_1, && [e_3, e_4] = \tfrac{1}{2}\sqrt{2(1 + \alpha^2)}\, e_2.
\end{aligned}$$

Case (1.ii): $b = 0$, $C \neq 0$, $D = 0$ In this case, $\widetilde{\mathfrak{C}}_{313} = -AC(\lambda_2 - 2)\lambda_2$ and $\widetilde{\mathfrak{C}}_{314} = (A^2 - R^2)C(\lambda_2 - 2)\lambda_2$. This implies that $\lambda_2 = 2$ and the cyclic-parallel condition reduces to $C^2 - 2(A^2 + R^2) = 0$. Hence $C = \varepsilon\sqrt{2(A^2 + R^2)}$, with $\varepsilon^2 = 1$. Observe that $e_1 \mapsto -e_1$, $e_2 \mapsto -e_2$ defines an isometry between the Lie groups corresponding to $\varepsilon = \pm 1$ as in Case (1.i), thus giving

$$\begin{array}{lll} [e_2, e_3] = e_1, & [e_3, e_1] = 2e_2, & \alpha \in \mathbb{R}, \\ [e_1, e_4] = 2\alpha e_2, & [e_2, e_4] = -\alpha e_1, & [e_3, e_4] = \sqrt{2(1+\alpha^2)}e_1. \end{array}$$

Now, setting $v_1 = e_1$, $v_2 = e_2$, $v_3 = \frac{1}{1+\alpha^2}(\alpha e_3 + e_4)$, $v_4 = \frac{1}{1+\alpha^2}(e_3 - \alpha e_4)$, both in Case (1.i) and Case (1.ii), one has that $[v_i, v_j] = 0$ for $i, j \in \{1, 2, 3\}$ and thus the left-invariant metrics above are isometric to some left-invariant metric on $\mathbb{R} \ltimes \mathbb{R}^3$. Therefore they are homothetic to the metric in Assertion (c).

Case (1.iii): $b = 0$, $CD \neq 0$, $\lambda_2 = -1$ There is no solution in this case since $\widetilde{\mathfrak{C}}_{313} = -3AC$ and $\widetilde{\mathfrak{C}}_{314} = 3C(A^2 - R^2)$.

Case (2): $b \neq 0$, $C = D = 0$ Since $\widetilde{\mathfrak{C}}_{114} = -bR^2(\lambda_2^2 - 1)$ it follows that either $\lambda_2 = 1$ or $\lambda_2 = -1$. If $\lambda_2 = 1$ then the manifold is cyclic-parallel and locally symmetric. If $\lambda_2 = -1$ then the cyclic-parallel condition $\{\widetilde{\mathfrak{C}}_{ijk} = 0\}$ reduces to $\widetilde{\mathfrak{C}}_{213} = 4(b^2 - A^2 - R^2)$ and $\widetilde{\mathfrak{C}}_{214} = -4A(b^2 - A^2 - R^2)$. Hence $b = \varepsilon\sqrt{A^2 + R^2}$, with $\varepsilon^2 = 1$, and a straightforward calculation shows that the manifold is symmetric.

- **Left-invariant metrics on $\widetilde{SL(2, \mathbb{R})} \times \mathbb{R}$ and $SU(2) \times \mathbb{R}$.** There exists an orthonormal basis $\{e_1, \dots, e_4\}$ of $\mathfrak{sl}(2, \mathbb{R}) \oplus \mathbb{R}$ or $\mathfrak{su}(2) \oplus \mathbb{R}$ such that

$$\begin{array}{lll} [e_1, e_2] = \lambda_3 e_3, & [e_2, e_3] = \lambda_1 e_1, & [e_3, e_1] = \lambda_2 e_2, \\ [e_1, e_4] = \frac{1}{R}(k_3\lambda_2 e_2 - k_2\lambda_3 e_3), & & [e_2, e_4] = \frac{1}{R}(k_1\lambda_3 e_3 - k_3\lambda_1 e_1), \\ [e_3, e_4] = \frac{1}{R}(k_2\lambda_1 e_1 - k_1\lambda_2 e_2), & & R > 0. \end{array} \tag{8}$$

where $\lambda_1\lambda_2\lambda_3 \neq 0$. A long but straightforward calculation shows that the cyclic-parallel condition, considering the structure constants in Eq. (8), is equivalent to the vanishing of the polynomials $\widetilde{\mathfrak{C}}_{ijk}$ given by

$$\begin{aligned}
\widetilde{\mathfrak{C}}_{112} &= R^2\mathfrak{C}_{112} = -(\lambda_1 - \lambda_2)(\lambda_1 - 2\lambda_2 + \lambda_3)\lambda_3 k_1 k_3, \\
\widetilde{\mathfrak{C}}_{113} &= R^2\mathfrak{C}_{113} = (\lambda_1 - \lambda_3)(\lambda_1 + \lambda_2 - 2\lambda_3)\lambda_2 k_1 k_2, \\
\widetilde{\mathfrak{C}}_{114} &= R^3\mathfrak{C}_{114} = -(\lambda_2 - \lambda_3)(\lambda_1 + \lambda_2 + \lambda_3)\lambda_1 k_1 k_2 k_3, \\
\widetilde{\mathfrak{C}}_{212} &= R^2\mathfrak{C}_{212} = (\lambda_1 - \lambda_2)(2\lambda_1 - \lambda_2 - \lambda_3)\lambda_3 k_2 k_3, \\
\widetilde{\mathfrak{C}}_{223} &= R^2\mathfrak{C}_{223} = -(\lambda_2 - \lambda_3)(\lambda_1 + \lambda_2 - 2\lambda_3)\lambda_1 k_1 k_2, \\
\widetilde{\mathfrak{C}}_{224} &= R^3\mathfrak{C}_{224} = (\lambda_1 - \lambda_3)(\lambda_1 + \lambda_2 + \lambda_3)\lambda_2 k_1 k_2 k_3, \\
\widetilde{\mathfrak{C}}_{313} &= R^2\mathfrak{C}_{313} = -(\lambda_1 - \lambda_3)(2\lambda_1 - \lambda_2 - \lambda_3)\lambda_2 k_2 k_3, \\
\widetilde{\mathfrak{C}}_{323} &= R^2\mathfrak{C}_{323} = -(\lambda_2 - \lambda_3)(\lambda_1 - 2\lambda_2 + \lambda_3)\lambda_1 k_1 k_3, \\
\widetilde{\mathfrak{C}}_{213} &= R^2\mathfrak{C}_{213} = -(\lambda_2 - \lambda_3)((\lambda_2 + \lambda_3)\lambda_1 - 2\lambda_2\lambda_3)k_1^2 \\
&\quad + (\lambda_1 - \lambda_3)((\lambda_2 - 2\lambda_3)\lambda_1 + \lambda_2\lambda_3)k_2^2 \\
&\quad - (\lambda_1 - \lambda_2)((\lambda_1 + \lambda_2)\lambda_3 - 2\lambda_1\lambda_2)k_3^2 \\
&\quad + 2(\lambda_1 - \lambda_2)(\lambda_1 - \lambda_3)(\lambda_2 - \lambda_3)R^2,
\end{aligned}$$

$$
\begin{aligned}
\widetilde{\mathfrak{C}}_{214} = 2R^3\mathfrak{C}_{214} = \{&((2\lambda_2 - 3\lambda_3)\lambda_1 + (2\lambda_2 - \lambda_3)\lambda_3)\lambda_2 k_1^2 \\
&- (2(\lambda_2 + \lambda_3)\lambda_1 - (3\lambda_2 + \lambda_3)\lambda_3)\lambda_1 k_2^2 \\
&- 4(\lambda_1 - \lambda_2)\lambda_1\lambda_2 k_3^2 \\
&- (\lambda_1 - \lambda_2)(2\lambda_1 - \lambda_3)(2\lambda_2 - \lambda_3)R^2\}k_3, \\
\widetilde{\mathfrak{C}}_{314} = 2R^3\mathfrak{C}_{314} = \{&((3\lambda_1 + \lambda_2)\lambda_2 - 2(\lambda_1 + \lambda_2)\lambda_3)\lambda_3 k_1^2 \\
&+ 4(\lambda_1 - \lambda_3)\lambda_1\lambda_3 k_2^2 \\
&+ (2(\lambda_2 + \lambda_3)\lambda_1 - (\lambda_2 + 3\lambda_3)\lambda_2)\lambda_1 k_3^2 \\
&- (2\lambda_1 - \lambda_2)(\lambda_1 - \lambda_3)(\lambda_2 - 2\lambda_3)R^2\}k_2, \\
\widetilde{\mathfrak{C}}_{324} = 2R^3\mathfrak{C}_{324} = -\{&4(\lambda_2 - \lambda_3)\lambda_2\lambda_3 k_1^2 \\
&+ ((\lambda_1 + 3\lambda_2)\lambda_1 - 2(\lambda_1 + \lambda_2)\lambda_3)\lambda_3 k_2^2 \\
&- ((\lambda_1 - 2\lambda_2)\lambda_1 + (3\lambda_1 - 2\lambda_2)\lambda_3)\lambda_2 k_3^2 \\
&+ (\lambda_1 - 2\lambda_2)(\lambda_1 - 2\lambda_3)(\lambda_2 - \lambda_3)R^2\}k_1,
\end{aligned}
$$

where $\mathfrak{C}_{ijk} = \mathfrak{C}(e_i, e_j, e_k)$. Since the structure constants of $\mathfrak{sl}(2, \mathbb{R})$ or $\mathfrak{su}(2)$ satisfy $\lambda_1\lambda_2\lambda_3 \neq 0$, one may work with a homothetic basis $\widetilde{e}_k = \frac{1}{\lambda_1}e_k$ so that we may assume $\lambda_1 = 1$ in the rest of the proof. Note that, without loss of generality, it is sufficient to analyze the following cases:

$$
\begin{aligned}
&(1)\ k_1 = k_2 = k_3 = 0, &&(2)\ k_1 = k_2 = 0, k_3 \neq 0, \\
&(3)\ k_1 = 0, k_2k_3 \neq 0, &&(4)\ k_1k_2k_3 \neq 0\,.
\end{aligned}
$$

Case (1): $k_1 = k_2 = k_3 = 0$ The cyclic-parallel condition $\{\widetilde{\mathfrak{C}}_{ijk} = 0\}$ reduces to $\widetilde{\mathfrak{C}}_{213} = 2(\lambda_2 - 1)(\lambda_3 - 1)(\lambda_2 - \lambda_3)R^2 = 0$. If $\lambda_2 = 1$ or $\lambda_3 = 1$ and $\lambda_2 \neq \lambda_3$, then the manifold corresponds to Assertion (a) in Theorem 1 since $\ker \mathrm{Ric}$ is parallel. If $\lambda_2 = \lambda_3$ then the manifold is either locally symmetric or locally isometric to a product $\mathbb{R} \times N$ as in Theorem 1-(a), depending on $\lambda_3 = 1$ or $\lambda_3 \neq 1$, respectively.

Case (2): $k_1 = k_2 = 0, k_3 \neq 0$ In this case, the cyclic-parallel condition reduces to

$$
\begin{aligned}
\widetilde{\mathfrak{C}}_{213} &= (\lambda_2 - 1)\{((\lambda_3 - 2)\lambda_2 + \lambda_3)k_3^2 + 2(\lambda_2 - \lambda_3)(\lambda_3 - 1)R^2\} = 0, \\
\widetilde{\mathfrak{C}}_{214} &= (\lambda_2 - 1)\{4\lambda_2 k_3^2 - (2\lambda_2 - \lambda_3)(\lambda_3 - 2)R^2\}k_3 = 0\,.
\end{aligned}
$$

If $\lambda_2 = 1$ then the manifold is symmetric or it corresponds to Theorem 1-(a), depending on $\lambda_3 = 1$ or $\lambda_3 \neq 1$, respectively. If $\lambda_2 \neq 1$ then a straightforward calculation shows that

$$
\lambda_3 = \frac{2(\lambda_2^2 - \lambda_2 + 1)}{\lambda_2 + 1}, \qquad k_3 = \varepsilon R\sqrt{\frac{(\lambda_2 - 2)(2\lambda_2 - 1)}{(\lambda_2 + 1)^2}},
$$

where $\varepsilon^2 = 1$ and $\lambda_2 \in (-\infty, \frac{1}{2}) \cup (2, +\infty)$, $\lambda_2 \notin \{-1, 0\}$. Moreover, note that replacing e_4 by $-e_4$ defines an isometry wich interchanges $\varepsilon = 1$ by $\varepsilon = -1$, which corresponds to Theorem 1-(d).

Case (3): $k_1 = 0, k_2k_3 \neq 0$ In this case, $\widetilde{\mathfrak{C}}_{212} = (\lambda_2 - 1)(\lambda_2 + \lambda_3 - 2)\lambda_3 k_2 k_3$ and $\widetilde{\mathfrak{C}}_{313} = -(\lambda_3 - 1)(\lambda_2 + \lambda_3 - 2)\lambda_2 k_2 k_3$. Hence, necessarily $\lambda_3 = 2 - \lambda_2$ and we get $\widetilde{\mathfrak{C}}_{214} = (\lambda_2 - 1)((3\lambda_2 - 2)R^2 - 2k_2^2 + 4k_3^2)\lambda_2 k_3$. If $\lambda_2 = 1$ then the manifold is cyclic-parallel and locally symmetric. Otherwise, if $\lambda_2 \neq 1$, then $\lambda_2 = \frac{2(k_2^2 - 2k_3^2 + R^2)}{3R^2}$ and the condition $\{\widetilde{\mathfrak{C}}_{ijk} = 0\}$ reduces to $(2k_2^2 - 4k_3^2 - R^2)(k_2^2 - 2(k_3^2 + R^2)) = 0$. As a consequence, $(\lambda_2, \lambda_3) = (1, 1)$ or $(\lambda_2, \lambda_3) = (2, 0)$ and therefore there is no solution.

Case (4): $k_1k_2k_3 \neq 0$ Considering $\widetilde{\mathfrak{C}}_{112} = -(\lambda_2 - 1)(2\lambda_2 - \lambda_3 - 1)\lambda_3 k_1 k_3$ and $\widetilde{\mathfrak{C}}_{212} = (\lambda_2 - 1)(\lambda_2 + \lambda_3 - 2)\lambda_3 k_2 k_3$, it follows that necessarily $\lambda_2 = 1$. Now, $\widetilde{\mathfrak{C}}_{113} = 2(\lambda_3 - 1)^2 k_1 k_2$ and therefore $\lambda_3 = 1$. For $\lambda_1 = \lambda_2 = \lambda_3 = 1$ we get a cyclic-parallel manifold which is locally symmetric. This finishes the proof. □

3 Weakly-Einstein Homogeneous Spaces

Let (M, g) be a four-dimensional symmetric space. Since the Ricci tensor is parallel, the Ricci curvatures are constant and the corresponding eigenspaces are parallel. Let $\{\kappa_i\}$ be the Ricci curvatures. If (M, g) is non-Einstein, then there are at least two-distinct Ricci curvatures. Let κ be a Ricci curvature appearing with multiplicity one. Then the associated eigenspace splits off a one-dimensional manifold so that (M, g) is locally isometric to $\mathbb{R} \times N$ for some three-dimensional symmetric manifold N. Now $\mathbb{R} \times N$ is $\check{R}$-Einstein, $\check{\rho}$-Einstein or $R[\rho]$-Einstein if and only if it is flat. Hence the only possibility is that the Ricci operator has exactly two-distinct Ricci curvatures with multiplicity two. Moreover, a standard calculation shows that the Ricci curvatures κ_1 and κ_2 satisfy $\kappa_1 = -\kappa_2$ which corresponds to the locally conformally flat product of two surfaces of constant opposite sectional curvature. The non-symmetric case allows other examples of left-invariant metrics on Lie groups as the following shows [3, 19].

Theorem 2 *Let (M, g) be a four-dimensional homogeneous manifold. Then*

(a) (M, g) is $\check{R}$-Einstein and non-symmetric if and only if it is homothetic to the Lie group $\mathbb{R} \ltimes \mathbb{R}^3$ with left-invariant metric determined by the Lie algebra

$$[e_4, e_1] = e_1, \quad [e_4, e_2] = -e_2, \quad [e_4, e_3] = -e_3.$$

(b) (M, g) is $\check{\rho}$-Einstein and non-symmetric if and only if it is homothetic to one of the following:

(b.1) The Lie group $SU(2) \times \mathbb{R}$ with left-invariant metric determined by the Lie algebra

$$[e_1, e_2] = (4 \pm 2\sqrt{2})e_3, \quad [e_2, e_3] = (3 \pm 2\sqrt{2})e_1, \quad [e_3, e_1] = e_2,$$
$$[e_4, e_1] = -e_2, \quad [e_4, e_2] = (3 \pm 2\sqrt{2})e_1.$$

(b.2) The Lie group $\mathbb{R} \ltimes H^3$ with left-invariant metric determined by the Lie algebra

$$[e_1, e_2] = e_3, \quad [e_1, e_4] = \frac{1}{2}e_1, \quad [e_2, e_4] = -\frac{1}{2}e_2.$$

(b.3) The Lie group $\mathbb{R} \ltimes \mathbb{R}^3$ with left-invariant metric determined by the Lie algebra

$$\begin{aligned} [e_4, e_1] &= e_1 - \frac{\alpha(1+\alpha+\alpha^2)}{(\alpha+1)^2(\alpha-1)}e_2 + \frac{\alpha(1+\alpha+\alpha^2)}{2\alpha+1}e_3, \\ [e_4, e_2] &= \frac{\alpha(1+\alpha+\alpha^2)}{(\alpha+1)^2(\alpha-1)}e_1 + \alpha e_2 + \frac{(1+\alpha+\alpha^2)^2}{\alpha(\alpha+2)}e_3, \\ [e_4, e_3] &= -\frac{\alpha(1+\alpha+\alpha^2)}{2\alpha+1}e_1 - \frac{(1+\alpha+\alpha^2)^2}{\alpha(\alpha+2)}e_2 - \frac{\alpha}{\alpha+1}e_3, \end{aligned}$$

where $\{e_1, \ldots, e_4\}$ is an orthonormal basis and $\alpha \in (-1, 1)$, $\alpha \neq -\frac{1}{2}$, $\alpha \neq 0$.

(c) If (M, g) is $R[\rho]$-Einstein then it is symmetric.

Remark 2 The left-invariant metrics on $\mathbb{R} \ltimes H^3$ given in Theorem 2-(b.2) have cyclic-parallel Ricci tensor, and are homothetic to the metrics corresponding to Assertion (b) in Theorem 1.

Remark 3 Metrics corresponding to Assertion (a) in Theorem 2 are algebraic Ricci solitons determined by the derivation $\mathfrak{D} = \operatorname{Ric} + 3\operatorname{Id}$. None of the other weakly-Einstein homogeneous four-dimensional manifolds admits a Ricci soliton structure.

Besides its importance in Mathematics, the Ricci flow is also central in Physics, where it arises as the first-order approximation of the renormalization group flow. Recently the second-order approximation of this renormalization group flow (RG-2 flow) has naturally appeared by considering the expansion truncated at the second term

$$\frac{\partial}{\partial t}g = -2\rho - \frac{\alpha}{2}\check{R}.$$

The fixed points of this nonlinear perturbation of the Ricci flow are those geometries for which $\rho + \frac{\alpha}{4}\check{R}$ vanishes [21]. It follows from Eq. (1) that any four-dimensional Einstein metric satisfies $\rho + \frac{\alpha}{4}\check{R} = 0$. Furthermore, the constant α is determined by tracing in $\rho + \frac{\alpha}{4}\check{R} = 0$, so that $\tau + \frac{\alpha}{4}\|R\|^2 = 0$.

While Ricci-flat homogeneous manifolds are flat, there are non-flat examples with vanishing $\rho - \frac{\tau}{\|R\|^2}\check{R}$. For instance, a locally symmetric four-dimensional manifold satisfies $\rho - \frac{\tau}{\|R\|^2}\check{R} = 0$ if and only if it is Einstein or a product of the form $\mathbb{R} \times N^3(c)$, or $\mathbb{R}^2 \times N^2(c)$. The non-symmetric homogeneous geometries satisfying $\rho + \frac{\alpha}{4}\check{R} = 0$ are given as follows [20].

Theorem 3 *Let (M, g) be a complete and simply connected four-dimensional homogeneous manifold satisfying $\rho + \frac{\alpha}{4}\check{R} = 0$. Then (M, g) is either symmetric or homothetic to the Lie group $SU(2) \times \mathbb{R}$ with the left-invariant metric*

$$[e_1, e_2] = e_3, \quad [e_3, e_1] = \frac{3}{4}e_2, \quad [e_2, e_3] = \frac{3}{4}e_1$$

where $\{e_1, \ldots, e_4\}$ is an orthonormal basis of $\mathfrak{su}(2) \times \mathbb{R}$.

Remark 4 The only non-symmetric three-dimensional homogeneous geometry satisfying $\rho + \frac{\alpha}{4}\check{R} = 0$ is homothetic to the one given by the left-invariant metric on $SU(2)$ determined by

$$[e_1, e_2] = e_3, \quad [e_3, e_1] = \frac{3}{4}e_2, \quad [e_2, e_3] = \frac{3}{4}e_1$$

where $\{e_1, e_2, e_3\}$ is an orthonormal basis. Furthermore, in this case $\alpha = -8$.

Remark 5 Clearly any metric satisfying $\rho + \frac{\alpha}{4}\check{R} = 0$ is a steady soliton of the RG-2 flow. While homogenenous metrics of the form $\mathbb{R}^k \times N(c)$ are also rigid gradient Ricci solitons, the metrics in Theorem 3 are not Ricci solitons.

4 Conformally Einstein Homogeneous Spaces

Conformally Einstein homogeneous four-manifolds were described in [11] as follows.

Theorem 4 *Let (M, g) be a four-dimensional complete and simply connected conformally Einstein homogeneous manifold. Then (M, g) is symmetric or otherwise it is homothetic to one of the following:*

(a) The Euclidean space $\mathbb{R}^4$ with coordinates (x, y, z, t) and metric

$$g_\alpha = e^{-2t}dx^2 + e^{-\frac{1}{2}t}(dy - \alpha z dt)^2 + e^{-\frac{1}{2}t}(dz + \alpha y dt)^2 + dt^2 .$$

The conformal metric $\widetilde{g}_\alpha = e^{\frac{3}{2}t} g_\alpha$ is Ricci-flat, and moreover, all metrics in this class are homothetic.

(b) The Euclidean space $\mathbb{R}^4$ with coordinates (x, y, z, t) and metric

$$\begin{aligned} g_\alpha = {} & e^{-2t}(dx + \alpha y dt)^2 + e^{-2t}(dy - \alpha x dt)^2 \\ & + e^{-4t}(dz + \tfrac{1}{2}(x dy - y dx) - \tfrac{1}{2}\alpha(x^2 + y^2)dt)^2 + dt^2 . \end{aligned}$$

The conformal metric $\widetilde{g}_\alpha = e^{3t} g_\alpha$ is Ricci-flat. All metrics in this class are half conformally flat and homothetic.

(c) *The Euclidean space* $\mathbb{R}^4$ *with coordinates* (x, y, z, t) *and metric*

$$g_\alpha = e^{-2t}dx^2 + e^{-2(\alpha+1)^2 t}dy^2 + e^{-2\alpha^2 t}dz^2 + dt^2 .$$

The conformal metric $\widetilde{g}_\alpha = e^{2(\alpha^2+\alpha+1)t} g_\alpha$ *is Ricci-flat. Moreover, for all* $\alpha > 1$ *the metrics above correspond to different homothety classes.*

Remark 6 Metrics in Theorem 4 are isometric to the left-invariant metrics determined by the following solvable Lie algebras:

(a) The Lie algebra $\mathfrak{g}_\alpha = \mathbb{R}e_4 \ltimes \mathfrak{r}^3$ given by

$$[e_4, e_1] = e_1, \quad [e_4, e_2] = \frac{1}{4}e_2 + \alpha e_3, \quad [e_4, e_3] = -\alpha e_2 + \frac{1}{4}e_3 .$$

(b) The Lie algebra $\mathfrak{g}_\alpha = \mathbb{R}e_4 \ltimes \mathfrak{h}^3$ given by

$$[e_1, e_2] = e_3 \quad [e_4, e_1] = e_1 - \alpha e_2, \quad [e_4, e_2] = \alpha e_1 + e_2, \quad [e_4, e_3] = 2e_3 .$$

(c) The Lie algebra $\mathfrak{g}_\alpha = \mathbb{R}e_4 \ltimes \mathfrak{r}^3$ given by

$$[e_4, e_1] = e_1, \quad [e_4, e_2] = (\alpha + 1)^2 e_2, \quad [e_4, e_3] = \alpha^2 e_3, \qquad \alpha > 1 .$$

Here $\{e_1, \ldots, e_4\}$ is an orthonormal basis. Further observe that although the metrics corresponding to the Lie algebras $\mathfrak{g}_\alpha$ in (a) and (b) are homothetic, the underlying Lie groups are not isomorphic.

Remark 7 Let (M, g) be a four-dimensional complete and simply connected conformally Einstein homogeneous Ricci soliton which is not Einstein. If the soliton is a gradient one, then it is rigid and isometric to a locally conformally flat product $N^3(c) \times \mathbb{R}$ [35]. Otherwise the Ricci soliton is expanding and isometric to a metric solvsoliton [5]. A straightforward calculation shows that they correspond to one of the following:

(a) The Lie algebra $\mathfrak{g}_\alpha = \mathbb{R}e_4 \ltimes \mathfrak{r}^3$ given by

$$[e_4, e_1] = e_1, \quad [e_4, e_2] = \frac{1}{4}e_2 + \alpha e_3, \quad [e_4, e_3] = -\alpha e_2 + \frac{1}{4}e_3 ,$$

where the soliton is determined by the derivation $\mathfrak{D} = \mathrm{Ric} + \frac{9}{8}\,\mathrm{Id}$.

(b) The Lie algebra $\mathfrak{g}_\alpha = \mathbb{R}e_4 \ltimes \mathfrak{r}^3$ given by

$$[e_4, e_1] = e_1, \quad [e_4, e_2] = (\alpha + 1)^2 e_2, \quad [e_4, e_3] = \alpha^2 e_3, \qquad \alpha > 1 ,$$

where the soliton is determined by the derivation $\mathfrak{D} = \mathrm{Ric} + 2(\alpha^2 + \alpha + 1)^2\,\mathrm{Id}$.

Remark 8 The following two strictly Bach-flat Lie groups have been constructed in [1] (where $\{e_1, \dots, e_4\}$ is an orthonormal basis):

(a) The Lie algebra $\mathfrak{g} = \mathbb{R}e_4 \ltimes \mathfrak{e}(1,1)$ given by

$$[e_2, e_3] = e_1, \qquad [e_1, e_3] = (2 + \sqrt{3})\, e_2,$$
$$[e_4, e_1] = \sqrt{6 + 3\sqrt{3}}\, e_1, \; [e_4, e_2] = \sqrt{6 + 3\sqrt{3}}\, e_2.$$

(b) The Lie algebra $\mathfrak{g} = \mathbb{R}e_4 \ltimes \mathfrak{h}_3$ given by

$$[e_1, e_2] = e_3, \qquad [e_4, e_1] = \tfrac{1}{4}\sqrt{7 - 3\sqrt{5}}\, e_1,$$
$$[e_2, e_4] = \tfrac{1}{4}\sqrt{7 + 3\sqrt{5}}\, e_2, \quad [e_3, e_4] = \frac{\sqrt{5}}{2\sqrt{2}}\, e_3.$$

It was shown in [11] that the two cases above are indeed the only strictly Bach-flat four-dimensional homogeneous spaces. Furthermore, observe that the corresponding Lie groups are essentially different, since the case (b) is also a solvsoliton, while the Lie group corresponding to Assertion (a) is not.

5 Conformal *C*-Spaces

Recall that a Riemannian manifold is a conformal C-space if there is a vector field ξ so that $C + i_\xi W = 0$ (see Eq. (3)).

Let (M, g) be a symmetric space which is not Einstein. Then either it is a product $\mathbb{R} \times N^3(c)$ or $N_1^2(c_1) \times N_2^2(c_2)$. In the first case the metric is locally conformally flat and thus conformally Cotton-flat. Moreover, if M is a product $N_1^2(c_1) \times N_2^2(c_2)$, then it is a C-space if and only if $c_1 = -c_2$ in which case it is also locally conformally flat. The non-symmetric case is given by the following [12]:

Theorem 5 *Let (M, g) be a four-dimensional homogeneous conformal C-space. Then it is conformally Einstein or homothetic to a Lie group G with a left-invariant metric determined by the Lie algebra $\mathfrak{g}_\alpha = \mathbb{R}e_4 \ltimes \mathfrak{r}^3$ given by*

$$[e_1, e_4] = e_1, \quad [e_2, e_4] = e_2, \quad [e_3, e_4] = \alpha e_3$$

where $\{e_1, \dots, e_4\}$ is an orthonormal basis and $\alpha \in \mathbb{R} \setminus \{0, 1, 4\}$.

Remark 9 The conformal C-space metrics in Theorem 5 satisfy the equation $C + i_\xi W = 0$ for a vector field $\xi = -6e_4$. Since $de^4 = 0$, ξ is a gradient and thus the homogeneous conformal C-space is indeed conformally Cotton-flat.

Remark 10 The Bach-flat left-invariant structures in Remark 8 are not conformal C-spaces (see [1]). Therefore, one has that a four-dimensional homogeneous conformal C-space is Bach-flat if and only if it is conformally Einstein.

Remark 11 Let $\{e^1, \ldots, e^4\}$ be the dual basis of $\{e_1, \ldots, e_4\}$. Then the structure equations corresponding to the Lie algebras in Theorem 5 are given by

$$de^4 = 0, \quad de^1 = -e^1 \wedge e^4, \quad de^2 = -e^2 \wedge e^4, \quad de^3 = -\alpha e^3 \wedge e^4.$$

Setting

$$e^4 = dt, \quad e^1 = e^t dx, \quad e^2 = e^t dy, \quad e^3 = e^{\alpha t} dz,$$

one has that the manifolds in Theorem 5 are isometric to the doubly warped product metric $\mathbb{R} \times_{exp(t)} \mathbb{R}^2 \times_{exp(\alpha t)} \mathbb{R}$ on $\mathbb{R}^4$ given by

$$g = dt^2 + e^{2t}(dx^2 + dy^2) + e^{2\alpha t} dz^2.$$

Moreover, a straightforward calculation shows that the conformal metric $\hat{g} = e^{-6t} g$ is Cotton-flat. Finally, observe that the Bach tensor $\mathfrak{B}$, when expressed in the above coordinates, is diagonal and given by

$$\mathfrak{B} = \frac{1}{6}\alpha(\alpha - 1)(\alpha - 4)\,\mathrm{diag}[\alpha - 1, -(\alpha + 1)e^{2t}, -(\alpha + 1)e^{2t}, (\alpha + 3)e^{2\alpha t}].$$

Hence the Bach tensor vanishes if and only if $\alpha = 0$, $\alpha = 1$, or $\alpha = 4$. For $\alpha = 0$, the underlying structure corresponds to a product $\mathbb{R} \times \mathbb{H}^3$ while it corresponds to the hyperbolic space $\mathbb{H}^4$ for $\alpha = 1$. The case $\alpha = 4$ is conformally Einstein since $\hat{g} = e^{-6t} g$ is Ricci-flat.

Remark 12 The metrics corresponding to Theorem 5 are never weakly-Einstein, unless the case given by $\alpha = -1$ which corresponds to the only homogeneous $\check{R}$-Einstein four-dimensional space (cf. Assertion (1) in Theorem 2).

Remark 13 Let $\{e^i\}$ be the dual basis of $\{e_i\}$, and set $E_1^\pm = \frac{1}{\sqrt{2}}\left(e^1 \wedge e^2 \pm e^3 \wedge e^4\right)$, $E_2^\pm = \frac{1}{\sqrt{2}}\left(e^1 \wedge e^3 \mp e^2 \wedge e^4\right)$, and $E_3^\pm = \frac{1}{\sqrt{2}}\left(e^1 \wedge e^4 \pm e^2 \wedge e^3\right)$.

The self-dual and anti-self-dual Weyl curvature operator of any left-invariant metric in Theorem 5 satisfy $W^\pm = -\frac{1}{6}\alpha(\alpha - 1)\,\mathrm{diag}[2, -1, -1]$ when expressed on the basis $\{E_1^+, E_2^+, E_3^+\}$ (resp., $\{E_1^-, E_2^-, E_3^-\}$) of self-dual (resp., anti-self-dual) two-forms. The distinguised eigenvalue of W^+ and W^- defines one-dimensional subspaces of Λ_+^2 and Λ_-^2. The corresponding sections determine two-forms given by $E_1^\pm$. A straightforward calculation shows that $dE_1^\pm = e^4 \wedge E_1^\pm$, where $de^4 = 0$. Hence the two forms $E_1^\pm$ are conformally symplectic and opposite conformally symplectic. Furthermore, observe that none of the corresponding almost complex structures $J^\pm$ (determined by $J^\pm e_1 = e_2$, and $J^\pm e_3 = \pm e_4$) is integrable.

A straightforward calculation shows that the two-forms $\Omega_\pm = e^{-2t} E_1^\pm$ determine a symplectic pair [6] (i.e., they are non-degenerate closed two-forms such that $\Omega_+ \wedge \Omega_- = 0$, and $\Omega_+ \wedge \Omega_+ = -\Omega_- \wedge \Omega_-$).

Remark 14 Homogeneous conformally C-spaces given in Theorem 5 are expanding Ricci solitons. Indeed, a straightforward calculation shows that they are algebraic Ricci solitons determined by the derivation $\mathfrak{D} = \mathrm{Ric} + (\alpha^2 + 2)\,\mathrm{Id}$.

Acknowledgement Supported by projects MTM2016-75897-P, and ED431C 2019/10 with FEDER Funds (Spain).

References

1. E. Abbena, S. Garbiero, S. Salamon, Bach-flat Lie groups in dimension 4. C. R. Acad. Sci. Paris Ser. I **351**, 303–306 (2013)
2. D.V. Alekseevski, B.N. Kimel'fel'd, Structure of homogeneous Riemann spaces with zero Ricci curvature. Funktsional'nyi Analiz i Ego Prolozheniya **9**, 5–11 (1975)
3. T. Arias-Marco, O. Kowalski, Classification of 4-dimensional homogeneous weakly Einstein manifolds. Czech. Math. J. **65**, 21–59 (2015)
4. T. Arias-Marco, O. Kowalski, Classification of 4-dimensional homogeneous D'Atri spaces. Czech. Math. J. **58**, 203–239 (2008)
5. R.M. Arroyo, R. Lafuente, Homogeneous Ricci solitons in low dimensions. Int. Math. Res. Not. IMRN **13**, 4901–4932 (2015)
6. G. Bande, D. Kotschick, The geometry of symplectic pairs. Trans. Am. Math. Soc. **358**, 1643–1655 (2006)
7. L. Bérard-Bergery, Les spaces homogènes Riemanniens de dimension 4. Riemannian Geom. Dimens. 4 (Paris 1978/1979) **3**, 40–60 (1981)
8. M. Berger, Quelques formules de variation pour une structure riemannienne. Ann. Sci. Éc. Norm. Super. (4) **3**, 285–294 (1970)
9. A.L. Besse, *Einstein Manifolds*. Ergebnisse der Mathematik, vol. 10 (Springer, Berlin, 1987)
10. X. Cao, H. Tran, The Weyl tensor of gradient Ricci solitons. Geom. Topol. **20**, 389–436 (2016)
11. E. Calviño-Louzao, X. García-Martínez, E. García-Río, I. Gutiérrez-Rodríguez, R. Vázquez-Lorenzo, Conformally Einstein and Bach-flat four-dimensional homogeneous manifolds. J. Math. Pures Appl. (9) **130**, 347–374 (2019)
12. E. Calviño-Louzao, E. García-Río, I. Gutiérrez-Rodríguez, R. Vázquez-Lorenzo, Four-dimensional homogeneous conformal C-spaces. Osaka J. Math. **58**, 413–433 (2021)
13. G. Catino, Generalized quasi-Einstein manifolds with harmonic Weyl tensor. Math. Z. **271**, 751–756 (2012)
14. G. Catino, P. Mastrolia, D.D. Monticelli, M. Rigoli, Analytic and geometric properties of generic Ricci solitons. Trans. Am. Math. Soc. **368**, 7533–7549 (2016)
15. D. Cox, D. Little, D. O'Shea, *Ideals, Varieties, and Algorithms. An Introduction to Computational Algebraic Geometry and Commutative Algebra*. Undergraduate Texts in Mathematics (Springer, Cham, 2015)
16. V. de Smedt, S. Salamon, Anti-self-dual metrics on Lie groups, in *Proc. Conf. Integrable Systems and Differential Geometry*, vol. 308 (Amer. Math. Soc., Providence, 2002), pp. 63–75
17. Y. Euh, J. Park, K. Sekigawa, A curvature identity on a 4-dimensional Riemannian manifold. Result. Math. **63**, 107–114 (2013)
18. M. Fernández-López, E. García-Río, D.N. Kupeli, B. Ünal, A curvature condition for a twisted product to be a warped product. Manuscripta Math. **106**, 213–217 (2001)
19. E. García-Río, A. Haji-Badali, R. Mariño-Villar, M. E. Vázquez-Abal, Four-dimensional homogeneous manifolds satisfying some Einstein-like conditions. Kodai Math. J. **43**(3), 465–488 (2020)

20. E. García-Río, R. Mariño-Villar, M. E. Vázquez-Abal, R. Vázquez-Lorenzo, Fixed points and steady solitons for the two-loop renormalization group flow, to appear
21. K. Gimre, Chr. Guenther, J. Isenberg, A geometric introduction to the two-loop renormalization group flow. J. Fixed Point Theory Appl. **14**, 3–20 (2013)
22. A.R. Gover, P. Nurowski, Obstructions to conformally Einstein metrics in n dimensions. J. Geom. Phys. **56**, 450–484 (2006)
23. R. Gover, P.-A. Nagy, Four-dimensional conformal C-spaces. Q. J. Math. **58**, 443–462 (2007)
24. A. Gray, Einstein-like manifolds which are not Einstein. Geom. Dedicata **7**, 259–280 (1978)
25. M. Jablonski, Homogeneous Ricci solitons. J. Reine Angew. Math. **699**, 159–182 (2015)
26. G.R. Jensen, Homogeneous Einstein spaces of dimension four. J. Differ. Geom. **3**, 309–349 (1969)
27. O. Kowalski, Spaces with volume preserving symmetries and related classes of Riemannian manifolds. Rend. Sem. Math. Univ. Politec. Tor. Fascicolo Speciale 131–158 (1983)
28. O. Kowalski, F. Prüfer, L. Vanhecke, D'Atri spaces, in *Topics in Geometry*. Progr. Nonlinear Differential Equations Appl., vol. 20 (Birkhäuser, Boston, 1996), pp. 241–284
29. O. Kowalski, L. Vanhecke, Four-dimensional naturally reductive homogeneous spaces, in *Conference on Differential Geometry on Homogeneous Spaces* (Turin, 1983). Rend. Sem. Mat. Univ. Politec. Torino 1983, Special Issue (1984), pp. 223–232
30. C.N. Kozameh, E.T. Newman, K.P. Tod, Conformal Einstein spaces. Gen. Rel. Grav. **17**, 343–352 (1985)
31. W. Kühnel, H.-B. Rademacher, Conformal transformations of pseudo-Riemannian manifolds, in *Recent Developments in Pseudo-Riemannian Geometry. ESI Lect. Math. Phys., Eur. Math. Soc.* (2008), pp. 261–298
32. J. Lauret, Ricci soliton homogeneous nilmanifolds. Math. Ann. **319**, 715–733 (2001)
33. A. Naber, Noncompact shrinking four solitons with nonnegative curvature. J. Reine Angew. Math. **645**, 125–153 (2010)
34. H. Pedersen, P. Tod, The Ledger curvature conditions and D'Atri geometry. Differ. Geom. Appl. **11**, 155–162 (1999)
35. P. Petersen, W. Wylie, On gradient Ricci solitons with symmetry. Proc. Am. Math. Soc. **137**, 2085–2092 (2009)
36. F. Podestà, A. Spiro, Four-dimensional Einstein-like manifolds and curvature homogeneity. Geom. Dedicata **54**, 225–243 (1995)

Conformal and Isometric Embeddings of Gravitational Instantons

Maciej Dunajski and Paul Tod

Abstract We construct isometric and conformally isometric embeddings of some gravitational instantons in $\mathbb{R}^7$ and $\mathbb{R}^8$. In particular we show that the embedding class of the Einstein–Maxwell instanton due to Burns is equal to 3. For $\mathbb{CP}^2$, Eguchi–Hanson and anti-self-dual Taub-NUT we obtain upper and lower bounds on the embedding class.

1 Introduction

Gravitational instantons are solutions to the four-dimensional Einstein equations in Riemannian signature which give complete metrics and are compact, or asymptotically look like flat space. Some instantons (for example the Euclidean Schwarzchild metric) arise as analytic continuations of Lorentzian metrics, but those with anti-self-dual conformal curvature do not have Lorentzian analogues [5, 11]. In an attempt to visualise gravitational instantons one may isometrically embed them in a flat higher-dimensional ambient space as surfaces.

In this paper we shall consider such embeddings as well as conformal embeddings (see Definition 2.1) where an instanton embeds in a conformally flat space. We shall exhibit explicit embeddings of Eguchi–Hanson, anti-self-dual Taub NUT, and the Fubini-Study metrics. In all three cases the embedding class (see Definition 2.2) is at most 4, and at least 3, and the conformal embedding class is at most 3 and at least 2. The only case (apart from the Schwarzschild instanton, where the embedding class is known to be 2) where we could establish a sharp result about the embedding class is an Einstein–Maxwell instanton known as the Burns metric [2]. It is a scalar-

M. Dunajski (✉)
Department of Applied Mathematics and Theoretical Physics, University of Cambridge, Cambridge, UK
e-mail: m.dunajski@damtp.cam.ac.uk

P. Tod
The Mathematical Institute, Oxford University, Oxford, UK
e-mail: tod@maths.ox.ac.uk

S. Hervik et al. (eds.), *Geometry, Lie Theory and Applications*, Abel Symposia 16,
https://doi.org/10.1007/978-3-030-81296-6_2

flat Kähler metric on the total space of the line bundle $\mathcal{O}(-1) \to \mathbb{CP}^1$ (see (3.28) for the coordinate formula)

Theorem 1.1 *The isometric embedding class of the Burns metric is* 3.

The paper is organised as follows. In the next section we shall recall the basic theory of isometric embeddings of class 2, and follow [1, 21] to give (in Theorem 2.3) necessary conditions for an existence of class 2 embeddings of four-manifolds. In Sect. 3 we shall introduce the Burns metric, and establish Theorem 1.1. In Sect. 4 we shall use the conformal equivalence between the Fubini-Study metric and the Burns metric to rule out the existence of a radially symmetric conformal embedding of $\mathbb{CP}^2$ in $\mathbb{R}^6$, and show (Theorem 4.2) that the conformal embedding of $\mathbb{CP}^2$ in $\mathbb{R}^7$ via Burns is induced by the canonical Veronese embedding $\mathbb{CP}^2 \to S^7$. In Sect. 5 we shall construct explicit conformal embeddings of the Eguchi–Hanson and anti-self-dual (ASD) Taub-NUT gravitational instantons in $\mathbb{R}^7$. The resulting embedding is global in the case of ASD Taub-NUT. We shall prove (Theorem 5.1) that the Burns metric is the unique non-flat locally rotationally symmetric Bianchi IX metric which is ASD and scalar-flat and which admits a radial isometric embedding in $\mathbb{R}^7$. In 5.2 we shall construct isometric embeddings of Eguchi–Hanson and ASD Taub–NUT in $\mathbb{R}^8$.

2 Isometric Embeddings of Class 2

Let $\mathbb{R}^{r,s}$ be an $(r+s)$-dimensional pseudo-Euclidean space with a flat metric η of signature (r, s). We shall start off with a definition

Definition 2.1 An isometric embedding of a pseudo-Riemannian n-dimensional manifold (M, g) as a surface in $\mathbb{R}^{r,s}$ is a map $\iota : M \to \mathbb{R}^{r,s}$ such that $\iota^*(\eta) = g$ and $\iota(M) \subset \mathbb{R}^{r,s}$ is diffeomorphic to M.

In the real analytic category an n-dimensional manifold is always locally embeddable in $\mathbb{R}^N$, where $N = n(n+1)/2$ dimensions [3].

Definition 2.2 The isometric embedding class of (M, g) is the smallest integer k such that there exists an isometric embedding of (M, g) into $\mathbb{R}^{r,s}$ with $r+s-n = k$.

An example of a gravitational instanton with embedding class 2 is the Euclidean Schwarzchild metric

$$g = \left(1 - \frac{2m}{r}\right)d\tau^2 + \left(1 - \frac{2m}{r}\right)^{-1}dr^2 + r^2(d\theta^2 + \sin\theta^2 d\phi^2) \quad r > 2m.$$

If τ is periodic with the period $8\pi m$, then g can be globally and isometrically embedded in $\mathbb{R}^6$ with the flat metric

$$\eta = d\rho^2 + \rho^2 d\psi^2 + dZ^2 + dR^2 + R^2(d\Theta^2 + \sin\Theta^2 d\Phi^2).$$

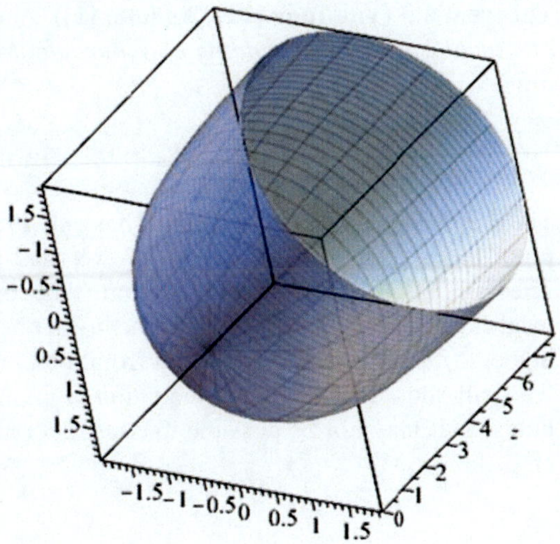

Fig. 1 The surface M (2.1) in (ρ, ψ, Z) coordinates. Each point on M corresponds to a two-sphere

The embedding is given by a modification of the Fronsdal construction [9]

$$R = r, \quad \rho = 4m\sqrt{1 - \frac{2m}{r}},$$
$$Z = \int \sqrt{\left(\frac{2m}{r}\right)^3 + \left(\frac{2m}{r}\right)^2 + \left(\frac{2m}{r}\right)} dr, \quad \psi = \frac{\tau}{4m}, \quad \Theta = \theta, \quad \Phi = \phi. \tag{2.1}$$

The corresponding surface (Fig. 1) in $\mathbb{R}^6$ is approximated by a paraboloid $8mZ = \sqrt{3}\rho^2$ near $r = 2m$, and for large r it asymptotically approaches the flat cylinder $\rho = 4m$. Moreover, as g is Ricci-flat, there does not exist an embedding in $\mathbb{R}^5$, even locally [18]. The embedding class of the Euclidean Schwarschild metric is therefore 2.

In the rest of this section we shall review some necessary conditions for existence of class 2 isometric embeddings of an oriented Riemannian four-manifold (M, g). Let ϵ_{abcd} and $R_{abc}{}^d$ be the volume form and the Riemann tensor of g respectively. Set

$$P_{ab}{}^{cd} := -\frac{1}{2}\epsilon_{abpq} R^{pqcd}, \tag{2.2}$$

and define

$$\Phi_3 \equiv \mathrm{Tr}(P^3) = P^{ab}{}_{cd} P^{cd}{}_{ef} P^{ef}{}_{ab}, \quad \Phi_5 \equiv \mathrm{Tr}(P^5). \tag{2.3}$$

Theorem 2.3 (Yakupov [21], Agaoka [1]) *Necessary conditions for an existence of a local isometric embedding of a four-manifold* (M, g) *into* $\mathbb{R}^{r,s}$ *with* $r + s = 6$ *are*

$$\Phi_3 = 0, \quad \Phi_5 = 0. \tag{2.4}$$

The sketch of proof given in [21] applies only to Einstein manifolds, while Agaoka's proof [1] is relatively long. In § 2.1 we shall give a self-contained proof of this Theorem based on the Gauss equation. The conditions (2.4) are not sufficient for local class 2 embeddability, and further local obstructions arise. Some of these are of higher differential order, and result from the Codazzi equations [14] but in Sect. 2.3 we shall show that the Gauss equation alone may not be satisfiable even if (2.4) holds, as it may not be possible to construct real second-fundamental forms of the embeddings.

2.1 Basic Theory

Let $\iota : M \to \mathbb{R}^{r,s}$ be a class two isometric embedding, with $r \geq 4$ and $r + s = 6$. The embedded surface M has two normals so two second fundamental forms but also a vector defining the connection on the normal bundle. Basic theory can be found for example in [18], but here is another way: Choose an orthonormal basis in 6 dimensions containing the 2 normals along with 4 vectors tangent to M and write the 6dim coordinate vector as

$$X^\alpha = (X^a, Y, Z),$$

so $\alpha = 0, \ldots, 5$ while $a = 0, \ldots, 3$, and

$$\delta_{\alpha\beta} X^\alpha X^\beta = g_{ab} X^a X^b + Y^2 + Z^2,$$

with $\delta_{\alpha\beta}$ the flat metric of $\mathbb{R}^6$ and g_{ab} the induced metric of M.

Proposition 2.4 *There exists symmetric tensors* K_{ab}, L_{ab}, *and a one-form* V_a *such that*

$$R_{abcd} = -2K_{c[a}K_{b]d} - 2L_{c[a}L_{b]d}, \quad \textit{Gauss equation} \tag{2.5}$$

$$\nabla_{[a}K_{b]c} = V_{[a}L_{b]c}, \quad \nabla_{[a}L_{b]c} = -V_{[a}K_{b]c}, \quad \textit{Codazzi equations} \tag{2.6}$$

$$\nabla_{[a}V_{b]} = L_{c[a}K_{b]}{}^{c} \quad \textit{Ricci equation.} \tag{2.7}$$

Proof The symmetric tensors (the *second fundamental forms*) K_{ab}, L_{ab} on M arise from

$$\nabla_a X_b = g_{ab} - Y K_{ab} - Z L_{ab}, \tag{2.8}$$

and a vector (the *torsion vector*) V_a satisfies

$$\nabla_a Y = K_{ab} X^b + Z V_a, \tag{2.9}$$

$$\nabla_a Z = L_{ab} X^b - Y V_a. \tag{2.10}$$

The formalism invites a complexification which emphasises the role of V_a in defining a connection on the normal bundle: set

$$\zeta = Y + iZ, \quad \Sigma_{ab} = K_{ab} + iL_{ab},$$

when

$$\nabla_a \zeta = \Sigma_{ab} X^b - i\zeta V_a. \tag{2.11}$$

Now a rotation in the normal bundle has the effect

$$\zeta \to \hat{\zeta} = e^{i\Theta(X)}\zeta, \quad \Sigma_{ab} \to \hat{\Sigma}_{ab} = e^{i\Theta}\Sigma_{ab}, \quad V_a \to \hat{V}_a = e^{i\Theta}(V_a - \nabla_a \Theta).$$

Commute derivatives on X_a, Y, Z to obtain (2.5, 2.6, 2.12), and hence also

$$\nabla_{[a}\Sigma_{b]c} = -i V_{[a}\Sigma_{b]c}. \tag{2.12}$$

Note that the consistency conditions for (2.6) are satisfied by virtue of themselves and (2.7), and that for (2.7) is automatically satisfied by virtue of (2.6). The Bianchi identity from (2.5) is automatic given (2.6). □

If we want embedding into signatures other than Riemannian we may replace the above by

$$\delta_{\alpha\beta} X^\alpha X^\beta = g_{ab} X^a X^b + \epsilon_1 Y^2 + \epsilon_2 Z^2,$$

with $\epsilon_i = \pm 1$. Then the second fundamental forms occur in the modified expression

$$\nabla_a X_b = g_{ab} - \epsilon_1 Y K_{ab} - \epsilon_2 Z L_{ab}, \tag{2.13}$$

and the vector V_a in

$$\nabla_a Y = K_{ab} X^b + Z V_a,$$

$$\nabla_a Z = L_{ab} X^b - \epsilon_1 \epsilon_2 Y V_a.$$

The Gauss equation becomes

$$R_{abcd} = -2\epsilon_1 K_{c[a}K_{b]d} - 2\epsilon_2 L_{c[a}L_{b]d}, \tag{2.14}$$

the Codazzi equations can be taken to be

$$\nabla_{[a}K_{b]c} = V_{[a}L_{b]c}, \quad \nabla_{[a}L_{b]c} = -\epsilon_1\epsilon_2 V_{[a}K_{b]c}, \tag{2.15}$$

and the Ricci equation becomes

$$\nabla_{[a}V_{b]} = \epsilon_2 L_{c[a}K_{b]}{}^{c}. \tag{2.16}$$

2.2 Necessary Conditions

We shall mostly be concerned with solving the Gauss equation (2.5) in what follows, and it will be sufficient to establish Theorem 2.3

***Proof of Theorem* 2.3** First note an identity for any symmetric T^{ab} in 4 dimensions:

$$\epsilon_{abpq}T^{pm}T^{qn}\epsilon_{mnef}T^{er}T^{fs}\epsilon_{rscd} = 4(\det T)\epsilon_{abcd}, \tag{2.17}$$

(this is in Riemannian signature; there will be sign modifications in Lorentzian signature). The determinant is defined by contraction of this formula with ϵ^{abcd}. The identity is easy to see for diagonal T^{ab}, and in Riemannian signature all symmetric T can be diagonalised.

Introduce a matrix notation: for symmetric T^{ab} the tensor $X_{ab}{}^{cd} := \epsilon_{abpq}T^{pc}T^{qd}$ defines an endomorphism on 2-forms, which is necessarily trace-free. Since

$$X_I{}^J = X_{ab}{}^{cd} = \epsilon_{abef}T^{ec}T^{fd},$$

we need

$$X_{ab}{}^{cb} = 0 \tag{2.18}$$

by the symmetry of T^{ab}. Suppress the indices and use bold font to indicate a 6×6-matrix, then (2.17) can be written

$$\mathbf{X}^2 = 4(\det T)\mathbf{I}, \tag{2.19}$$

where, as a tensor, $\mathbf{I} = \delta_{[a}{}^{c}\delta_{b]}{}^{d}$, and recall that $\mathrm{tr}\mathbf{X} = 0$ by the symmetry of T^{ab}, so that (2.19) is in fact the minimum polynomial of $\mathbf{X}$ (assuming $\mathbf{X}$ is nontrivial).

The Gauss equation (2.5) shows that, as a matrix,

$$\mathbf{P} = \mathbf{X} + \mathbf{Y},$$

where $\mathbf{X}$, $\mathbf{Y}$ are made from K^{ab}, L^{ab} respectively in the place of T^{ab}.

To obtain (2.4) we need to calculate

$$\Phi_3 = \mathrm{tr}((\mathbf{X} + \mathbf{Y})^3).$$

Note from (2.19) that

$$\mathrm{tr}(\mathbf{X}^3) = 4(\det K)\mathrm{tr}(\mathbf{X}) = 0,$$

and

$$\mathrm{tr}(\mathbf{X}^2\mathbf{Y}) = 4(\det K)\mathrm{tr}(\mathbf{Y}) = 0,$$

whence $\Phi_3 = 0$. The second condition $\Phi_5 = 0$ is the vanishing of $\mathrm{tr}((\mathbf{X} + \mathbf{Y})^5)$ and for this we need to consider

$$\mathrm{tr}(\mathbf{X}^5) = 16(\det K)^2\mathrm{tr}(\mathbf{X}) = 0,$$

$$\mathrm{tr}(\mathbf{X}^4\mathbf{Y}) = 16(\det K)^2\mathrm{tr}(\mathbf{Y}) = 0,$$

$$\mathrm{tr}(\mathbf{X}^3\mathbf{Y}^2) = 16(\det K \det L)\mathrm{tr}(\mathbf{X}) = 0,$$

$$\mathrm{tr}(\mathbf{X}^2\mathbf{YXY}) = 16(\det K)\mathrm{tr}(\mathbf{YXY}) = 0,$$

from which the result follows.

The Cayley-Hamilton Theorem for $\mathbf{P}$ shows the vanishing of $\mathrm{tr}(\mathbf{P}^{2k+1})$ inductively for all odd powers, since the characteristic polynomial of $\mathbf{P}$ must take the form

$$\mathbf{P}^6 + c_1\mathbf{P}^4 + c_2\mathbf{P}^2 + c_3\mathbf{I} = 0.$$

□

We can now rule out the existence of class 2 isometric embeddings for all anti-self-dual gravitational instantons, as well as for the Kerr solution

Corollary 2.5 *None of $\mathbb{CP}^2$ with the Fubini-Study metric, the (Lorentzian or Riemannian) Kerr solution, or any hyper-Kähler metric in four-dimensions with the J-invariant not equal to zero have isometric embedding class 2.*

Proof Recall [17] the spinor decomposition of the Weyl tensor

$$C_{abcd} = \psi_{ABCD}\epsilon_{A'B'}\epsilon_{C'D'} + \tilde{\psi}_{A'B'C'D'}\epsilon_{AB}\epsilon_{CD},$$

where ψ, $\tilde{\psi}$ are totally symmetric in their indices, and the ϵ spinors are skew. There are four algebraic invariants of the Weyl spinors:

$$I = \psi_{ABCD}\psi^{ABCD}, \quad J = \psi_{AB}{}^{CD}\psi_{CD}{}^{EF}\psi_{EF}{}^{AB}, \tag{2.20}$$
$$\tilde{I} = \psi_{A'B'C'D'}\psi^{A'B'C'D'}, \quad \tilde{J} = \psi_{A'B'}{}^{C'D'}\psi_{C'D'}{}^{E'F'}\psi_{E'F'}{}^{A'B'},$$

which are in general independent. For an Einstein metric

$$P_{ab}{}^{cd} = -\frac{1}{2}\epsilon_{abef}C^{efcd} - \frac{s}{12}\epsilon_{ab}{}^{cd},$$

where s is the scalar curvature. Therefore

$$\Phi_3 = 8(J - \tilde{J}) + 2s(I - \tilde{I}). \tag{2.21}$$

We can now rule out several possibilities: The Fubini-Study metric on $\mathbb{CP}^2$ is conformaly ASD (see Sect. 4), and we find $\Phi_3 = -s^3/9$. This was established in [1]. Any hyper-Kähler metric in four-dimensions is ASD and Ricci-flat. Therefore $\tilde{I} = \tilde{J} = s = 0$. Therefore $\Phi_3 \neq 0$ iff $J \neq 0$. For the Lorentzian Kerr solutions J is complex and Lorentzian signature $\tilde{J} = \overline{J}$, so $\Phi_3 \neq 0$. In Riemannian signature $J \sim (r - \alpha\cos\theta)^{-9}$ and $\widetilde{J} \sim (r + \alpha\cos\theta)^{-9}$ in the natural Kerr coordinates [8], therefore $J - \widetilde{J} \neq 0$. □

2.3 *Failure of Sufficiency*

Assume that the Riemann curvature of (M, g) satisfies the conditions (2.4) of Theorem (2.3). In order to construct the embedding we should be able to

1. Find $\mathbf{X}$ and $\mathbf{Y}$ such that

$$\mathbf{P} = \mathbf{X} + \mathbf{Y}, \quad \mathbf{X}^2 = \lambda\mathbf{I}, \quad \mathbf{Y}^2 = \mu\mathbf{I}, \quad \mathrm{tr}(\mathbf{X}) = 0 = \mathrm{tr}(\mathbf{Y}), \tag{2.22}$$

   where λ, μ are also to be found.
2. Assuming that $\mathbf{X}$ and $\mathbf{Y}$ satisfying (2.22) have been found, find symmetric tensors K and L such that

$$X_{ab}{}^{cd} := \epsilon_{abpq}K^{pc}K^{qd}, \quad Y_{ab}{}^{cd} := \epsilon_{abpq}L^{pc}L^{qd}. \tag{2.23}$$

3. Assuming that K and L satisfying (2.23) have been found, check that they satisfy the Codazzi equations (2.6).

In fact each of these steps may fail. We shall first show that the second step can fail in general, that is real second fundamental forms need not arise from real **X** and **Y**. In Sect. 3 we shall then use the example of the Burns metric to show that the first step can also fail.

2.4 Obtaining K_{ab} from X in the General Case

Proposition 2.6 *Conditions (2.4) are not sufficient for the existence of symmetric tensors K and L such that the Gauss equation (2.5) holds.*

Proof If we had found real **X**, **Y**, we would still need to find real K_{ab}, L_{ab} and there are new complex issues here. Suppose then we have a candidate real **X** with

$$\mathrm{tr}(\mathbf{X}) = 0, \quad \mathbf{X}^2 = 4\sigma_4\mathbf{I}, \quad X_{acb}{}^{c} = 0,$$

for real constant σ_4, which will be $\det K$, can we find real symmetric K_{ab} satisfying (2.23)? Introduce

$$T_{ab} = \epsilon_{adef} X_b{}^{def}.$$

Then T_{ab} is symmetric by virtue of the conditions on **X** and (2.23) becomes

$$T_{ab} = K_a{}^{c} K_{bc} - \sigma_1 K_{ab}, \tag{2.24}$$

with $\sigma_1 = K_c{}^{c}$. Again introduce a matrix notation, but now for 4×4-matrices:

$$\mathbf{K} := (K_{ab}), \quad \mathbf{T} = (T_{ab}),$$

and let σ_i for $i = 1, \ldots, 4$ be the symmetric polynomials of **K**, so that

$$\sigma_1 = \mathrm{tr}\mathbf{K}, \sigma_4 = \det\mathbf{K},$$

$$\mathrm{tr}(\mathbf{K}^2) = \sigma_1^2 - 2\sigma_2,$$

$$\mathrm{tr}(\mathbf{K}^3) = \sigma_1^3 - 3\sigma_1\sigma_2 + 3\sigma_3,$$

$$\mathrm{tr}(\mathbf{K}^4) = \sigma_1^4 - 4\sigma_1^2\sigma_2 + 2\sigma_2^2 + 4\sigma_1\sigma_3 - 4\sigma_4,$$

and the Cayley-Hamilton Theorem for **K** is

$$\mathbf{K}^4 - \sigma_1\mathbf{K}^3 + \sigma_2\mathbf{K}^2 - \sigma_3\mathbf{K} + \sigma_4\mathbf{I} = 0.$$

Given the Cayley-Hamilton Theorem we may calculate

$$\mathrm{tr}(\mathbf{K}^5) = \sigma_1^5 - 5\sigma_1^3\sigma_2 + 5\sigma_1^2\sigma_3 + 5\sigma_1\sigma_2^2 - 5\sigma_2\sigma_3 - 5\sigma_1\sigma_4,$$

and

$$\mathrm{tr}(\mathbf{K}^6) = \sigma_1^6 - 6\sigma_1^4\sigma_2 + 6\sigma_1^3\sigma_3 + 9\sigma_1^2\sigma_2^2 - 6\sigma_1^2\sigma_4 - 12\sigma_1\sigma_2\sigma_3 + 3\sigma_3^2 + 6\sigma_2\sigma_4.$$

In matrix notation, (2.24) becomes

$$\mathbf{T} = \mathbf{K}^2 - \sigma_1\mathbf{K}, \tag{2.25}$$

and we wish to solve this for real symmetric **K** given real symmetric **T**. Since **K** is real symmetric, there will be real orthogonal **O** and real diagonal **Δ** with

$$\mathbf{K} = \mathbf{O}^T\mathbf{\Delta}\mathbf{O},$$

so (2.25) becomes

$$\mathbf{T} = \mathbf{O}(\mathbf{\Delta}^2 - \sigma_1\mathbf{\Delta})\mathbf{O}^T,$$

and therefore the same **O** diagonalises **T**. Given **T** we find **O**, then we just need the σ_i to fix **Δ**. We already have

$$\sigma_4 = \frac{1}{24}\mathrm{tr}(\mathbf{X}^2).$$

From (2.25) by taking the trace we obtain

$$\sigma_2 = -\frac{1}{2}\mathrm{tr}(\mathbf{T}).$$

By squaring and tracing (2.25) we find

$$\sigma_1\sigma_3 = \sigma_2^2 - 2\sigma_4 - \frac{1}{2}\mathrm{tr}(\mathbf{T}^2), \tag{2.26}$$

and by cubing and tracing (2.25)

$$\mathrm{tr}(\mathbf{T}^3) = 5\sigma_1^2\sigma_4 + 3\sigma_1\sigma_2\sigma_3 + 3\sigma_3^2 + 6\sigma_2\sigma_4. \tag{2.27}$$

Eliminate σ_1 in favour of σ_3 with the aid of (2.26), then (2.27) gives a quadratic for σ_3^2 (What if the expression on the right in (2.26) vanishes? Then one of σ_1, σ_3 vanishes; one chooses which and then (2.27) gives the other, up to sign—the solution is non-unique but possibly complex). There is a sign ambiguity in σ_3 and therefore also in σ_1 but this is to be expected as K_{ab} can only be known up to sign, but, starting from **X**, there is no reason for σ_3 or σ_1 to be real, so reality of these is an extra condition on **T** and therefore on **X**. To see that this is a genuine issue, note that $\mathbf{K} = i\mathbf{I}$ leads to a real **T**, so indeed a real **T** is not obliged to give a real **K**. □

3 The Burns Metric

In this section we shall introduce the Burns metric [2]. It is Kähler with vanishing scalar curvature (and therefore has anti-self-dual Weyl tensor), and it is one of the few metrics where the embedding class can be established as in Theorem 1.1.

The Burns metric is the unique scalar-flat Kähler metric on the total space of the line bundle $\mathcal{O}(-1) \to \mathbb{CP}^1$. It can also be regarded as an Einstein–Maxwell gravitational instanton, with the self-dual part of the Maxwell field strength given by the Kähler form [4].

There are several ways to present it in coordinates, and we chose to write it as

$$g_B = d\rho^2 + \frac{1}{4}\rho^2(\sigma_1^2 + \sigma_2^2 + \sigma_3^2) + \frac{1}{4}(\sigma_1^2 + \sigma_2^2), \tag{3.28}$$

where $\sigma_1, \sigma_2, \sigma_3$ are left-invariant one forms on $SU(2)$, such that $d\sigma_1 + \sigma_2 \wedge \sigma_3 = 0$ etc. We shall verify by explicit computation that the conditions 2.4 of Theorem 2.3 are satisfied for this metric, and yet there does not exist a class 2 isometric embedding as real **X** does not exist. We shall also establish a (rather obvious) embedding in $\mathbb{R}^7$. This will complete the proof of Theorem 1.1.

Proof of Theorem 1.1 We will first show that the (2.4) hold for the Burns metric, and then demonstrate that these conditions are not sufficient for the existence of class 2 embedding. The vanishing of Φ_3 and Φ_5 could be verified by a MAPLE computation given the form of the metric (3.28) but we find it instructive to present a computation which can be directly checked by readers, and exhibits the expicit form of **P**.

To calculate the curvature, use Cartan calculus with the orthonormal tetrad:

$$\theta^0 = d\rho, \;\; \theta^1 = \frac{1}{2}(1+\rho^2)^{1/2}\sigma^1, \;\; \theta^2 = \frac{1}{2}(1+\rho^2)^{1/2}\sigma^2, \;\; \theta^3 = \frac{\rho}{2}\sigma^3. \tag{3.29}$$

The connection 1-forms are:

$$\omega^0{}_1 = -\frac{\rho}{1+\rho^2}\theta^1, \;\; \omega^0{}_2 = -\frac{\rho}{1+\rho^2}\theta^2, \;\; \omega^0{}_3 = -\frac{1}{\rho}\theta^3,$$

$$\omega^1{}_2 = \frac{\rho^2+2}{\rho(1+\rho^2)}\theta^3, \;\; \omega^3{}_1 = \frac{\rho}{(1+\rho^2)}\theta^2, \;\; \omega^2{}_3 = \frac{\rho}{(1+\rho^2)}\theta^1,$$

leading to Riemann tensor components

$$R_{0101} = R_{0202} = -R_{0123} = -R_{0231} = R_{3131} = R_{2323} = -(1+\rho^2)^{-2},$$

$$R_{0312} = -2(1+\rho^2)^{-2}, \quad R_{1212} = 4(1+\rho^2)^{-2}.$$

The Ricci components are

$$R_{00} = -R_{11} = -R_{22} = R_{33} = -2(1+\rho^2)^{-2},$$

and the scalar curvature is zero (as expected).

Now consider $P_I{}^J \equiv -P_{ab}{}^{cd} = \frac{1}{2}\epsilon_{abpq}R^{pqcd}$ as a 6×6 matrix with the indices ordered

$$I, J = 01, 23, 02, 31, 03, 12.$$

Claim this is

$$\mathbf{P} = \begin{pmatrix} -1 & 1 & 0 & 0 & 0 & 0 \\ 1 & -1 & 0 & 0 & 0 & 0 \\ 0 & 0 & -1 & 1 & 0 & 0 \\ 0 & 0 & 1 & -1 & 0 & 0 \\ 0 & 0 & 0 & 0 & 2 & -4 \\ 0 & 0 & 0 & 0 & 0 & 2 \end{pmatrix}, \tag{3.30}$$

omitting the factor $(1+R^2)^{-2}$ which is common to all terms. This can be written

$$\mathbf{P} = \mathrm{diag}\left(\begin{pmatrix} -1 & 1 \\ 1 & -1 \end{pmatrix}, \begin{pmatrix} -1 & 1 \\ 1 & -1 \end{pmatrix}, \begin{pmatrix} 2 & -4 \\ 0 & 2 \end{pmatrix}\right),$$

and this form is easier for calculation. One finds

$$\mathbf{P}^3 = \mathrm{diag}\left(\begin{pmatrix} -4 & 4 \\ 4 & -4 \end{pmatrix}, \begin{pmatrix} -4 & 4 \\ 4 & -4 \end{pmatrix}, \begin{pmatrix} 8 & -48 \\ 0 & 8 \end{pmatrix}\right),$$

$$\mathbf{P}^5 = \mathrm{diag}\left(\begin{pmatrix} -16 & 16 \\ 16 & -16 \end{pmatrix}, \begin{pmatrix} -16 & 16 \\ 16 & -16 \end{pmatrix}, \begin{pmatrix} 32 & -320 \\ 0 & 32 \end{pmatrix}\right),$$

Now it is clear that Agaoka's necessary conditions $\mathrm{tr}\mathbf{P}^3 = 0 = \mathrm{tr}\mathbf{P}^5$ both hold.

If the conditions (2.4) of Theorem 2.3 were sufficient we should now be able to find the isometric embedding of the Burns metric and so in particular to solve the Gauss equation. This requires finding $\mathbf{X}, \mathbf{Y}$ satisfying (2.22). However, there is another condition on $\mathbf{X}, \mathbf{Y}$, as (2.18) holds automatically for $\mathbf{P} = \mathbf{X} + \mathbf{Y}$ but would not hold necessarily for the summands $\mathbf{X}, \mathbf{Y}$ separately unless we imposed it.

We may seek **X** in partitioned form as

$$\mathbf{X} = \begin{pmatrix} A & B & C \\ D & E & F \\ J & K & L \end{pmatrix}$$

where all blocks are 2×2, and decompose **P** for brevity as

$$\mathbf{P} = \begin{pmatrix} M & 0 & 0 \\ 0 & M & 0 \\ 0 & 0 & N \end{pmatrix},$$

with

$$M = \begin{pmatrix} -1 & 1 \\ 1 & -1 \end{pmatrix}, \quad N = \begin{pmatrix} 2 & -4 \\ 0 & 2 \end{pmatrix}.$$

Write $\mathbf{Y} = \mathbf{P} - \mathbf{X}$ and square

$$\mathbf{Y}^2 = \mu\mathbf{I} = \mathbf{P}^2 - (\mathbf{PX} + \mathbf{XP}) + \lambda\mathbf{I},$$

or

$$\mathbf{PX} + \mathbf{XP} = \mathbf{P}^2 + (\lambda - \mu)\mathbf{I} = \mathbf{P}^2 + k\mathbf{I} \text{ say},$$

with $k = \lambda - \mu$. Now look at this in blocks:

$$\begin{aligned}
\mathbf{PX} + \mathbf{XP} &= \begin{pmatrix} AM + MA & BM + MB & CN + MC \\ DM + MD & EM + ME & FN + MF \\ JM + NJ & KM + NK & LN + NL \end{pmatrix} \\
&= \mathbf{P}^2 + k\mathbf{I} \\
&= \begin{pmatrix} M^2 + kI & 0 & 0 \\ 0 & M^2 + kI & 0 \\ 0 & 0 & N^2 + kI \end{pmatrix}.
\end{aligned}$$

This leads to constraints on the blocks in **X** which we can solve successively. First solve $BM + MB = 0$: set $B = \begin{pmatrix} \alpha & \beta \\ \gamma & \delta \end{pmatrix}$ then

$$MB + BM = \begin{pmatrix} -2\alpha + \beta + \gamma & -2\beta + \alpha + \delta \\ -2\gamma + \alpha + \delta & -2\delta + \beta + \gamma \end{pmatrix} = 0.$$

This rapidly leads to

$$B = \begin{pmatrix} \beta & \beta \\ \beta & \beta \end{pmatrix},$$

for some β, and similarly

$$D = \begin{pmatrix} \delta & \delta \\ \delta & \delta \end{pmatrix},$$

for some δ. Next solve $CN + MC = 0$: set $C = \begin{pmatrix} \alpha & \rho \\ \gamma & \sigma \end{pmatrix}$ then

$$CN + MC = \begin{pmatrix} \alpha + \gamma & -4\alpha + \rho + \sigma \\ \alpha + \gamma & -4\gamma + \rho + \sigma \end{pmatrix} = 0,$$

whence

$$C = \begin{pmatrix} 0 & \gamma \\ 0 & -\gamma \end{pmatrix},$$

for some γ, and similarly

$$F = \begin{pmatrix} 0 & \phi \\ 0 & -\phi \end{pmatrix},$$

for some ϕ. Next solve $JM + NJ = 0$: set $J = \begin{pmatrix} \rho & \sigma \\ \tau & \nu \end{pmatrix}$ then

$$JM + NJ = \begin{pmatrix} -4\tau + \rho + \sigma & -4\nu + \rho + \sigma \\ \tau + \nu & \tau + \nu \end{pmatrix} = 0,$$

whence

$$J = \begin{pmatrix} \rho & -\rho \\ 0 & 0 \end{pmatrix},$$

for some ρ and

$$K = \begin{pmatrix} \kappa & -\kappa \\ 0 & 0 \end{pmatrix},$$

for some κ. For A, E, L we have inhomogeneous equations (and we will re-use Greek letters). With $A = \begin{pmatrix} \alpha & \beta \\ \gamma & \delta \end{pmatrix}$ we want

$$AM + MA = \begin{pmatrix} -2\alpha + \beta + \gamma & -2\beta + \alpha + \delta \\ -2\gamma + \alpha + \delta & -2\delta + \beta + \gamma \end{pmatrix} = \begin{pmatrix} 2+k & -2 \\ -2 & 2+k \end{pmatrix},$$

which only has solutions with $k = 0$ i.e. $\lambda = \mu$, and then

$$A = \begin{pmatrix} \alpha & \alpha+1 \\ \alpha+1 & \alpha \end{pmatrix},$$

for some α, when also

$$E = \begin{pmatrix} \epsilon & \epsilon+1 \\ \epsilon+1 & \epsilon \end{pmatrix},$$

for some ϵ. Finally we need $L = \begin{pmatrix} \alpha & \beta \\ \gamma & \delta \end{pmatrix}$ satisfying

$$LN + NL = \begin{pmatrix} -4\gamma + 4\alpha & -4\alpha - 4\delta + 4\beta \\ 4\gamma & -4\gamma + 4\delta \end{pmatrix} = \begin{pmatrix} 4 & -16 \\ 0 & 4 \end{pmatrix}$$

since we now know that $k = 0$, and this leads to

$$L = \begin{pmatrix} 1 & -2 \\ 0 & 1 \end{pmatrix}.$$

We have found $\mathbf{X}$ in terms of parameters $\alpha, \beta, \gamma, \delta, \epsilon, \kappa, \phi$ and ρ, explicitly

$$\mathbf{X} = \begin{pmatrix} \alpha & \alpha+1 & \beta & \beta & 0 & \gamma \\ \alpha+1 & \alpha & \beta & \beta & 0 & -\gamma \\ \delta & \delta & \epsilon & \epsilon+1 & 0 & \phi \\ \delta & \delta & \epsilon+1 & \epsilon & 0 & -\phi \\ \kappa & -\kappa & \rho & -\rho & 1 & -2 \\ 0 & 0 & 0 & 0 & 0 & 1 \end{pmatrix}.$$

For the vanishing trace we need $\alpha + \epsilon = -1$, which we impose, then we impose $\mathbf{X}^2 = \lambda \mathbf{I}$ for some λ. By inspection of $\mathbf{X}$ (consider the bottom right entry of $\mathbf{X}^2$) only $\lambda = 1$ is possible. For the rest we obtain the conditions

$$\alpha(\alpha+1) + \beta\delta = 0 = \epsilon(\epsilon+1) + \beta\delta,$$

and

$$\kappa\gamma + \rho\phi = 1. \tag{3.31}$$

Given $\alpha+\epsilon+1 = 0$ from the vanishing trace, only one of the first pair is independent, so there are 6 free parameters remaining in **X**, and **Y** is uniquely determined given **X**.

However we still need to impose (2.18). Claim

$$X_{0b}{}^{1b} = \phi + \rho = 0,$$

$$X_{0b}{}^{2b} = -\gamma - \kappa = 0,$$

$$X_{0b}{}^{3b} = \beta - \delta = 0,$$

and the rest give nothing new. Now we have $\phi = -\rho, \kappa = -\gamma$ which with (3.31) gives

$$1 = \kappa\gamma + \rho\phi = -\kappa^2 - \phi^2,$$

which can't be satisfied with reals: there are no suitable real **X**, **Y**. In particular the Burns metric is not isometric embedding class 2 into Riemannian $\mathbb{R}^6$. By (2.14), embedding into other signatures still requires real **X** and so these are also ruled out.

To complete the proof we should establish an isometric embedding of the Burns metric in $\mathbb{R}^7$, but this is readily done: the first factor in (3.28) just flat $\mathbb{R}^4$, and the second factor is the round two-sphere embeddable in $\mathbb{R}^3$. To write this condition explicitly, as intersection of algebraic surfaces consider a flat metric on $\mathbb{R}^7$

$$\eta = |d\zeta_1|^2 + |d\zeta_2|^2 + dx_1{}^2 + dx_2{}^2 + dx_3{}^2,$$

where

$$x_1 = Z\cos\theta\cos\phi, \quad x_2 = Z\cos\theta\sin\phi, \quad x_3 = Z\sin\theta$$

and

$$\zeta_1 = \rho\cos\left(\frac{\theta}{2}\right)e^{i(\psi+\phi)/2}, \quad \zeta_2 = \rho\sin\left(\frac{\theta}{2}\right)e^{i(\psi-\phi)/2}$$

so one condition is

$$\left(\frac{|\zeta_1|^2 - |\zeta_2|^2}{|\zeta_1|^2 + |\zeta_2|^2}\right)^2 = \frac{x_3{}^2}{x_1{}^2 + x_2{}^2 + x_3{}^2}. \tag{3.32}$$

We also note that $\zeta_1/\zeta_2 = |\zeta_1/\zeta_2|e^{i\phi}$ so the second condition is

$$i\frac{\zeta_1\bar{\zeta}_2 + \bar{\zeta}_1\zeta_2}{\zeta_1\bar{\zeta}_2 - \bar{\zeta}_1\zeta_2} = \frac{x_1}{x_2}$$

which looks like an orbifold in $\mathbb{R}^7$. The final condition is

$$x_1{}^? + x_2{}^? + x_3{}^? = 1/4.$$

□

4 Conformal Isometric Embeddings of the Fubini-Study Metric

Recall Definition 2.1 of an isometric embedding, and make the following definition [6, 16]

Definition 4.1 A conformally isometric embedding of a pseudo-Riemannian n-dimensional manifold (M, g) as a surface in $\mathbb{R}^{r,s}$ is a map $\iota : M \to \mathbb{R}^{r,s}$ such that $\iota^*(\eta) = \Omega^2 g$ for some $\Omega : M \to \mathbb{R}^+$ and $\iota(M) \subset \mathbb{R}^{r,s}$ is diffeomorphic to M.

We shall also define the conformal embedding class of (M, g) to be the smallest integer k, such that (M, g) can be conformally isometrically embedded in $\mathbb{R}^{r,s}$ with $r + s - n = k$.

In this section we shall use the conformal equivalence between the Burns metric and the Fubini-Study metric on $\mathbb{CP}^2$ to show that the later does not admit a conformal embedding of class 2, under an additional assumption that the conformal factor is constant on the $SU(2)$ orbits (the existence of the general conformal embedding of class 1 for $\mathbb{CP}^2$ was ruled out in [6]).

The local form of the Fubini-Study metric is [10]

$$g = \frac{dr^2}{(1+r^2)^2} + \frac{1}{4}\frac{r^2\sigma_3^2}{(1+r^2)^2} + \frac{1}{4}\frac{r^2}{1+r^2}(\sigma_1^2 + \sigma_2^2). \tag{4.1}$$

The metric is regular everywhere on $\mathbb{CP}^2$, and the apparent singularity at $r = 0$ results from using spherical polars. Computing Φ_3 given by (2.3) for (4.1) gives -1536 (the Ricci scalar is 24 - if the Ricci scalar was s, then we would get $-s^3/9$). We are *assuming* that a global conformal embedding (if one exists) must have a conformal factor depending only of r. If we do not assume this, then we are just restricting the class of conformal factors. Set $\hat{g} = \Omega^2 g$, where (for convenience) we set $\Omega = \exp G(r)$. Now we compute Φ_3, and find that it vanishes iff

$$-r\left(-3+\left(r^3+r\right)\frac{d}{dr}G(r)\right)\left(r^2+1\right)^2\frac{d^2}{dr^2}G(r)+r^2\left(r^2+1\right)^3\left(\frac{d}{dr}G(r)\right)^3+$$

$$\left(-2r^7+6r^3+4r\right)\left(\frac{d}{dr}G(r)\right)^2+\left(3r^4+12r^2+9\right)\frac{d}{dr}G(r)-16r=0 \tag{4.2}$$

which is a first order ODE for $G'(r)$. We have shown that $\Phi_5 \equiv \mathrm{Tr}(P^5)$ is another obstruction. Computing this for $\hat{g}$ gives another second order ODE for $G(r)$ (this one also does not contain $G(r)$, but takes more space so we do not write it down). We now solve (4.2) for G'' and substitute to $\Phi_5 = 0$. This gives five candidates for $G'(r)$, and we find that only $G' = -2/((r^2+1)r)$ satisfies both $\Phi_3 = 0$ and $\Phi_5 = 0$. The resulting conformal factor is remarkably simple:

$$\Omega = 1 + r^{-2}, \tag{4.3}$$

and makes $\hat{g}$ scalar flat. To examine $\hat{g}$ set $r = 1/\rho$, so that

$$\hat{g} = g_B = d\rho^2 + \frac{1}{4}\rho^2(\sigma_1^2+\sigma_2^2+\sigma_3^2) + \frac{1}{4}(\sigma_1^2+\sigma_2^2)$$

which is the Burns metric (3.28). We have shown in Theorem 1.1 that this metric does not admit an isometric embedding in $\mathbb{R}^6$. This rules out the existence of the conformal embedding of $\mathbb{CP}^2$ in $\mathbb{R}^6$, at least if Ω is only allowed to depend on the radial coordinate.

The Burns metric isometrically embeds in $\mathbb{R}^7$. In the next section we shall show that the resulting conformal embedding of $\mathbb{CP}^2$ is equivalent to the standard one which results from the isometric embedding [15] (see also [13] for a more recent application of this embedding) of $\mathbb{CP}^2$ in S^7.

4.1 *Canonical Embedding of* $\mathbb{CP}^2$ *in* S^7

Let $H \cong \mathbb{R}^8$ be the 8 dimensional space of 3 by 3 Hermitian matrices with trace one. The pairing

$$(A, B) = \mathrm{Tr}(AB), \quad A, B \in H \tag{4.4}$$

induces a flat Euclidean metric[1] on $\mathbb{R}^8$, as $(A, A) = A_{ij}\overline{A}_{ij}$.

[1] Alternatively, this flat metric is induced on the hyper-plane $\mathrm{Tr}(A) = 1$ on the space $\mathbb{R}^9$ of all 3 by 3 Hermitian matrices.

Let $Z = [Z^1, Z^2, Z^3]$ be homogeneous coordinates of a point in $\mathbb{CP}^2$. A distance between two points $Z, W \in \mathbb{CP}^2$ measured along geodesics of the Fubini-Study metric is

$$d(Z, W) = 2 \arccos \sqrt{\kappa(Z, W)}, \quad \text{where} \quad \kappa(Z, W) = \frac{|<Z, W>|^2}{|Z|^2 |W|^2},$$

$$\text{and} \quad < Z, W > \equiv \overline{Z}^\alpha W^\alpha.$$

Using this notation, the Fubini-Study metric is the quadratic part of the expression

$$\kappa(Z, Z + dZ) = 1 - \frac{|Z|^2 |dZ|^2 - | < Z, dZ > |^2}{|Z|^4} + O(dZ^3).$$

From now on we shall impose a normalisation condition $< Z, Z > \equiv \overline{Z}^\alpha Z^\alpha = 1$. Consider the embedding $\phi : \mathbb{CP}^2 \to H$ given by

$$\phi(Z) = \begin{pmatrix} |Z^1|^2 & Z^1\overline{Z}^2 & Z^1\overline{Z}^3 \\ Z^2\overline{Z}^1 & |Z^2|^2 & Z^2\overline{Z}^3 \\ Z^3\overline{Z}^1 & Z^3\overline{Z}^2 & |Z^3|^2 \end{pmatrix}. \tag{4.5}$$

This embedding is $SU(3)$ invariant, and isometric (in flat $\mathbb{R}^8$) as

$$\mathrm{Tr}(\phi(Z)\phi(W)) = | < W, Z > |^2.$$

The normalisation implies $\mathrm{Tr}(\phi(Z)^2) = 1$, so this is also an isometric embedding of $\mathbb{CP}^2$ in S^7. Combining this result with the stereographic projection form S^7 to $\mathbb{R}^7$ gives a conformal embedding of $\mathbb{CP}^2$ in $\mathbb{R}^7$, and a natural question arises whether this conformal embedding is different to the one we established in Sect. 4 via conformal equivalence with the Burns metric.

Theorem 4.2 *Let $\pi : S^7 \to \mathbb{R}^7$ be the stereographic projection. The map*

$$\pi \circ \phi : \mathbb{CP}^2 \to \mathbb{R}^7$$

is a conformal isommetric embedding of $\mathbb{CP}^2$, and the diagram

$$\begin{array}{ccc} (\mathbb{CP}^2, g_{FS}) & \xrightarrow{\phi} & S^7 \subset \mathbb{R}^8 \\ \downarrow \Omega^2 & & \downarrow \pi \\ \textit{Burns metric} & \longrightarrow & \mathbb{R}^7 \end{array} \tag{4.6}$$

is commutative.

Proof Consider a metric on S^7 induced from the flat $\mathbb{R}^8$

$$g_{S^7} = dR^2 + R^2 g_{S^6} + dY^2 = \frac{dR^2}{1-R^2} + R^2 g_{S^6}, \quad \text{where} \quad R^2 + Y^2 = 1, \tag{4.7}$$

and g_{S^6} is the round metric on the six-sphere. The metric $F^2 g_{S^7}$ with

$$F^2 = \frac{1}{(1+\sqrt{1-R^2})^2} \tag{4.8}$$

is flat. To construct the embedding explicitly in the Bianchi-IX coordinates set

$$Z^1 = \frac{r}{\sqrt{1+r^2}} \sin(\theta/2) e^{-i\phi/2}, \quad Z^2 = \frac{r}{\sqrt{1+r^2}} \cos(\theta/2) e^{i\phi/2}, \quad Z^3 = \frac{1}{\sqrt{1+r^2}} e^{-i\psi}. \tag{4.9}$$

The corresponding element of $H \cong \mathbb{R}^8$ is

$$\phi(Z) = A = \begin{pmatrix} x & p & q \\ \overline{p} & y & s \\ \overline{q} & \overline{s} & 1-x-y, \end{pmatrix}$$

where real coordinates (x, y) and complex coordinates (p, q, s) are now explicit functions of (r, ϕ, θ, ψ) and we verify that $(d\phi(Z), d\phi(Z))$ is twice the Fubini-Study metric (4.1). The sphere equation $\text{Tr}(A^2) = 1$ takes the form

$$3(|p|^2 + |q|^2 + |s|^2) + \left(\frac{3}{2}(x+y) - 1\right)^2 + \left(\frac{\sqrt{3}}{2}(x-y)\right)^2 = 1 \tag{4.10}$$

and (a constant rescaling of) the metric is

$$\frac{3}{2}\text{Tr}(dA^2) = |\sqrt{3}dp|^2 + |\sqrt{3}dq|^2 + |\sqrt{3}dr|^2 + d\left(\frac{3}{2}(x+y) - 1\right)^2 + d\left(\frac{\sqrt{3}}{2}(x-y)\right)^2. \tag{4.11}$$

A choice of a coordinate Y in (4.7) is equivalent to a choice of a point on S^7 from which the projection $\pi : S^7 \to \mathbb{R}^7$ is made. Every such choice gives rise to a conformal factor (4.8). To reproduce the Burns metric as the pull-back $(\pi \circ \phi)^* g_{\mathbb{R}^7}$ consider[2]

$$Y = \frac{3}{2}(x+y) - 1 = \frac{3r^2}{2(1+r^2)} - 1, \quad \text{which gives} \quad F = \frac{1}{1+Y} = \frac{2}{3}(1+r^{-2})$$

[2]We note that other choices of Y give rise to non-radial conformal factors, and conformal embeddings of $\mathbb{CP}^2$ in S^7.

in agreement with (4.3) up to a constant conformal factor. We now need to show that not only the conformal factors, but also the embeddings agree in this case. To do that, recall that a flat metric on $\mathbb{R}^8$

$$\delta_{ij} dX^i dX^j + dY^2, \quad i, j = 1, \ldots, 7$$

gives rise to a metric on the sphere $\delta_{ij} X^i X^j + Y^2 = 1$ given by

$$g_{S^7} = \frac{d\xi_1{}^2 + \cdots + d\xi_7{}^2}{(1 + \xi_1{}^2 + \cdots + \xi_7{}^2)^2}, \quad \text{where} \quad \xi_i = \frac{X^i}{1+Y}. \tag{4.12}$$

This is conformally flat, with the conformal factor $(1 + Y)^{-2}$ as before. The flat Cartesian coordinates on $\mathbb{R}^7$ are given by (4.12) with

$$X^1+iX^2 = \sqrt{3}Z^1\overline{Z}^2, \quad X^3+iX^4 = \sqrt{3}Z^1\overline{Z}^3, \quad X^5+iX^6 = \sqrt{3}Z^2\overline{Z}^3, \quad X^7 = \frac{\sqrt{3}}{2}(|Z^1|^2-|Z^2|^2).$$

This gives

$$\begin{aligned}
\xi_1 + i\xi_2 &= \frac{1}{\sqrt{3}} \sin\theta e^{-i\phi}, \quad \xi_7 = \frac{1}{\sqrt{3}} \cos\theta, \\
\xi_3 + i\xi_4 &= \frac{1}{\sqrt{3}}\rho \sin(\theta/2) e^{i(\psi-\phi/2)}, \quad \xi_5 + i\xi_6 = \frac{1}{\sqrt{3}}\rho \cos(\theta/2) e^{i(\psi+\phi/2)}
\end{aligned} \tag{4.13}$$

where $\rho = r^{-1}$. The metric $d\xi_3{}^2 + d\xi_4{}^2 + d\xi_5{}^2 + d\xi_6{}^2$ is the flat $\mathbb{R}^4$ factor in (3.28) and $d\xi_1{}^2 + d\xi_2{}^2 + d\xi_7{}^2$ is the S^2 factor. They add up to the Burns metric (3.28). □

5 Conformal Embeddings of LRS Bianchi-Type IX in $\mathbb{R}^7$

The name *LRS Bianchi-type IX* means there is an isometry group locally isomorphic to $SU(2) \times U(1)$ transitive on 3-surfaces, with isotropy $U(1)$ at each point (so LRS stands for *locally-rotationally symmetric*). Einstein or vacuum examples are, without loss of generality, diagonal in a basis of invariant 1-forms [20] so we consider only the metric form: Let

$$g = a dr^2 + b(\sigma_1^2 + \sigma_2^2) + c\sigma_3^2, \tag{5.1}$$

and let

$$\eta = dR^2 + \frac{1}{4}R^2(\sigma_1^2 + \sigma_2^2 + \sigma_3^2) + \epsilon(dZ^2 + Z^2(\sigma_1^2 + \sigma_2^2)) \tag{5.2}$$

be a pull back of a flat metric from $\mathbb{R}^7$ (if $\epsilon = 1$) or $\mathbb{R}^{4,3}$ (if $\epsilon = -1$) to a five-manifold obtained by identifying the angles of the two-sphere factors in $\mathbb{R}^4$ and $\mathbb{R}^3$. The radial conformal embedding condition

$$\Omega^2 g = \eta, \quad \Omega = \Omega(r), \quad R = R(r), \quad Z = Z(r) \tag{5.3}$$

gives

$$R^2 = 4\Omega^2 c, \quad Z^2 = \epsilon\Omega^2(b - c) \geq 0 \quad \text{(this determines the sign of } \epsilon)$$

and

$$\Omega = \exp\Big(\int \frac{-\dot{c} - \epsilon\dot{h} \pm \sqrt{a(c + \epsilon h) - \frac{(c\dot{h} - h\dot{c})^2}{\epsilon ch}}}{2(c + \epsilon h)} dr\Big), \quad \text{where} \quad h = \frac{1}{4}(b - c). \tag{5.4}$$

Therefore in general there are two conformal factors which make the embedding possible. They are related by the involution

$$\Omega \to \frac{1}{(c + \epsilon h)\Omega}.$$

Now look at some ASD examples which will all have $\epsilon = 1$. We know (Corollary 2.4 in [6]) that all these must have a conformal embedding class at least 2. The calculations below show that for each of these examples the conformal embedding class is at most 3

- The Burns metric has

$$a = 1, \quad b = \frac{1}{4}(r^2 + 1), \quad c = \frac{1}{4}r^2,$$

  which gives

$$\Omega = 1, \quad \text{or} \quad \Omega = \frac{1}{1 + 4r^2}$$

  depending on the choice of sign in the integral (5.4). This embedding is global.
- The Eguchi–Hanson metric [7] has

$$a = \Big(1 - \frac{a^4}{r^4}\Big)^{-1}, \quad b = \frac{1}{4}r^2, \quad c = \frac{1}{4}r^2\Big(1 - \frac{a^4}{r^4}\Big), \quad \text{where} \quad r > a$$

which gives

$$\Omega = \exp\Big(\int \frac{3a^4 + 4r^4 \pm 2a^4\sqrt{\frac{7a^4-4r^4}{a^4-r^4}}}{r(3a^4 - 4r^4)} dr\Big).$$

This only covers the $r > (7/4)^{1/4}a$ range of the EH-manifold, and even in this range the embedding is only local, as the regularity of the Eguchi–Hanson metric at $r = a$ requires the range of ψ to be $0 \leq \psi \leq 2\pi$ (rather than $0 \leq \psi \leq 4\pi$ which is used in to cover $\mathbb{R}^7$ in (5.2)). A global isometric embedding of the Eguchi–Hanson manifold in $\mathbb{R}^{11}$ has been presented in [12].

- The Fubini-Study is conformal to Burns. It has

$$a = \frac{1}{(r^2+1)^2}, \quad b = \frac{1}{4}\frac{r^2}{r^2+1}, \quad c = \frac{1}{4}\frac{r^2}{(r^2+1)^2}$$

and we find

$$\Omega = 1 + r^{-2}, \quad \text{or} \quad \Omega = \frac{r^2+1}{r^2+4},$$

the first of which we already knew from (4.3).

- The Taub–NUT metric has

$$a = \frac{1}{4}\frac{r+m}{r-m}, \quad b = \frac{1}{4}(r^2 - m^2), \quad c = m^2\frac{r-m}{r+m}, \quad r > m$$

which gives

$$\Omega = \exp\Big(\int \frac{(3m+r)(4m^2 - mr + r^2) \pm (r+m)^2\sqrt{\frac{55m^3+3m^2r+5mr^2+r^3}{3m+r}}}{(m^2 - r^2)(13m^2 + 2mr + r^2)} dr\Big).$$

This covers the whole range $r > m$, and MAPLE computes Ω in terms of elliptic integrals. To analyse the apparent singularity at $r = m$ set $r = m + \rho^2/(2m)$ so that near $r = m$ the metric is flat

$$g \sim d\rho^2 + \frac{\rho^2}{4}\Big(\sigma_1^2 + \sigma_2^2 + \sigma_3^2\Big).$$

Making the same coordinare change in Ω yelds two conformal factors: $\Omega_1 \sim 1$, and $\Omega_2 \sim \rho^{-1}$ near $\rho = 0$. Using Ω_1 for conformal rescalling gives a flat metric near the NUT point $r = m$. The second conformal factor moves the NUT to ∞, which can be seen by replacing making a coordinate transformation $\rho = 1/\hat{\rho}$ in $\hat{g} = {\Omega_2}^2 g$. The resulting metric is also flat near $\hat{\rho} = 0$. Therefore the resulting conformal embedding is global.

- The Agaoka obstructions (2.3) do not vanish for the ASD Taub–NUT and the Eguchi–Hanson metrics. Moreover, there does not exist a conformal factor depending only on r and such that conformal rescallings of Eguchi–Hanson or ASD Taub NUT have vanishing (2.3). In fact a MAPLE aided computation shows that the only non-conformally flat LRS Bianchi IX metric with ASD conformal curvature and vanishing Agaoka's obstructions is the Burns metric.

5.1 *Isometric Embeddings in* $\mathbb{R}^7$

In Sect. 5 we listed a number of examples of conformally anti-self-dual metrics of the form (5.1) which admit a radial conformal embedding in $\mathbb{R}^7$. Of these only the Burns metric (3.28) admits an embedding with $\Omega = 1$, that is an isometric embedding. We shall now show that under the additional assumption of zero scalar curvature the Burns metric is the only metric with this property

Theorem 5.1 *Let (M, g) be a non-flat LRS Bianchi IX Riemannian manifold with anti-self-dual conformal curvature and zero Ricci scalar, and let $\iota : M \to \mathbb{R}^7$ be a radial isometric embedding (5.2), (5.3) such that $\iota^*(\eta) = g$. Then g is isometric to a constant multiple of the Burns metric (3.28).*

Proof First recall [19] that any diagonal Bianchi IX metric can be put in the form

$$g = w_1 w_2 w_3 dt^2 + \frac{w_2 w_3}{w_1}{\sigma_1}^2 + \frac{w_1 w_3}{w_2}{\sigma_2}^2 + \frac{w_1 w_2}{w_3}{\sigma_3}^2 \tag{5.5}$$

for some $w_j = w_j(t)$ where $j = 1, 2, 3$. Define three more functions $a_j = a_j(t)$ by

$$\dot{w}_1 = -w_2 w_3 + w_1(a_2 + a_3) \quad \text{(and cyclic permutations).} \tag{5.6}$$

Then the scalar-flat anti-self-duality equations are

$$\dot{a}_1 = -a_2 a_3 + a_1(a_2 + a_3) \quad \text{(and cyclic permutations).} \tag{5.7}$$

We restrict to the LRS Bianchi IX metrics (5.1), and set

$$w_1 = w_2 = w, \quad w_3 = u, \quad a_1 = a_2 = a, \quad a_3 = A.$$

Equations (5.7) and (5.6) simplify to

$$\dot{w} = -uw + (a + A)w \tag{5.8a}$$

$$\dot{u} = -w^2 + 2au \tag{5.8b}$$

$$\dot{a} = a^2 \tag{5.8c}$$

$$\dot{A} = -a^2 + 2aA. \tag{5.8d}$$

Comparing expressions (5.2) with (5.5) we find

$$R^2 = 4\frac{w^2}{u}, \quad Z^2 = u - \frac{w^2}{u},$$

when now R and Z are regarded as functions of t.

The isometric embedding condition $\dot{R}^2 + \dot{Z}^2 = w^2 u$ takes the form

$$\left(2\dot{w} - w\frac{\dot{u}}{u}\right)^2 + \frac{\left(\dot{u} - \frac{2w\dot{w}}{u} + \frac{w^2\dot{u}}{u^2}\right)^2}{4(u^2 - w^2)} = u^2 w^2. \tag{5.9}$$

Equation (5.8c) is the Riccati equation with the general solution $a = (t_0 - t)^{-1}$, and a singular integral $a = 0$. The constant t_0 can be set to 0 by a translation of the t-coordinate, and the remaining equations in (5.8a–5.8d) can be readily solved: □

The Singular Case The coupled system of Eqs. (5.8a–5.8d) can be solved to give

$$a = 0, \quad A = c, \quad w = \frac{4e^{t/c_1}}{e^{2t/c_1}{c_1}^2 - 4}, \quad u = \frac{e^{2t/c_1}{c_1}^2(c_1 c + 1) + 4(1 - c_1 c)}{c_1(e^{2t/c_1}{c_1}^2 - 4)} \tag{5.10}$$

where one of the integration constants has been elliminated by translating the t-coordinate, and c, c_1 are the remaining constants of integration. The embedding condition (5.9) forces $c_1 = \pm c^{-1}$, and w. l. g. we can choose $c_1 = c^{-1}$ by changing the sign of t if neccessary. This gives

$$u = \frac{2ce^{2tc}}{e^{trc} - 4c^2}, \quad w = 2ce^{-tc}u,$$

with the range of c now restricted by $0 < c < 1/2$ for regularity.

We claim that the resulting metric is a constant rescalling of the Burns solution (3.28). To see this, define ρ by

$$\frac{w^2}{u} = \frac{1}{4}\rho^2,$$

so that

$$t = \frac{1}{2c}\ln\left(\frac{4c^2(\rho^2 + 8c)}{\rho^2}\right), \quad u = \frac{\rho^2}{4} + 2c, \quad w = \frac{\rho}{4}\sqrt{\rho^2 + 8c}.$$

We verify that $uw^2 dt^2 = d\rho^2$, so that finally

$$g = d\rho^2 + \frac{\rho^2}{4}({\sigma_1}^2 + {\sigma_2}^2 + {\sigma_3}^2) + 2c({\sigma_1}^2 + {\sigma_2}^2) \tag{5.11}$$

which agrees with the Burns metric (3.28) if $c = 1/8$. For any other non-zero value of c, the coordinate transformation $\rho \to \sqrt{8c}\rho$ gives g as $8c$ times the Burns metric. If $c = 0$ then the metric (5.11) is flat.

The Generic Case Setting $t_0 = 0$ in the general solution of the Riccati equation (5.8c), and solving the remaining equations in (5.8a–5.8d) gives

$$a = -\frac{1}{t}, \quad A = -\frac{1}{t} + \frac{c}{t^2}, \quad w = \frac{4e^{\frac{c_2 t+1}{c_1}}}{t^2\left(c_1{}^2 e^{\frac{2}{c_1 t}} - 4e^{\frac{2c_2}{c_1}}\right)}, \quad u = -\frac{\dot{w}}{w} + \frac{c}{r^2} - \frac{2}{r}. \tag{5.12}$$

The embedding condition (5.9) does not hold for any values of the integration constants c, c_1, c_2, which can be seen by looking at the coefficients of various powers of $\gamma \equiv \exp(-1/(c_1 t))$ in (5.9). The coefficient of γ^{20} vanishes iff $c_1 = 1/c$, but then the coefficent of γ^2 is non-zero.

5.2 LRS Bianchi-IX Isometrically Embedded into Flat $\mathbb{R}^8$

This can be done with signatures depending on the example considered. Consider the flat metrics

$$dR^2 + \frac{R^2}{4}(\sigma_1^2 + \sigma_2^2 + \sigma_3^2) + \epsilon_1(dZ^2 + Z^2(\tilde{\sigma}_1^2 + \tilde{\sigma}_2^2)) + \epsilon_2 dF^2,$$

with $\epsilon_i = \pm 1$ and embedding

$$\tilde{\theta} = \theta, \quad \tilde{\phi} = \phi, \quad R(t), \quad F(t), \quad Z(t).$$

This gives (5.1) if we choose

$$R^2 = 4c, \quad \epsilon_1 Z^2 = b - c, \quad (R')^2 + \epsilon_1(Z')^2 + \epsilon_2(F')^2 = a.$$

The choices of ϵ_1, ϵ_2 are dictated by the requirement of reality for F and Z.

5.2.1 $\mathbb{CP}^2$

With $\epsilon_1 = \epsilon_2 = 1$ set

$$R = \frac{r}{1+r^2}, \quad Z = \frac{r^2}{2(1+r^2)}, \quad F = \frac{\sqrt{3}}{2(1+r^2)},$$

then the metric is Fubini-Study. Since for this embedding we have

$$2R^2 + 4Z^2 + 4F^2/3 = 1,$$

some rescaling of the flat coordinates gives this as the familiar embedding of $\mathbb{CP}^2$ into S^7.

5.2.2 Eguchi-Hanson

With $\epsilon_1 = 1 = -\epsilon_2$ set

$$R = \frac{(r^4 - a^4)^{1/2}}{r}, \quad Z = \frac{a^2}{2r}$$

and solve

$$F' = \frac{\sqrt{3}a^2}{2r^2} \left(\frac{3r^4 + a^4}{r^4 - a^4} \right)^{1/2},$$

for $F(r)$ to obtain the Eguchi-Hanson metric.

5.2.3 Anti-Self-Dual Taub-NUT

With $\epsilon_1 = \epsilon_2 = 1$ set

$$R = 2m \left(\frac{r - m}{r + m} \right)^{1/2}, \quad Z = \frac{(r - m)}{2} \left(\frac{r + 3m}{r + m} \right)^{1/2}$$

and solve

$$F' = \left(\frac{2mr^3 + 9m^2r^2 + 24m^3r + 29m^4}{4(r + m)^3(r + 3m)} \right)^{1/2}$$

to obtain the ASD Taub-NUT metric.

Acknowledgments We are grateful to Robert Bryant for correspondence, and pointing out the reference [1] to us. The work of M.D. has been partially supported by STFC consolidated grant no. ST/P000681/1.

References

1. Y. Agaoka, On the curvature of Riemannian submanifolds of codimension 2. Hokkaido Math. J. **14**, 107135 (1985)
2. D. Burns, Twistors and harmonic maps, in *Lecture, Amer. Math. Soc. Conference* (Char lotte, NC, 1986)
3. E. Cartan, Sur la possibilite de plonger un espace Riemannien donne dans espace Euclidean. Ann. Soc. Polon. Math. **6** 1–7 (1927)
4. M. Dunajski, S.A. Hartnoll, Einstein-Maxwell gravitational instantons and five dimensional solitonic strings. Class. Quantum. Grav. **24**, 1841–1862 (2007). hep-th/0610261
5. M. Dunajski, in *Solitons, Instantons, and Twistors. Oxford Graduate Texts in Mathematics* (Oxford University Press, Oxford, 2009)
6. M. Dunajski, K.P. Tod, Conformally isometric embeddings and Hawking temperature. Class. Quant. Grav. **36**, 125005 (2019). arXiv:1812.05468
7. T. Eguchi, A. Hanson, Selfdual solutions to Euclidean gravity. Ann. Phys. **120**, 82–105 (1979)
8. T. Eguchi, P.B. Gilkey, A.J. Hanson, Gravitation, gauge theories and differential geometry. Phys. Rep. **66**, 213–393 (1980)
9. C. Fronsdal, Completion and embedding of the Schwarzschild solution. Phys. Rev. **116**, 778 (1959)
10. G.W. Gibbons, C. Pope, $\mathbb{CP}^2$ As a Gravitational Instanton. Commun. Math. Phys. **61** 239 (1978)
11. G.W. Gibbons, S.W. Hawking, gravitational multi-instantons. Phys. Lett. **B78**, 430 (1978)
12. A.J. Hanson, J.P. Sha, Isometric embedding of the A_1 gravitational instanton, in *Memorial Volume for Kerson Huang* (World Scientific, Singapore, 2017)
13. L.P. Hughston, S.M. Salamon, Surveying points in the complex projective plane. Adv. Math. **286**, 1017–1052 (2016)
14. E. Kaneda, On the Gauss-Codazzi equations. Hokkaido Math. J **19**, 189–213 (1990)
15. S. Kobayashi, Isometric imbeddings of compact symmetric spaces. Tohoku Math. J. **20**, 21–25 (1968)
16. H. Jacobowitz, J.D. Moore, The Cartan–Janet theorem for conformal embeddings. Indiana Univ. Math. J. **23**, 187–203 (1973)
17. R. Penrose, W. Rindler, in *Spinors and Space-Time. Two-Spinor Calculus and Relativistic Fields.* Cambridge Monographs on Mathematical Physics (Cambridge University Press, Cambridge, 1987, 1988)
18. H. Stephani, D. Kramer, M. MacCallum, C. Hoenselaars, E. Herlt, *Exact Solutions of Einstein's Field Equations*, 2nd edn. (CUP, Cambridge, 2003)
19. K.P. Tod, Self-dual Einstein metrics from the Painleve VI equation. Phys. Lett. **A 190**, 221–224 (1994)
20. K.P. Tod, Cohomogeneity-one metrics with self-dual Weyl tensor, in *Twistor Theory*, Lecture Notes in Pure and Appl. Math., vol. 169 (Dekker, New York, 1995)
21. M.Sh Yakupov, On Einstein spaces of second embedding class. Sov. Phys. Dokl. **13**, 5857 (1968)

Recent Results on Closed G_2-Structures

Anna Fino and Alberto Raffero

Abstract We review recent results concerning closed G_2-structures on seven-dimensional manifolds. In particular, we discuss the construction of examples and some related problems.

1 Introduction

Let M be a seven-dimensional manifold. A G_2-reduction of its frame bundle, i.e., a G_2- *structure*, is characterised by the existence of a nowhere vanishing differential form $\varphi \in \Omega^3(M)$ with pointwise stabiliser isomorphic to the exceptional Lie group G_2. Any such 3-form φ gives rise to a Riemannian metric g_φ and to an orientation on M, and its covariant derivative $\nabla^{g_\varphi}\varphi$ with respect to the Levi Civita connection of g_φ represents the obstruction to having $\mathrm{Hol}(g_\varphi) \subseteq G_2$. By [18], a G_2-structure φ is *parallel*, i.e., $\nabla^{g_\varphi}\varphi = 0$, if and only if both $d\varphi = 0$ and $d *_\varphi \varphi = 0$, where $*_\varphi$ is the Hodge operator associated with g_φ and the given orientation. A G_2-structure satisfying the former condition is called *closed*, and a G_2-structure satisfying the latter is called *co-closed*.

All known and potentially effective methods to obtain parallel G_2-structures on compact 7-manifolds require the existence of a closed G_2-structure, possibly satisfying some additional properties [4, 7, 32, 34, 36, 37, 43]. However, currently there are no known results guaranteeing the existence of closed G_2-structures on compact manifolds, neither it is known whether the existence of this type of G_2-structures imposes any constraint on the properties of the manifold, e.g. on its topology.

In this survey, we review various known examples of compact and non-compact 7-manifolds admitting closed G_2-structures, and we discuss the techniques used

A. Fino (✉) · A. Raffero
Dipartimento di Matematica "G. Peano", Università degli Studi di Torino, Torino, Italy
e-mail: annamaria.fino@unito.it; alberto.raffero@unito.it

S. Hervik et al. (eds.), *Geometry, Lie Theory and Applications*, Abel Symposia 16,
https://doi.org/10.1007/978-3-030-81296-6_3

to obtain them. Moreover, we give a new example satisfying some remarkable properties, and we present some related open problems.

2 Preliminaries

A G_2-structure on a seven-dimensional manifold M is a reduction of the structure group of its frame bundle to the exceptional Lie group G_2. Any such reduction occurs if and only if M is orientable and spin [29], and it is characterised by the existence of a 3-form $\varphi \in \Omega^3(M)$ which can be written at each point x of M as

$$\varphi|_x = e^{127} + e^{347} + e^{567} + e^{135} - e^{146} - e^{236} - e^{245}, \tag{1}$$

with respect to some basis $\{e^1, \ldots, e^7\}$ of the cotangent space T_x^*M. Here and henceforth, the notation $e^{ijk\cdots}$ is used as a shorthand for the wedge product of covectors $e^i \wedge e^j \wedge e^k \wedge \cdots$. Any differential 3-form φ defining a G_2-structure is a section of an open subbundle $\Lambda^3_+(T^*M) \subset \Lambda^3(T^*M)$ with typical fibre isomorphic to $\mathrm{GL}(7, \mathbb{R})/G_2$. Therefore, small perturbations of φ still define a G_2-structure. We let $\Omega^3_+(M) := \Gamma(\Lambda^3_+(T^*M))$.

Since $G_2 \subset \mathrm{SO}(7)$, every G_2-structure φ gives rise to a Riemannian metric g_φ and to an orientation on M. In particular, g_φ and the corresponding Riemannian volume form dV_φ are related to φ as follows

$$g_\varphi(X, Y)\, dV_\varphi = \frac{1}{6}\, \iota_X\varphi \wedge \iota_Y\varphi \wedge \varphi,$$

for all $X, Y \in \Gamma(TM)$. We denote by ∇^{g_φ} the Levi Civita connection of g_φ, and by $*_\varphi$ the Hodge operator associated with g_φ and the given orientation.

Given a G_2-structure $\varphi \in \Omega^3_+(M)$, the obstruction to having $\mathrm{Hol}(g_\varphi) \subseteq G_2$ is represented by the *intrinsic torsion* T_φ. The latter is a section of a vector bundle over M with typical fibre $(\mathbb{R}^7)^* \otimes \mathfrak{g}_2^\perp$, where $\mathfrak{g}_2^\perp$ is the orthogonal complement in $\mathfrak{so}(7)$ of the Lie algebra $\mathfrak{g}_2$ of G_2. It is well-known that T_φ can be identified with the covariant derivative $\nabla^{g_\varphi}\varphi$, so that one has $\mathrm{Hol}(g_\varphi) \subseteq G_2$ if and only if $\nabla^{g_\varphi}\varphi = 0$. When this happens, the G_2-structure is said to be *parallel* or *torsion-free*. A further characterisation for parallel G_2-structures was obtained by Fernández and Gray in [18]. We recall it in the following.

Theorem 1 ([18]) *A G_2-structure φ is parallel if and only if both $d\varphi = 0$ and $d *_\varphi \varphi = 0$.*

As proved by Bonan in [3], the metric induced by a parallel G_2-structure is Ricci-flat. Thus, the construction of compact examples is not an easy task. For instance, compact homogeneous examples are exhausted by flat tori and, more generally, compact 7-manifolds with a parallel G_2-structure φ such that $\mathrm{Hol}(g_\varphi) = G_2$ cannot have continuous symmetries. Indeed, using the Bochner-Weitzenböck technique, it is possible to show that the Lie algebra of the automorphism group $\mathrm{Aut}(M, \varphi)$ of a G$_2$-structure φ

$$\mathfrak{aut}(M, \varphi) := \{X \in \Gamma(TM) \mid \mathcal{L}_X\varphi = 0\} \subseteq \mathfrak{isom}(M, g_\varphi)$$

is abelian with $\dim(\mathfrak{aut}(M, \varphi)) = 0, 1, 3, 7$, whenever φ is parallel and M is compact. The possible dimensions depend on having $\mathrm{Hol}(g_\varphi) = \mathrm{G}_2, \mathrm{SU}(3), \mathrm{SU}(2), \{1\}$, respectively.

Compact examples of Riemannian manifolds with holonomy equal to G$_2$ were constructed in [7, 32, 34, 36, 37] using techniques such as orbifolds resolutions, generalised connected sums, and geometric glueing. All these manifolds admit a parallel G$_2$-structure obtained by perturbing a G$_2$-structure φ satisfying the condition $d\varphi = 0$ and whose intrinsic torsion is *small* in a suitable sense. The manifold topology plays a role in this construction, too, as the metric induced by a parallel G$_2$-structure on a compact manifold M has holonomy equal to G$_2$ if and only if $\pi_1(M)$ is finite [32].

3 Closed G$_2$-Structures

From Theorem 1, we know that a G$_2$-structure φ is parallel if and only if $d\varphi = 0$ and $d *_\varphi \varphi = 0$. A G$_2$-structure satisfying the former condition is called *closed*, and one satisfying the latter is called *co-closed*. As we have previously recalled, closed G$_2$-structures play a central role in the known constructions of compact 7-manifolds with holonomy G$_2$. However, currently there are no general results ensuring the existence of closed G$_2$-structures on compact manifolds, neither it is known whether the existence of such type of structure imposes any constraint on the manifold structure, e.g. on its topology. On the other hand, Crowley and Nordström proved that the existence of co-closed G$_2$-structures on compact 7-manifolds is only a topological matter. Indeed, they always exist on any compact manifold admitting G$_2$-structures [10].

By Harvey and Lawson [30], a closed G$_2$-structure $\varphi \in \Omega^3_+(M)$ defines a calibration on M, that is, for every oriented tangent 3-plane $V \subset T_xM$ one has $\varphi|_V \leq \mathrm{vol}_V$, where vol_V denotes the volume form of the inner product $g_\varphi|_V$ on V. A three-dimensional oriented submanifold $N \hookrightarrow M$ is said to be *associative* if it is calibrated by φ, i.e., if $\varphi|_{T_xN} = \mathrm{vol}_{T_xN}$ for all $x \in N$. Recall that every compact associative 3-fold is volume minimising in its homology class (see [30, Thm. II.4.2]). Similarly, if a G$_2$-structure φ is co-closed, then the 4-form $*_\varphi\varphi$ defines a calibration on M, and one can consider four-dimensional oriented submanifolds of M calibrated by it. These submanifolds are called *co-associative*. It is possible to show that a four-dimensional oriented submanifold U in M is co-associative if and only if $\varphi|_U \equiv 0$ (cf. [30, Cor. IV.1.20]). Thus, when a G$_2$-structure φ is closed but not co-closed, one can define an oriented 4-fold U of M to be co-associative if $\varphi|_U \equiv 0$. However, U is not necessarily volume minimising in its homology class, as $d *_\varphi \varphi \neq 0$. For more details on associative and co-associative submanifolds, we refer the reader to [30, 33].

Let us now focus on a 7-manifold M endowed with a closed G_2-structure φ. In this case, the intrinsic torsion T_φ can be identified with a unique 2-form $\tau \in \Omega^2_{14}(M) := \{\alpha \in \Omega^2(M) \mid \alpha \wedge \varphi = - *_\varphi \alpha\}$ such that

$$d *_\varphi \varphi = \tau \wedge \varphi.$$

We shall refer to τ as the *intrinsic torsion form* of φ. Notice that τ vanishes identically if and only if $d *_\varphi \varphi = 0$, that is, if and only if the closed G_2-structure is parallel.

The Ricci tensor and the scalar curvature of the Riemannian metric g_φ induced by φ can be expressed in terms of the intrinsic torsion form τ as follows (cf. [4]):

$$\mathrm{Ric}(g_\varphi) = \frac{1}{4}|\tau|^2 g_\varphi - \frac{1}{4} j_\varphi \left(d\tau - \frac{1}{2} *_\varphi (\tau \wedge \tau) \right),$$

$$\mathrm{Scal}(g_\varphi) = -\frac{1}{2}|\tau|^2,$$

where the map $j_\varphi : \Omega^3(M) \to \mathcal{S}^2(M)$ assigns to each 3-form γ the symmetric 2-covariant tensor $j_\varphi(\gamma)(X, Y) = *_\varphi(\iota_X\varphi \wedge \iota_Y\varphi \wedge \gamma)$. In particular, the scalar curvature is always non-positive, and it vanishes identically if and only if φ is parallel. Further properties of $\mathrm{Ric}(g_\varphi)$ and $\mathrm{Scal}(g_\varphi)$ when M is compact are summarised in the next theorem.

Theorem 2 ([4, 8]) *Let M be a compact 7-manifold endowed with a closed* G_2-*structure φ. Then,*

1. *g_φ is Einstein if and only if φ is parallel;*
2. $\int_M [\mathrm{Scal}(g_\varphi)]^2 dV_\varphi \leq 3 \int_M |\mathrm{Ric}(g_\varphi)|^2 dV_\varphi$.

Remark 1 Currently, it is still not known whether there exist (even incomplete) closed non-parallel G_2-structures inducing an Einstein metric. A negative answer when the 7-manifold is a simply connected solvable Lie group endowed with a left-invariant closed G_2-structure was given in [15].

In [4], Bryant proved that the inequality in point 2. of Theorem 2 reduces to an equality if and only if

$$d\tau = \frac{|\tau|^2}{6}\varphi + \frac{1}{6} *_\varphi (\tau \wedge \tau).$$

A closed G_2-structure φ satisfying the above equation is called *extremally Ricci pinched* (*ERP* for short). Notice that this definition makes sense also in the non-compact setting.

The existence of an ERP G_2-structure on a compact 7-manifold imposes strong constraints on its geometry. In detail:

Proposition 1 ([4]) *Let M be a compact 7-manifold endowed with an ERP G_2-structure φ with intrinsic torsion form $\tau \in \Omega^2_{14}(M)$ not identically vanishing. Then*

1. *τ has constant (non-zero) norm and $\tau \wedge \tau \wedge \tau = 0$;*
2. *$\tau \wedge \tau$ and $*_\varphi(\tau \wedge \tau)$ are non-zero closed simple forms of constant norm. Consequently, the tangent bundle of M splits into the orthogonal direct sum of two integrable subbundles $TM = P \oplus Q$, with $P := \ker(\tau \wedge \tau)$, $Q := \ker(*_\varphi(\tau \wedge \tau))$. Moreover, the P-leaves are associative submanifolds calibrated by $-|\tau|^{-2} *_\varphi (\tau \wedge \tau)$, and the Q-leaves are coassociative submanifolds calibrated by $-|\tau|^{-2}(\tau \wedge \tau)$;*
3. *the Ricci tensor of g_φ is given by $\mathrm{Ric}(g_\varphi) = \frac{1}{12}\, j_\varphi(*_\varphi(\tau \wedge \tau)) = -\frac{1}{6}\, |\tau|^2\, g_\varphi|_P$. Hence, it is non-positive with eigenvalues $-\frac{1}{6}\,|\tau|^2$ of multiplicity three and 0 of multiplicity four.*

So far, the only known compact examples were obtained in [1, 4, 35]. The example in [4] consists of the quotient $\Gamma\backslash M$, where $M = \mathrm{SL}(2, \mathbb{C}) \ltimes \mathbb{C}^2/\mathrm{SU}(2)$ is a non-compact homogeneous space endowed with an invariant ERP G_2-structure, and $\Gamma \subset \mathrm{Aut}(M, \varphi)$ is a co-compact discrete subgroup. The example in [35] is the compact quotient of a simply connected solvable Lie group endowed with the left-invariant ERP G_2-structure obtained in [39, Ex. 4.7]. Further examples are given in [2, 25, 40, 41].

4 Examples of Manifolds Admitting Closed G_2-Structures

Since there are no general results ensuring the existence of closed G_2-structures on compact 7-manifolds, one should in principle try to solve the PDEs arising from the equation $d\varphi = 0$ for a generic 3-form $\varphi \in \Omega^3_+(M)$ on a given compact manifold M. In some cases, this problem may be reduced to a simpler one. In this section, we give an overview of various known examples and we discuss some new ones.

4.1 Closed G_2-Structures with Symmetry

The first examples of compact 7-manifolds admitting closed non-parallel G_2-structures were obtained by Fernández in [12, 13]. These examples are of the form $M = \Gamma\backslash \mathrm{G}$, where G is a seven-dimensional simply connected nilpotent or solvable Lie group, $\Gamma \subset \mathrm{G}$ is a co-compact discrete subgroup (lattice), and the closed G_2-structure on M is induced by a left-invariant one on G. In particular, (M, φ) is locally homogeneous. Further examples of this type have been obtained in [6, 28, 38]. We emphasise that in all these examples the Lie group G, being diffeomorphic to $\mathbb{R}^7$, is not compact. Moreover, any left-invariant G_2-structure φ on G is completely determined by the 3-form $\varphi|_{1_G}$ on the Lie algebra $\mathfrak{g} \cong T_{1_G}G$, and the condition $d\varphi = 0$ is equivalent to a system of algebraic equations for the coefficients of $\varphi|_{1_G}$.

As this last observation suggests, symmetry considerations might help to reduce the PDEs arising from the condition $d\varphi = 0$ to equations of a simpler type. Therefore, one may ask whether there exist examples of compact 7-manifolds M endowed with a closed non-parallel G_2-structure φ that are acted on with low cohomogeneity by the automorphism group $\mathrm{Aut}(M, \varphi)$ or a subgroup thereof. In [9], Cleyton and Swann considered 7-manifolds endowed with a G_2-structure and acted on with cohomogeneity one by a simple group of automorphisms, showing in particular that if φ is closed then only complete non-compact examples occur. Moreover, the existence problem of compact homogeneous 7-manifolds endowed with an invariant closed non-parallel G_2-structure, i.e., having a transitive group of automorphisms $G \subset \mathrm{Aut}(M, \varphi)$, has been open for some time (see [39, Question 3.1]).

In [45], Podestà and the second named author of the present survey studied the properties of $\mathrm{Aut}(M, \varphi)$ when M is compact and the G_2-structure φ is closed and non-parallel, obtaining the following result.

Theorem 3 ([45]) *Let M be a compact 7-manifold endowed with a closed non-parallel* G_2*-structure φ. Then,* $\mathfrak{aut}(M, \varphi)$ *is abelian with dimension bounded above by* $\min\{6, b_2(M)\}$.

The above result immediately implies that there are no compact homogeneous 7-manifolds endowed with an invariant closed non-parallel G_2-structure, and that the only cohomogeneity one examples occur on manifolds diffeomorphic to the 7-torus.

Notice that the examples of seven-dimensional simply connected Lie groups G endowed with a left-invariant closed G_2-structure φ mentioned earlier ([6, 12, 13, 28, 38]) correspond to the case when M is not compact and $G \subset \mathrm{Aut}(M, \varphi)$ acts simply transitively on it. Recently, the complete classification of non-compact 7-manifolds admitting closed non-parallel G_2-structures and acted on transitively by a reductive group of automorphisms has been obtained in [48].

We remark that Theorem 3 gives some insights on a currently open problem concerning the existence of a closed G_2-structure on the 7-sphere $\mathbb{S}^7$, as it implies that $\mathrm{Aut}(\mathbb{S}^7, \varphi)$ must be finite for any hypothetical closed G_2-structure $\varphi \in \Omega^3_+(\mathbb{S}^7)$.

4.2 A Compact Example Obtained via the Resolution of an Orbifold

Besides symmetry considerations, other techniques may be used to construct examples of compact 7-manifolds admitting closed G_2-structures. For instance, all known examples of compact manifolds admitting parallel G_2-structures are endowed with a closed one. The techniques used to obtain these examples include orbifolds resolutions [32], the twisted connected sum construction [7, 36, 37], and analysis on families of Eguchi-Hanson spaces [34].

Recently, Fernández, Kovalev, Muñoz and the first named author of the present survey obtained a new example of a compact 7-manifold M with $b_1(M) = 1$ and admitting closed G$_2$-structures but no parallel G$_2$-structures [14]. This example is constructed resolving the singularities of a certain orbifold, and the techniques involved in this construction are motivated by those used by Joyce to obtain compact 7 manifolds with holonomy G$_2$ via the resolution of orbifolds of the form $\mathbb{T}^7/\mathrm{F}$, where $\mathrm{F} \subset \mathrm{G}_2$ is a suitable finite subgroup [32].

The idea of [14] consists in starting with a compact nilmanifold $M = \Gamma\backslash\mathrm{N}$ endowed with an invariant closed G$_2$-structure in place of the 7-torus $\mathbb{T}^7 \cong \mathbb{R}^7/\mathbb{Z}^7$ endowed with the flat G$_2$-structure induced by the canonical one on $\mathbb{R}^7$. Here, N is the simply-connected nilpotent Lie group corresponding to the 3-step nilpotent Lie algebra $\mathfrak{n} = \langle e_1, \ldots, e_7\rangle$ with the following structure equations

$$[e_1, e_2] = -e_4, \ [e_1, e_3] = -e_5, \ [e_1, e_4] = -e_6, \ [e_1, e_5] = -e_7,$$

$\Gamma \cong 2\mathbb{Z} \times \mathbb{Z}^6$ is a lattice in N, and the invariant closed G$_2$-structure on M is induced by a left-invariant one on N. One then considers the action of the finite group $\mathrm{F} = \mathbb{Z}_2$ on N generated by

$$\rho : \mathrm{N} \to \mathrm{N}, \quad (x_1, \ldots, x_7) \mapsto (-x_1, -x_2, x_3, x_4, -x_5, -x_6, x_7).$$

This action satisfies $\rho(ab) = \rho(a)\rho(b)$, for all $a, b \in \mathrm{N}$, and $\rho(\Gamma) = \Gamma$. Thus, ρ induces an action of $\mathbb{Z}_2$ on $M = \Gamma\backslash\mathrm{N}$, and it is possible to consider the quotient space $\widehat{M} = M/\mathbb{Z}_2$. The following result can then be proved.

Theorem 4 ([14])

- $\widehat{M} = M/\mathbb{Z}_2$ *is a compact 7-orbifold with first Betti number* $b_1(\widehat{M}) = 1$ *and admitting an orbifold closed* G$_2$*-form* $\widehat{\varphi}$. *The singular locus* S *of* $\widehat{M}$ *is the disjoint union of 16 copies of* $\mathbb{T}^3$.
- *There exists a resolution* $\pi : (\widetilde{M}, \widetilde{\varphi}) \to (\widehat{M}, \widehat{\varphi})$, *where* $\widetilde{M}$ *is a compact smooth manifold,* $b_1(\widetilde{M}) = 1$, *and* $\widetilde{\varphi} = \pi^*\widehat{\varphi}$ *in the complement of a neighborhood of the exceptional locus* $\pi^{-1}(S)$. *Moreover,* $\pi_1(\widetilde{M}) = \mathbb{Z}$, $\widetilde{M}$ *is formal and it does not admit any parallel* G$_2$*-structure.*

4.3 A Comparison with Lie Algebras Admitting Symplectic Structures

As we recalled in Sect. 4.1, there are many (non-compact) examples of simply connected seven-dimensional Lie groups admitting left-invariant closed G$_2$-structures, and for any such example the properties of the G_2-structure may be studied at the Lie algebra level. This leads one to focusing on seven-dimensional Lie algebras $\mathfrak{g}$ endowed with a G_2-structure $\varphi \in \Lambda^3_+(\mathfrak{g}^*)$ which is closed with respect to the Chevalley-Eilenberg differential d of $\mathfrak{g}$. In this setting, there are some problems motivated by symplectic geometry that it is worth investigating.

It is known that symplectic structures and closed G_2-structures share similar properties. Indeed, they are both defined by closed differential forms satisfying the same non-degeneracy condition, namely at each point of the manifold their orbit under the action of the general linear group is open (see [31]). Moreover, the existence of a symplectic structure on an even-dimensional Lie algebra imposes some constraints on it. For instance, a unimodular Lie algebra admitting a symplectic structure must be solvable [5, 42], and any unimodular Lie algebra cannot admit symplectic structures defined by an exact 2-form [11]. Thus, it is natural to investigate whether similar results hold for seven-dimensional unimodular Lie algebras admitting closed G_2-structures, too. In our works [17, 23], we obtained a negative answer to these problems.

In [23], we studied the existence of closed G_2-structures on non-solvable unimodular Lie algebras, showing that examples actually occur and obtaining a complete classification of the possible Lie algebras up to isomorphism.

Theorem 5 ([23]) *Let $\mathfrak{g}$ be a seven-dimensional non-solvable unimodular Lie algebra admitting closed G_2-structures, and denote by $\mathfrak{g} = \mathfrak{s} \ltimes \mathfrak{r}$ its Levi decomposition, where $\mathfrak{r}$ is the radical of $\mathfrak{g}$ and $\mathfrak{s}$ is a semisimple subalgebra. Then, $\mathfrak{s} \cong \mathfrak{sl}(2,\mathbb{R})$ and*

- *if the Levi decomposition is trivial, namely $\mathfrak{g} = \mathfrak{s} \oplus \mathfrak{r}$, then $\mathfrak{r}$ is centerless, and there exists a basis $\{e^1, \dots, e^7\}$ of $\mathfrak{g}^*$ for which its structure equations $(de^1, \dots, de^7)$ are in the following list*

$$(-e^{23}, -2e^{12}, 2e^{13}, 0, -e^{45}, \tfrac{1}{2}e^{46}, -e^{47}, \tfrac{1}{2}e^{47});$$
$$(-e^{23}, -2e^{12}, 2e^{13}, 0, -e^{45}, -\mu e^{46}, (1+\mu)e^{47}), \quad -1 < \mu \leq \tfrac{1}{2};$$
$$(-e^{23}, -2e^{12}, 2e^{13}, 0, -\mu e^{45}, \tfrac{\mu}{2}e^{46} - e^{47}, e^{46} + \tfrac{\mu}{2}e^{47}), \ \mu > 0.$$

- *if the Levi decomposition is not trivial, then $\mathfrak{r} \cong \mathbb{R} \ltimes \mathbb{R}^3$, and there is a basis $\{e^1, \dots, e^7\}$ of $\mathfrak{g}^*$ for which its structure equations are the following*

$$(-e^{23}, -2e^{12}, 2e^{13}, -e^{14} - e^{25} - e^{47}, e^{15} - e^{34} - e^{57}, 2e^{67}, 0).$$

In our recent joint work with M. Fernández [17], we focused on closed G_2-structures defined by an exact 3-form φ on a unimodular Lie algebra $\mathfrak{g}$, showing that there exist examples when the third Betti number $b_3(\mathfrak{g})$ of $\mathfrak{g}$ vanishes and when $\mathfrak{g}$ is (2,3)-trivial, i.e., when $b_2(\mathfrak{g}) = 0 = b_3(\mathfrak{g})$.

4.4 Products, Warped Products, and Lie Algebras Extensions

Further examples not mentioned so far can be obtained on products of suitable manifolds. For instance, it is possible to define closed G_2-structures on the product of a 4-manifold endowed with a hypersymplectic structure and the 3-torus $\mathbb{T}^3$ (see e.g. [20]), and on the product of a 6-manifold endowed with a suitable SU(3)-structure and the circle $\mathbb{S}^1$.

Recall that an SU(3)-structure on a 6-manifold N is the data of an almost Hermitian structure (g, J), with corresponding fundamental 2-form $\omega := g(J\cdot, \cdot)$, and a complex (3,0)-form $\Psi = \psi + i\widehat{\psi}$ satisfying the normalisation condition $3\psi \wedge \widehat{\psi} = 2\omega^3$. Moreover, the whole structure is completely determined by the real 2-form ω and the real 3-form ψ, provided that they satisfy suitable conditions (see [31] for details). Given a positive function $f \in \mathcal{C}^\infty(N)$, it is possible to endow the product $N \times \mathbb{S}^1$ with the 3-form

$$\varphi = f\omega \wedge ds + \psi, \tag{2}$$

which is easily seen to define a G_2-structure with associated Riemannian metric $g_\varphi = g + f^2 ds^2$, where s denotes the coordinate on the circle. Notice that the manifold $N \times \mathbb{S}^1$ endowed with this metric is none other than the warped product of N and $\mathbb{S}^1$ with warping function f and fibre $\mathbb{S}^1$. The G_2-structure φ on $N \times \mathbb{S}^1$ is closed if and only if the SU(3)-structure on N satisfies

$$d\omega = -d\log(f) \wedge \omega, \quad d\psi = 0. \tag{3}$$

Thus, the problem of finding a closed G_2-structure of the form (2) on $N \times \mathbb{S}^1$ boils down to the construction of a 6-manifold endowed with an SU(3)-structure satisfying (3). An example on $N = \mathbb{T}^6$ can be found in [24].

When the warping function f is constant, the manifold reduces to a Riemannian product, and it is always possible to assume that $f \equiv 1$ up to scaling the coordinate s. In this case, the G_2-structure $\varphi = \omega \wedge ds + \psi$ is closed if and only if both $d\omega = 0$ and $d\psi = 0$. An SU(3)-structure satisfying these conditions is called *symplectic half-flat*, and its intrinsic torsion can be identified with the unique real 2-form w_2 of type (1,1) such that $w_2 \wedge \omega^2 = 0$ and $d\widehat{\psi} = w_2 \wedge \omega$. A simple computation shows that the intrinsic torsion form of the closed G_2-structure $\varphi = \omega \wedge ds + \psi$ coincides with w_2. We refer the reader to [19, 24, 46, 47] and the references therein for more details and examples.

The last example of closed G_2-structures on $N \times \mathbb{S}^1$ that we would like to mention is obtained as follows. Let us consider a *coupled* SU(3)-structure on N, i.e., an SU(3)-structure satisfying the condition $d\omega = c\psi$ for some non-zero real number c. The 3-form ψ is clearly closed, while the 3-form $\widehat{\psi}$ satisfies

$$d\widehat{\psi} = -\frac{2}{3}c\,\omega^2 + w_2 \wedge \omega, \tag{4}$$

for a unique real 2-form w_2 of type (1,1) and satisfying $w_2 \wedge \omega^2 = 0$. The pair (c, w_2) completely determines the intrinsic torsion of the coupled SU(3)-structure (ω, ψ). Now, the G_2-structure $\varphi = \omega \wedge ds + \psi$ on $N \times \mathbb{S}^1$ considered before is no longer closed, but there exists a global conformal change making it into a closed one, namely

$$\widetilde{\varphi} = e^{cs} (\omega \wedge ds + \psi) .$$

The Riemannian metric induced by $\widetilde{\varphi}$ is given by $g_{\widetilde{\varphi}} = e^{\frac{2}{3}cs} g_{\varphi}$, and its intrinsic torsion form is $\widetilde{\tau} = e^{\frac{cs}{3}} w_2$. For more details and examples regarding coupled SU(3)-structures and G_2-structures that are (locally) conformal to closed ones, the reader may refer to [16, 22].

As for Lie algebras, similar constructions can be made considering rank-one extensions of six-dimensional Lie algebras endowed with an SU(3)-structure.

We recall that the *rank-one extension* of an n-dimensional Lie algebra $\mathfrak{h}$ induced by a derivation $D \in \mathrm{Der}(\mathfrak{h})$ is the $(n + 1)$-dimensional Lie algebra given by the vector space $\mathfrak{h} \oplus \mathbb{R}$ endowed with the Lie bracket

$$[(X, a), (Y, b)] := \big([X, Y]_{\mathfrak{h}} + a\, D(Y) - b\, D(X), 0\big) ,$$

where $(X, a),\ (Y, b) \in \mathfrak{h} \oplus \mathbb{R}$, and $[\cdot, \cdot]_{\mathfrak{h}}$ is the Lie bracket on $\mathfrak{h}$. We shall denote this Lie algebra by $\mathfrak{h} \rtimes_D \mathbb{R}$. Moreover, we let $\xi := (0, 1)$, and we denote by η the 1-form on $\mathfrak{h} \rtimes_D \mathbb{R}$ such that $\eta(\xi) = 1$ and $\mathfrak{h} = \ker(\eta)$.

Let d and $\hat{d}$ denote the Chevalley-Eilenberg differential on $\mathfrak{h} \rtimes_D \mathbb{R}$ and on $\mathfrak{h}$, respectively. Then, for every k-form $\gamma \in \Lambda^k(\mathfrak{h}^*)$ the following identity holds:

$$d\gamma = \hat{d}\gamma + (-1)^{k+1} D^*\gamma \wedge \eta,$$

where the natural action of an endomorphism $A \in \mathrm{End}(\mathfrak{h})$ on $\Lambda^k(\mathfrak{h}^*)$ is given by

$$A^*\gamma(X_1, \dots, X_k) = \gamma(AX_1, \dots, X_k) + \cdots + \gamma(X_1, \dots, AX_k),$$

for all $X_1, \dots, X_k \in \mathfrak{h}$. If $D = 0 \in \mathrm{Der}(\mathfrak{h})$ is the trivial derivation, then the rank-one extension of $\mathfrak{h}$ reduces to the product Lie algebra $\mathfrak{h} \oplus \mathbb{R}$.

Assume now that $\mathfrak{h}$ is six-dimensional and that it is endowed with an SU(3)-structure (ω, ψ). Then, the rank-one extension $\mathfrak{h} \rtimes_D \mathbb{R}$ is always endowed with a G_2-structure defined by the 3-form $\varphi = \omega \wedge \eta + \psi$. Clearly, the condition $d\varphi = 0$ constrains the possible types of SU(3)-structures and the properties of the derivation D that one can consider. Indeed, it holds if and only if

$$\hat{d}\omega = -D^*\psi, \quad \hat{d}\psi = 0. \tag{5}$$

In the next proposition, whose proof immediately follows from (5), we recall some relevant cases related to the SU(3)-structures mentioned above.

Proposition 2 *Let* $\mathfrak{h}$ *be a six-dimensional Lie algebra endowed with an* SU(3)*-structure* (ω, ψ)*, and consider the rank-one extension* $\mathfrak{h}\rtimes_D\mathbb{R}$ *endowed with the* G$_2$*-structure* $\varphi = \omega \wedge \eta + \psi$*. Then,* φ *is closed in the following cases*

- *the* SU(3)*-structure is symplectic half-flat, i.e.,* $\hat{d}\omega = 0$*,* $\hat{d}\psi = 0$*, and either* $D = 0$ *or* $D^*\psi = 0$*;*
- *the* SU(3)*-structure is coupled with* $\hat{d}\omega = c\psi$*, and* $D^*\psi = -c\psi$*.*

The classification of six-dimensional solvable Lie algebras admitting symplectic half-flat SU(3)-structures was given in [19]. Examples of closed G$_2$-structures on rank-one extensions of these Lie algebras are discussed in [44].

In [22], we proved that a six-dimensional nilpotent Lie algebra admitting coupled SU(3)-structures is isomorphic to one of the following

$$\mathfrak{n}_1 = (0, 0, 0, e^{13}, e^{14}+e^{23}, e^{13}-e^{15}-e^{24}), \quad \mathfrak{n}_2 = (0, 0, 0, 0, e^{14}+e^{23}, e^{13}-e^{24}).$$

In both cases, there exists a coupled SU(3)-structure with *adapted* basis $\{e^1, \dots, e^6\}$, namely

$$\omega = e^{12} + e^{34} + e^{56}, \quad \Psi = (e^1 + ie^2) \wedge (e^3 + ie^4) \wedge (e^5 + ie^6), \tag{6}$$

and it is easy to check that $\hat{d}\omega = -\psi$, i.e., $c = -1$. Moreover, for the coupled SU(3)-structure on $\mathfrak{n}_2$, the unique 2-form w_2 fulfilling (4) is given by $w_2 = \frac{4}{3}e^{12} + \frac{4}{3}e^{34} - \frac{8}{3}e^{56}$ and its differential $\hat{d}w_2$ is proportional to ψ.

Remark 2 More generally, if (ω, ψ) is a coupled SU(3)-structure on a 6-manifold with dw_2 proportional to ψ, then one has $dw_2 = \frac{|w_2|^2}{4}\psi$ with $|w_2|$ constant (see e.g. [21]). Furthermore, on the nilpotent Lie algebra $\mathfrak{n}_1$ there are no coupled SU(3)-structures satisfying this condition (cf. [21, Prop. 12]).

The nilpotent Lie algebra $\mathfrak{n}_2$ endowed with the coupled SU(3)-structure (ω, ψ) given in (6) admits rank-one extensions for which the G$_2$-structure $\varphi = \omega\wedge\eta+\psi$ is closed. A one-parameter family of examples was given by Lauret in [39, Ex. 4.10], and it consists of the extension $\mathfrak{g}_a := \mathfrak{n}_2 \rtimes_{D_a} \mathbb{R}$ with respect to the derivation

$$D_a = \mathrm{diag}\left(a, a, \frac{1}{2} - a, \frac{1}{2} - a, \frac{1}{2}, \frac{1}{2}\right) \in \mathrm{Der}(\mathfrak{n}_2), \quad a \geq \frac{1}{4}.$$

Notice that D_a fulfills the condition $D_a^*\psi = \psi$, whence $d\varphi = 0$. The Lie algebra $\mathfrak{g}_a$ is completely solvable, and the bound on the real parameter a arises requiring the simply connected Lie groups G$_a$ with Lie algebra $\mathfrak{g}_a$ to be pairwise non-isomorphic for different values of a.

The Lie algebra $\mathfrak{n}_1$ endowed with the coupled SU(3)-structure given in (6) admits derivations satisfying the condition $D^*\psi = \psi$, too. It is not difficult to prove that the most general one has the following expression with respect to the basis $\{e_1, \ldots, e_6\}$ of $\mathfrak{n}_1$

$$D_{a,b} = \begin{pmatrix} 0 & 0 & 0 & 0 & 0 & 0 \\ 0 & 0 & 0 & 0 & 0 & 0 \\ a & 0 & \frac{1}{2} & 0 & 0 & 0 \\ 0 & a & 0 & \frac{1}{2} & 0 & 0 \\ b & 0 & 0 & 0 & \frac{1}{2} & 0 \\ 0 & b & 0 & 0 & 0 & \frac{1}{2} \end{pmatrix}, \quad a, b \in \mathbb{R}.$$

Consequently, any Lie algebra belonging to the two-parameter family $\mathfrak{g}_{a,b} := \mathfrak{n}_1 \rtimes_{D_{a,b}} \mathbb{R}$ admits a closed G_2-structure defined by $\varphi = \omega \wedge \eta + \psi$.

We now discuss a new example of a solvable non-nilpotent Lie algebra endowed with a coupled SU(3)-structure, and we show that there exist rank-one extensions admitting closed G_2-structures.

Example 1 Consider the two parameter family of six-dimensional unimodular solvable Lie algebras $\mathfrak{s}_{a,b}$, $a, b \in \mathbb{R}$, whose structure equations with respect to a basis $\{e^1, \ldots, e^6\}$ of $\mathfrak{s}_{a,b}^*$ are the following

$$\begin{cases} de^1 = -a\, e^{26}, \\ de^2 = a\, e^{16}, \\ de^3 = b\, e^{16} + b\, e^{25} + a\, e^{46}, \\ de^4 = b\, e^{15} - b\, e^{26} - a\, e^{36}, \\ de^5 = de^6 = 0. \end{cases}$$

Notice that $\mathfrak{s}_{a,b} \cong \mathbb{R}^4 \rtimes \mathbb{R}^2$, where the subalgebra $\mathbb{R}^2$ is spanned by the vectors e_5 and e_6. We endow $\mathfrak{s}_{a,b}$ with the SU(3)-structure (ω, ψ) given by (6).

If $a = 0$, then the Lie algebra $\mathfrak{s}_{0,b}$ is isomorphic to the nilpotent Lie algebra $\mathfrak{n}_2$ and the SU(3)-structure is the coupled one described before. If $b = 0$, then $\mathfrak{s}_{a,0}$ is solvable non-nilpotent, the SU(3)-structure satisfies $\hat{d}\omega = 0$, $\hat{d}\psi = 0 = \hat{d}\widehat{\psi}$, and the associated inner product $g = \sum_{k=1}^{6} (e^k)^2$ is flat.

From now on, we assume that $a \neq 0$ and $b \neq 0$. Under this assumption, the Lie algebra is solvable non-nilpotent and the SU(3)-structure is coupled with

$$\hat{d}\omega = b\,\psi.$$

Moreover, it has $w_2 = -\frac{4}{3}be^{12} + \frac{8}{3}be^{34} - \frac{4}{3}be^{56}$, which is easily seen to satisfy the condition $\hat{d}w_2 = \frac{8}{3}b^2\psi$.

There exists a one-parameter family of derivations $D_k \in \mathrm{Der}(\mathfrak{s}_{a,b})$ for which the 3-form $\varphi = \omega \wedge \eta + \psi$ defines a closed G_2-structure on the rank-one extension $\mathfrak{g}_{a,b,k} := \mathfrak{s}_{a,b} \rtimes_{D_k} \mathbb{R}$. It has the following expression

$$D_k = \begin{pmatrix} -\frac{b}{2} & k & 0 & 0 & 0 & 0 \\ -k & -\frac{b}{2} & 0 & 0 & 0 & 0 \\ 0 & 0 & -\frac{b}{2} & -k & 0 & 0 \\ 0 & 0 & k & -\frac{b}{2} & 0 & 0 \\ 0 & 0 & 0 & 0 & 0 & 0 \\ 0 & 0 & 0 & 0 & 0 & 0 \end{pmatrix}, \qquad k \in \mathbb{R},$$

and it is easily seen to satisfy the equation $D_k^* \psi = -b\,\psi$, for any $k \in \mathbb{R}$.

5 Some Remarks on Closed G_2-Structures Induced by Coupled SU(3)-Structures on Lie Algebras

An interesting property of Lauret's example [39, Ex. 4.10], reviewed in Sect. 4.4, is that the left-invariant closed G_2-structure induced by $\varphi = \omega \wedge \eta + \psi$ on the simply connected Lie group G_a is always a *Laplacian soliton*, that is, its Hodge Laplacian $\Delta_\varphi \varphi = -d *_\varphi d *_\varphi \varphi = d\tau$ satisfies the equation

$$\Delta_\varphi \varphi = \lambda \varphi + \mathcal{L}_X \varphi, \tag{7}$$

for $\lambda = \lambda(a) := 8a^2 - 4a - 4$, and for a suitable vector field X on G_a. Moreover, this Laplacian soliton is *algebraic* according to [38, Def. 3.9], namely there exists a derivation $B \in \mathrm{Der}(\mathfrak{g}_a)$ such that the vector field X is induced by the unique one-parameter group of automorphisms $\Phi_t \in \mathrm{Aut}(\mathrm{G}_a)$ satisfying $d\Phi_t|_{1_{\mathrm{G}_a}} = \exp(tB) \in \mathrm{Aut}(\mathfrak{g}_a)$. Finally, depending on the value of a, the Laplacian soliton can be *shrinking* ($\lambda(a) < 0$), *steady* ($\lambda(a) = 0$), or *expanding* ($\lambda(a) > 0$).

Remark 3

1. So far, the shrinking Laplacian solitons on G_a, for $\frac{1}{4} \leq a < 1$, and the example in [27] constitute the only known examples of such solitons.
2. The left-invariant steady Laplacian soliton, corresponding to the value $a = 1$, is also an ERP G_2-structure. More generally, Lauret and Nicolini proved that every left-invariant ERP G_2-structure on a simply connected Lie group is a steady Laplacian soliton [40]. On the other hand, there exists an example of a left-invariant steady Laplacian soliton that is not ERP [26].

On a (compact) 7-manifold M, all closed G_2-structures satisfying (7) have a distinguished behaviour when they evolve under the G_2-*Laplacian flow*. This flow, which was introduced by Bryant in [4], prescribes the evolution of a closed G_2-structure $\varphi \in \Omega^3_+(M)$ as follows

$$\begin{cases} \frac{\partial}{\partial t}\varphi(t) = \Delta_{\varphi(t)}\varphi(t), \\ d\varphi(t) = 0, \\ \varphi(0) = \varphi. \end{cases} \tag{8}$$

The initial datum φ is a Laplacian soliton, i.e., it satisfies (7) for some real number λ and some vector field $X \in \Gamma(TM)$, if and only if the solution $\varphi(t)$ of (8) starting from it at $t = 0$ is *self-similar*, that is

$$\varphi(t) = \left(1 + \frac{2}{3}\lambda t\right)^{\frac{3}{2}} F_t^*\varphi,$$

where $F_t \in \mathrm{Diff}(M)$ is the one-parameter family of diffeomorphisms generated by the vector field $X(t) = \left(1 + \frac{2}{3}\lambda t\right)^{-1} X$ and satisfying $F_0 = \mathrm{Id}_M$. In particular, $\varphi(t)$ exists on the maximal time interval $\left(-\infty, -\frac{3}{2\lambda}\right)$ when $\lambda < 0$, $(-\infty, +\infty)$ when $\lambda = 0$, and $\left(-\frac{3}{2\lambda}, +\infty\right)$ when $\lambda > 0$. We refer the reader to [4, 43] for more details on the G_2-Laplacian flow and its solitons.

In [39], Lauret computed the solution of the G_2-Laplacian flow on $\mathfrak{g}_a = \mathfrak{n}_2 \rtimes_{D_a} \mathbb{R}$ starting from the closed G_2-structure $\varphi = \omega \wedge \eta + \psi$. When $a \neq 1$, it is given by

$$\varphi(t) = A_t{}^{q_1(a)}\, e^{127} + A_t{}^{q_2(a)}\, e^{347} + A_t{}^{q_3(a)} \left(e^{567} + e^{135} - e^{146} - e^{236} - e^{245}\right),$$

where $e^7 := \eta$, $A_t = \frac{2}{3}\,\lambda(a)\, t + 1$, and

$$q_1(a) = \frac{3a}{2(2a+1)}, \qquad q_2(a) = \frac{3(2a-1)}{8(a-1)}, \qquad q_3(a) = \frac{9}{8(2a+1)(a-1)}.$$

The maximal interval of existence of $\varphi(t)$ is $\left(-\infty, -\frac{3}{2\lambda(a)}\right)$, when $\frac{1}{4} \leq a < 1$, and $\left(-\frac{3}{2\lambda(a)}, +\infty\right)$, when $a > 1$.

An examination of the solution $\varphi(t)$ allows us to notice the following remarkable property.

Proposition 3 *Consider the seven-dimensional Lie algebra $\mathfrak{g}_a = \mathfrak{n}_2 \rtimes_{D_a} \mathbb{R}$, for $a \geq \frac{1}{4}$ with $a \neq 1$, and let $\varphi = \omega \wedge \eta + \psi$ be the closed G_2-structure on it induced by the coupled $SU(3)$-structure (ω, ψ) on $\mathfrak{n}_2$ given in (6). Then, the solution of the G_2-Laplacian flow starting from φ at $t = 0$ can be written as follows*

$$\varphi(t) = \omega(t) \wedge f(t)\,\eta + \psi(t),$$

where $f(t) = (A_t)^{\frac{1}{2}}$ *and the pair*

$$\omega(t) = A_t{}^{q_1(a)-\frac{1}{2}} e^{127} + A_t{}^{q_2(a)-\frac{1}{2}} e^{347} + A_t{}^{q_3(a)-\frac{1}{2}} e^{567},$$
$$\psi(t) = A_t{}^{q_3(a)} \left(e^{135} - e^{146} - e^{236} - e^{245}\right) = A_t{}^{q_3(a)} \psi,$$

defines a family of coupled SU(3)*-structures on* $\mathfrak{n}_2$ *satisfying*

$$\hat{d}\omega(t) = c(t)\,\psi(t), \quad c(t) = -(A_t)^{-\frac{1}{2}},$$

and having $\hat{d}w_2(t)$ *proportional to* $\psi(t)$*. Finally,* $\varphi(t)$ *induces the metric* $g_\varphi(t) = g(t) + [f(t)]^2\eta^2$*, where* $g(t)$ *denotes the metric associated with the coupled* SU(3)*-structure* $(\omega(t), \psi(t))$*.*

Let $\mathfrak{h}$ denote a six-dimensional Lie algebra endowed with a coupled SU(3)-structure (ω, ψ) satisfying $\hat{d}\omega = c\psi$, and assume that there exists a derivation $D \in \mathrm{Der}(\mathfrak{h})$ such that $D^*\psi = -c\psi$. The observations given in Proposition 3 lead to the question whether the G_2-Laplacian flow on $\mathfrak{h} \rtimes_D \mathbb{R}$ starting from $\varphi = \omega\wedge\eta + \psi$ can be reinterpreted as a flow on $\mathfrak{h}$ evolving the coupled SU(3)-structure (ω, ψ). The examination of the flow on the Lie algebras $\mathfrak{g}_{a,b}$ and $\mathfrak{g}_{a,b,k}$ defined in Sect. 4.4 may give some insights on this problem. For the sake of convenience, henceforth we let $e^7 := \eta$ and $e_7 := \xi$, so that the closed G_2-structure φ on these Lie algebras can be written as in (1) with respect to the basis $\{e^1, \dots, e^6, e^7\}$ we are considering.

Let us begin with the Lie algebra $\mathfrak{g}_{a,b,k}$. Notice that the coupled SU(3)-structure (ω, ψ) on $\mathfrak{s}_{a,b}$ satisfies the additional property $\hat{d}w_2 \propto \psi$, which is also satisfied by the coupled SU(3)-structure considered on $\mathfrak{n}_2$ and is never satisfied by any coupled SU(3)-structure on $\mathfrak{n}_1$ (cf. Remark 2).

We consider the following Ansatz for the solution

$$\varphi(t) = C_1e^{127} + C_2e^{347} + C_3e^{567} + C_4e^{135} - C_5e^{146} - C_6e^{236} - C_7e^{245}, \tag{9}$$

where $C_i = C_i(t)$, $i = 1, \dots, 7$, are unknown positive functions depending only on t and satisfying $C_i(0) = 1$. Then, $\varphi(t)$ still defines a G_2-structure, and it is closed if and only if $C_7 = C_6 = C_5 = C_4 = C_2$. Now, some straightforward computations show that the intrinsic torsion form $\tau(t)$ of $\varphi(t)$ has the following expression

$$\tau(t) = -b\left(\frac{C_1{C_2}^2}{{C_3}^2}\right)^{\frac{1}{3}} e^{12} + 3b\left(\frac{{C_2}^5}{{C_1}^2{C_3}^2}\right)^{\frac{1}{3}} e^{34} - 2b\left(\frac{{C_2}^2C_3}{{C_1}^2}\right)^{\frac{1}{3}} e^{56}.$$

Consequently, recalling that $\Delta_{\varphi(t)}\varphi(t) = d\tau(t)$, we have

$$\Delta_{\varphi(t)}\varphi(t) = -b^2\left(\frac{C_1{C_2}^2}{{C_3}^2}\right)^{\frac{1}{3}} e^{127} + 3b^2\left(\frac{{C_2}^5}{{C_1}^2{C_3}^2}\right)^{\frac{1}{3}} \left(e^{347} + e^{135} - e^{146} - e^{236} - e^{245}\right),$$

whence we see that the Laplacian flow Eq. (8) is equivalent to the following system of ODEs for the functions C_1, C_2, C_3:

$$\begin{cases} \frac{d}{dt}C_1 = -b^2\left(\frac{C_1 C_2{}^2}{C_3{}^2}\right)^{\frac{1}{3}}, \\ \frac{d}{dt}C_2 = 3b^2\left(\frac{C_2{}^5}{C_1{}^2 C_3{}^2}\right)^{\frac{1}{3}}, \\ \frac{d}{dt}C_3 = 0. \end{cases}$$

Combining these ODEs with the initial conditions $C_1(0) = 1 = C_2(0) = C_3(0)$, we immediately get $C_3(t) = 1$ and $C_1 = C_2{}^{-\frac{1}{3}}$, with C_2 solving the Cauchy problem

$$\begin{cases} \frac{d}{dt}C_2 = 3b^2 C_2{}^{\frac{17}{9}}, \\ C_2(0) = 1. \end{cases}$$

Thus, we have $C_2(t) = \left(-\frac{8}{3}b^2 t + 1\right)^{-\frac{9}{8}}$, and we see that the solution of the G_2-Laplacian flow starting from φ is defined in the interval $\left(-\infty, \frac{3}{8b^2}\right)$. Moreover, it can be written as

$$\varphi(t) = \omega(t) \wedge f(t)\,\eta + \psi(t),$$

where $f(t) = \left(-\frac{8}{3}b^2 t + 1\right)^{\frac{1}{2}}$, and $(\omega(t), \psi(t))$ is a family of coupled SU(3)-structures on $\mathfrak{s}_{a,b}$ satisfying $\hat{d}\omega(t) = c(t)\psi(t)$ with $c(t) = b\left(-\frac{8}{3}b^2 t + 1\right)^{-\frac{1}{2}}$ and having $\hat{d}w_2(t)$ proportional to $\psi(t)$.

Comparing this discussion with Proposition 3, we see that the solution we have just found shares similar properties with the solution of the Laplacian flow obtained by Lauret on $\mathfrak{g}_a$, for $\frac{1}{4} \leq a < 1$. However, we claim that the closed G_2-structure we are considering is never an algebraic soliton. The proof goes as follows. From [38], we know that if φ gives rise to a left-invariant algebraic soliton on the simply connected Lie group with Lie algebra $\mathfrak{g}_{a,b,k}$, then there exist a derivation $B \in \mathrm{Der}(\mathfrak{g}_{a,b,k})$ and a real number λ such that

$$\Delta_\varphi \varphi = d\tau = \lambda\varphi + B^*\varphi.$$

Since $a \neq 0$ and $b \neq 0$, the space $\mathrm{Der}(\mathfrak{g}_{a,b,k})$ is eight-dimensional, and we can write a generic derivation with respect to a basis $\{B_1, \dots, B_8\}$ of $\mathrm{Der}(\mathfrak{g}_{a,b,k})$ as $B = \sum_{i=1}^{8} A^i B_i$, for some real parameters A_i. We can then compute the 3-form $B^*\varphi$, and see that the above equation reduces to a system of polynomial equations in the unknowns A_i and λ. It turns out that this system has no solutions under the assumption $b \neq 0$.

On the other hand, obtaining the explicit solution of the G$_2$-Laplacian flow on $\mathfrak{g}_{a,b}$ satisfying $\varphi(0) = \varphi$ seems challenging. For instance, the Ansatz (9) is not helpful in this case, as the flow equation in (8) implies that some of the functions $C_i(t)$ must be identically vanishing, a contradiction. Notice that in this example the coupled SU(3)-structure does not satisfy the condition $\hat{d}w_2 \propto \psi$.

We now collect some observations and open problems motivated by the results presented in this last section.

1. Is it possible to determine the explicit expression of the solution of the G$_2$-Laplacian flow on $\mathfrak{g}_{a,b}$ starting from the closed G$_2$-structure φ?
2. Albeit the Lie algebras $\mathfrak{g}_a$, $\frac{1}{4} \leq a < 1$, and $\mathfrak{g}_{a,b,k}$, $a \neq 0$, $b \neq 0$, are not isomorphic, they admit coupled SU(3)-structures sharing similar properties. Are the corresponding simply connected Lie groups endowed with the left-invariant coupled SU(3)-structures equivalent in a suitable sense?
3. Consider a six-dimensional Lie algebra $\mathfrak{h}$ endowed with a coupled SU(3)-structure (ω, ψ) with $\hat{d}\omega = c\psi$ and admitting a derivation satisfying $D^*\psi = -c\psi$. Is it possible to reinterpret the G$_2$-Laplacian flow on $\mathfrak{g} = \mathfrak{h} \rtimes_D \mathbb{R}$ evolving the closed G$_2$-structure $\varphi = \omega \wedge \eta + \psi$ as a flow evolving the coupled SU(3)-structure on $\mathfrak{h}$? Does the condition $\hat{d}w_2 \propto \psi$ play some role in this problem?
4. Is it possible to classify all six-dimensional solvable Lie algebras admitting a coupled SU(3)-structure up to isomorphism, at least when the condition $\hat{d}w_2 \propto \psi$ holds?

Acknowledgments The authors were supported by G.N.S.A.G.A. of I.N.d.A.M. and by the project P.R.I.N. 2017 "Real and Complex Manifolds: Topology, Geometry and Holomorphic Dynamics". A. F. would like to thank the organisers of the Abel Symposium, J. Figueroa-O'Farrill, S. Hervik, B. Kruglikov, I. Markina, J. Slovák, D. The, and B. Ørsted, for the wonderful conference and their hospitality.

References

1. G. Ball, Quadratic closed G_2-structures. arXiv:2006.14155
2. G. Ball, Seven-Dimensional Geometries With Special Torsion. Ph.D. dissertation, Duke University
3. E. Bonan, Sur des variétés riemanniennes á groupe d'holonomie G$_2$ ou Spin(7). C. R. Acad. Sci. Paris Sér. A-B **262**, A127–A129 (1966)
4. R. Bryant, Some remarks on G$_2$-structures, in *Proceedings of Gökova Geometry-Topology Conference 2005, Gökova Geometry/Topology Conference (GGT)*, Gökova (2006), pp. 75–109
5. B.Y. Chu, Symplectic homogeneous spaces. Trans. Am. Math. Soc. **197**, 145–159 (1974)
6. D. Conti, M. Fernández, Nilmanifolds with a calibrated G$_2$-structure. Differ. Geom. Appl. **29**(4), 493–506 (2011)
7. A. Corti, M. Haskins, J. Nordström, T. Pacini, G$_2$-manifolds and associative submanifolds via semi-Fano 3-folds. Duke Math. J. **164**(10), 1971–2092 (2015)
8. R. Cleyton, S. Ivanov, On the geometry of closed G$_2$-structures. Commun. Math. Phys. **270**(1), 53–67 (2007)
9. R. Cleyton, A. Swann Cohomogeneity-one G$_2$-structures. J. Geom. Phys. **44**(2–3), 202–220 (2002)

10. D. Crowley, J. Nordström, New invariants of G_2-structures. Geom. Topol. **19**(5), 2949–2992 (2015)
11. A. Diatta, B. Manga, On properties of principal elements of Frobenius Lie algebras. J. Lie Theory **24**(3), 849–864 (2014)
12. M. Fernández, An example of a compact calibrated manifold associated with the exceptional Lie group G_2. J. Differ. Geom. **26**(2), 367–370 (1987)
13. M. Fernández, A family of compact solvable G_2-calibrated manifolds. Tohoku Math. J. **39**(2), 287–289 (1987)
14. M. Fernández, A. Fino, A. Kovalev, V. Munoz, A compact G_2-calibrated manifold with first Betti number $b_1 = 1$. Adv. Math. **381**, 37 (2021)
15. M. Fernández, A. Fino, V. Manero, G_2-structures on Einstein solvmanifolds, Asian J. Math. **19**(2), 321–342 (2015)
16. M. Fernández, A. Fino, A. Raffero, Locally conformal calibrated G_2-manifolds. Ann. Mat. Pura Appl. **195**(5), 1721–1736 (2016)
17. M. Fernández, A. Fino, A. Raffero, Exact G_2-structures on unimodular Lie algebras. Monatsh. Math. **193**(1), 47–60 (2020)
18. M. Fernández, A. Gray, Riemannian manifolds with structure group G_2. Ann. Mat. Pura Appl. **32**, 19–45 (1982)
19. M. Fernández, V. Manero, A. Otal, L. Ugarte, Symplectic half-flat solvmanifolds. Ann. Global Anal. Geom. **43**(4), 367–383 (2013)
20. J. Fine, C. Yao, Hypersymplectic 4-manifolds, the G_2-Laplacian flow, and extension assuming bounded scalar curvature. Duke Math. J. **167**, 3533-3589 (2018)
21. A. Fino, A. Raffero, Coupled SU(3)-structures and supersymmetry. Symmetry **7**(2), 625–650 (2015)
22. A. Fino, A. Raffero, Einstein locally conformal calibrated G_2-structures. Math. Z. **280**(3–4), 1093–1106 (2015)
23. A. Fino, A. Raffero, Closed G_2-structures on non-solvable Lie groups. Rev. Mat. Complut. **32**(3), 837–851 (2019)
24. A. Fino, A. Raffero, Closed warped G_2-structures evolving under the Laplacian flow. Ann. Sc. Norm. Sup. Pisa Cl. Sci. **20**(1), 315–348 (2020)
25. A. Fino, A. Raffero, A class of eternal solutions to the G_2-Laplacian flow. J. Geom. Anal. **31**(5), 4641–4660 (2021)
26. A. Fino, A. Raffero, Remarks on homogeneous solitons of the G_2-Laplacian flow. C. R. Math. Acad. Sci. Paris. **358**(4), 401–406 (2020)
27. U. Fowdar, S^1-invariant Laplacian flow. J. Geom. Anal. (2020, to appear). arXiv:2007.05130
28. M. Freibert Calibrated and parallel structures on almost Abelian Lie algebras. arXiv:1307.2542
29. A. Gray, Vector cross products on manifolds. Trans. Am. Math. Soc. **141**, 465–504 (1969)
30. R. Harvey, H.B. Lawson Jr., Calibrated geometries. Acta Math. **148**, 47–157 (1982)
31. N. Hitchin, Stable forms and special metrics, in *Global Differential Geometry: The Mathematical Legacy of Alfred Gray (Bilbao, 2000)*, vol. 288. Contemp. Math. (Amer. Math. Soc., Providence, 2001), pp. 70–89
32. D.D. Joyce, Compact Riemannian 7-manifolds with holonomy G_2. I, II. J. Differ. Geom. **43**, 291–328, 329–375 (1996)
33. D.D. Joyce, in *Riemannian Holonomy Groups and Calibrated Geometry*. Oxford Graduate Texts in Mathematics, vol. 12 (Oxford University Press, Oxford, 2007)
34. D.D. Joyce, S. Karigiannis, A new construction of compact torsion-free G_2-manifolds by gluing families of Eguchi-Hanson spaces. J. Differ. Geom. arXiv:1707.09325
35. I. Kath, J. Lauret, A new example of a compact ERP G_2-structure. Bull. Lond. Math. Soc. To appear. https://doi.org/10.1112/blms.12520
36. A. Kovalev, Twisted connected sums and special Riemannian holonomy. J. Reine Angew. Math. **565**, 125–160 (2003)
37. A. Kovalev, N.-H. Lee, K3 surfaces with non-symplectic involution and compact irreducible G_2-manifolds. Math. Proc. Camb. Philos. Soc. **151**, 193–218 (2011)

38. J. Lauret, Laplacian flow of homogeneous G_2-structures and its solitons. Proc. Lond. Math. Soc. **114**(3), 527–560 (2017)
39. J. Lauret, Laplacian solitons: questions and homogeneous examples. Differ. Geom. Appl. **54**(B), 345–360 (2017)
40. J. Lauret, M. Nicolini, Extremally Ricci pinched G_2-structures on Lie groups. Commun. Anal. Geom. To appear
41. J. Lauret, M. Nicolini, The classification of ERP G_2-structures on Lie groups. Ann. Mat. Pura Appl. **199**, 2489–2510 (2020)
42. A. Lichnerowicz, A. Medina, On Lie groups with left-invariant symplectic or Kählerian structures. Lett. Math. Phys. **16**(3), 225–235 (1988)
43. J.D. Lotay, Y. Wei, Laplacian flow for closed G_2 structures: Shi-type estimates, uniqueness and compactness. Geom. Funct. Anal. **27**(1), 165–233 (2017)
44. V. Manero, Compact solvmanifolds with calibrated and cocalibrated G_2-structures. Manuscripta Math. **162**, 315–339 (2020)
45. F. Podestà, A. Raffero, On the automorphism group of a closed G_2-structure. Q. J. Math. **70**(1), 195–200 (2019)
46. F. Podestà, A. Raffero, Homogeneous symplectic half-flat 6-manifolds. Ann. Global Anal. Geom. **55**(1), 1–15 (2019)
47. F. Podestà, A. Raffero, On the automorphism group of a symplectic half-flat 6-manifold. Forum Math. **31**(1), 265–273 (2019)
48. F. Podestà, A. Raffero, Closed G_2-structures with a transitive reductive group of automorphisms. Asian J. Math. (to appear)

Almost Zoll Affine Surfaces

Peter B. Gilkey

Abstract An affine surface is said to be a Zoll surface if all affine geodesics close smoothly. It is said to be an almost Zoll surface if thru any point, every affine geodesic but one closes smoothly (the exceptional geodesic is said to be alienated as it does not return). We exhibit an affine structure on the cylinder which is almost Zoll. This structure is geodesically complete, affine Killing complete, and affine symmetric.

Keywords Affine manifold · Zoll surface

1 Introduction

Our purpose in this brief note is two-fold. We shall present some new research results related to the theory of Zoll surfaces in the affine context. But enroute to that goal, we shall take the opportunity to survey some of the relevant results in affine geometry.

A Riemannian manifold is said to be a *Zoll manifold* if the unit speed geodesics are all simple closed curves of the same length. Zoll [24] showed that the sphere S^2 admits many such metrics in addition to the round one. Later authors called these surfaces *auf Wiedersehensflächen* or "until we meet again" surfaces. In Spanish, this becomes *te veo de nuevo superficie*.

The only compact surfaces which admit Zoll Riemannian metrics are the sphere S^2 and real projective space $\mathbb{RP}^2$. Green [11] showed that the only Zoll metric on $\mathbb{RP}^2$ is the standard homogeneous metric, up to isometry and rescaling; this was later extended by Pries [20] to show that if all the geodesics of a metric on $\mathbb{RP}^2$

Research partially supported by Project MTM2016-75897-P (AEI/FEDER, Spain).

P. B. Gilkey (✉)
Math. Dept., University of Oregon, Eugene, OR, USA
e-mail: gilkey@uoregon.edu

S. Hervik et al. (eds.), *Geometry, Lie Theory and Applications*, Abel Symposia 16,
https://doi.org/10.1007/978-3-030-81296-6_4

are closed, then the metric is the standard homogeneous metric, up to isometry and rescaling. One can consider analogous questions in the Lorentzian setting—see, for example, the work of Mounoud and Suhr [17]. Zoll surfaces have been used in many contexts; see, for example, the work on Balacheff et al. [3] concerning a geodesic length conjecture. Zoll surfaces also have been considered in the orbifold context; see, for example, Lange [13]. We refer to Besse [4] for a discussion of more of the history of the subject than is available here and also to the references cited above.

An affine surface $\mathcal{M}$ is a pair (M, ∇) where M is a 2-dimensional manifold and ∇ is a torsion free connection on the tangent bundle of M. We assume henceforth that M is connected. A diffeomorphism Φ from (M_1, ∇_1) to (M_2, ∇_2) is said to be *affine* if Φ intertwines ∇_1 and ∇_2. An affine surface $\mathcal{M}$ is said to be *homogeneous* if the group of affine diffeomorphisms acts transitively on $\mathcal{M}$. A vector field X on $\mathcal{M}$ is said to be an *affine Killing vector field* if the (locally defined) flows of X are (locally defined) affine diffeomorphisms of $\mathcal{M}$ or, equivalently by Kobayashi and Nomizu [12], the Lie derivative $\mathcal{L}_X\nabla = 0$. The Lie bracket makes the set $\mathfrak{K}\mathcal{M})$ of affine Killing vector fields into a Lie algebra. We say that $\mathcal{M}$ is *affine Killing complete* if the integral curves of any affine Killing vector field are defined for $t \in \mathbb{R}$; if $\mathcal{M}$ is affine Killing complete, then $\mathfrak{K}(\mathcal{M})$ is the Lie algebra of the Lie group of affine diffeomorphisms of $\mathcal{M}$. A curve $\sigma(t)$ in M is said to be an *geodesic* if it is *auto parallel*, i.e. if $\nabla_{\dot\sigma}\dot\sigma = 0$. A geodesic is defined by its initial position $\sigma(0)$ and initial direction $\dot\sigma(0)$. $\mathcal{M}$ is said to be *geodesically complete* if every geodesic is defined for $t \in \mathbb{R}$. $\mathcal{M}$ is said to be a *Zoll affine surface* if all geodesics close smoothly. We refer, for example, to Contatto and Dunajski [8] where affine Zoll surfaces are constructed arising from two-dimensional Frobenius manifolds and to McKay [16] for a discussion in relation to complete projective connections.

Let $R(X, Y) := \nabla_X\nabla_Y - \nabla_Y\nabla_X - \nabla_{[X,Y]}$ be the curvature operator. The Ricci tensor is given by $\rho(X, Y) := \operatorname{Tr}(Z \to R(X, Y)Z)$. We say that $\mathcal{M}$ is *symmetric* if and only if $\nabla R = 0$ or, equivalently as we are in the 2-dimensional setting and the R can be expressed in terms of the Ricci tensor, $\nabla\rho = 0$. Since ρ need not be symmetric in the affine context, let $\rho_s(X, Y) := \frac{1}{2}\{\rho(X, Y) + \rho(Y, X)\}$ be the *symmetric Ricci tensor*. We say two affine structures ∇ and $\tilde\nabla$ on the same underlying surface M are *projectively equivalent* if there exists a smooth 1-form ω on M so that we have $\tilde\nabla_X Y = \nabla_X Y + \omega(X)Y + \omega(Y)X$. Two affine structures are projectively equivalent if and only if their unparametrized geodesics coincide (see for example, Schouten [21]). Thus it is natural to pass to projective equivalence in considering Zoll structures. Note that projective equivalence does not preserve geodesic completeness since the parametrized geodesics will in general be different. A *projective structure* is an equivalence class of projectively equivalent structures on M. There is a question of which connections or more generally which projective structures are *metrizable*, i.e. which are projectively equivalent to the Levi-Civita connection of some Riemannian or pseudo-Riemannian metric; we refer to Bryant et al. [7] and Vanžurová [22, 23] for further details concerning this question. The geometries are said to be *strongly projectively equivalent* if $\omega = d\phi$ is exact, i.e.

$$\tilde{\nabla}_X Y = \nabla_X Y + X(\phi)Y + Y(\phi)X \quad \text{for some} \quad \phi \in C^\infty(M)\,.$$

In this setting, we shall say that ϕ provides a *strong projective equivalence* from $\mathcal{M} = (M, \nabla)$ to ${}^\phi\mathcal{M} := (M, \tilde{\nabla})$. We say that $\mathcal{M}$ is *strongly projectively flat* if $\mathcal{M}$ is strongly projectively equivalent to a flat connection.

As noted above, the question of whether or not $\mathcal{M}$ is a Zoll surface is a projective question. LeBrun and Mason [14] (see also later related work in [15]) showed that the only compact surfaces which admit Zoll affine structures are $\mathbb{S}^2$ and $\mathbb{RP}^2$. More generally, they show that any *tame* affine manifold of arbitrary dimension is compact, has finite fundamental group, and the exponential map is surjective; we shall discuss this further in Sect. 5; the requirement that $\mathcal{M}$ be "tame" is a mild additional technical assumption. Consequently, it is natural to weaken the Zoll condition just a little. We say that an affine surface is *almost Zoll* if for every point of the surface, there is a single geodesic, which will be called the *alienated geodesic*, which does not close and such that all other geodesics thru that point are simple closed curves which return to the initial point; as the manifold in question need not be geodesically complete, we assume each geodesic is defined on its maximal domain. In this brief note, we present several examples of this phenomena.

Here is a brief guide to the remainder of this paper. In Sect. 2, following [9], we define the solution space of the *affine quasi-Einstein equation* and present its basic properties in Theorem 1. In Sect. 3, we introduce the affine surfaces $\mathcal{M}(c)$ that will form the focus of our investigation; $\mathcal{M}(c) = (\mathbb{R}^2, \nabla)$ where the (possibly) non-zero Christoffel symbols are $\Gamma_{22}{}^1 = (1 + c^2)x^1$ and $\Gamma_{22}{}^2 = 2c$ for $c \geq 0$. In Theorem 2, we show that $\mathcal{M}(c)$ is affine homogeneous, determine the Lie algebra of affine Killing vector fields of $\mathcal{M}(c)$, show that $\mathcal{M}(c)$ is affine Killing complete, and determine the connected component $\mathcal{G}_0(c)$ of the 4-dimensional Lie group of affine diffeomorphisms. In Theorem 3, we show the geometries $\mathcal{M}(c)$ and $\mathcal{M}(\tilde{c})$ are neither strongly projectively equivalent nor locally affine equivalent for $c \neq \tilde{c}$. In Sect. 4, we show in Theorem 4 that $\mathcal{M}(0)$ is geodesically complete and that $\mathcal{M}(c)$ is geodesically incomplete for $c > 0$. In Sect. 5, we use $\mathcal{M}(0)$ to construct an almost Zoll structure C on the cylinder. We also construct a related almost Zoll affine structure on the Möbious strip. We state the fundamental Lemma of LeBrun and Mason [14] concerning Zoll affine surfaces and use C to show why this does not extend to the almost Zoll setting.

2 The Quasi-Einstein Equation

The solution space to the quasi-Einstein equation is a fundamental invariant in the theory of affine surfaces. Let $\mathcal{H}\phi = (\partial_{x^i}\partial_{x^j}\phi - \Gamma_{ij}{}^k\partial_{x^k}\phi)dx^i \otimes dx^j \in S^2(M)$ be the *Hessian operator* of an affine surface $\mathcal{M}$ where we adopt the *Einstein convention* and sum over repeated indices. Let $Q(\mathcal{M})$ be the solution space of the *quasi-Einstein equation*: $Q(\mathcal{M}) := \{\phi \in C^\infty(M) : \mathcal{H}\phi + \phi\rho_s = 0\}$.

There is a close relationship between strong projective equivalence and the solutions of the quasi-Einstein equation. We refer to Gilkey and Valle-Regueiro [9] for the proof of the following result and to Brozos-Vázquez et al. [6] for a further discussion of the quasi-Einstein equation and its use through the modified Riemannian extension to study conformally Einstein Walker metrics of signature (2, 2) on the cotangent bundle of an affine surface.

Theorem 1 *Let $\mathcal{M} = (M, \nabla)$ be a simply connected affine surface.*

1. $\dim\{Q(\mathcal{M})\} \leq 3$.
2. $\dim\{Q(\mathcal{M})\} = 3$ *if and only if $\mathcal{M}$ is strongly projectively flat.*
3. $Q({}^\phi\mathcal{M}) = e^\phi Q(\mathcal{M})$.
4. *Let $\mathcal{M}_1$ and $\mathcal{M}_2$ be two strongly projectively flat affine surfaces and let Φ be a diffeomorphism from M_1 to M_2. If $\Phi^*Q(\mathcal{M}_2) = Q(\mathcal{M}_1)$, then Φ is an affine morphism from $\mathcal{M}_1$ to $\mathcal{M}_2$.*

3 The Affine Structures $\mathcal{M}(c)$

The following family of surfaces was introduced by D'Ascanio et al. [2] in their study of geodesically complete homogeneous affine surfaces. Let $\mathcal{M}(c) := (\mathbb{R}^2, \nabla)$ where the only (possibly) non-zero Christoffel symbols of ∇ are $\Gamma_{22}{}^1 = (1 + c^2)x^1$ and $\Gamma_{22}{}^2 = 2c$. Replacing x^2 by $-x^2$ interchanges the roles of $\mathcal{M}(c)$ and $\mathcal{M}(-c)$ so we shall assume $c \geq 0$ henceforth. A direct computation shows

$$\rho = (1 + c^2)dx^2 \otimes dx^2 \text{ and } \nabla\rho = -4c(1 + c^2)dx^2 \otimes dx^2 \otimes dx^2 . \quad (1)$$

Let $T(\alpha, \beta, \gamma, \delta) : (x^1, x^2) \to (e^\alpha x^1 + \beta e^{cx^2}\cos(x^2) + \gamma e^{cx^2}\sin(x^2), x^2 + \delta)$ for $(\alpha, \beta, \gamma, \delta) \in \mathbb{R}^4$. Let $\mathfrak{g}(c) := \text{Span}\{x^1\partial_{x^1}, e^{cx^2}\cos(x^2)\partial_{x^1}, e^{cx^2}\sin(x^2)\partial_{x^1}, \partial_{x^2}\}$. Patera et al. [19] classified the 4-dimensional Lie algebras. We follow their notation. Let $A_{4,12}$ be the 4-dimensional Lie algebra with bracket relations $[e_1, e_3] = e_1$, $[e_2, e_3] = e_2$, $[e_1, e_4] = -e_2$, $[e_2, e_4] = e_1$.

Theorem 2 *Let $\mathcal{G}_0(c) := \{T(\alpha, \beta, \gamma, \delta) : (\alpha, \beta, \gamma, \delta) \in \mathbb{R}^4\}$.*

1. $Q(\mathcal{M}(c)) = \text{Span}\{e^{cx^2}\cos(x^2), e^{cx^2}\sin(x^2), x^1\}$.
2. *$\mathcal{M}(c)$ is strongly projectively flat.*
3. *Composition makes $\mathcal{G}_0(c)$ into a 4-dimensional Lie group.*
4. *$\mathcal{G}_0(c)$ acts transitively on $\mathcal{M}(c)$ by affine diffeomorphisms.*
5. *$\mathcal{M}(c)$ is a homogeneous space that is symmetric if and only if $c = 0$.*
6. *$\mathfrak{g}(c)$ is the Lie algebra of $\mathcal{G}_0(c)$. The vector fields of $\mathfrak{g}(c)$ are complete.*
7. *$\mathfrak{K}\mathcal{M}(c) = \mathfrak{g}(c)$, $\mathcal{M}$ is affine Killing complete, and $\mathfrak{g}(c)$ is isomorphic to $A_{4,12}$.*

Proof We follow the discussion in Gilkey et al. [10]. A direct computation shows that $\{e^{cx^2}\cos(x^2), e^{cx^2}\sin(x^2), x^1\} \subset \mathcal{Q}(\mathcal{M}(c))$. Thus $\dim\{\mathcal{Q}(\mathcal{M}(c)) \geq 3$. Assertions 1 and 2 now follow from Theorem 1. We show that $\mathcal{G}_0(c)$ is closed under composition and inverse and verify Assertion 3 by computing that

$$\begin{aligned}
&T(\alpha, \beta, \gamma, \delta) \circ T(\tilde{\alpha}, \tilde{\beta}, \tilde{\gamma}, \tilde{\delta}) \\
&= T(e^{\alpha+\tilde{\alpha}}, e^{\alpha}\tilde{\beta} + \beta e^{c\tilde{\delta}}\cos(\tilde{\delta}) + \gamma e^{c\tilde{\delta}}\sin(\tilde{\delta}), \\
&\quad e^{\alpha}\tilde{\gamma} - \beta e^{c\tilde{\delta}}\sin(\tilde{\delta}) + \gamma e^{c\tilde{\delta}}\cos(\tilde{\delta}), \delta + \tilde{\delta}), \\
&T(\alpha, \beta, \gamma, \delta)^{-1} = T(-\alpha, -e^{-\alpha-c\delta}(\beta\cos(\delta) - \gamma\sin(\delta)), \\
&\quad -e^{-\alpha-c\delta}(\beta\sin(\delta) + \gamma\cos(\delta)), -\delta).
\end{aligned}$$

A direct computation shows that $T(\alpha, \beta, \gamma, \delta)^*\mathcal{Q}(\mathcal{M}(c)) = \mathcal{Q}(\mathcal{M}(c))$. Theorem 1 shows that $T(\alpha, \beta, \gamma, \delta)$ is an affine map; it is immediate that the action is transitive. This verifies Assertion 4. It then follows that $\mathcal{M}(c)$ is homogeneous; Eq. (1) shows that $\mathcal{M}(c)$ is symmetric if and only if $c = 0$. This establishes Assertion 5. A direct computation verifies that $\mathfrak{g}(c)$ is the associated Lie algebra which establishes Assertion 6. The one parameter subgroups of $\mathcal{G}_0(c)$ are the flows of the vector fields in $\mathfrak{g}(c)$ so all these vector fields are complete.

Following the notation of Opozda [18] (who classified the geometry of locally homogeneous affine surfaces), an affine structure on $\mathbb{R}^2$ is said to be Type $\mathcal{A}$ if the Christoffel symbols are constant; work of Arias-Marco and Kowalski [1] and of Vanzurova [23] show that if the Ricci tensor of such a structure is non-zero, then it is not metrizable, i.e. it does not arise as the Levi-Civita connection of a pseudo-Riemannian metric. Let $\mathcal{M}_0(c)$ be the affine structure on $\mathbb{R}^2$ where the only (possibly) non-zero Christoffel symbols are $\Gamma_{11}{}^1 = 1$, $\Gamma_{22}{}^1 = 1 + c^2$, and $\Gamma_{22}{}^2 = 2c$. This is a Type $\mathcal{A}$ geometric structure which is linearly equivalent to the surfaces $\mathcal{M}_5^c$ of [5] but is in a slightly more convenient form for our purposes. Let (u^1, u^2) be coordinates on $\mathbb{R}^2$. We have $\{e^{cu^2}\cos(u^2), e^{cu^2}\sin(u^2), e^{u^1}\} \subset \mathcal{Q}(\mathcal{M}_0(c))$. Consequently, by Theorem 1, these elements form a basis for $\mathcal{M}_0(c)$. Consider the diffeomorphism $\Phi : (u^1, u^2) \to (e^{u^1}, u^2)$. We then verify that $\Phi^*\mathcal{Q}(\mathcal{M}(c)) = \mathcal{M}_0(c)$ so Φ is an affine embedding of $\mathcal{M}_0(c)$ in $\mathcal{M}(c)$ by Theorem 1. Consequently, since the Ricci tensor of $\mathcal{M}(c)$ has rank 1, the Ricci tensor of $\mathcal{M}_0(c)$ has rank 1 as well. Results of Brozos-Vázquez et al. [5] show that if $\mathcal{M}$ is a Type $\mathcal{A}$ structure on $\mathbb{R}^2$ and if $\text{Rank}\{\rho\} = 1$, then $\dim\{\mathfrak{K}(\mathcal{M}(c)\} = 4$. Consequently, the Lie algebra of Killing vector fields $\mathfrak{K}\mathcal{M}_0)$ is 4 -dimensional and, consequently, $\mathfrak{K}(\mathcal{M}(c))$ is 4-dimensional as well. Since $\mathfrak{g}(c)$ is 4-dimensional, we have $\mathfrak{K}(\mathcal{M}(c)) = \mathfrak{g}(c)$. Assertion 7 now follows from Assertion 6. Results of [5] show $\mathfrak{K}(\mathcal{M}_0(c))$ is isomorphic to $A_{4,12}$ and, consequently, $\mathfrak{K}(\mathcal{M}(c))$ is isomorphic to $A_{4,12}$ as well. Assertion 7 follows. □

Let $\mathcal{G}(c)$ be the full Lie group of affine diffeomorphisms of $\mathcal{M}(c)$. The analysis performed above shows $\mathcal{G}_0(c)$ is the connected component of the identity in $\mathcal{G}(c)$.

However, $\mathcal{G}(c)$ is in fact larger. The map $(x^1, x^2) \to (-x^1, x^2)$ preserves $\mathcal{Q}(\mathcal{M}(c))$ and hence is an affine diffeomorphism of $\mathcal{M}(c)$. This involution will play a central role in the analysis of Sect. 5. Furthermore, if $c = 0$, then the map $(x^1, x^2) \to (x^1, -x^2)$ also preserves $\mathcal{Q}(\mathcal{M}(0))$ and hence is an affine diffeomorphism. In fact, $\mathcal{G}(c)$ has 2 arc components if $c \neq 0$ and 4 arc components if $c = 0$. We omit details in the interests of brevity as this will play no role in our analysis.

Theorem 3

1. *The geometries $\mathcal{M}(c)$ and $\mathcal{M}(\tilde{c})$ are not locally affine equivalent for $c \neq \tilde{c}$.*
2. *The geometries $\mathcal{M}(c)$ and $\mathcal{M}(\tilde{c})$ are not strongly projectively equivalent for $c \neq \tilde{c}$.*
3. *Let $\Phi_c(x^1, x^2) = (e^{cx^2}x^1, x^2)$ define a diffeomorphism of $\mathbb{R}^2$. Then $\Phi^*\mathcal{M}(c)$ is strongly projectively equivalent to $\mathcal{M}(0)$.*

Proof Let $\mathcal{M} = (M, \nabla)$ be an affine surface. One says that the Ricci tensor of $\mathcal{M}$ is *recurrent* if $\nabla\rho = \omega \otimes \rho$ for some suitably chosen 1-form ρ. Assume in addition that the Ricci tensor of $\mathcal{M}$ is symmetric and has rank 1. Chose a (locally defined) vector field X on M with $\rho(X, X) \neq 0$. Let $\alpha_X := \nabla\rho(X, X, X)^2\rho(X, X)^{-3}$. It is an easy exercise to show α_X is independent of the particular X which is chosen and defines an affine invariant $\alpha(\mathcal{M})$ (see, for example, [5]). One uses Eq. (1) to see that this analysis applies to $\mathcal{M}(c)$ and to compute that $\alpha(\mathcal{M}(c)) = \frac{16c^2}{1+c^2}$. This function is strictly monotonically increasing for $c \geq 0$. Consequently, $\mathcal{M}(c)$ is locally affine equivalent to $\mathcal{M}(\tilde{c})$ if and only if $c = \tilde{c}$; these geometries are all distinct.

Suppose $c \neq \tilde{c}$. There is no function h so that $\mathcal{Q}(\mathcal{M}(c)) = e^h\mathcal{Q}(\mathcal{M}(\tilde{c}))$. Consequently, by Theorem 1, these geometries are not strongly projectively equivalent. However, we verify that $\Phi_c^*\mathcal{Q}(\mathcal{M}(c)) = e^{cx^2}\mathcal{Q}(\mathcal{M}(0))$. Consequently, by Theorem 1 $\Phi_c^*\mathcal{M}(c)$ is strongly projectively equivalent to $\mathcal{M}(0)$. □

4 The Geodesic Structure of $\mathcal{M}(c)$

The geometry $\mathcal{M}(0)$ is geodesically complete; the geometries $\mathcal{M}(c)$ for $c > 0$ are geodesically incomplete. The following theorem follows from a direct computation.

Theorem 4

1. *The curves* $\sigma_{u,v;a,b}(t) = \left\{\begin{array}{ll}(u\cos(bt) + \frac{a}{b}\sin(bt), bt + v) & \text{if } b \neq 0\\ (at + u, 0) & \text{if } b = 0\end{array}\right\}$ *are geodesics for the geometry $\mathcal{M}(0)$ with initial position (u, v) and initial velocity (a, b). $\mathcal{M}(0)$ is geodesically complete.*
2. *For $c > 0$, let $\tau(t) = \log(1 + 2bct)$. Then the curves*

$$\sigma_{u,v;a,b}(t) = \left\{\begin{array}{ll}(\frac{\sqrt{1+2bct}}{b}(bu\cos(\frac{\tau(t)}{2c}) + (a - bcu)\sin(\frac{\tau(t)}{2c}), v + \frac{\tau(t)}{2c}) & \text{if } b \neq 0\\ (at + u, 0) & \text{if } b = 0\end{array}\right\}$$

are geodesics of $\mathcal{M}(c)$ with initial position (u, v) and initial velocity (a, b). $\mathcal{M}(c)$ is geodesically incomplete since these curves are only defined for the parameter range $1 + 2bct > 0$.

Proof One verifies directly the curves $\sigma_{u,v;a,b}$ are geodesics with the given initial conditions. Since they are defined for all time if $c = 0$, $\mathcal{M}(0)$ is geodesically complete. On the other hand, if $c > 0$, they fail to be defined when $1 + 2bct = 0$ and thus $\mathcal{M}(c)$ is geodesically incomplete for $c > 0$. □

Remark 1 We give below a picture of the geodesic structure of $\mathcal{M}(0)$ at various points. The geodesics fall into 2 families; those with $b > 0$ all meet at the point $(0, \pi)$; those with $b < 0$ all meet at the point $(0, -\pi)$, and those with $b = 0$ lie along the x^1 axis and are alienated.

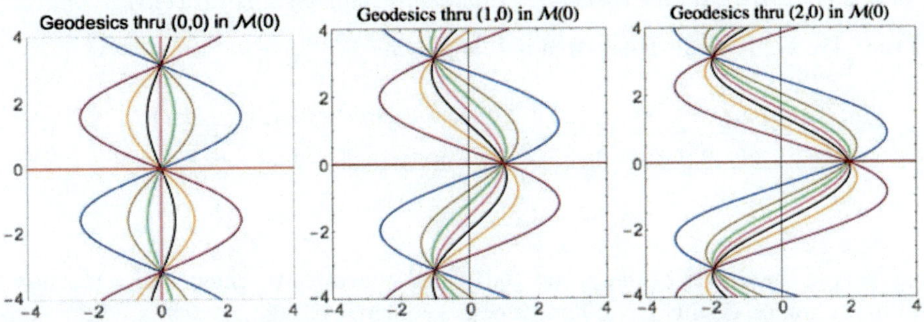

Remark 2 Let $\Phi(x^1, x^2) = (e^{cx^2}x^1, x^2)$. By Theorem 2, $\Phi^*\mathcal{Q}(0) = e^{cx^2}\mathcal{Q}(c)$. Theorem 1 then implies that $\Phi^*\mathcal{M}(0)$ is strongly projectively equivalent to $\mathcal{M}(c)$. Thus modulo the diffeomorphism Φ, the picture of the geodesic structure for $\mathcal{M}(c)$ is the same as that for $\mathcal{M}(0)$; Φ takes unparametrized geodesics in $\mathcal{M}(c)$ to unparameterized geodesics in $\mathcal{M}(0)$. Φ does not, preserve, the parametrization; $\mathcal{M}(c)$ is geodesically incomplete for $c > 0$.

Remark 3 We have $\rho = (1 + c^2)dx^2 \otimes dx^2$. We use ρ to define a positive semi-definite inner product and let $\rho(X, X)$ be the "speed". We suppose $c \neq 0$. We compute

$$\rho(\dot{\sigma}_{u,v,a,b}, \dot{\sigma}_{u,v,a,b}) = \left\{ \begin{array}{ll} (1 + c^2)\frac{2b}{(1+2bct)^2} & \text{if } b \neq 0 \\ 0 & \text{if } b = 0 \end{array} \right\}.$$

Thus the alienated geodesics are the null geodesics in the geometry $\mathcal{M}(c)$. The speed of the remaining geodesics goes to ∞ as $t \to -\frac{1}{2bc}$. Thus this geometry is essentially geodesically incomplete, it can not be embedded as an open subset of a geodesically complete geometry. On the other hand, if $c = 0$, then $\rho(\dot{\sigma}_{u,v,a,b}, \dot{\sigma}_{u,v,a,b}) = b^2$, Thus once again, the alienated geodesics are the null geodesics, but the remaining geodesics all have constant non-zero speed.

5 An Almost Zoll Structure on the Cylinder

In light of Remark 2, we take $c = 0$ henceforth. Let $\Phi(x^1, x^2) = (x^1, x^2 + 2\pi)$ generate a fixed point free action of $\mathbb{Z}$ on $\mathbb{R}^2$; this corresponds to taking $\alpha = 0$, $\beta = 0$, $\gamma = 0$, and $\delta = \pi$ in Theorem 2. We divide by this action to define an affine structure on the cylinder $C := (\mathbb{R} \times S^1, \nabla)$. Similarly, let $\Psi(x^1, x^2) := (-x^1, x^2 + \pi)$ generate a fixed point free action of $\mathbb{Z}$ on $\mathbb{R}^2$. Let $\mathcal{L} := \mathbb{R}^2/\Psi(\mathbb{Z})$ be the quotient; this is the Möbius strip. In a purely formal sense, this corresponds to taking $\alpha = \pi\sqrt{-1}$, $\beta = \gamma = 0$, and $\delta = \pi$ in Theorem 2.

Theorem 5 *C and $\mathcal{L}$ are homogeneous almost Zoll affine surfaces which are geodesically complete and affine Killing complete. The associated exponential map on C is not surjective; the associated exponential map on $\mathcal{L}$ is surjective.*

Proof We verify immediately that if $c = 0$, then

$$\begin{aligned} T(\alpha, \beta, \gamma, \delta)\Phi_k(x^1, x^2) &= (e^\alpha x^1 + \beta \cos(x^2 + 2k\pi) + \gamma \sin(x^2 + 2k\pi), x^2 + 2k\pi) \\ &= (e^\alpha x^1 + \beta \cos(x^2) + \gamma \sin(x^2), x^2 + 2k\pi) \\ &= \Phi_k T(\alpha, \beta, \gamma, \delta)(x^1, x^2) \end{aligned}$$

so Φ is in the center of the group $\mathcal{G}_0(0)$ and hence $\mathcal{G}_0(0)$ extends to a transitive affine action on C; this is a homogeneous geometry. All the geodesics with a non-trivial vertical component close smoothly. The horizontal geodesic is the alienated geodesic. We use Theorem 2 to see C is affine Killing complete. We use Theorem 4 to see C is geodesically complete. It follows from Theorem 4 that the alienated geodesic thru $(0, \pi)$ is not in the range of the exponential map—see Remark 1.

Similarly, to examine $\mathcal{L}$, we compute

$$\begin{aligned} T(\alpha, \beta, \gamma, \delta)\Psi(x^1, x^2) &= (-e^\alpha x^1 + \beta \cos(x^2 + \pi) + \gamma \sin(x^2 + \pi), x^2 + \pi) \\ &= (-e^\alpha x^1 - \beta \cos(x^2) - \gamma \sin(x^2), x^2 + \pi) = \Psi T(\alpha, \beta, \gamma, \delta)(x^1, x^2). \end{aligned}$$

Thus this is a homogeneous affine structure as well and we have a double cover $\mathbb{Z}_2 \to C \to \mathcal{L}$ on which $\mathcal{G}_0(0)$ acts equivariantly. With this structure, the Möbius strip is affine homogeneous, geodesically complete and affine Killing complete. All geodesics but the alienated geodesic close smoothly owing to the twist in x^1. Since the structure is π periodic in x^2, the exponential map is surjective. □

We now relate these examples to the results of LeBrun and Mason [14]. If M is a smooth manifold, let $\mathbb{P}(M)$ be the projective tangent bundle. If ∇ is an affine structure on M, the tangent of lifted geodesics defines a natural foliation on $\mathbb{P}(M)$; the affine structure is said to be *tame* if this structure on $\mathbb{P}(M)$ is locally trivial. LeBrun and Mason showed (see Lemma 2.7 and Lemma 2.8)

Theorem 6 *Suppose that $\mathcal{M}$ is an affine tame Zoll manifold.*

1. *The universal cover of M with the induced affine structure is tame Zoll.*
2. *M is compact and any two points of M can be joined by an affine geodesic.*
3. *M has finite fundamental group.*

Our examples show that these results fail for almost Zoll affine structures. Let $\mathcal{M}$ is be an almost Zoll surface. Associating to any point of M the tangent to the alienated geodesic through that point defines a natural section to $\mathbb{P}(M)$; let $\mathbb{P}(M)$ be the complement of this section. The alienated geodesics of C are the horizontal geodesics; $\sigma_{u,v;a,b}$ is not alienated if and only if $b \neq 0$. Since we are working projectively, we may set $b = 1$. Let $\chi(r,s,t) := \dot{\sigma}_{r,0,s,1}(t)$ parametrize $\tilde{\mathbb{P}}(C)$; this identifies $\tilde{\mathbb{P}}(C)$ with $\mathbb{R} \times \mathbb{R} \times S^1$ since, of course, $\sigma_{r,0,s,1}(t)$ is periodic with period 2π in t. This shows the foliation of $\tilde{\mathbb{P}}(TM)$ by lifted geodesics is a trivial circle bundle over $\mathbb{R}^2$ and hence is tame. Similarly, if we lift to the universal cover, the foliation of $\mathbb{P}(T\mathbb{R}^2) - S(\mathbb{R}^2)$ by lifted geodesics is a trivial $\mathbb{R}$ bundle over $\mathbb{R}^2$ and hence tame. Clearly, however, the affine structure on $\mathbb{R}^2$ is no longer almost Zoll and we can not join any two points of $\mathbb{R}^2$ by geodesics. And the cylinder does not have finite fundamental group. On the other hand, the cylinder is the oriented double cover of the Möbius strip.

As noted above, the tangent to the alienated geodesic thru any point of an almost Zoll manifold is a section to $\mathbb{P}(TM)$. Consequently, if M is compact, then the Euler-Poincare characteristic of M vanishes. Thus, in particular, the only compact surfaces which could potentially admit an almost Zoll structure are the torus and the Klein bottle. The structure on $\mathbb{R}^2$ given by $\mathcal{M}(0)$ passes to the cylinder and the Möbius strip; it does not pass to the torus or the Klein bottle. We do not know if the torus or Klein bottle admit an almost Zoll affine structure but we suspect the answer is no.

Acknowledgments Research partially supported by Project MTM2016-75897-P (AEI/FEDER, Spain). We acknowledge facilities provided by the Universidade de Santiago de Compostela (Spain) where much of this research was performed.

References

1. T. Arias-Marco, O. Kowalski, Classification of locally homogeneous linear connections with arbitrary torsion on 2-dimensional manifolds. Monatsh. Math. **153**, 1–18 (2008)
2. D. D'Ascanio, P. Gilkey, P. Pisani, Geodesic completeness for type $\mathcal{A}$ surfaces. Differ. Geom. Appl. **54**, 31–43 (2017)
3. F. Balacheff, C. Croke, M. Katz, A Zoll counter example to a geodesic length conjecture. Geom. Funct. Anal. **19**, 1–10 (2009)
4. A.L. Besse, *Manifolds All of Whose Geodesics Are Closed* (Springer, Berlin, 1978)
5. M. Brozos-Vázquez, E. García-Río, P. Gilkey, Homogeneous affine surfaces: affine Killing vector fields and Gradient Ricci solitons. J. Math. Soc. Jpn **70**, 25–70 (2018)
6. M. Brozos-Vázquez, E. García-Río, P. Gilkey, X. Valle-Regueiro, A natural linear equation in affine geometry: the affine quasi-Einstein equation. Proc. Amer. Math. Soc. **146**, 3485–3497 (2018)

7. R. Bryant, M. Dunajski, M. Eastwood, Metrisability of two-dimensional projective structures. J. Differ. Geom. **83**, 465–499 (2009)
8. F. Contatto, M. Dunajski, First integrals of affine connections and Hamiltonian systems of hydrodynamic type. J. Integr. Syst. (2016). https://doi.org/10.1093/integr/xyw009
9. P. Gilkey, X. Valle-Regueiro, Applications of PDEs to the study of affine surface geometry. Mat. Vesnik **71**, 45–62 (2019)
10. P. Gilkey, J.H. Park, X. Valle-Regueiro, Affine Killing complete and geodesically complete homogeneous affine surfaces. J. Math. Anal. Appl. **474**, 179–193 (2019)
11. L.W. Green, Auf Wiedersehensflächen. Ann. Math. **78**, 289–299 (1963)
12. S. Kobayashi and K. Nomizu, in *Foundations of Differential Geometry. Vols. I and II*. Wiley Classics Library (Wiley, New York, 1996)
13. C. Lange, On metrics on 2-orbifolds all of whose geodesics are closed. J. Reine Angew. Math. https://doi.org/10.1515/crelle-2017-0050
14. C. LeBrun, L.J. Mason, Zoll manifolds and complex surfaces. J. Differential Geom. **61**, 453–535 (2002)
15. C. LeBrun, L.J. Mason, Zoll metrics, branched covers, and holomorphic disks. Commun. Anal. Geom. **18**, 475–502 (2010)
16. B. McKay, Complete projective connections. ArXiv 0504082
17. P. Mounoud, S. Suhr, On spacelike Zoll surfaces with symmetries. J. Differential Geom. **102**, 243–284 (2016)
18. B. Opozda, A classification of locally homogeneous connections on 2-dimensional manifolds. Differ. Geom. Appl. **21**, 173–198 (2004)
19. J. Patera, R.T. Sharp, P. Winternitz, H. Zassenhaus, Invariants of real low dimension Lie algebras. J. Math. Phys. **17**, 986–994 (1976)
20. C. Pries, Geodesics closed on the projective plane. Geom. Funct. Anal. **18**, 1774–1785 (2009)
21. J.A. Schouten. *Ricci Calculus: An Introduction to Tensor Analysis and Its Geometrical Applications* (Springer, Berlin, 1954)
22. A. Vanžurová, Metrizability of connections on two-manifolds. Acta Univ. Palack. Olomuc. Fac. Rerum Natur. Math. **48**, 157–170 (2009)
23. A. Vanžurová, On metrizability of a class of 2-manifolds with linear connection. Miskolc Math. Notes **14**, 621–627 (2013)
24. O. Zoll, Über Fläschen mit Scharen geschlossener geodätischer Linen". Math. Ann. **57**, 108–133 (1903)

Distinguished Curves and First Integrals on Poincaré–Einstein and Other Conformally Singular Geometries

A. Rod Gover and Daniel Snell

Abstract We treat the problem of defining, and characterising in a practical way, an appropriate class of distinguished curves for Poincaré-Einstein manifolds, and other conformally singular geometries. These "generalised geodesics" agree with geodesics away from the conformal singularity set and are shown to satisfy natural "boundary conditions" at points where they meet or cross the metric singularity set. We also characterise when they coincide with conformal circles. In the case of (Poincaré-)Einstein manifolds, we are able to provide a very general theory of first integrals for these distinguished curves. As well as the general procedure outlined, a specific example is given.

1 Introduction

Geodesics and other distinguished curves play a basic and essential role in differential geometry and its applications [1, 14, 21, 24, 31]. The determination and study of these can be enormously simplified if one has available curve first integrals [1, 6, 8, 14, 25, 29]. Certain conformally singular geometries such as Poincaré-Einstein manifolds have proved to have a central place in mathematical physics, geometric scattering, the AdS/CFT correspondence of physics, as well as in conformal geometry itself [10–12, 20, 22, 27, 32]. It is thus important to study the distinguished curves, and their possible first integrals, for such geometries. However classical theory is not directly applicable, as for these structures there is only a well-defined metric on a dense open subset of the geometry.

We recall the notion of a conformally compact manifold. Let $\mathring{M}$ be the interior of a n-manifold with boundary M. So $M = \mathring{M} \cup \partial M$ and the boundary ∂M has dimension $n-1$. The metric $\mathring{g}$ on the interior $\mathring{M}$ of M is said to be *conformally compact* if $\mathring{g} = u^{-2} g$ where g is a metric on M (and so is nondegenerate up to

A. Rod Gover (✉) · D. Snell
Department of Mathematics, The University of Auckland, Auckland, New Zealand
e-mail: r.gover@auckland.ac.nz; daniel.snell@auckland.ac.nz

S. Hervik et al. (eds.), *Geometry, Lie Theory and Applications*, Abel Symposia 16,
https://doi.org/10.1007/978-3-030-81296-6_5

∂M) and u is a defining function for ∂M, i.e. the zero locus $\mathcal{Z}(u)$ of u is exactly the boundary, so $\mathcal{Z}(u) = \partial M$, and $du \neq 0$ at all points of ∂M. If in addition $\mathring{g}$ is Einstein, then M is said to be *Poincaré-Einstein*. In Riemannian signature a Poincaré-Einstein manifold is necessarily of negative scalar curvature, and more generally this condition on the curvature holds asymptotically for conformally compact manifolds as defined here. There are variations of these statements for metrics of other signatures. There is also a suitable notion of conformal conformally compact for the case of zero scalar curvature. These and the related constraints, may all be understood as special cases of the notion of an almost (pseudo-)Riemannian manifold given in Definition 1. See [7].

Note that for conformally compact manifolds a first problem is to describe the right classes of distinguished curves. The usual geodesic equation is only applicable away from the singularity of the metric. However because the metric is singular in a conformal way, the entire manifold M has a well-defined conformal structure $[g]$. This means that conformal circles, as defined in [2, 13, 28, 30, 33] are well-defined and can play an important role. However conformal circles provide a class of curves that is, in a sense, "too large". It requires a point, velocity, *and acceleration*, at that point, to determine a conformal circle.

A second problem is to treat first integrals. Suppose that one understands that, on a conformally compact manifold, a certain class of unparametrised geodesics meet the boundary. Then can these sometimes be understood, or at least simplified using first integrals? The problem now is that geodesic first integrals are related to the existence of so called Killing tensors [29], but the Killing tensor equation is not conformally invariant and is also not well defined where the metric is conformally singular. Thus the standard theory is insufficient.

The aim of this short note is to attack and, to a considerable extent, treat these problems. We first recall some recent advances in the treatment of conformal circles, and the construction of conserved quantities for these. This is done in Sects. 3 and 6 respectively, see especially Theorem 1, and follows [19]. That source provides a new definition of conformal circles that is especially useful for constructing the corresponding first integrals, and also provides a rather general theory of the latter. That new machinery is then used here to understand the links between conformal circles and geodesics. This leads us to a way to define a class of distinguished curves for conformally compact manifolds and their generalisations. The result is that for a large class of conformally singular geometries we provide a distinguished curve equation that is well defined at all points and specialises to exactly the usual unparametrised geodesic equation at points where the metric is non-singular, see Definition 2. The situation is especially beautiful and simple in the case of Einstein and Poincaré-Einstein manifolds, and in those settings we explain how to use conformal geometry to construct first integrals for these distinguished curves.

In more detail the main results are then as follows: Theorem 2 gives the "conformal characterisation" of metric geodesics that leads to the definition of generalised geodesics in Definition 2. Proposition 2 then shows that the generalised geodesics extend geodesics. Proposition 3 shows that in the case of conformally compact manifolds the generalised geodesics with interior points and that meet the boundary must meet it orthogonally. Theorem 3 gives a characterisation of (Poincaré-)Einstein manifolds as an agreement between (generalised) geodesics and conformal circles. This then leads to a main application, namely that all the machinery from [19] for proliferating first integrals of conformal circles may be applied (by specialisation) to provide first integrals of the (generalised) geodesics of (Poincaré-)Einstein manifolds. This is taken up briefly in Sect. 6.

Some remarks on the notation and conventions used in this paper are in order. We work on a (pseudo-)Riemannian or conformal manifold of dimension $n \geq 3$. We will always assume that our manifold is connected. The Riemannian or conformal structure is allowed to have any signature (p, q), although some aspects of the paper will only be relevant when the signature is strictly pseudo-Riemannian (namely $pq \neq 0$). We will use Penrose's abstract index notation [26] for sections of vector bundles. In this notation, lower case Latin indices are used to indicate the type of a section of a tensor bundle in the following way. The tangent bundle is denoted by $\mathcal{E}^a$ and the cotangent bundle is denoted by $\mathcal{E}_b$. Tensor products are then denoted by an appropriate combination of indices. So for example, v^a is a vector field, ω_{bcd} is a 3-form, and $T^a{}_b$ is an endomorphism of the tangent (or cotangent) bundle. The notation $\mathcal{E}$ is used for the trivial bundle, so that $\Gamma(\mathcal{E}) = C^\infty(M)$. Square brackets around indices denotes the completely anti-symmetric part of the given tensor, and round brackets around indices denotes the completely symmetric part. Given a vector bundle $\mathcal{V}$, $\Lambda^k\mathcal{V}$ will denote the subbundle of $\otimes^k\mathcal{V}$ consisting of the totally anti-symmetric tensors. Similarly, $S^\ell\mathcal{V}$ denotes the subbundle of totally symmetric tensors. Finally we will occasionally employ a somewhat informal "wedge" notation for convenience. If $S^{a_1\cdots a_k} \in \Gamma(\Lambda^k\mathcal{V})$ and $T^{b_1\cdots b_\ell} \in \Gamma(\Lambda^\ell\mathcal{V})$, we define

$$S \wedge T := S^{[a_1\cdots a_k}T^{b_1\cdots b_\ell]}. \tag{1}$$

All objects and functions are assumed smooth (C^∞).

2 Conformal Geometry and Conformal Tractor Calculus

We review the basic theory of conformal tractor calculus that will be required for our purposes.

Conformal density bundles are a family of line bundles which arise naturally in conformal geometry. Recall that any manifold M is equipped with the oriented line bundle $\mathcal{K} := (\Lambda^n TM)^2$. Conformal density bundles are then various roots of this bundle. For $w \in \mathbb{R}$, let

$$\mathcal{E}[w] := \mathcal{K}^{\frac{w}{2n}}. \tag{2}$$

Given a vector bundle $\mathcal{V}$, we define $\mathcal{V}[w] := \mathcal{V} \otimes \mathcal{E}[w]$.

The conformal structure $(M, \boldsymbol{c})$ determines a distinguished section of $\mathcal{E}_{(ab)}[2]$ which we denote $\mathbf{g}_{ab}$ and call the *conformal metric*. This section is characterised by the fact that any choice of metric g in the conformal class may be realised as $g = \sigma_g^{-2}\mathbf{g}$ for some $\sigma_g \in \Gamma(\mathcal{E}_+[1])$ depending on g. Conversely, given a non-vanishing 1-density $\sigma \in \Gamma(\mathcal{E}_+[1])$, $g := \sigma^{-2}\mathbf{g}$ is a metric in the conformal class. In light of this bijective correspondence between metrics in the conformal class and sections of the bundle $\mathcal{E}_+[1]$, we call sections of this bundle *scales*, and will often refer to a choice of metric to facilitate computations as a *choice of scale*. We shall use $\mathbf{g}$ to raise and lower indices on a conformal manifold. This is very similar to the way that the Riemannian metric is used, since the conformal metric is preserved by *any* Levi-Civita connection from the conformal class. There is however the small caveat that raising and lowering indices via the conformal metric and its inverse does change the weight of the section on which it acts:

$$\mathbf{g}_{ab} : \mathcal{E}^a \to \mathcal{E}_b[2] \quad \text{by} \quad v^a \mapsto \mathbf{g}_{ab}v^a. \tag{3}$$

There are many ways to realise the standard conformal tractor bundle; here we shall present one that has the advantage of introducing an equation we will use again later.

It is a well-known fact that the following equation is conformally invariant

$$\nabla_{(a}\nabla_{b)_0}\sigma + P_{(ab)_0}\sigma = 0, \tag{4}$$

where $\sigma \in \mathcal{E}[1]$ is a conformal density. This equation will be called the *Almost-Einstein equation* (A.E.), since a solution σ defines an Einstein metric everywhere $\sigma \neq 0$. The above equation may be *prolonged* to give an equivalent first order closed system; see [7] for a more detailed description of this process. The prolongation moreover yields a linear connection on a certain vector bundle. The solutions of the A.E. equation are in bijective correspondence with sections of this vector bundle which are parallel for this connection. This vector bundle is the *tractor bundle*, denoted $\mathcal{T}$ and the connection is the *tractor connection*, denoted simply ∇ or sometimes $\nabla^{\mathcal{T}}$ to avoid ambiguity.

A choice of metric $g \in \boldsymbol{c}$ in the conformal class induces an isomorphism

$$\mathcal{T} \stackrel{g}{\cong} \mathcal{E}[1] \oplus \mathcal{E}_a[1] \oplus \mathcal{E}[-1]. \tag{5}$$

In terms of this, the connection acts as

$$\nabla_a^{\mathcal{T}} \begin{pmatrix} \sigma \\ \mu_b \\ \rho \end{pmatrix} \stackrel{g}{=} \begin{pmatrix} \nabla_a\sigma - \mu_a \\ \nabla_a\mu_b + \mathbf{g}_{ab}\rho + P_{ab}\sigma \\ \nabla_a\rho - P_{ab}\mu^b \end{pmatrix}, \tag{6}$$

where we are again using a choice of metric to explicitly write down the formula.

The tractor bundle also comes equipped with a symmetric non-degenerate bilinear form, defined by

$$h(V, V') \stackrel{g}{=} \begin{pmatrix} \sigma & \mu_a & \rho \end{pmatrix} \begin{pmatrix} 0 & 0 & 1 \\ 0 & \mathbf{g}^{ab} & 0 \\ 1 & 0 & 1 \end{pmatrix} \begin{pmatrix} \sigma' \\ \mu'_b \\ \rho' \end{pmatrix}. \tag{7}$$

In abstract indices, tractors will be denoted with upper case Latin indices. Hence the tractor metric will be written h_{AB} and is a section of $\mathcal{E}_{(AB)}$. The tractor metric identifies $\mathcal{E}^A$, the tractor bundle, with its dual $\mathcal{E}_A$.

There is a conformally-invariant bundle map $\mathcal{E}[-1] \hookrightarrow \mathcal{T}$. We denote this map by X and refer to it as the *canonical tractor* or *position tractor*. It plays a central role in our characterisation of conformal circles. The map X may be thought of as a section of $\mathcal{E}_A[1]$. Raising the tractor index via the tractor metric, we obtain a section $X^A \in \Gamma(\mathcal{E}^A[1])$, which coincides with the conformally invariant projection $X^A : \mathcal{E}_A \to \mathcal{E}[1]$.

This idea can be extended to give a convenient notation for working with tractors. We have already seen that there is a map which inserts the subbundle $\mathcal{E}[-1]$ into the tractor bundle. Given a choice of scale there are corresponding sections mapping the other two slots into the triple. We emphasise these are not conformally invariant. For a section $V^A \in \Gamma(\mathcal{E}^A)$, we write $V^A = \sigma Y^A + \mu^a Z_a^A + \rho X^A$ following [17]. Here, $Y^A \in \Gamma(\mathcal{E}^A[-1])$ and $Z_a^A \in \Gamma(\mathcal{E}_a^A[1])$.

The final element of the standard theory of tractor calculus we shall require is the second order differential operator $D : \mathcal{E}[1] \to \mathcal{T}$ such that $\nabla_a^{\mathcal{T}} D_A \sigma = 0$ if, and only if, σ solves the A.E. equation (4). It is a way of encoding a choice of scale in a tractor. In a choice of scale and using the tractor projector notation of the introduced in the previous paragraph, one has

$$D_A \sigma \stackrel{g}{=} \sigma Y_A + \nabla_b \sigma Z_A^b - \frac{1}{n}(\Delta + J)\sigma X_A, \tag{8}$$

where J is the conformal metric trace of the Schouten tensor of g. We say that a tractor $I^A \in \Gamma(\mathcal{E}^A)$ is a *scale tractor* if $I^A = D^A \sigma$ for some $\sigma \in \Gamma(\mathcal{E}_+[1])$ and I^A is nowhere zero.

The canonical tractor X^A mentioned previously recovers the section σ from such a scale tractor. Explicitly, $X^A D_A \sigma = \sigma$.

3 Distinguished Curves in Conformal Geometry

For us, a curve will be a smooth, regular map $\gamma : I \to M$, where $I \subset \mathbb{R}$ is an interval. The distinguished curves of a conformal manifold are divided into two classes: the null geodesics (if the manifold is strictly pseudo-Riemannian) and the (necessarily non-null) conformal circles.

We recall here the main results that we will require from [19]. Rather than working directly with the velocity and acceleration of the curve, it turns out to be more natural to work with weighted versions of these. For non-null curves these are constructed as follows. The velocity u^a of the curve γ determines a scale $\sigma_u \in \Gamma(\mathcal{E}_+[1]|_\gamma)$ along the curve via

$$\sigma_u := \begin{cases} \sqrt{\mathbf{g}_{ab}u^a u^b} & \text{if } u^a \text{ is spacelike,} \\ \sqrt{-\mathbf{g}_{ab}u^a u^b} & \text{if } u^a \text{ is timelike.} \end{cases} \tag{9}$$

Using this, we define the weighted velocity, $\mathbf{u}^a \in \Gamma(T\gamma[-1])$ by $\mathbf{u}^a := \sigma_u^{-1}u^a$. This weighted velocity does not depend on the parametrisation of the curve. Moreover, we define the weighted version of the acceleration, $\mathbf{a}^b \in \Gamma(T\gamma[-2])$ by $\mathbf{a}^b := \mathbf{u}^a\nabla_a\mathbf{u}^b$. Note that the while weighted velocity is conformally invariant, the weighted acceleration is not.

The original conformal circle equation is a second-order ODE in the velocity of the curve; see e.g. [2]. With the above definitions, we can give an equivalent presentation of the conformal circle equation, which is essentially a weighted version of equation (7) from [2].

Lemma 1 *Let γ be an oriented nowhere-null curve on $(M, \boldsymbol{c})$. Then γ is a conformal circle if and only if its weighted velocity $\mathbf{u}^a$ and acceleration $\mathbf{a}^a$ satisfy the conformally invariant equation*

$$\left(\mathbf{u}^c\nabla_c\mathbf{a}^{[a}\right)\mathbf{u}^{b]} = \pm\mathbf{u}^c\mathrm{P}_c{}^{[a}\mathbf{u}^{b]}\,, \qquad \textit{whenever } \mathbf{u}^a\mathbf{u}_a = \pm 1, \tag{10}$$

or equivalently,

$$\mathbf{u}^b\nabla_b\mathbf{a}^a = \pm\mathbf{u}^b\mathrm{P}_b{}^a - (\mathrm{P}_{bc}\mathbf{u}^b\mathbf{u}^c \pm \mathbf{a}\cdot\mathbf{a})\mathbf{u}^a\,, \quad \textit{whenever } \mathbf{u}^a\mathbf{u}_a = \pm 1, \tag{11}$$

for any $g \in \boldsymbol{c}$ with Levi-Civita connection ∇.

Proof See [19]. □

The previous lemma may be reformulated in terms of tractors, using a moving incidence relation; this is a main result from [19] for conformal circles.

Theorem 1 *On a pseudo-Riemannian or conformal manifold a nowhere null curve γ is an oriented conformal circle if and only if along γ there is a parallel 3-tractor $0 \neq \Sigma \in \Gamma(\Lambda^3\mathcal{T}|_\gamma)$ such that*

$$X \wedge \Sigma = 0. \tag{12}$$

For a given oriented conformal circle γ the 3-tractor Σ_γ satisfying (12) is unique up to multiplication by a positive constant, and unique if we specify $|\Sigma_\gamma|^2 = -1$ when γ is spacelike, or $|\Sigma_\gamma|^2 = 1$ when γ is timelike.

Proof We sketch here the proof, but for details see [19]. First, following [3], for a non-null curve γ, we define the *velocity tractor* and *acceleration tractor* respectively by

$$U^A := \sigma_u \mathbf{u}^a \nabla_a (\sigma_u^{-1} X^A), \quad \text{and} \quad A^B := \sigma_u \mathbf{u}^a \nabla_a U^B. \tag{13}$$

These are conformally invariant by construction. But they are not parametrisation independent. Then form

$$\Sigma^{ABC} := 6\sigma_u^{-1} X^{[A} U^B A^{C]}, = \pm 6\mathbf{u}^c X^{[A} Y^B Z^{C]}{}_c + 6\mathbf{u}^b \mathbf{a}^c X^{[A} Z^B{}_b Z^{C]}{}_c. \tag{14}$$

This is parametrisation independent and conformally invariant. Moreover it is parallel along γ if, and only if, (11) holds.

Now if $\Sigma \in \Gamma(\Lambda^3 \mathcal{T})$ takes the form (14), then $X \wedge \Sigma = 0$. On the other hand, if $\Sigma \in \Gamma(\Lambda^3 \mathcal{T})$ is parallel along the curve and satisfies $X \wedge \Sigma = 0$, then Σ must take the form (14). This completes the proof. □

We wish to remark here that the velocity and acceleration tractors, and therefore the 3-tractor Σ as defined above, can be defined for *any* non-null curve γ; one does not require that the curve is a conformal circle. Thus Σ is a fundamental conformal invariant of such unparametrised curves. This observation will be relevant when we consider the distinguished curves of Riemannian geometry in the following section.

We shall also require the derivative of such a Σ along the curve, and so we record that formula here for later reference:

$$\mathbf{u}^d \nabla_d \Sigma^{ABC} = 6 \left(\mathbf{u}^d \nabla_d \mathbf{a}^c \mp \mathbf{u}^d \mathrm{P}_d{}^c \right) \mathbf{u}^b \, X^{[A} Z^B{}_b Z^{C]}{}_c \,, \text{ whenever } \mathbf{u}^a \mathbf{u}_a = \pm 1, \tag{15}$$

As a consequence, we obtain a parametrisation-independent version of Proposition 3.3 from [2].

Proposition 1 *Let $(M, \boldsymbol{c})$ be a conformal manifold and γ a non-null, oriented curve in M. Then γ is an unparametrised conformal circle if, and only if, γ is an unparametrised geodesic for some metric in the conformal class and $\mathbf{u}^a P_a{}^b \propto \mathbf{u}^b$, where P is the Schouten tensor for the given metric.*

Proof First, suppose that γ is a conformal circle. Locally, there exists a metric in the conformal class for which γ is an affine geodesic [2, 9]. Let σ be the scale corresponding to this metric. Now, working in the scale σ, (15) becomes

$$\mathbf{u}^d \nabla_d \Sigma^{ABC} = \mp 6 \mathbf{u}^d \mathrm{P}_d{}^c \mathbf{u}^b \, X^{[A} Z^B{}_b Z^{C]}{}_c \,, \tag{16}$$

where $\mathrm{P}_d{}^c$ is the Schouten for this metric. But since γ is a conformal circle, the left side of the previous display must be zero by Theorem 1. Hence $\mathbf{u}^d P_d{}^c \propto \mathbf{u}^c$. Conversely, if there is a scale in the conformal class for which γ is a geodesic, then working in the assumed scale, $\mathbf{a} \propto \mathbf{u}$ and hence (15) will again take the form (16).

Hence if $\mathbf{u}^d P_d{}^c \propto \mathbf{u}^c$ then $\mathbf{u}^d \nabla_d \Sigma^{ABC} = 0$, and so γ is a conformal circle by Theorem 1. □

4 Metric Geodesics via Conformal Tractor Machinery

Here we give first a characterisation of geodesics that, in its nature, is "as conformal as possible" and which also clarifies the relation between geodesics and conformal circles. We have already noted that the section $\Sigma \in \Gamma(\Lambda^3\mathcal{T}|_\gamma)$ from display (14), and defined by $\Sigma^{ABC} := 6\sigma_u^{-1} X^{[A} U^B A^{C]}$, is determined by *any* non-null unparametrised curve γ in a conformal manifold. While initially defined and used in our work characterising conformal circles, the following theorem shows that this 3-tractor also turns out to provide a way to characterise Riemannian distinguished curves.

Theorem 2 *Let (M, g) be a pseudo-Riemannian manifold. A non-null curve γ is an unparametrised geodesic of g if, and only if, the scale tractor $I = D\sigma$ (where $g = \sigma^{-2}\mathbf{g}$) satisfies $I \wedge \Sigma = 0$.*

Proof Suppose that γ is an unparametrised geodesic for g. Then we may form Σ^{ABC}. Let σ be the scale determined by g, i.e. $g = \sigma^{-2}\mathbf{g}$, and let $I := D\sigma$. Then, working in the scale σ, we have that $\nabla\sigma = 0$ (see e.g. [7]) and so $I^A = \sigma Y^A + \rho X^A$, for some $\rho \in \mathcal{E}[-1]$ whose explicit form will not be needed. Then

$$I \wedge \Sigma = \left(\sigma Y^A + \rho X^A\right) \wedge \left(6\mathbf{u}^a X^{[B} Y^C Z_a^{D]} \mp 6\mathbf{u}^a \mathbf{a}^b X^{[B} Z_a^C Z_b^{D]}\right)$$
$$= \mp 6\sigma \mathbf{u}^a \mathbf{a}^b Y^{[A} X^B Z_a^C Z_b^{D]}.$$

Note that in the scale σ, $\mathbf{a}^c = \sigma^{-2} a^c$ and $\mathbf{u}^c = \sigma^{-1} u^b$, and hence $\mathbf{u} \propto \mathbf{a}$ if, and only if $a \propto u$. But the latter is the case, since γ is an unparametrised geodesic, so $I \wedge \Sigma = 0$.

Conversely, suppose that the scale tractor I satisfies $I \wedge \Sigma = 0$. Then in the scale σ, $I \wedge \Sigma$ again takes the form of the above display. Since the tractor projectors are linearly independent pointwise, $I \wedge \Sigma = 0$ implies that $\mathbf{a} \propto \mathbf{u}$, which in turn implies that $a^b \propto u^b$ in the scale σ. Hence γ is an unparametrised geodesic for the metric g. □

Theorem 2 suggests a way to generalise the geodesic equation to a larger class of structures, which includes conformally compact manifolds.

Recall that on an Einstein manifold with Einstein scale $\sigma \in \Gamma(\mathcal{E}_+[1])$, the corresponding scale tractor $I_A := D_A\sigma$ is parallel. More generally, such a scale tractor may be used to define a slight weakening of the notion of a pseudo-Riemannian manifold. Note that if a scale tractor I_A is nowhere zero, then $\sigma = X^A I_A$ is non-zero on an open dense set. Therefore that same open dense set possesses a pseudo-Riemannian metric defined by $\mathring{g} := \sigma^{-2}\mathbf{g}$.

Definition 1 We shall call $(M, \boldsymbol{c}, \sigma)$ an *almost pseudo-Riemannian manifold* if $I := D\sigma$ is nowhere-zero and the 1-jet $j^1\sigma$ vanishes at most at isolated points.

This definition encompasses conformally compact structures; the open dense set where σ is nowhere zero being the interior, uncompactified manifold, and there $\sigma^{-2}\mathbf{g}$ recovers the original metric. Similar comments apply to suitable conformal type compactifications of metrics in any signature, and also the case where the scalar curvature is zero [7]. It is also interesting to consider closed manifolds with such an almost pseudo-Riemannian structure [7, 16]. For any such geometry there is a natural extension of the (unparametrised) geodesic equation to any points where the metric is singular, as follows.

Definition 2 Let $(M, \boldsymbol{c}, \sigma)$ be an almost pseudo-Riemannian manifold. We will say that an unparametrised non-null curve γ in M is a *generalised geodesic* if

$$I \wedge \Sigma = 0, \tag{17}$$

where $I := D\sigma$ and Σ is the 3-tractor defined in Sect. 4.

Observe that Eq. (17) is well defined globally on an almost pseudo-Riemannian manifold, and in particular through the zero locus $\mathcal{Z}(\sigma)$ (of the generalised scale σ) where the metric is singular. However in any neighbourhood where σ is non-vanishing the generalised geodesic equation recovers the usual geodesic equation. Thus generalised geodesics extend geodesics to(/through) the metric singularity set.

Proposition 2 *Let $(M, \boldsymbol{c}, \sigma)$ be an almost Riemannian manifold. Suppose that γ is a smooth unparametrised nowhere-null curve on M that is a geodesic for $g = \sigma^{-2}\mathbf{g}$ on $M \setminus \mathcal{Z}(\sigma)$. Then γ is a generalised geodesic of $(M, \boldsymbol{c}, \sigma)$.*

Proof Since γ satisfies the geodesic equation on $M \setminus \mathcal{Z}(\sigma)$, we have that $I \wedge \Sigma = 0$ on that open and dense set, as seen in Theorem 2. But by smoothness (and a slight extension of the argument in Proposition 3 below), this is then also true on the closure M, and hence $I \wedge \Sigma = 0$ everywhere, and γ is a generalised geodesic. □

Specialising to conformally compact manifolds we have the following result. It turns out that the generalised geodesics which extend smoothly to the boundary cannot do so arbitrarily. In fact, they may only meet the boundary in a very controlled way (that generalises the well known special case of compactified hyperbolic space).

Proposition 3 *On a conformally compact manifold, any generalised geodesic which extends to the boundary meets the boundary orthogonally.*

Proof Since $\partial M = \mathcal{Z}(\sigma)$, at the boundary,

$$I^A|_{\partial M} = \nabla^a \sigma Z^A{}_a - \frac{1}{n}\Delta\sigma X^A, \tag{18}$$

and $\nabla\sigma$ is non-vanishing at all boundary points. Thus

$$(I^A \wedge \Sigma^{BCD})|_{\partial M} = \left(\nabla^a\sigma Z_a{}^A - \frac{1}{n}\Delta\sigma X^A\right) \wedge \left(6\mathbf{u}^d X^{[B} Y^C Z^{D]}_d \mp 6\mathbf{u}^c\mathbf{a}^d X^{[B} Z^C_c Z^{D]}_d\right)$$
$$= 6\mathbf{u}^d \nabla^a\sigma Z^{[A}{}_a X^B Y^C Z^{D]}{}_d.$$

Thus $I \wedge \Sigma = 0$ implies that $\mathbf{u} \propto \nabla\sigma$. But $\nabla\sigma$ is a conormal to the boundary, and therefore $\mathbf{u}$ is orthogonal to the boundary as well. □

5 (Poincaré-)Einstein Structures

We have already noted that a solution to (4) determines an Einstein metric almost-everywhere. Hence following e.g. [15], we make the following definition.

Definition 3 We say that $(M, \boldsymbol{c}, \sigma)$ is an *Almost-Einstein structure* if σ is a nontrivial solution to (4).

In this case the scale tractor I is parallel, and thus nowhere zero. Thus the structure is almost pseudo-Riemannian and σ is non-vanishing on an open, dense set. On the set where σ is non-zero the metric defined by $g := \sigma^{-2}\mathbf{g}$ is Einstein. Proposition 1 now gives the following result.

Corollary 1 *Suppose $(M, \boldsymbol{c}, \sigma)$ is an almost-Einstein manifold. If γ is an generalised geodesic then γ is an unparametrised conformal circle.*

Proof Note that if γ is a geodesic and in addition $\mathbf{u}^b P_b{}^c \propto \mathbf{u}^c$, then γ is a conformal circle by Proposition 1. But this immediately implies the result on $M \setminus \mathcal{Z}(\sigma)$, since if M is Einstein, $P_{bc} \propto g_{bc}$. Then the full result follows by continuity. □

The above also admits a converse, yielding the following characterisation of Einstein manifolds and Poincaré-Einstein manifolds.

Theorem 3 *Let $(M, \boldsymbol{c}, \sigma)$ be a (conformally compact or) pseudo-Riemannian manifold. Then every (generalised) geodesic of M is a conformal circle if, and only if, $(M, \boldsymbol{c}, \sigma)$ is (Poincaré-)Einstein.*

Proof One direction is exactly Corollary 1. For the converse, we must show that if every geodesic of the manifold is a conformal circle, then the metric $g = \sigma^{-2}\boldsymbol{g}$ on $M \setminus \mathcal{Z}(\sigma)$ is Einstein. Note that for any $p \in M \setminus \mathcal{Z}(\sigma)$ and any $\mathbf{u}^a \in \mathcal{E}^a[-1]$, there is an unparametrised geodesic passing through p with velocity $\mathbf{u}^a$. The statement we wish to prove now essentially reduces to a question of linear algebra. Fix an orthonormal basis $\{e_1, \ldots e_n\}$ for $T_pM[-1]$, by which we mean $g(e_i, e_i) = \pm 1$ for all i according to the signature of the metric. Recalling corollary 1, each geodesic is an unparametrised conformal circle only if $\mathbf{u}^a P_a{}^b \propto \mathbf{u}^b$. Where $P_a{}^b$ is the Schouten tensor of the metric g. Working in the above basis, we see immediately that, $P(e_i, e_j) = 0$ when $i \neq j$, and hence when written as a matrix,

$P = \mathrm{diag}\{\lambda_1, \lambda_2, \ldots, \lambda_n\}$. But $\mathbf{u}^a P_a{}^b \propto \mathbf{u}^b$ for *all* $\mathbf{u} \in T_pM[-1]$. By taking $\mathbf{u} = e_i + \alpha e_j$, for $\alpha \in \mathbb{R}$ as i, j range over $1, 2, \ldots, n$, we conclude that $\lambda_i = \lambda_j$ for all i, j. Hence $P = \lambda\,\mathrm{Id}$ (where P is the matrix of $P_i{}^j$ in this basis) and the proposition follows. □

This theorem suggests one way to approach the study and construction of conserved quantities on Einstein manifolds. Namely, one may view the (generalised-)geodesics of an (Poincaré-)Einstein manifold as conformal circles that in addition satisfy $I \wedge \Sigma = 0$. Then, using the theory from [19], conformal BGG equations become available as a source of "symmetries" to construct first integrals. We will touch on this idea in the following section.

6 Conserved Quantities

One of the main applications of a tractor characterisation of distinguished curves is the development of a general theory of first integrals for such curves. We review a general method for producing conserved quantities for conformal circles, and refer the reader to [19] for a more detailed treatment. Suppose γ is a distinguished curve. Let $S \in \Gamma(\mathcal{V})$ where $\mathcal{V}$ is some (suitably) irreducible part of $S^k\Sigma$ or for $k \in \mathbb{N}$, where by $S^k\Sigma$ we mean $\Sigma \odot \Sigma \odot \cdots \odot \Sigma$, with Σ repeated k times. Then since S will only involve Σ and possibly the tractor metric, one has

$$\mathbf{u}^a \nabla_a S = 0.$$

Suppose moreover that T is a section of the dual bundle $\mathcal{V}^*$. Then, if $\mathbf{u}^a \nabla_a T = 0$, we will have

$$\mathbf{u}^a \nabla_a \langle S, T \rangle = 0,$$

where $\langle \cdot, \cdot \rangle$ denotes the dual pairing between sections of $\mathcal{V}$ and $\mathcal{V}^*$. Asking for such sections T that are parallel is restrictive, but far from fatally. Solutions to so-called BGG equations are in bijective correspondence with sections of tractor bundles which are parallel for a modified tractor connection—the *prolongation connection* of [23]. In the case that a solution is a *normal* solution [4, 5], this modified connection agrees with the standard tractor connection. Thus if T arises as the section of some tractor bundle corresponding to a normal solution to a BGG equation, we will have $\mathbf{u}^a \nabla_a T = 0$ as desired. For some BGG equations all solutions are normal. On conformally flat manifolds all solutions are normal. Most known superintegrable geometries are conformally flat. In fact, the examples in [19] prove that even in the general setting, surprisingly, the first integrals often do not require normality. We treat such an example here.

6.1 An Example in the Poincaré-Einstein Case

We have already seen that on a (Poincaré-)Einstein manifold any generalised geodesic is a conformal circle, that satisfies in addition $I \wedge \Sigma = 0$.

As a consequence of Theorem 3, we see that conformal geometry machinery may be used to proliferate first integrals of the generalised geodesics on (Poincaré-)Einstein manifolds. (Or more more generally almost-Einstein manifolds. For simplicity here we simply discuss the Poincaré-Einstein case.)

Proposition 4 *Let $(M, \boldsymbol{c}, \sigma)$ be a (Poincaré-)Einstein manifold. The first integrals of generalised geodesics of M are also first integrals of conformal geodesics, and may therefore be produced using solutions to BGG equations, c.f. Theorem 6.3 of [19].*

Recall the conformal Killing form equation on 2-forms:

$$\mathrm{tf}(\nabla F) \in \Gamma(\Lambda^3 T^* M), \tag{19}$$

where tf means the metric trace-free part of the given tensor. Normal solutions to this equation are in bijective correspondence with sections of $\Lambda^3 \mathcal{T}^*$ that are parallel for the usual tractor connection [18]. It turns out that the normality assumption is not required, and the following holds.

Theorem 4 *Let $(M, \boldsymbol{c}, \sigma)$ be a Poincaré-Einstein manifold. Suppose that γ is a generalised geodesic, and form Σ^{ABC} according to (14). Finally, suppose that $k_{ab} \in \Gamma(\mathcal{E}_{[bc]}[3])$ is a conformal Killing-Yano 2-form of the conformal manifold $(M, \boldsymbol{c})$, i.e. k_{ab} satisfies*

$$\nabla_a k_{bc} = \nabla_{[a} k_{bc]} - \frac{2}{n-1} \boldsymbol{g}_{a[b} \nabla^p k_{c]p} \,. \tag{20}$$

Write $\Gamma(\Lambda^3 \mathcal{T}^*) \ni \mathbb{K}_{ABC} := L(k)$, where the BGG splitting operator $L : \mathcal{E}_{[bc]}[3] \to \mathcal{E}_{[ABC]}$ maps a conformal Killing 2-form invariantly to its corresponding 3-tractor [5, 23]. Then the expression

$$\Sigma^{ABC} \mathbb{K}_{ABC}$$

equivalently,

$$\mathbf{u}^a \mathbf{a}^b k_{ab} \mp \frac{1}{n-1} \mathbf{u}^a \nabla^p k_{pa},$$

is a first integral of the generalised geodesic γ, where $\mathbf{u}^a$ denotes its weighted velocity, with $\mathbf{u}^a \mathbf{u}_a = \pm 1$, depending on whether the curve γ is spacelike or timelike respectively.

Proof According to Corollary 1, any generalised geodesic of the (Poincaré-)Einstein manifold is in fact a conformal circle for the underlying conformal manifold. Thus the curve γ is also a conformal circle. Having established this, the result follows at once from Theorem 6.8 of [19]. □

Note that on the interior of the manifold we may work in the scale $\mathring{g} = \sigma^{-2}\boldsymbol{g}$ whence the first integral simplifies to $\mathbf{u}^a \nabla^p k_{pa}$.

Acknowledgement ARG gratefully acknowledges support from the Royal Society of New Zealand via Marsden Grants 16-UOA-051 and 19-UOA-008.

References

1. L. Andersson, P. Blue, Hidden symmetries and decay for the wave equation on the Kerr spacetime. Ann. of Math. (2) **182**(3), 787–853 (2015). https://doi.org/10.4007/annals.2015.182.3.1
2. T.N. Bailey, M.G. Eastwood, Conformal circles and parametrizations of curves in conformal manifolds. Proc. Am. Math. Soc. **108**(1), 215–221 (1990). https://doi.org/10.2307/2047716
3. T.N. Bailey, M.G. Eastwood, A.R. Gover, Thomas's structure bundle for conformal, projective and related structures. Rocky Mountain J. Math. **24**(4), 1191–1217 (1994). https://doi.org/10.1216/rmjm/1181072333
4. A. Čap, A.R. Gover, M. Hammerl, Normal BGG solutions and polynomials. Int. J. Math. **23**(11) (2012)
5. A. Čap, A.R. Gover, M. Hammerl, Holonomy reductions of Cartan geometries and curved orbit decompositions. Duke Math. J. **163**(5), 1035–1070 (2014)
6. B. Carter, Global structure of the Kerr family of gravitational fields. Phys. Rev. **174**(5), 1559 (1968)
7. S. Curry, A.R. Gover, An introduction to conformal geometry and tractor calculus, with a view to applications in general relativity, in *Asymptotic Analysis in General Relativity*. London Mathematical Society Lecture Note Series, vol. 443 (2018)
8. M. Dunajski, P. Tod, Conformal geodesics on gravitational instantons, in Mathematical Proceedings of the Cambridge Philosophical Society, (2019).
9. M.G. Eastwood, L. Zalabová, Special metrics and scales in parabolic geometry. arXiv:2002.02199 [math.DG]
10. C. Fefferman, *The Ambient Metric*, in Annals of Mathematics Studies, vol. 178 (Princeton University Press, Princeton, 2012)
11. C. Fefferman, C. Graham, Q-curvature and Poincaré metrics. Math. Res. Lett. **9**(2–3), 139–151 (2002)
12. C. Fefferman, K. Hirachi, Ambient metric construction of Q-curvature in conformal and CR geometries. Math. Res. Lett. **10**(5–6), 819–831 (2003)
13. H. Friedrich, B.G. Schmidt, Conformal geodesics in general relativity. Proc. Roy. Soc. Lond. Ser. A **414**(1846), 171–195 (1987)
14. V.P. Frolov, P. Krtouš, D. Kubizňák, Black holes, hidden symmetries, and complete integrability. Living Rev. Relativ. **20**(1), 6 (2017). https://doi.org/10.1007/s41114-017-0009-9
15. A.R. Gover, Almost conformally Einstein manifolds and obstructions, in *Proceedings, 9th International Conference on Differential Geometry and Its Applications*, Prague, Czech Republic, August 30–September 3, 2004 (2005), pp. 247–260
16. A.R. Gover, Almost Einstein and Poincaré-Einstein manifolds in Riemannian signature. J. Geom. Phys. **60**(2), 182–204 (2010). https://doi.org/10.1016/j.geomphys.2009.09.016

17. A.R. Gover, L.J. Peterson, Conformally invariant powers of the Laplacian, Q-curvature, and tractor calculus. Commun. Math. Phys. **235**(2), 339–378 (2003). https://doi.org/10.1007/s00220-002-0790-4
18. A.R. Gover, J. Šilhan, The conformal Killing equation on forms—prolongations and applications. Differ. Geom. Appl. **26**(3), 244–266 (2008). https://doi.org/10.1016/j.difgeo.2007.11.014
19. A.R. Gover, D. Snell, A. Taghavi-Chabert, Distinguished curves and integrability in Riemannian, conformal, and projective geometry (2018). arXiv preprint arXiv:1806.09830
20. C.R. Graham, M. Zworski, Scattering matrix in conformal geometry. Invent. Math. **152**(1), 89–118 (2003)
21. C. Guillarmou, Scattering for the geodesic flow on surfaces with boundary, in *Geometric and Computational Spectral Theory*. Contemp. Math., vol. 700 (Amer. Math. Soc., Providence, 2017), pp. 267–284. https://doi.org/10.1090/conm/700/14189
22. C. Guillarmou, J. Qing, Spectral characterization of Poincaré-Einstein manifolds with infinity of positive Yamabe type (2009). http://search.proquest.com/docview/2087791691/
23. M. Hammerl, P. Somberg, V. Souček, J. Šilhan, On a new normalization for tractor covariant derivatives. J. Eur. Math. Soc. **14**(6), 1859–1883 (2012). https://doi.org/10.4171/JEMS/349
24. S. Holman, G. Uhlmann, On the microlocal analysis of the geodesic X-ray transform with conjugate points. J. Differ. Geom. **108**(3), 459–494 (2018). https://doi.org/10.4310/jdg/1519959623
25. T. Houri, K. Tomoda, Y. Yasui, On integrability of the Killing equation. Classical Quantum Gravity **35**(7), 075014, 22 (2018). https://doi.org/10.1088/1361-6382/aaa4e7
26. R. Penrose, W. Rindler, Spinors and space-time, vol. 1. Cambridge Monographs on Mathematical Physics (Cambridge University Press, Cambridge, 1987)
27. C. Robin Graham, E. Witten, Conformal anomaly of submanifold observables in AdS/CFT correspondence. Nucl. Phys. B **546**(1–2), 52–64 (1999)
28. J.A. Schouten, *Ricci-Calculus, an Introduction to Tensor Analysis and Its Geometrical Applications* (Springer, Berlin, 1954)
29. P. Sommers, On Killing tensors and constants of motion. J. Math. Phys. **14**, 787–790 (1973). https://doi.org/10.1063/1.1666395
30. P. Tod, Some examples of the behaviour of conformal geodesics. J. Geom. Phys. **62**(8), 1778–1792 (2012). https://doi.org/10.1016/j.geomphys.2012.03.010
31. R.M. Wald, *General Relativity* (University of Chicago Press, Chicago, 1984). https://doi.org/10.7208/chicago/9780226870373.001.0001
32. P. Yang, S. Chang, J. Qing, On the renormalized volumes for conformally compact Einstein manifolds. J. Math. Sci. **149**(6), 1755–1769 (2008)
33. K. Yano, Sur les circonférences généralisées dans les espaces à connexion conforme. Proc. Imp. Acad. **14**(9), 329–332 (1938). https://doi.org/10.3792/pia/1195579559

A Car as Parabolic Geometry

C. Denson Hill and Paweł Nurowski

Abstract We show that a car, viewed as a nonholonomic system, provides an example of a flat parabolic geometry of type $(\mathbf{SO}(2,3), P_{12})$, where P_{12} is a Borel parabolic subgroup in $\mathbf{SO}(2,3)$. We discuss the relations of this geometry of a car with the geometry of circles in the plane (a low dimensional Lie sphere geometry), the geometry of 3-dimensional conformal Minkowski spacetime, the geometry of 3-rd order ODEs, projective contact geometry in three dimensions, and the corresponding twistor fibrations. We indicate how all these classical geometries can be interpreted in terms of the nonholonomic kinematics of a car.

1 Car and Engel Distribution

1.1 Configuration Space and Nonholonomic Constraints

In this note we look at a car from the point of view of an observer that is situated in space over the plane on which the car is moving. We idealize the car as an interval of length ℓ in the plane $\mathbb{R}^2$. The car has two pairs of wheels; we idealize them to be attached at both ends of the interval. The rear wheels are always parallel to the interval, whereas the front wheels can be rotated around the line vertical to the

This work was supported by the Polish National Science Centre (NCN) via the grant number 2018/29/B/ST1/02583 and via the POLONEZ grant 2016/23/P/ST1/04148, which received funding from the European Union's Horizon 2020 research and innovation programme under the Marie Skł odowska-Curie grant agreement No. 665778.

C. Denson Hill (✉)
Department of Mathematics, Stony Brook University, Stony Brook, NY, USA
e-mail: Dhill@math.stonybrook.edu

P. Nurowski
Centrum Fizyki Teoretycznej, Polska Akademia Nauk, Warszawa, Poland
e-mail: nurowski@cft.edu.pl

S. Hervik et al. (eds.), *Geometry, Lie Theory and Applications*, Abel Symposia 16,
https://doi.org/10.1007/978-3-030-81296-6_6

plane passing through the point of their attachment to the car. At every moment the direction of the front wheels can assume any angle with respect to the direction of the headlights of the car. To describe the position of the car we need *four* numbers. One can define these four numbers in many ways; here we choose the setting depicted in the figure below:

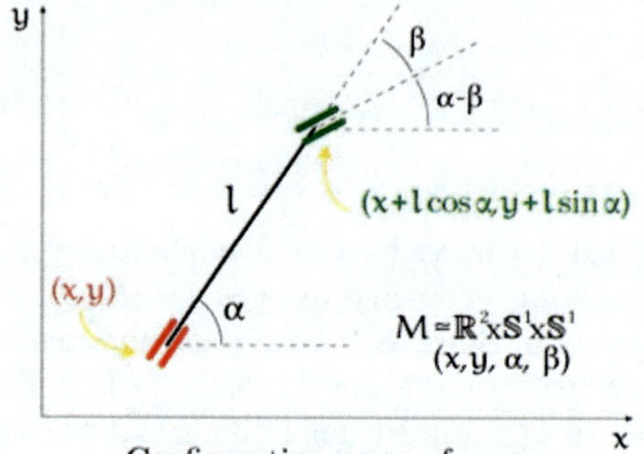

Configuration space of a car

We introduce a Cartesian coordinate system in the plane so that the position of the rear wheels of the car has coordinates (x, y). Then as a fixed line in the plane we choose the line $y = 0$, and to keep track of the orientation of the chassis of the car we take the angle α that the interval representing the car forms with this line. The orientation of the front wheels is the angle β, between the direction defined by the front wheels and the direction of the interval representing the chassis of the car. As a result we have four numbers (x, y, α, β) describing uniquely the position of the car as it moves. Thus the *configuration space* of the car is a 4-dimensional manifold M, locally diffeomorphic to

$$\mathbb{R}^2 \times \mathbb{S}^1 \times \mathbb{S}^1 = \{ \, (x, y, \alpha, \beta) \; : \; (x, y) \in \mathbb{R}^2; \; \alpha, \beta \in \mathbb{S}^1 \, \}.$$

1.2 Movement and the Role of the Tires

When the car is moving it traverses a curve $q(t) = (x(t), y(t), \alpha(t), \beta(t))$ in its configuration space M. The *velocity* of the car at time t is $\dot{q}(t) = (\dot{x}(t), \dot{y}(t), \dot{\alpha}(t), \dot{\beta}(t))$. It is a *vector* from the tangent space $T_{q(t)}M$.

A safe car has *tires*. Their role is to prevent the car from *skidding*. Our car will have *perfect* tires. They impose *nonholonomic constraints*. These are constraints on positions *and velocities*, that can not be integrated to constraints on positions only. Indeed, what is expected from a properly behaving car is that its rear wheels, i.e. the point (x, y) has its (x, y)-plane velocity parallel to the direction of the body of the car, and that the front wheels. i.e. the point $(x + \ell \cos \alpha, y + \ell \sin \alpha)$, has its (x, y)-plane velocity in the plane parallel to the orientation of the front wheels. Thus, the movement of a car, represented by the curve $q(t) = (x(t), y(t), \alpha(t), \beta(t)) \in M$, at every moment of time t, must satisfy

$$\tfrac{d}{dt}(x, y) \quad || \quad (\cos\alpha, \sin\alpha) \qquad \&$$
$$\tfrac{d}{dt}(x + \ell\cos\alpha, y + \ell\sin\alpha) \quad || \quad (\cos(\alpha - \beta), \sin(\alpha - \beta)),$$

or, what is the same

$$\dot{x}\sin\alpha - \dot{y}\cos\alpha = 0 \qquad \&$$
$$(\dot{x} - \ell\dot{\alpha}\sin\alpha)\,\sin(\alpha - \beta) - (\dot{y} + \ell\dot{\alpha}\cos\alpha)\,\cos(\alpha - \beta) = 0.$$

We emphasize that the above constraints are *linear* in velocities. Solving them we get the possible velocities as

$$\begin{pmatrix}\dot{x}\\ \dot{y}\\ \dot{\alpha}\\ \dot{\beta}\end{pmatrix} = A(t)\begin{pmatrix}0\\0\\0\\1\end{pmatrix} + B(t)\begin{pmatrix}\ell\cos\alpha\cos\beta\\ \ell\sin\alpha\cos\beta\\ -\sin\beta\\ 0\end{pmatrix}.$$

where $\alpha = \alpha(t)$, $\beta = \beta(t)$, $A = A(t)$ and $B = B(t)$ are *arbitrary* functions of time.

1.3 Velocity Distribution as an Engel Distribution

We can rephrase this by saying that at each point $q = (x, y, \alpha, \beta)^T$ in the tangent space T_qM, which is considered as the space of *all* possible velocities, there is a *distinguished* vector subspace

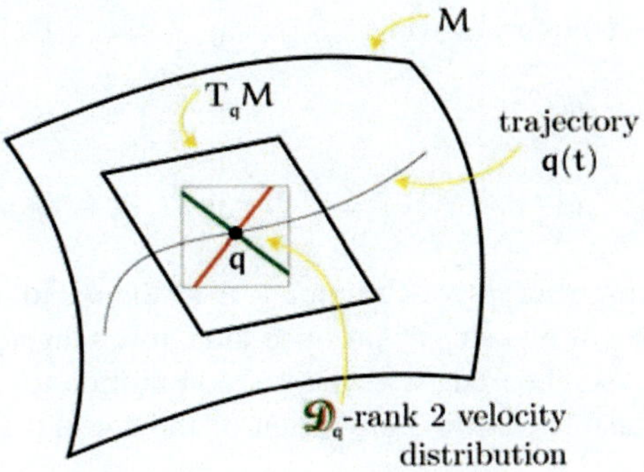

$\mathcal{D}_q = \text{Span}_{\mathbb{R}}(X_3, X_4)$ spanned at each point $q \in M$ by the vectors tangent to the vector fields

$$X_3 = \partial_\beta \quad \& \quad X_4 = -\sin\beta\partial_\alpha + \ell\cos\beta(\cos\alpha\partial_x + \sin\alpha\partial_y), \tag{1.1}$$

which is the space of *admissible* velocities of the car at q. The car with perfect tires moves always along the curves $q(t) = (x(t), y(t), \alpha(t), \beta(t))^T$ such that its velocity $\dot{q}$ in the configuration space satisfies

$$\dot{q} = AX_3 + BX_4.$$

The arbitrary functions $A = A(t)$ and $B = B(t)$ are called *controls* of the car.[1]

Thus on M there is a *rank 2* distribution $\mathcal{D}$ on M, describing the space of possible velocities, given by

$$\mathcal{D} = Span_{\mathcal{F}(M)}(X_3, X_4). \tag{1.2}$$

with $\mathcal{F}(M)$ being the ring of all differentiable functions on M. Therefore 'the structure of a car with perfect tires' is *up to now*

$$(M, \mathcal{D}),$$

i.e. a 4-manifold M with a rank 2 distribution $(M, \mathcal{D})$.

Now the fundamental question is: *Is $\mathcal{D}$ integrable?*

The answer is: Obviously *not*, since everybody knows that a car can be driven from any position in its configuration space to any other position (Chow-Raszewski theorem). One can also convince oneself about that by calculating the commutators of X_3 and X_4. We have:

$$\begin{aligned} [X_3, X_4] &= -\cos\beta\partial_\alpha - \ell\sin\beta(\sin\alpha\partial_y + \cos\alpha\partial_x) := X_2 \\ [X_4, X_2] &= \ell(\cos\alpha\partial_y - \sin\alpha\partial_x) := X_1, \end{aligned} \tag{1.3}$$

and it is easy to check that

$$X_1 \wedge X_2 \wedge X_3 \wedge X_4 = \ell^2\partial_x \wedge \partial_y \wedge \partial_\alpha \wedge \partial_\beta \neq 0.$$

This shows that taking successive commutators of the vectors from the car distribution $\mathcal{D}$ we quickly (in two steps!) produce the entire tangent bundle to M. This, by the Chow-Raszewski theorem, is a well know condition for curves tangent to the distribution to be capable reaching any point of the configuration space from any other point.

---

[1] Sometimes the vector fields X_3 and X_4 are also called controls.

We summarize this by defining three distributions $\mathcal{D}_{-1}$, $\mathcal{D}_{-2}$ and $\mathcal{D}_{-3}$ on M as in the table below:

		rank
$\mathcal{D}_{-1} := \mathcal{D}$	$\mathrm{Span}(X_4, X_3)$	2
$\mathcal{D}_{-2} := [\mathcal{D}_{-1}, \mathcal{D}_{-1}]$	$\mathrm{Span}(X_4, X_3, X_2)$	3
$\mathcal{D}_{-3} := [\mathcal{D}_{-1}, \mathcal{D}_{-2}]$	$\mathrm{Span}(X_4, X_3, X_2, X_1) = \mathrm{T}M$	4

Thus given the so far defined structure of the car $(M, \mathcal{D})$, we have a filtration $\mathcal{D}_{-1} \subset \mathcal{D}_{-2} \subset \mathcal{D}_{-3} = \mathrm{T}M$ of distributions with the *constant growth vector* $(2, 3, 4)$. These collective properties of the car distribution $\mathcal{D}$ make it an *Engel distribution*. Here we recall that an abstract *Engel distribution* is a rank 2 distribution on a 4-manifold such that its derived flag of distributions $\mathcal{D}_{-1} = \mathcal{D}$, $\mathcal{D}_{-2} := [\mathcal{D}_{-1}, \mathcal{D}_{-1}]$ and $\mathcal{D}_{-3} := [\mathcal{D}_{-1}, \mathcal{D}_{-2}]$ has respective *constant* ranks 2, 3 and 4.

1.4 Equivalence of Engel Distributions

Our discussion so far shows that the geometric structure associated with a car is $(M, \mathcal{D})$ with $\mathcal{D}$ being an Engel distribution on a manifold M.

A newcomer to this subject has an immediate question: are there nonequivalent Engel distributions? To answer this we need the notion of *equivalence* of distributions.

We say that two distributions $\mathcal{D}$ and $\bar{\mathcal{D}}$ of the same rank on manifolds M and $\bar{M}$ of the same dimension are *(locally) equivalent* iff there exists a (local) diffeomorphism $\phi : M \to \bar{M}$ such that $\phi_*\mathcal{D} = \bar{\mathcal{D}}$. (Local) self-equivalence maps $\phi : M \to M$, i.e. maps such that $\phi_*\mathcal{D} = \mathcal{D}$ are called (local) *symmetries* of $\mathcal{D}$. They form a *group of (local) symmetries of* $\mathcal{D}$. This notion has its infinitesimal version: we say that a vector field X on M is an *infinitesimal symmetry* of $\mathcal{D}$ if and only if $\mathcal{L}_X\mathcal{D} \subset \mathcal{D}$. Since the commutator $[X, Y]$ of two infinitesimal symmetries X and Y is also an infinitesimal symmetry, this leads to the notion of the *Lie algebra* $\mathfrak{g}_{\mathcal{D}}$ *of infinitesimal symmetries* of $\mathcal{D}$.

Now, one convinces herself that the distribution

$$\mathcal{D}_E = (\partial_q, \partial_x + p\partial_y + q\partial_p)$$

defined on an open set of $\mathbb{R}^4$ parametrized by (x, y, p, q) is an Engel distribution. We have the following classical theorem due to Friedrich Engel.

Theorem *Every Engel distribution is locally equivalent to the distribution* $\mathcal{D}_E$.

One may say that we are in trouble: Since the car structure $(M, \mathcal{D})$ is a structure of an Engel distribution, there is no geometry associated to the car. The wrong argument in this kind of criticisim is that an Engel distribution is not the only structure that a car with perfect tires has. It turns out that the geometry associated with a car is more subtle than just the geometry of an Engel distribution. The car features equip its Engel distribution with an additional structure.

2 Car and Engel Distribution with a Split

2.1 Two Distinguished Directions

To see this consider the vector field: $X_4 = -\sin\beta\partial_\alpha + \ell\cos\beta(\cos\alpha\partial_x + \sin\alpha\partial_y)$. When $\beta = 0$ it becomes $X_4 = \ell(\cos\alpha\partial_x + \sin\alpha\partial_y)$ and if the car chooses this direction of its velocity it makes a simple movement by going along a straight line in the direction $(\cos\alpha, \sin\alpha)$ in the (x, y) plane. On the other hand, if the car chooses its velocity in the direction of the vector fild $X_3 = \partial_\beta$, then although it does move in the configuration space, it does not perform any movement in the physical (x, y) plane, merely rotating the steering wheel/front wheels with the engine at idle.

Cars owners/producers perfectly know and *make use* of these two particular vector fields (X_3, X_4) in the distribution $\mathcal{D}$. In particular, car owners alternate using these two vector fields, each separately at proper instants/intervals of time, in parallel parking.

Indeed, if one wants to park a car one first approaches the parking spot by having its velocity aligned with X_4 vector field with $\beta = 0$. Then the car stops and rotates its front wheels towards the sidewalk passing from $\beta = 0$ to $\beta = \beta_0$=const. This is done by aligning its velocity with the vector field X_3. After this, the car velocity again becomes aligned with X_4, which now has $\beta = \beta_0$=const, so that the car goes backwards towards the sidewalk.

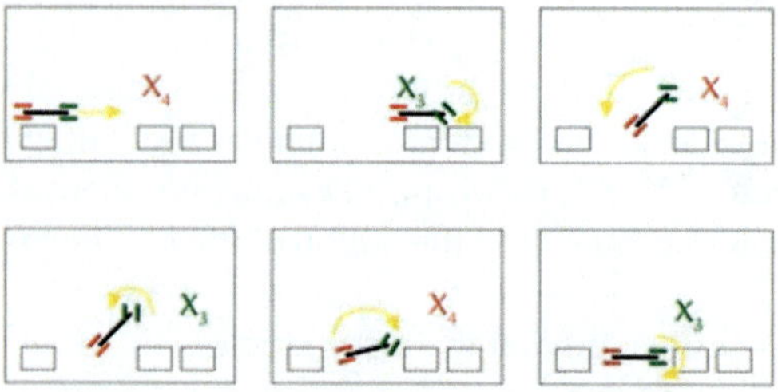

Parking by alternating vector fields X_3 and X_4

When the rear wheels are close to the sidewalk the car stops again, and aligns its velocity with X_3, going back from $\beta = \beta_0$ to $\beta = -\beta_0$. Again applying backwards X_4 with this constant $\beta = -\beta_0$ enables the driver to orient the rear wheels parallely to the sidewalk. If this happens, the car stops and applies X_3 to make $\beta = 0$ again. Finally the car aligns its velocity with X_4 having $\beta = 0$ to move parallely to the sidewalk and to take the midlle position between the two cars before and after it.

Thus the car's distribution $\mathcal{D}$ has an additional structure, which is its split

$$\mathcal{D} = \mathcal{D}_w \oplus \mathcal{D}_g,$$

onto rank one subdistributions

$$\mathcal{D}_w = \mathrm{Span}_{\mathcal{F}(M)}(X_3) \quad \text{and} \quad \mathcal{D}_g = \mathrm{Span}_{\mathcal{F}(M)}(X_4).$$

These subdistributions have a clear physical meaning:

The distribution $\mathcal{D}_w$ as spanned by $X_3 = \partial_\beta$, is responsible for the steering wheel control, and will be called the *steering wheel space*; on the other hand the distribution $\mathcal{D}_g$, as spanned by the generator of the forward-backward movement $X_4 = -\sin\beta\partial_\alpha + \ell\cos\beta(\cos\alpha\partial_x + \sin\alpha\partial_y)$ will be called the *gas space*.

This results in the statement that the car structure is actually $(M, \mathcal{D} = \mathcal{D}_w \oplus \mathcal{D}_g)$, with $\mathcal{D}$ being an Engel distribution with a *split* $\mathcal{D} = \mathcal{D}_w \oplus \mathcal{D}_g$ onto rank one, steering wheel and gas, subdistributions. So considering a car's geometry more thoroughly we land in a realm of the subtle geometry of Engel distributions *with a split*!

2.2 New Geometry: Engel Distributions with a Split

Thus we ultimately established that the geometry of a car with perfect tires, is given by a structure $(M, \mathcal{D} = \mathcal{D}_w \oplus \mathcal{D}_g)$, where $\mathcal{D}$ is an Engel distribution *with a (car's) split* $\mathcal{D} = \mathcal{D}_w \oplus \mathcal{D}_g$.

Abstractly, irrespectively of car's considerations, let us consider a geometry in the form $(M, \mathcal{D} = \mathcal{D}_1 \oplus \mathcal{D}_2)$, where $\dim M = 4$, $\mathcal{D}$ is an Engel distribution on M, and both subdistributions $\mathcal{D}_1$ and $\mathcal{D}_2$ in $\mathcal{D}$ have rank one. Let us call this an *Engel structure with a split*.

Such structures have their own equivalence problem, related to the following definitions:

Two Engel structures with a split $(M, \mathcal{D} = \mathcal{D}_1 \oplus \mathcal{D}_2)$ and $(\bar{M}, \bar{\mathcal{D}} = \bar{\mathcal{D}}_1 \oplus \bar{\mathcal{D}}_2)$ are *(locally) equivalent* if and only if there exists a (local) diffeomorphism $\phi : M \to \bar{M}$ such that $\phi_* \mathcal{D}_1 = \bar{\mathcal{D}}_1$ and $\phi_* \mathcal{D}_2 = \bar{\mathcal{D}}_2$. Infinitesimally, we consider vector fields S on M such that $\mathcal{L}_S \mathcal{D}_1 \subset \mathcal{D}_1$ and $\mathcal{L}_S \mathcal{D}_2 \subset \mathcal{D}_2$, and we call such vector fields *infinitesimal symmetries* of $(M, \mathcal{D} = \mathcal{D}_1 \oplus \mathcal{D}_2)$. This, as usual, leads to a notion of the *Lie algebra* $\mathfrak{g}_{\mathcal{D}}$ *of infinitesimal symmetries of* an Engel structure $(M, \mathcal{D} = \mathcal{D}_1 \oplus \mathcal{D}_2)$ with a split, as the Lie algebra of the vectors fields S as above.

We can now ask about the Lie algebra of infinitesimal symmetries of the Engel structure with a split $(M, \mathcal{D} = \mathcal{D}_w \oplus \mathcal{D}_g)$ of a car. In this case we have $\mathcal{D}_1 = \mathcal{D}_w$ and $\mathcal{D}_2 = \mathcal{D}_g$. As an answer we get a bit surprising result as below:

Theorem 2.1 *Consider the car structure $(M, \mathcal{D})$ consisting of its velocity distribution $\mathcal{D}$ and the split of $\mathcal{D}$ onto rank 1 distributions $\mathcal{D} = \mathcal{D}_w \oplus \mathcal{D}_g$ with $\mathcal{D}_w = \mathrm{Span}(\partial_\beta)$, $\mathcal{D}_g = \mathrm{Span}(-\sin\beta\partial_\alpha + \ell\cos\beta(\cos\alpha\partial_x + \sin\alpha\partial_y)$.*

The Lie algebra of infinitesimal symmetries of this Engel structure with a split is 10-dimensional, with the following generators

$$
\begin{aligned}
S_1 &= \partial_x \\
S_2 &= \partial_y \\
S_3 &= x\partial_y - y\partial_x + \partial_\alpha \\
S_4 &= \ell(\sin\alpha\partial_x - \cos\alpha\partial_y) + \sin^2\beta\partial_\beta \\
S_5 &= x\partial_x + y\partial_y - \sin\beta\cos\beta\partial_\beta \\
S_6 &= (x^2 - y^2)\partial_x + 2xy\partial_y + 2y\partial_\alpha - 2\cos\beta\Big(\ell\cos\beta\sin\alpha + x\sin\beta\Big)\partial_\beta \\
S_7 &= \ell\Big(x(\sin\alpha\partial_x - \cos\alpha\partial_y) - \cos\alpha\partial_\alpha\Big) + \sin\beta\Big(\ell\cos\beta\sin\alpha + x\sin\beta\Big)\partial_\beta \\
S_8 &= \ell\Big(y(\sin\alpha\partial_x - \cos\alpha\partial_y) - \sin\alpha\partial_\alpha\Big) - \sin\beta\Big(\ell\cos\beta\cos\alpha - y\sin\beta\Big)\partial_\beta \\
S_9 &= 2xy\partial_x + (y^2 - x^2)\partial_y - 2x\partial_\alpha + 2\cos\beta\Big(\ell\cos\beta\cos\alpha - y\sin\beta\Big)\partial_\beta \\
S_{10} &= \ell(x^2 + y^2)\Big(\sin\alpha\partial_x - \cos\alpha\partial_y\Big) - 2\ell\Big(x\cos\alpha + y\sin\alpha\Big)\partial_\alpha + \\
&\quad \Big(2\ell\sin\beta\cos\beta\big(x\sin\alpha - y\cos\alpha\big) + \sin^2\beta(x^2 + y^2) + 2\ell^2\cos^2\beta\Big)\partial_\beta
\end{aligned}
$$

It is isomorphic to the simple real Lie algebra $\mathfrak{so}(2,3) = \mathfrak{sp}(2,\mathbb{R})$. *Moreover, there are plenty of locally nonequivalent Engel distributions with a split, but the split* $\mathcal{D} = \mathcal{D}_w \oplus \mathcal{D}_g$ *on the (Engel) car distribution used by car owners and provided by cars' producers is* the most symmetric.

The fact that there are many locally nonequivalent Engel structures with a split is not surprising at all. What is surprising here, is that the split on the Engel distribution provided by the 'steering-wheel–gas' control of a car is the most symmetric. Moreover, the appearence of a *simple* Lie algebra $\mathfrak{so}(2,3) = \mathfrak{sp}(2,\mathbb{R})$ as the full algebra of symmetries of car's $\mathcal{D} = \mathcal{D}_w \oplus \mathcal{D}_g$ is also striking. Especially that $\mathfrak{so}(2,3)$ is the Lie algebra of the group of conformal symmetries of 3-dimensional *Minkowski space*. How on earth Minkowski space can be related to a car?

3 Explaining the $\mathfrak{so}(2,3) = \mathfrak{sp}(2,\mathbb{R})$ Symmetry

3.1 A Double Fibration

Consider integral curves of the two distinguished directions X_3 and X_4 defined by the split in the car's distribution $\mathcal{D}$. Let us call the integral curves of X_3 by q_3 and the integral curves of X_4 by q_4 respectively. They define two *foliations* of M, the first having q_3 as the leaves, and the second consisting of leaves given by q_4. Passing to the space of leaves of these two foliations, which we denote by P and by Q, respectively, we get a double fibration

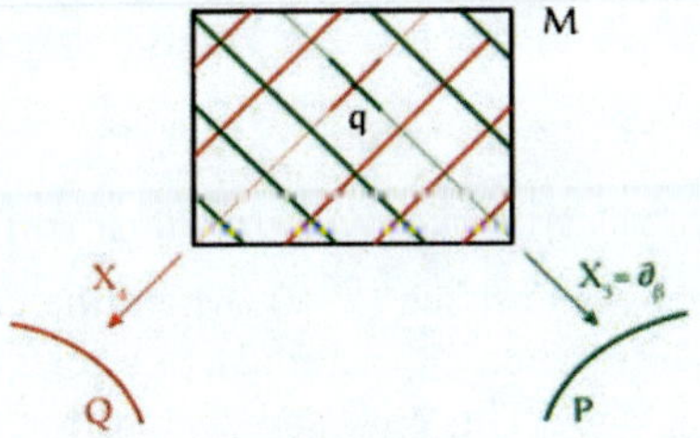

Double fibration

with the 4-dimensional configuration space M of a car on top, and the two 3-dimensional spaces P and Q at the bottom.

We will now analyze the geometry of each of the base spaces of this fibration, devoting a subsection to each of them.

3.2 *Conformal Structure on Q*

Points of Q are just the integral curves of X_4. What are these curves in M? In an appropriate parametrization they are:

$$q_4(t) = \begin{pmatrix} 2\ell \cot\beta_0 \cos(\alpha_0 - \frac{1}{2}t \sin\beta_0) \sin(\frac{1}{2}t \sin\beta_0) + x_0 \\ 2\ell \cot\beta_0 \sin(\alpha_0 - \frac{1}{2}t \sin\beta_0) \sin(\frac{1}{2}t \sin\beta_0) + y_0 \\ -t \sin\beta_0 + \alpha_0 \\ \beta_0 \end{pmatrix} \quad \text{when} \quad \beta_0 \neq 0, \tag{3.1}$$

or

$$q_4(t) = \begin{pmatrix} t\ell \cos\alpha_0 + x_0 \\ t\ell \sin\alpha_0 + y_0 \\ \alpha_0 \\ 0 \end{pmatrix} \quad \text{when} \quad \beta_0 = 0. \tag{3.2}$$

Here $(x_0, y_0, \alpha_0, \beta_0)$ are constants, corresponding to the position of the car at $t = 0$.

These curves $q_4(t)$ correspond to the movement of the car, when the β angle is fixed. Thus in the configuration space M, they are *helices* $(x(t), y(t), \alpha(t))$ in the 3-dimensional space $\beta = \beta_0 = \text{const}$, parametrized by (x, y, α). The axi of these helices are given by $(x_0 + \ell \cot\beta_0 \sin\alpha_0, y_0 - \ell \cot(\beta_0) \cos\alpha_0, t)$, their radii are $R = \ell \cot\beta_0$ and their pitch is 2π, for each choice of initial conditions (x_0, y_0, α_0).

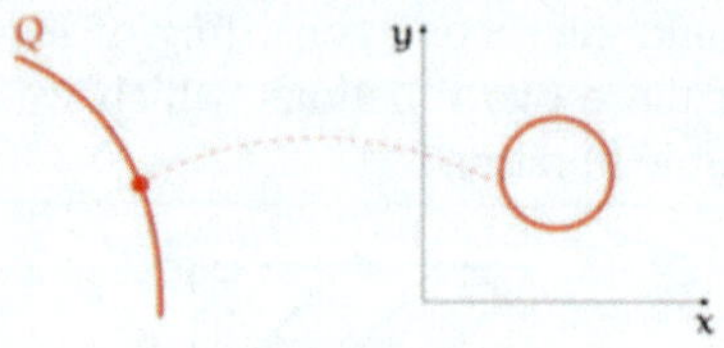

3 - dimensional space of all circles
on the plane (x,y)

In the physical 2-dimensional space (x, y), where the car is physically moving, these curves are either points (when $\beta_0 = \pm\pi/2$), or circles (when $0 < |\beta_0| < \pi/2$), or straight lines (when $\beta_0 = 0$). This corresponds to the simple fact that if one sets the steering wheel in a given position, or what is the same keeps the constant angle $\beta = \beta_0$ between the front wheels and the axis of the chasis of the car, the *rear* wheels of the car will go on a straight line if $\beta = 0$, will go on circles if $0 < |\beta| < \pi/2$, or will stay at a given point (x_0, y_0) if the front wheels are perpendicular to the axis of the car. It is important to note that, by setting the initial conditions $(x_0, y_0, \alpha_0, \beta_0)$ properly, one can obtain any point, line or a circle in the plane (x, y), as a trajectory of a physical movement of the car in the plane (x, y).

Thus there is a one-to one correspondence between the points q of the 3-dimensional space Q of the integral curves of the vector field X_4 (the helices at each plane $\beta = \beta_0$ in M) and the 3-dimensional *space* **Q** *of all points, circles and lines in* $\mathbb{R}^2$ coordinatized by (x, y).

3.2.1 Geometry of *Oriented* Circles on the Plane

Since two circles on the plane can be disjoint, or can intersect, or be tangent, and since these relations between any two circles are invariant with respect to diffeomorphisms of the plane, they should be used to further determine the geometry of the space **Q** and in turn the geometry of the leaf space Q.

The geometry of circles on the plane is a classical subject first considered by S. Lie (see e.g. [16]). Consider a set **Q** of all objects in the plane whose coordinates (x, y) satisfy

$$x^2 + y^2 - 2ax - 2by + c = 0,$$

with some real constants a, b, c. Introducing

$$R^2 = a^2 + b^2 - c,$$

and projective coordinates $[\xi : \eta : \zeta : \mu : \nu]$ in $\mathbb{R}P^4$ via

$$a = \frac{\xi}{\nu}, \quad b = \frac{\eta}{\nu}, \quad c = \frac{\mu}{\nu}, \quad R = \frac{\zeta}{\nu}, \tag{3.3}$$

we see that $\mathbf{Q}$ is a *projective quadric*

$$\mathbf{Q} = \{\, \mathbb{R}P^4 \ni [\xi : \eta : \zeta : \mu : \nu] \;:\; \xi^2 + \eta^2 - \zeta^2 - \mu\nu = 0 \,\} \tag{3.4}$$

in $\mathbb{R}P^4$. The objects (the points) of this set are stratified as follows. Generically they form the set $\mathbf{Q}_c$ of (all) circles in the plane; this occurs when $\xi^2 + \eta^2 - \mu\nu > 0$. When the radius R is infinite, i.e. when $\nu = 0$, the objects belong to $\mathbf{Q}_\ell$, the set of (all) lines in the plane; finally, when $\zeta = 0$, the objects belong to $\mathbf{Q}_p$, the set of (all) points on the plane. Thus we have

$$\mathbf{Q} = \mathbf{Q}_c \sqcup \mathbf{Q}_\ell \sqcup \mathbf{Q}_p,$$

i.e. $\mathbf{Q}$ is the set of *all circles*, *lines* and *points* on the plane. In addition we easily see that the *three dimensional* set $\mathbf{Q}$, as a *null* projective quadric in $\mathbb{R}P^4$, acquires a natural *conformal Lorentzian structure* $[g]$, coming from the quadratic form

$$Q(\xi, \eta, \zeta, \mu, \nu) = \xi^2 + \eta^2 - \zeta^2 - \mu\nu \tag{3.5}$$

in $\mathbb{R}^5$.

It is important to notice that by considering R as in formula (3.3) we *doubled* the number of circles in the plane. This is because, depending on the sign of $\zeta\nu$, the radius R of the circle may be positive or negative. This has an obvious interpretation: the space $\mathbf{Q}$ consists of all *oriented* circles/lines. We adapt the convention that a circle/line $(x - a)^2 + (y - b)^2 = R^2$ is oriented *counterclockwise* iff $R > 0$, and it is oriented *clockwise* iff $R < 0$.

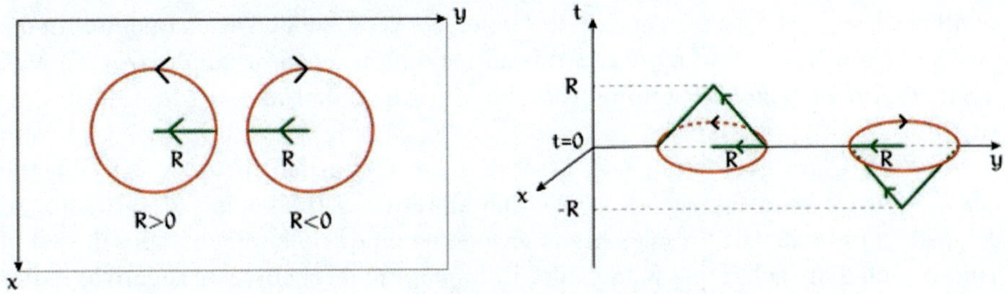

Lie has shown that the conformal structure in $\mathbf{Q}$, whose points are generically oriented circles in the plane, is identical with the structure defined by the *incidence* relation between the circles: *two circles* from $\mathbf{Q}$ *are incident if and only if they are tangent to each other in such a way that their orientations coincide when one of the circles is inside the other and are opposite when they are external to each other.*

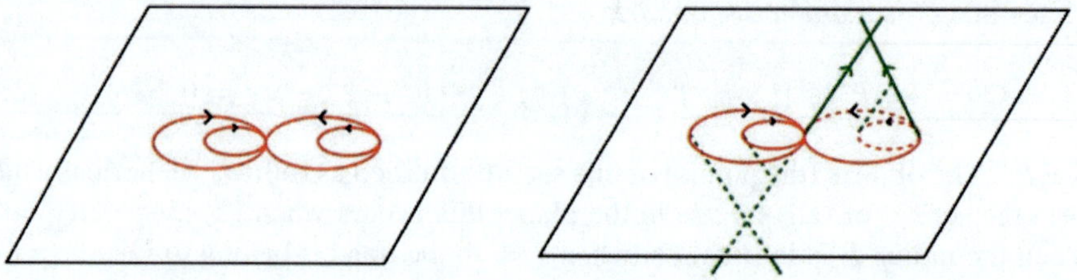

Circles in incidence: the two kissing circles on the left and the third one - the larger one on the right, are in incidence. The two circles on the right - are also in incidence. The point of contact of the first three defines an outgoing light ray in 3D Minkowski spacetime (the darkgreen line with an arrow going from left to right). Likewise the point of contact of the two kissing circles on the right defines a light ray (the darkgreen light with an arrow going from right to left).

Indeed, parametrizing the space of circles on the plane by (a, b, R), where (a, b) are the coordinates of their center in the plane, and R is their (negative or positive) radius, we see that close circles corresponding to (a, b, R) and $(a+\mathrm{d}a, b+\mathrm{d}b, R+\mathrm{d}R)$ have only one point of intersection iff the equations

$$(x-a)^2+(y-b)^2-R^2=0 \quad \& \quad (x-a-\mathrm{d}a)^2+(y-b-\mathrm{d}b)^2-(R+\mathrm{d}R)^2=0,$$

have a unique solution for (x, y). It is only possible if and only if

$$(\mathrm{d}a)^2+(\mathrm{d}b)^2-(\mathrm{d}R)^2=0,$$

i.e. when the circles corresponding to (a, b, R) and $(a+\mathrm{d}a, b+\mathrm{d}b, R+\mathrm{d}R)$ are *null* separated in the Lorentzian metric $g=(\mathrm{d}a)^2+(\mathrm{d}b)^2-(\mathrm{d}R)^2$ on the space of all circles $\mathbf{Q}_c$. Thus the space of all circles $\mathbf{Q}_c$ is embedded as an open set in the projective quadric $\mathbf{Q}$, and moreover this embedding is a *conformal embedding* with a *flat conformal structure* coming from the *Minkowski metric* $g=(\mathrm{d}a)^2+(\mathrm{d}b)^2-(\mathrm{d}R)^2$.

Another, more geometric, way of seeing the conformal metric $g=(\mathrm{d}a)^2+(\mathrm{d}b)^2-(\mathrm{d}R)^2$ on the space $\mathbf{Q}_c$ is to think about (a, b) plane as a $R=0$ slice of $\mathbb{R}^3$ with coordinates (a, b, R). This space can be uniquely equipped with the set of cones, such that each circle with center in (a_0, b_0) and (positive or negative) radius R_0 on the $R=0$ plane is an intersection of this plane with a cone having tip at (a_0, b_0, R_0). Then one declares $\mathbb{R}^3$ with such cones as a conformal 3-dimensional manifold on which these cones are light cones. By construction these cones are light cones in the metric $g=(\mathrm{d}a)^2+(\mathrm{d}b)^2-(\mathrm{d}R)^2$.

3.2.2 Conformal Minkowski Space in 3-Dimensions is SO(2, 3) Symmetric

Since, following Lie, we have shown that the space $\mathbf{Q}$ of all circles on the plane has a natural structure of 3-dimensional conformal Minkowski space which has $\mathbf{SO}(2, 3)$ as a group of symmetries, and since $\mathbf{Q}$ is in one to one correspondence with the base

Q of the fibration $M \to Q$, then also the space Q of all integral curves of the vector field X_4 in M has $\mathbf{SO}(2, 3)$ as a symmetry. But this is naturally associated with the configuration space M of a car equipped with the geometry of an (velocity) Engel distribution with car's split. This gives an argument why the Lie algebra $\mathfrak{so}(2, 3)$ is the algebra of infinitesimal symmetries of the car structure $(M, \mathcal{D} = \mathcal{D}_w \oplus \mathcal{D}_g)$.

3.3 Geometry of 3rd Order ODEs

It turns out that the double fibration of the type

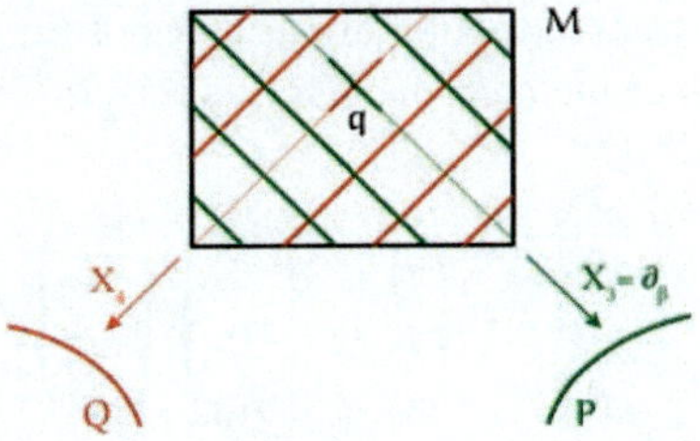

is also associated with the geometry of *3rd order ODEs considered modulo contact transformations of variables*. Indeed, in [7] S.S. Chern studied the geometry of an ordinary differential equation (ODE)

$$y''' = F(x, y, y', y'') \tag{3.6}$$

considered modulo contact transformation of variables, and established that the space M of second jets of the ODE, i.e. the *four*-dimensional space coordinatized by the jet coordinates (x, y, y', y''), is naturally equipped with two 1-dimensional foliations. These are given

- in terms of the integral curves of a vector field $X_3 = \partial_{y''}$ responsible for the projection $(x, y, y', y'') \to (x, y, y')$ from the space M of second jets to space P of the first jets, and
- in terms of the *total differential vector field* $X_4 = \partial_x + y'\partial_y + y''\partial_{y'} + F\partial_{y''}$ of the equation.

He has also shown that these two foliations on M do not change when the ODE undergoes contact transformation of variables. This led him to the study of a double fibration $Q \leftarrow M \to P$, with the 3-dimensional space Q beeing the leaf space of the foliation given by X_4.

In this section we recall Chern's considerations, and will show their relation to the geometry of the car fibration.

Equation (3.6) can be equivalently written as a system $y' = p$, $p' = q$, $q' = F(x, y, p, q)$ and as such is defined on the space of second jets $M = \mathcal{J}^2$ over the x-axis. This space is parameterized by (x, y, p, q) and every solution to (3.6) is a curve $\gamma(t) = (x(t), y(t), p(t), q(t))$ in $\mathcal{J}^2$ such that its tangent vector $\dot{\gamma}(t)$ annihilates the contact forms

$$\omega^1 = \mathrm{d}y - p\mathrm{d}x, \quad \omega^2 = \mathrm{d}p - q\mathrm{d}x, \quad \omega^3 = \mathrm{d}q - F(x, y, p, q)\mathrm{d}x. \tag{3.7}$$

These can be supplemented by

$$\omega^4 = \mathrm{d}x \tag{3.8}$$

to a coframe on $\mathcal{J}^2$.

Chern, inspired by the earlier work of E. Cartan's [5], (see also [6]), established that an arbitrary *contact* transformation of variables of Eq. (3.6) is equivalent to the following transformation of the *coframe* 1-forms $(\omega^1, \omega^2, \omega^3, \omega^4)$ in $\mathcal{J}^2$:

$$\begin{pmatrix}\omega^1\\ \omega^2\\ \omega^3\\ \omega^4\end{pmatrix} \to \begin{pmatrix} t_1 & t_2 & 0 & 0\\ t_3 & t_4 & 0 & 0\\ t_5 & t_6 & t_7 & 0\\ t_8 & t_9 & 0 & t_{10}\end{pmatrix}\begin{pmatrix}\omega^1\\ \omega^2\\ \omega^3\\ \omega^4\end{pmatrix}. \tag{3.9}$$

Here the t_i are arbitrary functions on $\mathcal{J}^2$ such that $(t_1t_4 - t_2t_3)t_7t_{10} \neq 0$. Thus the local equivalence of 3-rd order ODEs, considered modulo contact transformations, got reformulated by Chern into the local equivalence of coframes (3.7)–(3.8) given modulo transformations (3.9).

Looking at the transformation (3.9) defining a contact equivalence class of ODEs (3.6), we see that the frame vector fields (X_1, X_2, X_3, X_4), which on $\mathcal{J}^2$ are dual to $(\omega^1, \omega^2, \omega^3, \omega^4)$, $X_i \lrcorner\, \omega^j = \delta_i{}^j$, are given up to the transformations

$$\begin{pmatrix}X_1\\ X_2\\ X_3\\ X_4\end{pmatrix} \to \begin{pmatrix} * & * & * & *\\ * & * & * & *\\ 0 & 0 & \frac{1}{t_7} & 0\\ 0 & 0 & 0 & \frac{1}{t_{10}}\end{pmatrix}\begin{pmatrix}X_1\\ X_2\\ X_3\\ X_4\end{pmatrix}. \tag{3.10}$$

Thus a 3rd order ODE (3.6) considered modulo contact transformations distinguishes two well defined directions on $\mathcal{J}^2$. They are spanned by the respective vector fields

$$X_3 = \partial_q \quad \text{and} \quad X_4 = \partial_x + p\partial_y + q\partial_p + F\partial_q.$$

These in turn span a rank 2 distribution $\boldsymbol{\mathcal{D}} = \mathrm{Span}_{\mathcal{F}(\mathcal{J}^2)}(X_3, X_4)$ which happens to be an Engel distribution. Thus we have an Engel distribution $\boldsymbol{\mathcal{D}}$ with a natural split $\boldsymbol{\mathcal{D}} = \mathcal{D}_w \oplus \mathcal{D}_g$ given by $\mathcal{D}_w = \mathrm{Span}_{\mathcal{F}(\mathcal{J}^2)}(X_3)$ and $\mathcal{D}_g = \mathrm{Span}_{\mathcal{F}(\mathcal{J}^2)}(X_4)$. So the geometry of the jet space $\mathcal{J}^2$ with a 3rd order ODE considered modulo contact transformations of variables is very much like the geometry of car's configuration space!

Can we thus associate a 3rd order ODE to the car? If so, what is the ODE?

It turns out that the car structure geometry is a special case of geometries studied by N. Tanaka in [21], and by us in the paper [17]. In particular, in our paper [17] we considered manifolds M of dimension $k+n$ and the geometry of rank $n = r+s$ distributions $\mathcal{D}$ on M which had the split $\mathcal{D} = \mathcal{D}_r \oplus \mathcal{D}_s$ onto *integrable* subdistributions of respective ranks r and s. We called such structures *para-CR structures of type* (k, r, s). Since rank 1 distributions are always integrable then, in this sense, the geometry of car's structure $(M, \boldsymbol{\mathcal{D}} = \mathcal{D}_w \oplus \mathcal{D}_g)$ is a para-CR structure of type (2, 1, 1).

Actually, in Ref. [17], Sec. 4, Proposition 4.2, we have shown that the geometry of para-CR structures of type (2, 1, 1) is the same as the geometry of 3rd order ODEs considered modulo contact transformation of variables. Thus, according to this general result, there definitely exists a contact equivalence class of 3rd order ODEs associated with a car. So what is an ODE representing this class?

The car structure $(M, \boldsymbol{\mathcal{D}} = \mathcal{D}_w \oplus \mathcal{D}_g)$ defines a G-structure [11] on M, i.e. the reduction of the structure group $\mathbf{GL}(4, \mathbb{R})$ of the tangent bundle $\mathrm{T}M$ to its subgroup $G = \{\mathbf{GL}(4, \mathbb{R}) \ni A : AX_3 = \lambda_3 X_3, AX_4 = \lambda_4 X_4\}$, preserving $\boldsymbol{\mathcal{D}}$ and its split $\boldsymbol{\mathcal{D}} = \mathcal{D}_w \oplus \mathcal{D}_g$. It is more convenient to think about a G-structure dually: it is a G-subbundle of the bundle $F^*(M)$ of $\mathbf{GL}(4, \mathbb{R})$-coframes of M. The requirement that the G-structure is given by the car structure $(M, \boldsymbol{\mathcal{D}} = \mathcal{D}_w \oplus \mathcal{D}_g)$ is reflected in the G transformation of coframes as follows. We first consider the coframe $(\omega^1, \omega^2, \omega^3, \omega^4)$ dual to the car frame (X_1, X_2, X_3, X_4) on M given in (1.1), (1.3). We have:

$$
\begin{aligned}
\omega^1 &= \ell^{-1}(\cos\alpha \mathrm{d}y - \sin\alpha \mathrm{d}x) \\
\omega^2 &= -\cos\beta \mathrm{d}\alpha - \ell^{-1}\sin\beta\Big(\cos\alpha \mathrm{d}x + \sin\alpha \mathrm{d}y\Big) \\
\omega^3 &= \mathrm{d}\beta \\
\omega^4 &= -\sin\beta \mathrm{d}\alpha + \ell^{-1}\cos\beta\Big(\cos\alpha \mathrm{d}x + \sin\alpha \mathrm{d}y\Big),
\end{aligned}
\tag{3.11}
$$

and $X_i \lrcorner \omega^j = \delta_i{}^j$. Now, the coframe (ω^i), $i = 1, 2, 3, 4$, is given by the geometry of the car up to the transformation

$$
\omega^i \to \bar{\omega}^i = A^i{}_j \omega^j, \tag{3.12}
$$

with

$$
A = (A^i{}_j) = \begin{pmatrix} t_1 & t_2 & 0 & 0 \\ t_3 & t_4 & 0 & 0 \\ t_5 & t_6 & t_7 & 0 \\ t_8 & t_9 & 0 & t_{10} \end{pmatrix} \qquad \text{with} \qquad t_B \in \mathcal{F}(M), \quad \text{and} \quad \det A \neq 0. \tag{3.13}
$$

The G-structure group G of the car structure is therefore

$$G = \{A \in M_{4\times4}(\mathbb{R}) \ : \ A = \begin{pmatrix} t_1 & t_2 & 0 & 0 \\ t_3 & t_4 & 0 & 0 \\ t_5 & t_6 & t_7 & 0 \\ t_8 & t_9 & 0 & t_{10} \end{pmatrix} \text{ with } t_B \in \mathbb{R}, \text{ and } \det A \neq 0\}.$$

We now use transformations (3.12)–(3.13) to bring the coframe forms (3.11) to a form which is convenient to see a 3rd order ODE related to the car's geometry.

Taking

$$A_1 = \begin{pmatrix} \ell \sec\alpha & 0 & 0 & 0 \\ 0 & 1 & 0 & 0 \\ 0 & 0 & 1 & 0 \\ 0 & 0 & 0 & 1 \end{pmatrix} \tag{3.14}$$

we bring ω^1 into the form

$$\omega^1 = \mathrm{d}y - \tan\alpha \mathrm{d}x. \tag{3.15}$$

Now we observe that

$$\omega^2 = -\cos\beta\cos^2\alpha\Big(\mathrm{d}\tan\alpha + \ell^{-1}\tan\beta\sec^3\alpha\mathrm{d}x\Big) - \ell^{-1}\sin\beta\sin\alpha\omega^1,$$

where we have used the new ω^1 given by (3.15). This means that by taking

$$A_2 = \begin{pmatrix} 1 & 0 & 0 & 0 \\ -\ell^{-1}\tan\beta\tan\alpha\sec\alpha & -\sec\beta\sec^2\alpha & 0 & 0 \\ 0 & 0 & 1 & 0 \\ 0 & 0 & 0 & 1 \end{pmatrix} \tag{3.16}$$

we can bring the coframe 1-form ω^2 into the form

$$\omega^2 = \mathrm{d}\tan\alpha + \ell^{-1}\tan\beta\sec^3\alpha\mathrm{d}x. \tag{3.17}$$

We further observe that

$$\omega^3 = -\ell^{-1}\sec^3\alpha\sec^2\beta\Big(-\mathrm{d}(\ell^{-1}\sec^3\alpha\tan\beta) - 3\ell^{-2}\sec^5\alpha\sin\alpha\tan^2\beta\mathrm{d}x\Big) - \tfrac{3}{4}\sin 2\alpha\sin 2\beta\omega^2,$$

where we have used the new ω^2 given by (3.17). This means that by means of the matrix

$$A_3 = \begin{pmatrix} 1 & 0 & 0 & 0 \\ 0 & 1 & 0 & 0 \\ 0 & -3\ell^{-1} \sec\alpha \tan\alpha \tan\beta & -\ell^{-1} \sec^3\alpha \sec^2\beta & 0 \\ 0 & 0 & 0 & 1 \end{pmatrix} \tag{3.18}$$

we can bring the 1-form ω^3 into the form

$$\omega^3 = -\mathrm{d}\big(\ell^{-1} \sec^3\alpha \tan\beta\big) - 3\ell^{-2} \sec^5\alpha \sin\alpha \tan^2\beta \mathrm{d}x. \tag{3.19}$$

Finally, we also see that

$$\omega^4 = \ell^{-1} \sec\alpha \sec\beta \mathrm{d}x - \cos^2\alpha \sin\beta\omega^2 + \ell^{-1} \cos\beta \sin\alpha\omega^1,$$

with ω^1 and ω^2 as in (3.15), (3.17), which shows that the matrix

$$A_4 = \begin{pmatrix} 1 & 0 & 0 & 0 \\ 0 & 1 & 0 & 0 \\ 0 & 0 & 1 & 0 \\ -\frac{1}{2}\cos^2\beta \sin 2\alpha & \frac{1}{2}\ell \cos^3\alpha \sin 2\beta & 0 & \ell \cos\alpha \cos\beta \end{pmatrix} \tag{3.20}$$

brings the form ω^4 into

$$\omega^4 = \mathrm{d}x. \tag{3.21}$$

Summarizing what we have obtained so far we note that by a linear transformation

$$A = A_4 A_3 A_2 A_1,$$

with A_i as in (3.14), (3.16), (3.18), (3.20), which is of the form of (3.13), we can bring the car coframe (3.11) to the G-equivalent coframe

$$\begin{aligned} \omega^1 &= \mathrm{d}y - \tan\alpha \mathrm{d}x \\ \omega^2 &= \mathrm{d}\tan\alpha + \ell^{-1} \tan\beta \sec^3\alpha \mathrm{d}x \\ \omega^3 &= -\mathrm{d}\big(\ell^{-1} \sec^3\alpha \tan\beta\big) - 3\ell^{-2} \sec^5\alpha \sin\alpha \tan^2\beta \mathrm{d}x \\ \omega^4 &= \mathrm{d}x. \end{aligned} \tag{3.22}$$

Now we introduce the new coordinates (x, y, p, q) on M related to the coordinates (x, y, α, β) via

$$p = \tan\alpha, \qquad q = -\ell^{-1} \tan\beta \sec^3\alpha.$$

In these new coordinates the coframe 1-forms (3.22) read:

$$\omega^1 = \mathrm{d}y - p\mathrm{d}x$$
$$\omega^2 = \mathrm{d}p - q\mathrm{d}x$$
$$\omega^3 = \mathrm{d}q - F(x, y, p, q)\mathrm{d}x$$
$$\omega^4 = \mathrm{d}x,$$

with

$$F = 3\ell^{-2}\sec^5\alpha\sin\alpha\tan^2\beta = \frac{3pq^2}{1+p^2}.$$

Thus the car structure can equivalently be described in terms of coordinates (x, y, p, q) with the adapted coframe 1-forms

$$\begin{aligned}\omega^1 &= \mathrm{d}y - p\mathrm{d}x\\ \omega^2 &= \mathrm{d}p - q\mathrm{d}x\\ \omega^3 &= \mathrm{d}q - \frac{3pq^2}{1+p^2}\mathrm{d}x\\ \omega^4 &= \mathrm{d}x.\end{aligned} \tag{3.23}$$

The car velocity distribution

$$\mathcal{D} = \mathrm{Span}_{\mathcal{F}(M)}(X_3, X_4)$$

is in these coordinates spanned by the vector fields

$$X_3 = \partial_q \quad \text{and} \quad X_4 = \partial_x + p\partial_y + q\partial_p + \frac{3pq^2}{1+p^2}\partial_q. \tag{3.24}$$

They form a part of a frame (X_1, X_2, X_3, X_4) dual to $(\omega^1, \omega^2, \omega^3, \omega^4)$ given by (3.23). The 'steering wheel'-'gas' split,

$$\mathcal{D} = \mathcal{D}_w \oplus \mathcal{D}_g,$$

is given by

$$\mathcal{D}_w = \mathrm{Span}_{\mathcal{F}(M)}(X_3) \quad \text{and} \quad \mathcal{D}_g = \mathrm{Span}_{\mathcal{F}(M)}(X_4).$$

Since the coframe 1-forms (3.23) are just the standard *contact forms on the bundle of second jets* $\mathcal{J}^2$ with the standard jet coordinates $(x, y, p = y', q = y'')$ as in (3.7)–(3.8), we recognize here *the third ODE* $y''' = F(x, y, y', y'')$, with $F = \frac{3pq^2}{1+p^2}$. The possible transformations (3.12)–(3.13) of these forms, are equivalent to the *contact transformations* of variables for this equation (see [17], Sec. 4). Thus, the geometry of the car structure $(M, \mathcal{D} = \mathcal{D}_w \oplus \mathcal{D}_g)$ is locally diffeomorphically equivalent to the local differential geometry of the 3rd order ODE

$$y''' = \frac{3y'y''^2}{1+y'^2} \tag{3.25}$$

considered modulo contact transformation of variables.

What is this equation? This is the equation whose graphs of *general* solutions $(x, y(x))$ describe all circles on the plane (x, y). Indeed one can easilly check that the general solution to (3.25) is given by

$$\nu(x^2 + y^2) - 2\xi x - 2\eta y + \mu = 0,$$

where ν, ξ, η, μ are real constants. Since this formula is projective, the space of solutions $\mathbf{Q}$ is 3-dimensional. Taking $\nu = 1$ we get the space $\mathbf{Q}_c$ of all circles on the plane (with radius $R = \sqrt{\eta^2 + \xi^2 - \mu}$, centered at $x = a = \xi$ and $y = b = \eta$), taking $\nu = 0$ we get the space $\mathbf{Q}_\ell$ of all lines in the plane, and taking $\nu = 1$ and $\eta^2 + \xi^2 = \mu$ we get the space $\mathbf{Q}_p$ of all points in the plane.

What is the relation of the circles $(x^2 + y^2) - 2\xi x - 2\eta y + \mu = 0$ *and the lines* $2\xi x + 2\eta y - \mu = 0$ *to the car movement?* By construction, the vector field X_4 in (3.24) differs from the vector field X_4 in (1.1) by rescaling. Thus, *modulo a reparametrization*, both of them have *the same* integral curves in M. We know that the curves defined by X_4 from (1.1) are *helixes* ($\beta_0 \neq 0$) or *straight lines* ($\beta_0 = 0$), which when projected on the (x, y) plane, are *circles* or *straight lines* there. Likewise the integral curves of X_4 from (3.24) are helixes or straight lines which project to the circles or straight lines in the (x, y) plane. To see this one considers a curve $q_4(t) = (x(t), y(t), p(t), q(t))$ in M such that $\dot{q}_4$ is tangent to X_4 from (3.24). It satisfies the system of ODEs $(\dot{x}, \dot{y}, \dot{p}, \dot{q}) = (1, p, q, \frac{3pq^2}{1+p^2})$. This means that $x = t$, $p = \dot{y}$, $q = \dot{p} = \ddot{y}$, and finally $\dot{q} = \dddot{y} = \frac{3\dot{y}\ddot{y}^2}{1+\dot{y}^2}$. Thus, the graphs of solutions $y = y(t)$ of the last equation in the plane $(x = t, y)$, which are *circles* or *straight lines*, are just the circles or straight lines which the rear wheels of the car are performing in the plane (x, y) when the driver of a car applies a primitive 'gas control' only.

3.4 Contact Projective Geometry on P

We now pass to analyse the geometry of P, i.e. the base of the fibration $M \to P$, whose fibers are the steering wheel trajectories generated by the steering wheel vector field X_3 on the car's configuration space M. So what is the geometry on P?

To answer this question let us start with the interpretation of the configuration space M of the car as the second jet space for the car's ODE $y''' = \frac{3y'y''^2}{1+y'^2}$. In this interpretation, the unparametrized integral curves of $X_3 = \partial_\beta$ are the same as the unparametrized integral curves of $X_3 = \partial_q$, and they constitute natural fibres of the fibration $\pi : M = \mathcal{J}^2 \to P = \mathcal{J}^1$ of the second jet space $\mathcal{J}^2$ with coordinates (x, y, p, q) over the first jet space $\mathcal{J}^1$ with coordinates (x, y, p). Consider now the trajectories of X_4, which in the second jet interpretation of M, are just curves $(x, y, p, q) = (x, y(x), y'(x), y''(x))$ in $\mathcal{J}^2$ corresponding to *solutions* $y = y(x)$ of the ODE (3.25). There is a natural projection $\pi((x, y(x), y'(x), y''(x))) = (x, y(x), y'(x))$ of these curves to the 3-dimensional space $\mathcal{J}^1$ of the first jets. The important observation is that these projected curves $(x, y, p) = (x, y(x), y'(x))$ in $\mathcal{J}^1$, as curves corresponding to the solutions of (3.25), are *always tangent to the contact distribution* $\mathcal{C} = \{X \in \Gamma(\mathrm{T}\mathcal{J}^1) \ : \ X \lrcorner (\mathrm{d}y - p\mathrm{d}x) = 0\}$, which is a *natural* structure on $\mathcal{J}^1$. Moreover, since we have a solution to (3.25) for every choice of initial conditions $y(x_0) = y_0$, $y'(x_0) = p_0$, then at every point (x_0, y_0, p_0) in $\mathcal{J}^1$ the projection $\pi((x, y(x), y'(x), y''(x)))$ defines a curve tangent to $\mathcal{C}$ in *every* direction of $\mathcal{C}$. It follows that the projections $\pi((x, y(x), y'(x), y''(x)))$ of solution curves from $\mathcal{J}^2$ to $\mathcal{J}^1$ can be considered as *geodesics of* a certain class of *torsion free connections* on $P = \mathcal{J}^1$.

Indeed, consider a curve $\gamma(t) = x(t)\partial_x + y(t)\partial_y + p(t)\partial_p$ tangent to a distribution $\mathcal{C}$ in $\mathcal{J}^1$, and a frame (Z_1, Z_2, Z_3) in $\mathcal{J}^1$ with

$$Z_1 = \partial_y, \quad Z_2 = \partial_x + \partial_y, \quad Z_3 = \partial_p.$$

Since $\mathcal{C} = \mathrm{Span}_{\mathcal{F}(\mathcal{J}^1)}(Z_2, Z_3)$, the velocity of this curve,

$$\dot{\gamma} = \dot{x}\partial_x + \dot{y}\partial_y + \dot{p}\partial_p = \dot{\gamma}^1 Z_1 + \dot{\gamma}^2 Z_2 + \dot{\gamma}^3 Z_3 = (\dot{y} - p\dot{x})\partial_y + \dot{x} Z_2 + \dot{p} Z_3,$$

has the following components in the frame (Z_1, Z_2, Z_3):

$$\dot{\gamma}^1 = \dot{y} - p\dot{x} = 0, \quad \dot{\gamma}^2 = \dot{x}, \quad \dot{\gamma}^3 = \dot{p}. \tag{3.26}$$

If the curve $\gamma(t)$ is a geodesic of a torsion free connection, there should exist functional coefficients $\Gamma^i{}_{jk} = \Gamma^i{}_{kj}$—the connection coefficients in the frame (Z_1, Z_2, Z_3)—such that

$$\dot{\gamma}^i + \Gamma^i{}_{jk}\gamma^j\gamma^k = 0.$$

Thus, to interprete $\gamma(t)$ as a geodesic it is enough to find $\Gamma^i{}_{jk} = \Gamma^i{}_{kj}$ such that

$$\ddot{x}+\Gamma^2{}_{22}\dot{x}^2+2\Gamma^2{}_{23}\dot{x}\dot{p}+\Gamma^2{}_{33}\dot{p}^2=0 \qquad \& \qquad \ddot{p}+\Gamma^3{}_{22}\dot{x}^2+2\Gamma^3{}_{23}\dot{x}\dot{p}+\Gamma^3{}_{33}\dot{p}^2=0. \tag{3.27}$$

For this we eliminate t from both of these equations, by parametrizing $y = y(t)$ and $p = p(t)$ by x. Because of the first equation in (3.26) we have

$$p = \frac{\dot{y}}{\dot{x}} = \frac{\mathrm{d}y}{\mathrm{d}x} = y', \quad \dot{p} = \dot{x}y'', \quad \ddot{p} = \ddot{x}y'' + \dot{x}^2 y''',$$

and the last two of these equations compared with the second equation in (3.27) shows that

$$-(\Gamma^3{}_{22}\dot{x}^2+2\Gamma^3{}_{23}\dot{x}^2y''+\Gamma^3{}_{33}\dot{x}^2y''^2) = -y''(\Gamma^2{}_{22}\dot{x}^2+2\Gamma^2{}_{23}\dot{x}^2y''+\Gamma^2{}_{33}\dot{x}^2y''^2)+\dot{x}^2y'''.$$

Simplifying, we get:

$$y''' = \Gamma^2{}_{33}y''^3 + (2\Gamma^2_{23} - \Gamma^3{}_{33})y''^2 + (\Gamma^2{}_{22} - 2\Gamma^3{}_{23})y'' - \Gamma^3{}_{22},$$

where $\Gamma^i{}_{jk}$ are functions of x, y and $p = y'(x)$ only.

Thus, for an equation $y''' = F(x, y, y', y'')$ to define on the space of first jets $\mathcal{J}^1$ a structure of a contact manifold with geodesics passing through every point in every direction and such that they are tangent to the contact distribution, it is neccessary that the function $F = F(x, y, y', y'')$ is a polynomial of at most 3rd order in the variable y''. It follows that this condition for F is also sufficient for getting such a structure on $\mathcal{J}^1$.

Since the car's structure equation $y''' = \frac{3y'y''^2}{1+y'^2}$ depends on y'' quadratically, this implies that its 3-dimensional space P i.e. its space of first jets $\mathcal{J}^1$ is naturally equipped with the structure as in the following definition [12].

Definition 3.1 A *contact projective structure* on the first jet space $\mathcal{J}^1$ is given by the following data.

- The contact distribution $\mathcal{C}$, that is the distribution annihilated by $\omega^1 = dy - pdx$.
- A family of unparameterized curves everywhere tangent to $\mathcal{C}$ and such that:
  - for a given point and a direction in $\mathcal{C}$ there is exactly one curve passing through that point and tangent to that direction,
  - curves of the family are among unparameterized geodesics for some linear connection on $\mathcal{J}^1$.

To make the statement above the definition more explicit, we argue as follows:

We have the fibration $M \to P$, which on the one hand is a fibration of the second jet space $M = \mathcal{J}^2$ of a contact equivalence class of ODEs $y''' = \frac{3y'y''^2}{1+y'^2}$ over the space $P = \mathcal{J}^1$, and on the other hand the car fibration $M \to P$ of the configuration space M of a car and the space P of the possible movements of the car modulo the moves of a steering wheel. As we explained in Sect. 3.3 there is a natural bundle isomorphism between the car configuration space and the space of second jets of the contact equivalence classes of ODEs represented by $y''' = \frac{3y'y''^2}{1+y'^2}$, making an equivalence between the car's Engel geometry with a split and the contact geometry of this ODE. Since the ODE $y''' = \frac{3y'y''^2}{1+y'^2}$ has only quadratic dependence on y'' it belongs to the class of ODEs $y''' = A_3 y''^3 + A_2 y''^2 + A_1 y'' + A_0$. Thus the car's ODE first jet space $\mathcal{J}^1$ has a natural *contact projective structure*. This, via the bundle isomorphism $\mathcal{J}^2 \to P$, induced by the isomorphism between geometries on the car's configuration space and the bundle of the second jets, shows that the *car's 3-dimensional space P* of leaves generated by $X_3 = \partial_\beta$ has a natural *contact projective structure*. Such structures were in particular studied in [12–14]. Since the car's ODE $y''' = \frac{3y'y''^2}{1+y'^2}$ is contact equivalent to $y''' = 0$, it follows from these studies that this contact projective structure is *Cartan flat*. More precisely we have the following theorem:

Theorem 3.2 *The car's Engel structure with a split $(M, \mathcal{D} = \mathcal{D}_w \oplus \mathcal{D}_g)$ induces a natural contact projective structure on the car's space P of all possible positions of a car considered modulo orientation of the front wheels. This contact projective structure has a 10-dimensional Lie algebra of symmetries, which is isomorphic to the simple Lie algebra $\mathfrak{sp}(2, \mathbb{R})$. It is flat in the sense of having vanishing curvature of the natural normal $\mathfrak{sp}(2, \mathbb{R})$-valued Cartan connection uniquely defined by this contact projective structure.*

Since $\mathfrak{sp}(2, \mathbb{R})$ is isomorphic to $\mathfrak{so}(2, 3)$ (see Sect. 4.1.1) we again have an indication why the geometry of car's configuration space M has $\mathfrak{so}(2, 3)$ as its local symmetry.

3.5 Chern's Double Fibration $Q \leftarrow M \to P$, the Geometries on Q and P and a Problem About a Car on a Curved Terrain

If somebody inspired by this article would like to *curve* the geometry of a car, she will find useful the following information about the geometry of general third order ODEs, $y''' = F(x, y, y', y'')$, considered modulo contact transformations.

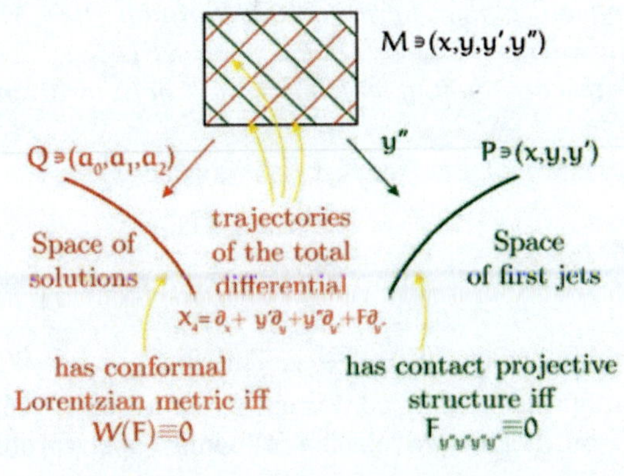

As we mentioned in Sect. 3.3 Chern in 1940 noticed the above double fibration $Q \leftarrow M \rightarrow P$ for any contact equivalence class of third order ODEs. If the class is defined by the equation $y''' = F(x, y, y', y'')$, and if the general solution of the defining the equation is written as $y = y(x, a_1, a_2, a_3)$, where a_1, a_2, a_3 are the three constants of integration, then the base space Q is the leaf space of the total differential $X_4 = \partial_x + y'\partial_y + y''\partial_{y'} + F\partial_{y''}$, which is parameterized by (a_1, a_2, a_3), and the base space P is the space of first jets—the leaf space of the integral curves of the vector field $X_3 = \partial_{y''}$, which is parameterized by (x, y, y'). We emphasize that this double fibration exists for any choice of the function F, and in turn is associated with *any* contact equivalence class of 3rd order ODEs. However, and this is the main observation of S.S. Chern in [7], the space of solutions Q has a natural conformal Lorentzian geometry on it, and/or the first jet space P has a natural contact projective structure on it, if and only if the function F satisfies certain conditions, which are invariant with respect to contact change of the variables of the equation.

We have the following theorem [7, 13, 14].

Theorem 3.3 *The space Q in Chern's double fibration $Q \leftarrow M \rightarrow P$ associated with a contact equivalence class of ODEs $y''' = F(x, y, y', y'')$ has a natural conformal Lorentzian structure on it, if and only if the Wünschmann invariant*

$$W[F] = 9\, X_4(X_4(X_3(F)) - 27\, X_4(F_{y'}) - 18\, X_3(F)\, X_4(X_3(F)) + 18\, X_3(F)\, F_{y'} + 4\, X_3(F)^3 + 54\, F_y$$

identically vanishes for F.

Similarly, the space P in the Chern's double fibration $Q \leftarrow M \rightarrow P$ associated with a contact equivalence class of ODEs $y''' = F(x, y, y', y'')$ has a natural contact projective structure on it, if and only if the Chern invariant

$$C[F] = X_3(X_3(X_3(X_3(F))))$$

identically vanishes for F.

Here, $X_3 = \partial_{y''}$, $X_4 = \partial_x + y'\partial_y + y''\partial_{y'} + F\partial_{y''}$, $F_y = \frac{\partial F}{\partial y}$ and $F_{y'} = \frac{\partial F}{\partial y'}$.

In the car's fibration we have $F = \frac{3y'y''^2}{1+y'^2}$. This function has $W[F] \equiv C[F] \equiv 0$. Thus the car fibration has a (flat) conformal structure on Q and a (flat) contact projective structure on P. This provokes the following (open) problem.

Problem *Generalize the car setting enabling the car to move on a curved terrain. This should lead to a nonflat Engel structure with a split $(M, \mathcal{D} = \mathcal{D}_w \oplus \mathcal{D}_g)$ on the car's configuration space. Characterize, in terms of Chern's invariants $W[F]$, $C[F]$, and possibly their derivatives, those Engel structures with a split, which are configuration space structures of cars on curved terrains. Which of the two geometries: the conformal Lorentzian one, or the contact projective one will survive for a car on a general terrain? Perhaps none?*

4 Lie's Correspondence

4.1 Lagrangian Planes in $\mathbb{R}^4$ and Oriented Circles in the Plane

It was S. Lie who understood the geometry of the projective quadric $\mathbf{Q}$, as in (3.4), in terms of the geometry of Lagrangian planes in a real 4-dimensional vector space. (see [2, 16] for more details). To talk about *Lagrangian planes* we need to have a *real 4-dimensional vector space* V and a *symplectic form* in V, i.e. a 2-form $\omega \in \bigwedge^2 V^*$ such that $\omega \wedge \omega \neq 0$. Now, a 2-plane $q = \mathrm{Span}(Y_1, Y_2)$, with $Y_1, Y_2 \in V$ and $Y_1 \wedge Y_2 \neq 0$, is Lagrangian in V if and only if $\omega(Y_1, Y_2) = 0$.

Given a symplectic form ω in V we consider the 5-dimensional vector space $\omega^\perp \subset \bigwedge^2 V$ consisting of elements $Y \in \bigwedge^2 V$ annihilating ω:

$$\omega^\perp = \{\textstyle\bigwedge^2 V \ni Y \;:\; Y \lrcorner\, \omega = 0\}.$$

It is now convenient to introduce a basis (e_1, e_2, e_3, e_4) in V, such that the symplectic form ω reads as

$$\omega = e^1 \wedge e^4 + e^2 \wedge e^3, \tag{4.1}$$

in its dual cobasis (e^1, e^2, e^3, e^4), $e_i \lrcorner e^j = \delta_i{}^j$, in V^*. Then the most general element $Y \in \omega^\perp$ is:

$$Y = (\eta+\zeta)\, e_1 \wedge e_2 + \mu\, e_1 \wedge e_3 + \nu\, e_4 \wedge e_2 + (\eta-\zeta)\, e_4 \wedge e_3 + \xi\, (e_1 \wedge e_4 - e_2 \wedge e_3), \tag{4.2}$$

where $(\xi, \eta, \zeta, \mu, \nu) \in \mathbb{R}^5$.

We now ask the question as to when such Y is a *simple* bivector. Recall that an element $0 \neq Y$ of $\bigwedge^2 V$ is *simple* if and only if $Y \wedge Y = 0$. In such case there exist vectors Y_1 and Y_2 in V such that $Y = Y_1 \wedge Y_2$. Thus such Y defines a 2-plane

$$q = \mathrm{Span}_{\mathbb{R}}(Y_1, Y_2),$$

in V. If in addition, a *simple* Y belongs to the 5-dimensional subspace $\omega^\perp$, then its *direction*,

$$\mathrm{dir}(Y) := \{\lambda Y,\ \lambda \in \mathbb{R}\},$$

defines a 2-plane which is *Lagrangian*. It turns out that *every* Lagrangian 2-plane in V is defined in terms of $0 \neq Y \in \omega^\perp$ such that $Y \wedge Y = 0$.

Simple algebra applied to a generic $Y \in \omega^\perp$ as in (4.2) gives:

$$Y \wedge Y = 2(\zeta^2 - \eta^2 + \mu\nu - \xi^2) e_1 \wedge e_2 \wedge e_3 \wedge e_4 = -\tfrac{1}{2} Q(\xi, \eta, \zeta, \mu, \nu) e_1 \wedge e_2 \wedge e_3 \wedge e_4. \tag{4.3}$$

Note the appearence of the quadratic form (3.5) in this formula! Thus such an Y is simple, $Y \wedge Y = 0$, if and only if the quintuple $[\xi : \eta : \zeta : \mu : \nu]$ belongs to the projective quadric $\mathbf{Q}$ considered in Sect. 3.2. Now, let us define

$$\mathbf{Q}' = \{P(\textstyle\bigwedge^2 V) \ni \mathrm{dir}(Y)\ :\ Y \lrcorner \omega = 0 \quad \& \quad Y \wedge Y = 0\},$$

where, as it is customary, we denoted the *projectivization* of $\bigwedge^2 V$ by $P(\bigwedge^2 V)$.

Since $Y \wedge Y = 0$ for $Y \in \omega^\perp$ is equivalent to $\zeta^2 - \eta^2 + \mu\nu - \xi^2 = 0$ for $[\xi : \eta : \zeta : \mu : \nu] \in \mathbb{R}P^4$, then

$$\mathbf{Q}' = \{\mathrm{dir}(Y)\ :\ Y \text{ as in (4.2) with } [\xi : \eta : \zeta : \mu : \nu] \in \mathbf{Q}\}.$$

This in turn establishes a diffeomorphism between $\mathbf{Q}$ and the space of all Lagrangian 2-planes in V. With some abuse of notation we will denote this space also by $\mathbf{Q}'$,

$$\mathbf{Q}' = \{\text{set of all Lagrangian } 2-\text{planes in } (V, \omega)\}.$$

Let us now parametrize those $\mathrm{dir}(Y)$ in $\mathbf{Q}'$ that correspond to all circles with a *finite* radius in the plane. Since such circles are points of the set $\mathbf{Q}_c \subset \mathbf{Q}$, with

$\nu \neq 0$, we can conveniently parametrize them by $\nu = 1$, $\mu = \xi^2 + \eta^2 - \zeta^2$. Thus, the corresponding bivectors $\mathrm{dir}(Y)$ in $\mathbf{Q}'$ may be represented by

$$Y = (\eta+\zeta)\, e_1\wedge e_2+(\xi^2+\eta^2-\zeta^2)\, e_1\wedge e_3+e_4\wedge e_2+(\eta-\zeta)\, e_4\wedge e_3+\xi\, (e_1\wedge e_4-e_2\wedge e_3), \tag{4.4}$$

or what is the same by $Y = \Big((\eta+\zeta)\, e_1 + e_4 + \xi\, e_3 \Big) \wedge \Big(-\xi\, e_1 + e_2 + (\eta-\zeta)\, e_3 \Big)$. Thus in the 3-dimensional space $\mathbf{Q}'$ there is an open set $\mathbf{Q}'_c$ of bivectors Y given by (4.4). This set, in turn, is diffeomorphic to the space of all Lagrangian 2-planes

$$q(\xi,\eta,\zeta) = \mathrm{Span}_{\mathbb{R}}(Y_1, Y_2), \tag{4.5}$$

spanned by

$$Y_1 = (\eta+\zeta)\, e_1 + e_4 + \xi\, e_3 \quad \& \quad Y_2 = -\xi\, e_1 + e_2 + (\eta-\zeta)\, e_3. \tag{4.6}$$

Again, with some abuse, we denote this space by $\mathbf{Q}'_c$.

In $\mathbf{Q}_c$ we had a nice interpretation of the incidence between two points (circles): two close circles in $\mathbf{Q}_c$ were incident if they were tangent to each other. The natural *incidence between the points of* $\mathbf{Q}'_c$, i.e. between two close Lagrangian 2-planes in V, *is their intersection along a line*. Let us see what such an incidence means:

If we take a Lagrangian 2-plane $q(\xi,\eta,\zeta)$ and its close neighbour $q(\xi+\mathrm{d}\xi, \eta+\mathrm{d}\eta, \zeta+\mathrm{d}\zeta)$, then they intersect in a line iff their corresponding bivectors

$$Y = \Big((\eta+\zeta)\, e_1 + e_4 + \xi\, e_3 \Big) \wedge \Big(-\xi\, e_1 + e_2 + (\eta-\zeta)\, e_3 \Big)$$

and

$$Y+\mathrm{d}Y = \Big((\eta+\zeta+\mathrm{d}\eta+\mathrm{d}\zeta)\, e_1+e_4+(\xi+\mathrm{d}\xi)\, e_3 \Big)\wedge\Big(-(\xi+\mathrm{d}\xi)\, e_1+ e_2+(\eta-\zeta+\mathrm{d}\eta-\mathrm{d}\zeta)\, e_3\Big)$$

$$Y \wedge (Y+\mathrm{d}Y) = 0.$$

A short algebra shows that

$$Y \wedge (Y+\mathrm{d}Y) = \big((\mathrm{d}\eta)^2 + (\mathrm{d}\xi)^2 - (\mathrm{d}\zeta)^2\big) e_1 \wedge e_2 \wedge e_3 \wedge e_4.$$

Hence the two Lagrangian planes from $\mathbf{Q}'_c$ intersect in a line if and only if the connecting vector $(\mathrm{d}\xi, \mathrm{d}\eta, \mathrm{d}\zeta)$ between the points (ξ,η,ζ) and $(\xi+\mathrm{d}\xi, \eta+\mathrm{d}\eta, \zeta+\mathrm{d}\zeta)$ in $\mathbf{Q}'_c$ is null in the 3-dimensional Minkowski metric $g = (\mathrm{d}\eta)^2+(\mathrm{d}\xi)^2-(\mathrm{d}\zeta)^2$. Comparing with (3.3) we see that in the present parametrization of $\mathbf{Q}_c$, we have

$$\xi = a, \qquad \eta = b \quad \text{and} \quad \zeta = R.$$

Hence $g = (\mathrm{d}a)^2 + (\mathrm{d}b)^2 - (\mathrm{d}R)^2$, and the condition that two neighbouring Lagrangian planes from $\mathbf{Q}'_c$ intersect in a line in V is then equivalent to the condition that the corresponding neighbouring circles from $\mathbf{Q}_c$ are kissing each other in the plane (x, y). This is the essence of *Lie's observation* :

Tangent circles in $\mathbb{R}^2$ with orientations as in moving gears correspond to Lagrangian planes in $\mathbb{R}^4$ intersecting in a line.

4.1.1 Double Cover of SO(2, 3) by Sp(2, $\mathbb{R}$)

It was Lie who established the isomorphism between the simple Lie algebras $\mathfrak{so}(2,3)$ and $\mathfrak{sp}(2,\mathbb{R})$. This is, for example, very nicely explained in [2]. Here we argue for this as follows:

The symplectic group $\mathbf{Sp}(2,\mathbb{R})$ is defined as

$$\mathbf{Sp}(2,\mathbb{R}) = \{\mathbf{GL}(V) \ni A \mid \omega(Av, Aw) = \omega(v, w),\ v, w \in V\},$$

where as before V is a real 4-dimensional vector space, and ω is a symplectic form on V. Note that $\mathbb{Z}_2 = \{I, -I\}$, where I is the identity in $\mathbf{GL}(V)$, is a subgroup of $\mathbf{Sp}(2,\mathbb{R})$, $\mathbb{Z}_2 \subset \mathbf{Sp}(2,\mathbb{R})$.

Introducing, $A^\mu{}_\nu$ via $A(e_\mu) = A^\nu{}_\mu e_\nu$, and $\omega_{\mu\nu} = \omega(e_\mu, e_\nu)$, we obtain that the matrix elements of those $A \in \mathbf{GL}(V)$ that are in $\mathbf{Sp}(2,\mathbb{R})$ satisfy

$$A^\mu{}_\alpha A^\nu{}_\beta \omega_{\mu\nu} = \omega_{\alpha\beta}. \tag{4.7}$$

Since $\dim V = 4$ we have

$$\tfrac14 \omega_{\mu\nu}\omega_{\rho\sigma} e^\mu \wedge e^\nu \wedge e^\rho \wedge e^\sigma = \omega\wedge\omega = 2e^1\wedge e^2\wedge e^2\wedge e^4 = \tfrac{1}{12}\epsilon_{\mu\nu\rho\sigma} e^\mu \wedge e^\nu \wedge e^\rho \wedge e^\sigma,$$

and hence

$$\omega_{[\mu\nu}\omega_{\rho\sigma]} = \tfrac13 \epsilon_{\mu\nu\rho\sigma}. \tag{4.8}$$

Here $\epsilon_{\mu\nu\rho\sigma}$ denotes the totally skew Levi-Civita symbol in $\mathbb{R}^4$.

Let us now take an element Y from $\omega^\perp$. We have $Y = \frac12 Y^{\mu\nu} e_\mu \wedge e_\nu$. Then, according to (4.3) we have

$$-\tfrac12 Q(Y) e_1\wedge e_2\wedge e_3\wedge e_4 = Y\wedge Y = \tfrac14 Y^{\mu\nu}Y^{\rho\sigma} e_\mu\wedge e_\nu\wedge e_\rho\wedge e_\sigma = \tfrac14 Y^{\mu\nu}Y^{\rho\sigma}\epsilon_{\mu\nu\rho\sigma} e_1\wedge e_2\wedge e_3\wedge e_4,$$

so the quadratic form $Q(Y)$ written in terms of the components $Y^{\mu\nu} = Y^{[\mu\nu]}$ of the bivector Y is

$$Q(Y) = -\tfrac12 Y^{\mu\nu}Y^{\rho\sigma}\epsilon_{\mu\nu\rho\sigma}.$$

There is a natural action of $\mathbf{Sp}(2,\mathbb{R})$ on the space $\omega^\perp$ induced by the action of $\mathbf{Sp}(2,\mathbb{R})$ in V. In components it reads

$$\mathbf{Sp}(2,\mathbb{R})\times\omega^\perp \quad\ni\quad (A, Y^{\mu\nu}) \longrightarrow (AY)^{\mu\nu} = A^{-1\mu}{}_\alpha A^{-1\nu}{}_\beta Y^{\alpha\beta} \quad\in\quad \omega^\perp.$$

If we now apply the form Q on the $\mathbf{Sp}(2,\mathbb{R})$ transformed bivector AY we get

$$\begin{aligned} Q(AY) = &-\tfrac{1}{2}A^{-1\mu}{}_\alpha A^{-1\nu}{}_\beta A^{-1\rho}{}_\gamma A^{-1\sigma}{}_\delta Y^{\alpha\beta}Y^{\gamma\delta}\epsilon_{\mu\nu\rho\sigma} = \\ &-\tfrac{3}{2}A^{-1\mu}{}_\alpha A^{-1\nu}{}_\beta A^{-1\rho}{}_\gamma A^{-1\sigma}{}_\delta Y^{\alpha\beta}Y^{\gamma\delta}\omega_{[\mu\nu}\omega_{\rho\sigma]} = \\ &-\tfrac{3}{2}Y^{\alpha\beta}Y^{\gamma\delta}\omega_{[\mu\nu}\omega_{\rho\sigma]} = -\tfrac{1}{2}Y^{\alpha\beta}Y^{\gamma\delta}\epsilon_{\mu\nu\rho\sigma} = Q(Y), \end{aligned}$$

where the expressions after the second and the fourth equality sign follow from (4.8), and the expression after the third equality sign follows from (4.7). Thus the symplectic transformation $v \mapsto Av$ in V induces a linear transformation $Y \mapsto AY$ in $\omega^\perp$ which *preserves the real quadratic form* Q of signature $(2, 3)$. This gives a homomorphism of $\mathbf{Sp}(2,\mathbb{R})$ onto $\mathbf{SO}(2,3)$. Its kernel is $\mathbb{Z}_2$, since

$$(\mathbf{Sp}(2,\mathbb{R})\supset\mathbb{Z}_2)\times\omega^\perp \quad\ni\quad (A=\pm I, Y^{\mu\nu}) \longrightarrow (\pm\delta^\mu{}_\alpha)(\pm\delta^\nu{}_\beta)Y^{\alpha\beta} = Y^{\mu\nu} \quad\in\quad \omega^\perp.$$

@ This gives the Lie's *double cover* of $\mathbf{SO}(2,3)$ by $\mathbf{Sp}(2,\mathbb{R})$,

$$\mathbb{Z}_2 \to \mathbf{Sp}(2,\mathbb{R}) \to \mathbf{SO}(2,3),$$

which has its local version in the *isomorphism of the Lie algebras* $\mathfrak{sp}(2,\mathbb{R})$ and $\mathfrak{so}(2,3)$.

4.2 *Lie's Twistor Fibration*

The relation between the groups $\mathbf{Sp}(2,\mathbb{R})$ and $\mathbf{SO}(3,2)$ recalled in the previous section is the basis for *Lie's correspondence* [2], Section 3. This can be described in yet another incarnation of the car's fibration $Q \leftarrow M \rightarrow P$, which is *Lie's twistor fibration*; see Section 4.4 in [3] for a general theory of these things.

To explain this we start with the space Q of all Lagrangian planes in V as before. This 3-dimensional space can be locally parameterized by (ξ, η, ζ) as in (4.5)–(4.6), with a Lagrangian plane $\mathrm{dir}(Y)$ spanned by

$$Y_1 = (\eta+\zeta)\,e_1 + e_4 + \xi\,e_3 \quad \& \quad Y_2 = -\xi\,e_1 + e_2 + (\eta-\zeta)\,e_3.$$

There is also another 3-dimensional space associated with V. This is

$$P := P(V) = \{\mathrm{dir}(v) \mid \lambda v,\ v\in V,\ \lambda\in\mathbb{R}\},$$

the projectivization of V. This can be locally parametetrized by (x^1, x^2, x^3), where a generic element of P is $\ell = \text{dir}(x^1 e_1 + x^2 e_2 + x^3 e_3 + e_4)$.

There is a third space, M, associated with our pair (V, ω). This is

$$M = \{P \times Q \ni (\ell, \text{dir}(Y)) \mid \ell \in \text{dir}(Y)\},$$

i.e. the *space of all pairs* (line ℓ, Lagrangian plane associated with Y) passing through zero in V with an incidence relation such that *a line ℓ is in the plane* $\text{dir}(Y)$. This space is *four* dimensional, as a generic such pair can be parametrized by (ξ, η, ζ, s), where (ξ, η, ζ) parametrizes the plane spanned by Y_1 and Y_2, and the parameter s comes from $\ell = \text{dir}(Y_1 + sY_2)$ and specifies a given line from the wealth of lines passing through zero in $\text{dir}(Y)$.

We again have a natural fibration $Q \leftarrow M \rightarrow P$:

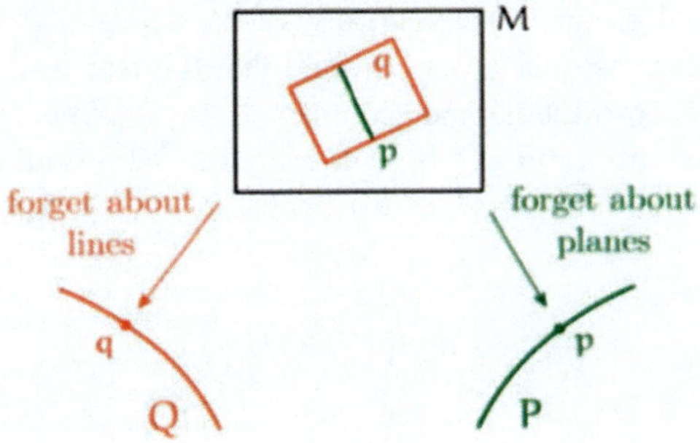

where the map $M \rightarrow Q$ is given by $(\ell, Y) \rightarrow Y$, and the map $M \rightarrow P$ is given by $(\ell, Y) \rightarrow \ell$. It is the *Lie's twistor fibration*.

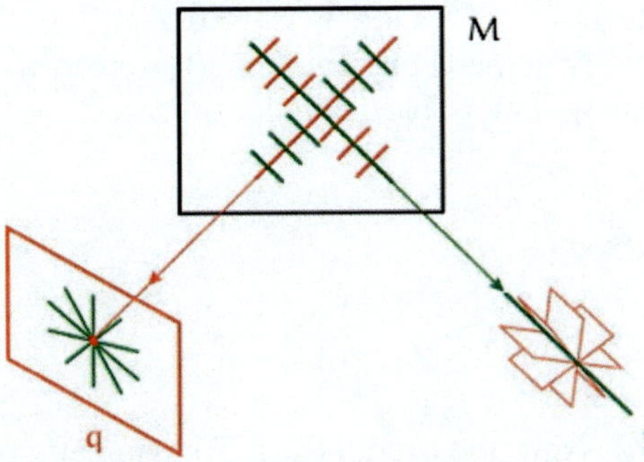

In it the fiber over a point $q \in Q$, i.e. over a Lagrangian plane $\text{dir}(Y)$ in V, consists of all lines ℓ passing through zero in this plane. Therefore the topology of such a fiber is the same as $\mathbb{R}P^1$. Likewise, the fiber over a point $p \in P$, i.e. over a line ℓ passing through zero in V, consists of all Lagrangian planes $\text{dir}(Y)$ containing the line ℓ. Such a fiber also has topology of $\mathbb{R}P^1$.

The group $\mathbf{Sp}(2,\mathbb{R})$ naturally acts on Q and P. These actions are given by

$$(A, \mathrm{dir}(Y)) \quad \to \quad \mathrm{dir}(AY) = \mathrm{dir}(\tfrac{1}{2}\, A^{-1\mu}{}_{\alpha} A^{-1\nu}{}_{\beta} Y^{\alpha\beta}\, e_\mu \wedge e_\nu) \tag{4.9}$$

and

$$(A, \mathrm{dir}(v)) \to \mathrm{dir}(\,Av), \tag{4.10}$$

where $Y = \frac{1}{2} Y^{\mu\nu} e_\mu \wedge e_\nu \in \omega^\perp$, $v \in V$ and $A \in \mathbf{Sp}(2,\mathbb{R})$. It also has an induced action on the elements $(\ell, Y) \in M$, via

$$(A, (\mathrm{dir}(v), \mathrm{dir}(Y)) \to (\mathrm{dir}(Av), \mathrm{dir}(AY)). \tag{4.11}$$

It is a matter of checking that the isotropy of the action (4.9) of $\mathbf{Sp}(2,\mathbb{R})$ on Q is a certain 7-dimensional group P_1, the isotropy of the action (4.10) of $\mathbf{Sp}(2,\mathbb{R})$ on P is also a certain 7-dimensional group P_2, and that the isotropy of the action (4.11) of $\mathbf{Sp}(2,\mathbb{R})$ on M is a 6-dimensional group $P_{12} = P_1 \cap P_2$.

Thus Lie's twistor fibration can be considered to be a double fibration of three $\mathbf{Sp}(2,\mathbb{R})$ *homogeneous spaces*: $M = \mathbf{Sp}(2,\mathbb{R})/P_{12}$, $Q = \mathbf{Sp}(2,\mathbb{R})/P_1$ and $P = \mathbf{Sp}(2,\mathbb{R})/P_2$.

$$\begin{array}{ccc} & M = \mathbf{Sp}(2,\mathbb{R})/P_{12} & \\ \swarrow & & \searrow \\ Q = \mathbf{Sp}(2,\mathbb{R})/P_1 & & P = \mathbf{Sp}(2,\mathbb{R})/P_2\,. \end{array} \tag{4.12}$$

Due to Lie's double cover of $\mathbf{SO}(2,3)$ by $\mathbf{Sp}(2,\mathbb{R})$, and due to the proper dimensions of the spaces in the above fibration, it is clear that this gives a global version of the car's configuration space fibration

$$\begin{array}{ccc} & (M, \mathcal{D} = \mathcal{D}_w \oplus \mathcal{D}_g) & \\ \swarrow & & \searrow \\ Q & & P. \end{array}$$

considered in Sect. 3.1. Now, the overall $\mathbf{Sp}(2,\mathbb{R})$ symmetry of all the ingredients of the fibration is obvious.

4.3 *The Picture in Terms of Parabolic Subgroups in* Sp(2, ℝ)

The double fibration (4.12) is a low dimensional example of the twistor correspondences discussed in [3], Section 4.4.6. The crucial point here is that the subgroups P_1, P_2 and P_{12} considered in the previous section are *parabolic* subgroups of a simple Lie group $\mathbf{Sp}(2,\mathbb{R})$; moreover they are such that P_1 and P_2 contain the same *Borel* subgroup, which happens to be P_{12}. To comment about this we need some preparations.

4.3.1 Car's Gradation in $\mathfrak{sp}(2,\mathbb{R})$

The elements E of the Lie algebra $\mathfrak{sp}(2,\mathbb{R})$ of $\mathbf{Sp}(2,\mathbb{R})$ can be considered as 4×4 real matrices $E=(E^\alpha{}_\beta)$ that preserve the symplectic form $\omega=\frac{1}{2}\omega_{\mu\nu}e^\mu\wedge e^\nu$, i.e.

$$E^\gamma{}_\alpha\omega_{\gamma\beta}+E^\gamma{}_\beta\omega_{\alpha\gamma}=0.$$

With our choice of a basis (e_μ) in V, in which the symplectic form ω is as in (4.1), the matrix E giving the generic element of the Lie algebra $\mathfrak{sp}(2,\mathbb{R})$ is given by

$$E=(E^\alpha{}_\beta)=\begin{pmatrix} a_5 & a_7 & a_9 & 2a_{10}\\ -a_4 & a_6 & a_8 & a_9\\ a_2 & a_3 & -a_6 & -a_7\\ -2a_1 & a_2 & a_4 & -a_5\end{pmatrix},$$

where the coefficients a_I, $I=1,2,\dots 10$, are real constants.

Now, viewing $\mathfrak{sp}(2,\mathbb{R})$ as a Lie algebra consisting of all 4×4 real matrices E as above, with the commutator in $\mathfrak{sp}(2,\mathbb{R})$ being the usual commutator $[E,E']=E\cdot E'-E'\cdot E$ of two matrices E and E', we get a convenient basis (E_I) in $\mathfrak{sp}(2,\mathbb{R})$ by

$$E_I=\frac{\partial E}{\partial a_I},\qquad I=1,2,\dots 10.$$

In this basis, modulo the antisymmetry, we have the following nonvanishing commutators: $[E_1,E_5]=2E_1$, $[E_1,E_7]=-2E_2$, $[E_1,E_9]=-2E_4$, $[E_1,E_{10}]=4E_5$, $[E_2,E_4]=E_1$, $[E_2,E_5]=E_2$, $[E_2,E_6]=E_2$, $[E_2,E_7]=2E_3$, $[E_2,E_8]=E_4$, $[E_2,E_9]=-E_5-E_6$, $[E_2,E_{10}]=-2E_7$, $[E_3,E_4]=-E_2$, $[E_3,E_6]=2E_3$, $[E_3,E_8]=-E_6$, $[E_3,E_9]=-E_7$, $[E_4,E_5]=E_4$, $[E_4,E_6]=-E_4$, $[E_4,E_7]=E_5-E_6$, $[E_4,E_9]=-2E_8$, $[E_4,E_{10}]=-2E_9$, $[E_5,E_7]=E_7$, $[E_5,E_9]=E_9$, $[E_5,E_{10}]=2E_{10}$, $[E_6,E_7]=-E_7$, $[E_6,E_8]=2E_8$, $[E_6,E_9]=E_9$, $[E_7,E_8]=E_9$, $[E_7,E_9]=E_{10}$.

What can be seen from this colorful mess?

First, it is useful to note that our choice of the basis E_I in $\mathfrak{sp}(2,\mathbb{R})$ is related to the following root diagram of the Lie algebra $\mathfrak{sp}(2,\mathbb{R})$:

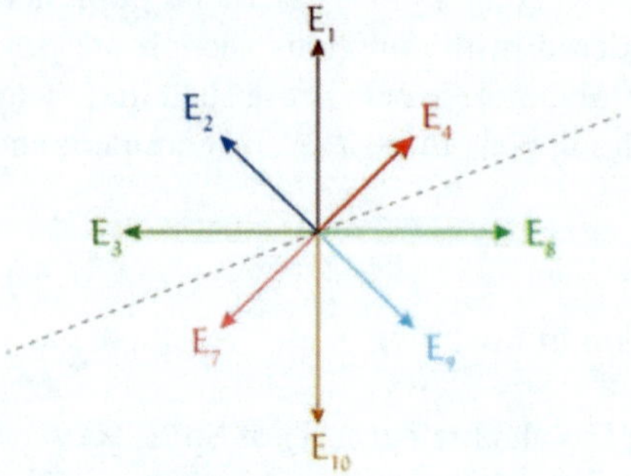

Root diagram for $\mathfrak{sp}(2,\mathbb{R})$

This gives a mnemonic technique on how to get the directions of the vectors representing the commutators: a commutator of two vectors E_I and E_K in $\mathfrak{sp}(2,\mathbb{R})$ either *vanishes* or *is along the direction of* $E_I + E_K$, where the sum is the usual sum of the vectors E_I and E_K in the plane of the diagram. The commutators are nonzero if and only if the sum $E_I + E_K$ of vectors in the diagram belongs to the diagram.

Morever, the commutation realtions above show, in particular, a certain *gradation* in $\mathfrak{sp}(2,\mathbb{R})$. Indeed, define

$$\begin{aligned}
\mathfrak{g}_{-3} &= \mathrm{Span}_{\mathbb{R}}(E_1)\\
\mathfrak{g}_{-2} &= \mathrm{Span}_{\mathbb{R}}(E_2)\\
\mathfrak{g}_{-1} &= \mathrm{Span}_{\mathbb{R}}(E_3, E_4)\\
\mathfrak{g}_{0} &= \mathrm{Span}_{\mathbb{R}}(E_5, E_6)\\
\mathfrak{g}_{1} &= \mathrm{Span}_{\mathbb{R}}(E_7, E_8)\\
\mathfrak{g}_{2} &= \mathrm{Span}_{\mathbb{R}}(E_9)\\
\mathfrak{g}_{3} &= \mathrm{Span}_{\mathbb{R}}(E_{10}),
\end{aligned}$$

and observe that due to the above commutation relations of the basis vectors E_I, these vector subspaces in $\mathfrak{sp}(2,\mathbb{R})$ satisfy

$$[\mathfrak{g}_i, \mathfrak{g}_j] \subset \mathfrak{g}_{i+j},$$

when $|i+j| \leq 3$, or

$$[\mathfrak{g}_i, \mathfrak{g}_j] = \{0\},$$

otherwise. This observation decomposes $\mathfrak{sp}(2,\mathbb{R})$ onto

$$\mathfrak{sp}(2,\mathbb{R}) = \mathfrak{g}_{-3} \oplus \mathfrak{g}_{-2} \oplus \mathfrak{g}_{-1} \oplus \mathfrak{g}_0 \oplus \mathfrak{g}_1 \oplus \mathfrak{g}_2 \oplus \mathfrak{g}_3,$$

and makes it into a *3-step graded Lie algebra*.

We further make a decomposition of $\mathfrak{g}_{-1}$ and $\mathfrak{g}_1$ onto

$$\mathfrak{g}_{-1} = \mathfrak{g}_{-1w} \oplus \mathfrak{g}_{-1g} \quad \text{and} \quad \mathfrak{g}_1 = \mathfrak{g}_{1g} \oplus \mathfrak{g}_{1w}$$

with

$$\mathfrak{g}_{-1w} = \mathrm{Span}_{\mathbb{R}}(E_3),\ \ \mathfrak{g}_{-1g} = \mathrm{Span}_{\mathbb{R}}(E_4),\ \ \mathfrak{g}_{1g} = \mathrm{Span}_{\mathbb{R}}(E_7),\ \ \text{and}\ \ \mathfrak{g}_{1w} = \mathrm{Span}_{\mathbb{R}}(E_8).$$

The commutation relations above show also that the following vector subspaces in $\mathfrak{sp}(2,\mathbb{R})$ are *Lie subalgebras*:

$$\begin{aligned}
\mathfrak{p}_1 &= \mathfrak{g}_{-1w} \oplus \mathfrak{g}_0 \oplus \mathfrak{g}_1 \oplus \mathfrak{g}_2 \oplus \mathfrak{g}_3\\
\mathfrak{p}_2 &= \mathfrak{g}_{-1g} \oplus \mathfrak{g}_0 \oplus \mathfrak{g}_1 \oplus \mathfrak{g}_2 \oplus \mathfrak{g}_3\\
\mathfrak{p}_{12} = \mathfrak{p}_1 \cap \mathfrak{p}_2 &= \mathfrak{g}_0 \oplus \mathfrak{g}_1 \oplus \mathfrak{g}_2 \oplus \mathfrak{g}_3\\
\mathfrak{n}_{12} &= \mathfrak{g}_1 \oplus \mathfrak{g}_2 \oplus \mathfrak{g}_3\\
\mathfrak{n}_1 &= \mathfrak{g}_{1g} \oplus \mathfrak{g}_2 \oplus \mathfrak{g}_3\\
\mathfrak{n}_2 &= \mathfrak{g}_{1w} \oplus \mathfrak{g}_2 \oplus \mathfrak{g}_3
\end{aligned}$$

$$\begin{aligned}
\mathfrak{m} &= \mathfrak{g}_{-3} \oplus \mathfrak{g}_{-2} \oplus \mathfrak{g}_{-1}\\
\mathfrak{q} &= \mathfrak{g}_{-3} \oplus \mathfrak{g}_{-2} \oplus \mathfrak{g}_{-1g}\\
\mathfrak{p} &= \mathfrak{g}_{-3} \oplus \mathfrak{g}_{-2} \oplus \mathfrak{g}_{-1w}.
\end{aligned} \tag{4.13}$$

4.3.2 Parabolic Subalgebras in $\mathfrak{sp}(2,\mathbb{R})$

We recall that a *Lie subalgebra* $\mathfrak{h}$ in the Lie algebra $\mathfrak{g}$ *is* (k-step) *nilpotent* if and only if the following sequence

$$\mathfrak{g}_{-1} = \mathfrak{h}, \quad \mathfrak{g}_{-\ell-1} = [\mathfrak{g}_{-1}, \mathfrak{g}_{-\ell}], \quad \ell = 1, 2, \ldots,$$

of vector subspaces in $\mathfrak{g}$ terminates at step $k+1$. Here the term 'terminates at step $k+1$' means that $\mathfrak{g}_{-k} \neq \{0\}$, and $\mathfrak{g}_{-k-1} = \{0\}$, for some finite $k \geq 1$. Note, that according to this definition, the *Lie subalgebras* $\mathfrak{n}_{12}$, $\mathfrak{n}_1$, $\mathfrak{n}_2$, $\mathfrak{m}$, $\mathfrak{p}$ and $\mathfrak{q}$, of respective dimensions 4,3,3,4,3,3, *are nilpotent* in $\mathfrak{sp}(2,\mathbb{R})$.

Using the structure constants $c^I{}_{JK}$, defined in our basis E_I of $\mathfrak{sp}(2,\mathbb{R})$ by $[E_I, E_J] = c^K{}_{IJ} E_K$, we find that the *Killing form* K of $\mathfrak{sp}(2,\mathbb{R})$ is

$$K = \tfrac{1}{12} K_{IJ} E^I \odot E^J = -4E^1 \odot E^{10} + 2E^2 \odot E^9 + E^3 \odot E^8 - 2E^4 \odot E^7 + E^5 \odot E^5 + E^6 \odot E^6,$$

where the coefficients K_{IJ} are calculated using $K_{IJ} = c^K{}_{IL} c^L{}_{JK}$. Here E^I, $I = 1, 2, \ldots, 10$, is the dual basis in $\mathfrak{sp}(2,\mathbb{R})^*$ to the basis E_I in $\mathfrak{sp}(2,\mathbb{R})$, $E_I \lrcorner E^J = \delta^J{}_I$.

Denoting by $\mathfrak{h}^\perp$ the subspace in $\mathfrak{sp}(2,\mathbb{R})$, which is *Killing-form-orthogonal* to $\mathfrak{h}$,

$$\mathfrak{h}^\perp = \{\mathfrak{sp}(2,\mathbb{R}) \ni E \mid K(H,E) = 0,\ \forall H \in \mathfrak{h}\},$$

we can now easily see that the nilpotent subalgebras $\mathfrak{n}_1$, $\mathfrak{n}_2$ and $\mathfrak{n}_{12}$ are Killing orthogonals to the respective Lie subalgebras $\mathfrak{p}_1$, $\mathfrak{p}_2$ and $\mathfrak{p}_{12}$,

$$\mathfrak{p}_1{}^\perp = \mathfrak{n}_1, \quad \mathfrak{p}_2{}^\perp = \mathfrak{n}_2, \quad \text{and} \quad \mathfrak{p}_{12}{}^\perp = \mathfrak{n}_{12}.$$

Now we recall the following definition:

Definition 4.1 A Lie subalgebra $\mathfrak{p}$ is a *parabolic* subalgebra of a (semi)simple Lie algebra $\mathfrak{g}$ if and only if its Killing orthogonal $\mathfrak{p}^\perp$ is a nilpotent subalgebra in $\mathfrak{g}$.

Thus according to this definition, we found *three parabolic subalgebras*, $\mathfrak{p}_1$, $\mathfrak{p}_2$ and $\mathfrak{p}_{12}$, in the simple Lie algebra $\mathfrak{sp}(2,\mathbb{R})$.[2]

4.3.3 Twistor Fibration and Three Flat Parabolic Geometries Associated with a Car

Consider now the simple Lie group $G = \mathbf{Sp}(2,\mathbb{R})$ and its three parabolic subgroups P_1, P_2 and $P_{12} = P_1 \cap P_2$ corresponding to the parabolic subalgebras p_1, p_2 and p_{12} is $\mathfrak{sp}(2,\mathbb{R})$. Accordingly we have *three* corresponding *homogeneous spaces* $M = G/P_{12}$, $Q = G/P_1$ and $P = G/P_2$. By construction all these three spaces are $\mathbf{Sp}(2,\mathbb{R})$ *symmetric*. Moreover, their tangent spaces at each point have the structure of the corresponding quotient vectors spaces $\mathfrak{m} = \mathfrak{sp}(2,\mathbb{R})/\mathfrak{p}_{12}$, $\mathfrak{q} = \mathfrak{sp}(2,\mathbb{R})/\mathfrak{p}_1$ and $\mathfrak{p} = \mathfrak{sp}(2,\mathbb{R})/\mathfrak{p}_2$. In particular, $\mathfrak{m}$, which can be identified with $\mathfrak{m} = \mathfrak{g}_{-3} \oplus \mathfrak{g}_{-2} \oplus \mathfrak{g}_{-1}$, has a well defined 2-dimensional vector space $\mathfrak{g}_{-1}$ with a well defined *split* $\mathfrak{g}_{-1} = \mathfrak{g}_{-1w} \oplus \mathfrak{g}_{-1g}$. This, point by point on $M = \mathfrak{sp}(2,\mathbb{R})$, defines an *Engel distribution with a split* $\mathcal{D} = \mathcal{D}_w \oplus \mathcal{D}_g$ on M, which *by construction* is $\mathbf{Sp}(2,\mathbb{R})$ symmetric. Therefore this M must be locally equivalent to the configuration space M of a car.

[2] It further follows that the 6-dimensional parabolic algebra $\mathfrak{p}_{12} = \mathfrak{p}_1 \cap \mathfrak{p}_2$ is a *Borel* subalgebra in $\mathfrak{sp}(2,\mathbb{R})$.

We leave to the reader to figure out, directly from the algebraic properties of $\mathfrak{sp}(2, \mathbb{R})$ and $\mathfrak{p}_1$ and $\mathfrak{p}_2$, how the spaces $Q = \mathbf{Sp}(2, \mathbb{R})/P_1$ and $P = \mathbf{Sp}(2, \mathbb{R})/P_2$ get equipped with the respective conformal Lorentzian structure, and the contact projective structure.

Anyhow, we can now write the global version of the car's double fibration as a *parabolic twistor fibration*

$$\begin{array}{ccc} & M = \mathbf{Sp}(2, \mathbb{R})/P_{12} & \\ \swarrow & & \searrow \\ Q = \mathbf{Sp}(2, \mathbb{R})/P_1 & & P = \mathbf{Sp}(2, \mathbb{R})/P_2 \end{array} \tag{4.14}$$

invoked in (4.12). Each space in this fibration is now a *(Cartan) flat model* for a parabolic geometry of the type $(\mathbf{Sp}(2, \mathbb{R}), P)$, where P is one of P_1, P_2 or P_{12}. In this sense the car's geometry falls in the realm of *parabolic geometries* [3].

Another simpler form of this fibration, can be obtained by taking the *simply connected nilpotent* Lie *groups* M, Q and P whose corresponding Lie algebras are $\mathfrak{m}$, $\mathfrak{q}$ and $\mathfrak{p}$ as in (4.13). These are *Carnot groups* [19] with additional structure, such as the Engel structure with a split on M. This enables us to interpret the car's double fibration as the following double *fibration of Carnot groups*:

$$\begin{array}{ccc} & M & \\ \swarrow & & \searrow \\ Q & & P\,. \end{array} \tag{4.15}$$

Although this fibration is made only in terms of Lie groups, and although it is, in a sense, a minimal fibration locally equivalent to the car's fibration, its disadvantage with comparison to the twistor parabolic fibration is that, similar to the cars fibration, the overall $\mathfrak{so}(2, 3) = \mathfrak{sp}(2, \mathbb{R})$ symmetry is not immediately visible in it.

4.4 Outlook: Parabolic Twistor Fibrations in Physics and in Nonholonomic Mechanics

The geometry of a car, which we discussed in this paper, is a *baby version* of the well known Penrose's twistor fibration [20]

$$\begin{array}{ccc} & M & \\ \swarrow & & \searrow \\ Q & & P\,, \end{array} \tag{4.16}$$

in which Q is a 4-dimensional conformal Minkowski spacetime, P is a 5-dimensional space of all null rays in Q, and M is the 6-dimensional bundle of null directions over Q (see [20]). Penrose's fibration is also known as a basis for the *Klein correspondence* (see [22], Part 1, Section 1). To explain this we again need some preparations:

The determination of how many different parabolic subgroups is in a given simple Lie algebra $\mathfrak{g}$, is obtained in terms of the Dynkin diagram of $\mathfrak{g}$: if $\mathfrak{g}$ is considered over the complex numbers, then the choice of a parabolic subgroup in $\mathfrak{g}$ is in one-to-one corespondence with the choice of a decoration of its Dynkin diagram with crosses marked at the nodes of the diagram. In this sense, in the Klein's-Penrose's case where the symmetry algebra is $\mathfrak{so}(2,4) = \mathfrak{su}(2,2)$, the twistor parabolic fibration looks like:

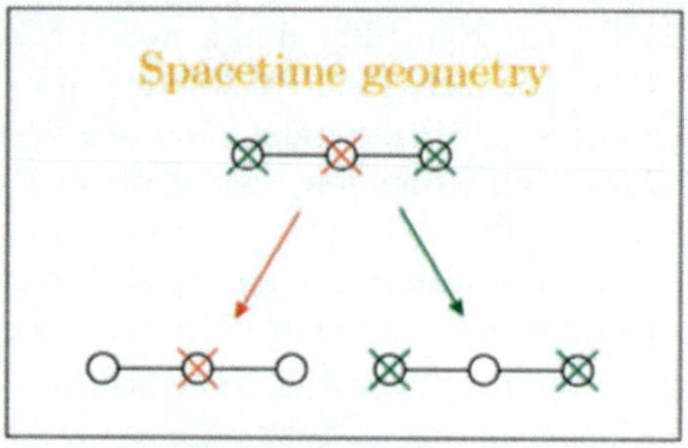

D_3 twistor correspondence of Klein and Penrose

Note that the symmetry Lie algebra here is a simple Lie algebra of rank 3—there are three nodes in each of the manifolds of the diagram.

The car's geometry is related to the symmetry algebra $\mathfrak{so}(2,3) = \mathfrak{sp}(2,\mathbb{R})$, which is a simple Lie algebra of rank 2. The corresponding twistor fibration, in terms of the Dynkin diagrams, looks like this:

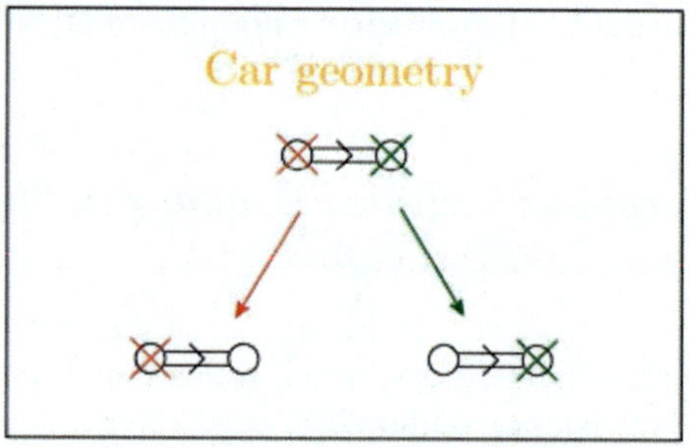

B_2 twistor correspondence of Lie

R. Bryant in the beautiful article [2] describes mathematically all the twistor parabolic fibrations associated with simple Lie algebras of rank 2. Since we interpreted the nonholonomic geometry of a car in terms of the twistor parabolic fibration related to the simple Lie algebra of Cartan-Killing type B_2, one can ask if there are similar physical—possibly related to nonholonomic mechanics—interpretations of the twistor parabolic fibrations related to the simple Lie algebras of type A_2 and G_2? The answer to this question is *yes*. It turns out that the A_2 fibration

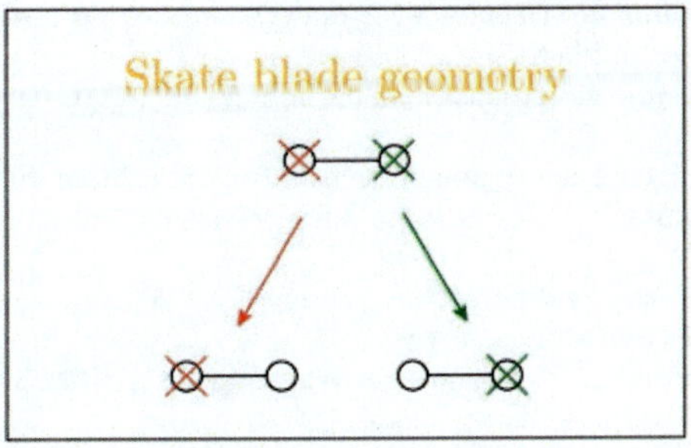

A_2 twistor correspondence of ...

corresponds to the nonholonomic movement of a *skate on an ice ring* [15, 18]. And this case, due to the dimension of M being equal to *three*, and dim$Q$=dim$P$=2, is really the simplest to describe. It is also very similar to the car's B_2 case, since M is really the configuration space of the physical object (a car, a skate) subject to the nonholonomic constraints.

The G_2 case [4, 10], corresponding to the twistor diagram

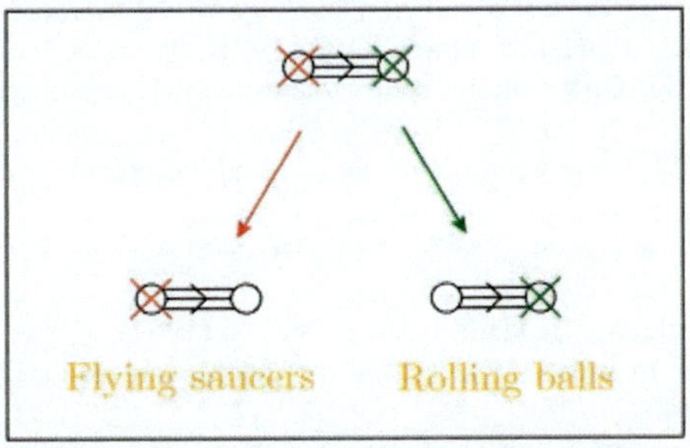

G_2 twistor correspondence of Cartan and Engel

is quite different. Here the dimension of M is 6, and the dimensions of Q and P are both 5. In this case however, the physical objects subjected to the nonholonomic constraints (rolling surfaces, a flying saucer) have their configuration spaces as Q and P [1, 8, 9] and M is merely the *correspondence space* enabling to translate nonholonomic movements between Q and P. If in this, G_2 case, an interpretation of M as a configuration space of some nonholonomic system exists we do not know.

References

1. D. An, P. Nurowski, Twistor space for rolling bodies. Commun. Math. Phys. **326**, 393–414 (2014)
2. R.L. Bryant, Élie Cartan and geometric duality. *Journees Elie Cartan 1998 et 1999*. Institut Elie Cartan, vol. 16 (2000), pp. 5–20. https://services.math.duke.edu/~bryant/Cartan.pdf
3. A. Čap, J. Slovák, *Parabolic Geometries I, Background and General Theory*. Math. Surveys Monogr., vol. 154 (Amer. Math. Soc., 2009)
4. É. Cartan, Sur la structure des groupes simples finis et continus. C. R. Acad. Sci. **116**, 784–786 (1893)
5. É. Cartan, Sur les varietes a connection projective. Bull. Soc. Math. France **52**, 205–241 (1924)
6. É. Cartan, La geometria de las ecuaciones diferenciales de tercer orden. Rev. Mat. Hispano-Amer. **4**, 1–31 (1941)
7. S.S. Chern, The geometry of differential equation $y''' = F(x, y, y', y'')$. Sci. Rep. Nat. Tsing Hua Univ. **4**, 97–111 (1940)
8. M. Eastwood, P. Nurowski, Aerobatics of flying saucers (2019). https://arxiv.org/abs/1810.04852
9. M. Eastwood, P. Nurowski, Aerodynamics of flying saucers (2019). https://arxiv.org/abs/1810.04855
10. F. Engel, Sur un groupe simple a quatorze parametres. C. R. Acad. Sci. **116**, 786–788 (1893)
11. S.S. Chern, The geometry of G-structures. Bull. Am. Math. Soc. **72**, 167–219 (1966)
12. D. Fox, Contact projective structures. Indiana Univ. Math. J. **54**, 1547 (2005). arXiv:math/0402332
13. M. Godliński, *Geometry of third order ordinary differential equations and its applications in General Relativity*, Ph.D. Thesis, Faculty of Physics, University of Warsaw (2008)
14. M. Godliński, P. Nurowski, Geometry of third order ODEs (2009). arXiv: 0902.4129
15. J. Gutt, Projective geometry of simple nonholonomic models, in *Colloquium at CFT PAS*, 24 January 2014. https://www.youtube.com/watch?v=4LV3J_mPUAI
16. S. Helgason, Sophus Lie, the mathematician, in *Proceedings Sophus Lie Memorial Conference, Oslo 1992*, ed. by O.A. Laudal, B. Jahren, Oslo (1994), pp. 3–21
17. C.D. Hill, P. Nurowski, Differential equations and para-CR structures. Bollettino dell'Unione Matematica Italiana (9) III, 25–91 (2010)
18. C.D. Hill, P. Nurowski, Nonholonomic mechanics and the exceptional simple Lie group G_2 (in preparation)
19. P. Pansu, Métriques de Carnot-Carathéodory et quasiisométries des espaces symétriques de rang un. Ann. Math. Second Series **129**(1), 1–60 (1989)
20. R. Penrose, Twistor algebra. J. Math. Phys. **8**, 345–66 (1967)
21. N. Tanaka, On affine symmetric spaces and the automorphism groups of product manifolds. Hokkaido Math. J. **14**, 277–351 (1985)
22. R. Ward, R. Wells, Jr., *Twistor Geometry and Field Theory*. Cambridge Monographs on Mathematical Physics (Cambridge University Press, Cambridge, 1990). https://doi.org/10.1017/CBO9780511524493

Legendrian Cone Structures and Contact Prolongations

Jun-Muk Hwang

Abstract We study a cone structure $C \subset \mathbb{P}D$ on a holomorphic contact manifold $(M, D \subset T_M)$ such that each fiber $C_x \subset \mathbb{P}D_x$ is isomorphic to a Legendrian submanifold of fixed isomorphism type. When C_x is a subadjoint variety, the associated contact G-structure is a parabolic contact structure, for which we have the theory of Tanaka connection. We study the case when C_x is not a subadjoint variety and show that the canonical distribution on the associated contact G-structure admits a holomorphic horizontal splitting by showing that a Legendrian submanifold different from subadjoint varieties has no nonzero contact prolongation.

1 Introduction

We will work in the complex analytic setting: all geometric objects refer to holomorphic ones.

A contact structure on a complex manifold M of dimension $2n + 1$ is a subbundle $D \subset T_M$ of rank $2n$ such that the Lie bracket of vector fields induces a nondegenerate bilinear form

$$\omega_x : \wedge^2 D_x \to L_x$$

for each $x \in M$, where L is the quotient line bundle T_M/D. The following conjecture is well-known.

Conjecture 1 Let X be a Fano manifold with a contact structure $D \subset T_X$. Assume that $\mathrm{Pic} X \cong \mathbb{Z} \cdot L$ for the line bundle $L = T_X/D$. Then X is homogeneous.

J.-M. Hwang (✉)
Institute for Basic Science, Daejon, Korea
e-mail: jmhwang@ibs.re.kr

S. Hervik et al. (eds.), *Geometry, Lie Theory and Applications*, Abel Symposia 16,
https://doi.org/10.1007/978-3-030-81296-6_7

Conjecture 1 has drawn much attention because it implies the LeBrun-Salamon conjecture on quaternion-Kähler manifolds in Riemannian holonomy theory, for which we may assume additionally that X is Kähler-Einstein (see [12]).

One approach to Conjecture 1 is via contact lines and their varieties of minimal rational tangents. To explain this approach, we recall the following definitions.

Definition 1 Let (V, σ) be a symplectic vector space, i.e., a vector space V equipped with a symplectic form $\sigma : \wedge^2 V \to \mathbb{C}$.

(1) The *conformal symplectic group* $\mathrm{CSp}(V) \subset \mathrm{GL}(V)$ is the direct sum of the symplectic group $\mathrm{Sp}(V)$ and the center $\mathbb{C}^\times \cdot \mathrm{Id}_V$.
(2) A projective subvariety $Z \subset \mathbb{P}V$ is *Legendrian* if its affine cone $\widehat{Z} \subset V$ satisfies $\dim \widehat{Z} = \frac{1}{2} \dim V$ and is isotropic with respect to σ, i.e.,

$$\omega(T_{\widehat{Z},v}, T_{\widehat{Z},v}) = 0$$

for each nonsingular point $v \in \widehat{Z}$. Here, we regard the tangent space $T_{\widehat{Z},z}$ as a subspace of V. We say that Z is a Legendrian submanifold if furthermore it is nonsingular.
(3) A Legendrian submanifold $Z \subset \mathbb{P}V$ is *nondegenerate* if the affine cone $\widehat{Z} \subset V$ spans the vector space V.
(4) A nondegenerate Legendrian submanifold $Z \subset \mathbb{P}V$ is a *subadjoint variety* if the group $G^Z \subset \mathrm{CSp}(V)$ of elements of $\mathrm{CSp}(V)$ preserving $\widehat{Z} \subset V$ acts transitively on Z.

The above definition of subadjoint varieties agrees with the standard one by Theorem 11 of [11]. Many examples of Legendrian submanifolds in $\mathbb{P}V$ other than subadjoint varieties are given in [1, 2] and [11].

In the setting of Conjecture 1, a rational curve $C \subset X$ is called a contact line if $C \cdot L = 1$ and C is tangent to $D \subset T_X$. It is known that contact lines exist through every point of X. Kebekus proved the following result in Theorem 4.1 and Theorem 4.4 of [10].

Theorem 1 *For X, D and L as in Conjecture 1, fix an irreducible component $\mathcal{K}$ of the space of contact lines such that members of $\mathcal{K}$ cover X. Let $x \in X$ be a general point and $\mathcal{K}_x \subset \mathcal{K}$ be the subscheme parametrizing members of $\mathcal{K}$ passing through x. Then*

(a) all members of $\mathcal{K}_x$ are nonsingular;
(b) the algebraic subset $\mathcal{C}_x \subset \mathbb{P}D_x$ consisting of tangent spaces to members of $\mathcal{K}_x$ is nonsingular; and
(c) each irreducible component of $\mathcal{C}_x$ is a Legendrian submanifold for the symplectic form ω_x on D_x.

The subset $\mathcal{C}_x \subset \mathbb{P}D_x$ in Theorem 1 is precisely the variety of minimal rational tangents of the Fano manifold X in the sense of [7] and [8]. By applying the theory of varieties of minimal rational tangents, Mok obtained the following result (Main Theorem in Section 2 of [13]).

Theorem 2 *In the setting of Theorem 1, if C_x is a subadjoint variety, then X is homogeneous.*

This seems to be a promising approach to Conjecture 1, as it does not assume the existence of vector fields on X. However, no good strategy has been suggested so far to prove C_x is a subadjoint variety in Theorem 1. This gives a motivation to study the geometry of general 'Legendrian cone structures' in the following sense.

Definition 2 Let $D \subset T_M$ be a contact structure on a complex manifold M and let $\omega_x : \wedge^2 D_x \to L_x$ be the associated symplectic form at each $x \in M$. A submanifold $C \subset \mathbb{P}D$ is a *Legendrian cone structure* on M if each fiber $C_x \subset \mathbb{P}D_x$ is a Legendrian submanifold in the sense of Definition 1. A Legendrian cone structure is *nondegenerate* if C_x is nondegenerate. For a fixed Legendrian submanifold $Z \subset \mathbb{P}V$ in the sense of Definition 1, a Legendrian cone structure $C \subset \mathbb{P}D$ on M is Z-*isotrivial* if the submanifold $C_x \subset \mathbb{P}D_x$ is equivalent to $Z \subset \mathbb{P}V$ modulo $\mathrm{CSp}(V)$ for each $x \in M$.

A theory of general Legendrian cone structures is yet to be developed. It seems reasonable to start with studying Z-isotrivial Legendrian cone structures for a general nondegenerate $Z \subset \mathbb{P}V$. When $Z \subset \mathbb{P}V$ is a subadjoint variety, a Z-isotrivial Legendrian cone structure is a parabolic contact structure in the sense of Section 4.2 of [3]. In fact, the proof of Theorem 2 uses the result of [6] which employed Tanaka connections for parabolic contact structures. As a first step toward the geometry of Z-isotrial Legendrian cone structures, one may ask whether there is an analog of Tanaka connection in this general setting. We prove the following Theorem, which says that the canonical distribution on the associated contact G-structure admits a holomorphic horizontal splitting. This can be viewed as an analog of Tanaka connection.

Theorem 3 *Let $Z \subset \mathbb{P}V$ be a nondegenerate Legendrian submanifold for a symplectic vector space (V, σ), which is not a subadjoint variety and let $G \subset$ $\mathrm{CSp}(V)$ the automorphism group of Z. Let M be a complex manifold with* $\dim M = \dim V + 1$, *equipped with a contact structure $D \subset T_M$ and a Z-isotrival Legendrian cone structure $C \subset \mathbb{P}D$. We have the associated G-principal fiber subbundle $\pi : \mathcal{P} \to M$ of the contact frame bundle of M. Let $T^{\pi} \dot{\subset} T_{\mathcal{P}}$ be the kernel of* $\mathrm{d}\pi$ *and let $\mathcal{D} \subset T_{\mathcal{P}}$ be the natural distribution satisfying $\mathcal{D}/T^{\pi} \cong \pi^* D$. Then there exists a subbundle $\mathcal{H} \subset \mathcal{D}$ such that $\mathcal{D} \cong T^{\pi} \oplus \mathcal{H}$.*

Theorem 3 is an immediate consequence of two results, Theorem 4 and Theorem 5 below. Theorem 4 is a general result on contact G-structures and contact prolongations. In principle, Theorem 4 is contained in Morimoto's general theory [14] of G-structures on filtered manifolds. As it can be stated and proved more explicitly in the contact setting, we present a self-contained account here. Theorem 5 is a characterization of subadjoint varieties in terms of contact prolongations, a contact analog of Theorem 7.13 in [4]. Its proof is simpler than that of [4], thanks to the work [11]. These theorems are expected to be useful in the approach to Conjecture 1 via varieties of minimal rational tangents of contact lines.

2 Contact Prolongation and Contact G-Structure

In this section, we look at the theory of contact G-structures. We will give a self-contained presentation, modifying the ordinary G-structure theory presented in [15]. Although this is a special case of the general theory of geometric structures on filtered manifolds developed by Morimoto in [13], the explicit presentation here could be useful in applications.

Definition 3 Let (V, σ) be a symplectic vector space.

(1) The *conformal symplectic algebra* is the Lie algebra $\mathfrak{csp}(V) \subset \mathfrak{gl}(V)$ of the conformal symplectic group $\mathrm{CSp}(V) \subset \mathrm{GL}(V)$ in Definition 1. Recall that $a \in \mathfrak{gl}(V)$ belongs to $\mathfrak{csp}(V)$ if and only if

$$\sigma(a(v), w) + \sigma(v, a(w)) = \frac{2}{\dim V}\mathrm{tr}(a) \cdot \sigma(v, w)$$

for all $v, w \in V$, where $\mathrm{tr}(a)$ is the trace of $a \in \mathrm{End}(V)$.

(2) For a Lie subalgebra $\mathfrak{g} \subset \mathfrak{csp}(V)$ and an element $A \in \mathrm{Hom}(V, \mathfrak{g})$, we will write $A(u) \in \mathfrak{g}, u \in V$, simply as $A_u \in \mathrm{End(V)}$. Denote by $\mathbf{A} \in V$ the unique vector satisfying

$$\sigma(\mathbf{A}, u) = \frac{2}{\dim V}\mathrm{tr}(A_u) \text{ for all } u \in V.$$

(3) For $A \in \mathrm{Hom}(V, \mathfrak{g})$, let δA be the element of $\mathrm{Hom}(\wedge^2 V, V)$ defined by

$$\delta A(u, v) = A_u(v) - A_v(u) - \sigma(u, v)\,\mathbf{A}$$

for $u, v \in V$. This defines a homomorphism $\delta : \mathrm{Hom}(V, \mathfrak{g}) \to \mathrm{Hom}(\wedge^2 V, V)$. An element $A \in \mathrm{Hom}(V, \mathfrak{g})$ is a *contact prolongation* of $\mathfrak{g}$ if $\delta A = 0$.

Definition 4 Let M be a complex manifold with a contact structure $D \subset M$ and the contact line bundle $L = T_M/D$. Denote by $\omega : \wedge^2 D \to L$ the homomorphism induced by the Lie bracket of local vector fields, equipping D_x with the L-valued symplectic form $\omega_x : \wedge^2 D_x \to L_x$. Fix a symplectic vector space (V, σ) with $\dim M = \dim V + 1$.

(1) A *contact frame* at $x \in M$ is a linear isomorphism $f : V \to D_x$ such that $\omega_x(f(u), f(v)) = 0$ for any $u, v \in V$ satisfying $\sigma(u, v) = 0$. There exists a unique nonzero vector $\mathbf{f} \in L_x$ such that

$$\omega_x(f(u), f(v)) = \sigma(u, v)\,\mathbf{f}$$

for all $u, v \in V$.

(2) Denote by $\mathrm{Frame}_x(M, D)$ the set of all contact frames at x. The union

$$\mathrm{Frame}(M, D) := \cup_{x \in M} \mathrm{Frame}_x(M, D)$$

is a principal $\mathrm{CSp}(V)$-bundle over M, called the *contact frame bundle* of (M, D). The right action

$$R_g : \mathrm{Frame}(M, D) \to \mathrm{Frame}(M, D)$$

of $g \in \mathrm{CSp}(V)$ is given by

$$f \in \mathrm{Frame}(M, D) \mapsto R_g(f) = f \circ g \in \mathrm{Frame}(M, D).$$

Note that the subgroup $\mathrm{Sp}(V) \subset \mathrm{CSp}(V)$ acts trivially on $\mathbf{f}$, while an element $c \cdot \mathrm{Id}_V, c \in \mathbb{C}^\times$, sends $\mathbf{f}$ to $c^2\mathbf{f}$.

(3) For a closed subgroup $G \subset \mathrm{CSp}(V)$, a G-principal subbundle $\mathcal{P} \subset \mathrm{Frame}(M, D)$ is called a *contact G-structure* on M. Denote by $\pi : \mathcal{P} \to M$ the natural projection.

(4) In (3), denote by λ the unique 1-form on $\mathcal{P}$ whose value $\lambda(\mathbf{w}) \in \mathbb{C}$ at $\mathbf{w} \in T_{\mathcal{P}, f}$ satisfies

$$\lambda(\mathbf{w})\mathbf{f} = \mathrm{d}\pi(\mathbf{w}) \mod D_x, \ x = \pi(f).$$

Let $\mathcal{D} \subset T_{\mathcal{P}}$ be the subbundle of corank 1 annihilated by λ. The bundle $T^\pi := \mathrm{Ker}(\mathrm{d}\pi)$ is contained in $\mathcal{D}$ and there is a natural isomorphism of vector bundles $\mathcal{D}/T^\pi \cong \pi^* D$.

The following is our main result on contact G-structures.

Theorem 4 *Let $G \subset \mathrm{CSp}(V)$ be a closed subgroup such that its Lie algebra $\mathfrak{g} \subset \mathfrak{csp}(V)$ has no nonzero contact prolongation. Let $\mathcal{P} \subset \mathrm{Frame}(M, D)$ be a contact G-structure on a contact manifold (M, D) with $\dim M = \dim V + 1$ and let $T^\pi \subset \mathcal{D} \subset T_{\mathcal{P}}$ be as in Definition 4 (4). Then there exists a subbundle $\mathcal{H} \subset \mathcal{D}$ inducing a splitting $\mathcal{D} \cong T^\pi \oplus \mathcal{H}$.*

Remark 1 The subbundle $\mathcal{H}$ in Theorem 4 is not necessarily G-equivariant. So it is not a (partial) connection on the principal bundle $\mathcal{P}$.

The rest of the section is devoted to the proof of Theorem 4. We start with some definitions.

Definition 5 For a contact G-structure $\mathcal{P} \subset \mathrm{Frame}(M, D)$, we have the bundles $T^\pi \subset \mathcal{D} \subset T_{\mathcal{P}}$ from Definition 4.

(1) For each $f \in \mathcal{P}$ and $x = \pi(f) \in M$, let $\theta_f : \mathcal{D}_f \to V$ be the homomorphism obtained by the composition

$$\mathcal{D}_f \xrightarrow{\mathrm{d}\pi} D_x \xrightarrow{f^{-1}} V$$

such that $\mathrm{Ker}(\theta_f) = T^{\pi}_f$. Let $\theta : \mathcal{D} \to \mathcal{P} \times V$ be the homomorphism of vector bundles on $\mathcal{P}$ given by $\{\theta_f,\ f \in \mathcal{P}\}$.

(2) Suppose there exist a line subbundle $\mathcal{L} \subset T_M$ such that $D \cap \mathcal{L} = 0$. Define a V-valued 1-form $\theta^{\mathcal{L}}$ on $\mathcal{P}$ by

$$\theta^{\mathcal{L}}(\mathbf{w}) = \theta(\mathbf{w}) \text{ if } \mathbf{w} \in \mathcal{D} \text{ and}$$

$$\theta^{\mathcal{L}}(\mathbf{w}) = 0 \text{ if } \mathrm{d}\pi(\mathbf{w}) \in \mathcal{L}.$$

Lemma 1 *In the notation of Definitions 4 and 5, denote by $\widetilde{a}$ the fundamental vector field on $\mathcal{P}$ generated by $a \in \mathfrak{g}$. Then for any $\mathbf{w} \in T_{\mathcal{P}}$, we have*

$$\mathrm{d}\lambda(\widetilde{a}, \mathbf{w}) = -\frac{2}{\dim V}\,\mathrm{tr}(a)\,\lambda(\mathbf{w}) \text{ and } \mathrm{d}\theta^{\mathcal{L}}(\widetilde{a}, \mathbf{w}) = -a \cdot \theta^{\mathcal{L}}(\mathbf{w}).$$

Proof It suffices to prove it when $G = \mathrm{CSp}(V)$, because the forms λ and $\theta^{\mathcal{L}}$ for a subbundle $\mathcal{P} \subset \mathrm{Frame}(M, D)$ are just restrictions of the corresponding forms on $\mathrm{Frame}(M, D)$.

We claim that

(i) $R_g^*\lambda = c^{-2}\lambda$ for $g = (g', c\,\mathrm{Id}_V) \in \mathrm{Sp}(V) \times \mathbb{C}^\times \cdot \mathrm{Id}_V$; and
(ii) $R_g^*\theta^{\mathcal{L}} = g^{-1} \cdot \theta^{\mathcal{L}}$ for any $g \in \mathrm{CSp}(V)$.

To prove (i), note that $\mathrm{d}\pi(\mathbf{w}) = \mathrm{d}\pi(R_{g*}\mathbf{w})$ for $\mathbf{w} \in T_{\mathcal{P},f}$. From Definition 4, we obtain, modulo D_x, $x = \pi(f)$,

$$\lambda(R_{g*}\mathbf{w})R_g(\mathbf{f}) \equiv \mathrm{d}\pi(R_{g*}\mathbf{w}) \equiv \mathrm{d}\pi(\mathbf{w}) \equiv \lambda(\mathbf{w})\mathbf{f}.$$

Since $R_g(\mathbf{f}) = c^2\,\mathbf{f}$, we obtain (i).

To prove (ii), if $\mathbf{w} \in \mathcal{D}_f$,

$$\theta^{\mathcal{L}}(R_{g*}\mathbf{w}) = (f \circ g)^{-1}(\mathrm{d}\pi(\mathbf{w})) = g^{-1} \circ f^{-1}(\mathrm{d}\pi(\mathbf{w})) = g^{-1} \cdot \theta^{\mathcal{L}}(\mathbf{w}).$$

On the other hand, if $\mathrm{d}\pi(\mathbf{w}) \in \mathcal{L}$, then $\mathrm{d}\pi(R_{g_*}\mathbf{w}) \in \mathcal{L}$. Thus

$$\theta^{\mathcal{L}}(R_{g*}\mathbf{w}) = 0 = \theta^{\mathcal{L}}(\mathbf{w}).$$

This proves (ii).

The derivative of (i) shows

$$\mathrm{Lie}_{\widetilde{a}}\lambda = -\frac{2}{\dim V}\mathrm{tr}(a)\,\lambda$$

where $\text{Lie}_{\widetilde{a}}$ denotes the Lie derivative with respect to the vector field $\widetilde{a}$. Since $\lambda(\widetilde{a}) = 0$, Cartan's formula for the Lie derivative of forms (e.g, Equation (1.9) in Chapter 3 of [15]) gives $d\lambda(\widetilde{a}, \mathbf{w}) = -\frac{2}{\dim V} \operatorname{tr}(a)\, \lambda(\mathbf{w})$. Similarly, the derivative of (ii) shows

$$\text{Lie}_{\widetilde{a}}\theta^{\mathcal{L}} = -a \cdot \theta^{\mathcal{L}}.$$

Since $\theta^{\mathcal{L}}(\widetilde{a}) = 0$, Cartan's formula gives $d\theta^{\mathcal{L}}(\widetilde{a}, \mathbf{w}) = -a \cdot \theta^{\mathcal{L}}(\mathbf{w})$. □

Lemma 2 *In the notation of Definition 4 (4), let $H \subset \mathcal{D}_f$ be a subspace such that $\mathcal{D}_f = T^{\pi}_f \oplus H$. For $u, v \in V$, let $u^H, v^H \in H$ be the corresponding vectors by the two isomorphisms*

$$V \xrightarrow{f} D_x \xleftarrow{d\pi} H.$$

Then $d\lambda(u^H, v^H) = -\sigma(u, v)$.

Proof Choose local sections $\widetilde{u}$ and $\widetilde{v}$ of $\mathcal{D}$ near $f \in \mathcal{P}$ whose values at f are u^H and v^H, respectively. Then

$$\begin{aligned} d\lambda(\widetilde{u}, \widetilde{v}) &= \widetilde{u}(\lambda(\widetilde{v})) - \widetilde{v}(\lambda(\widetilde{u})) - \lambda([\widetilde{u}, \widetilde{v}]) \\ &= -\lambda([\widetilde{u}, \widetilde{v}]). \end{aligned}$$

Thus

$$\begin{aligned} d\lambda(u^H, v^H)\, \mathbf{f} &= d\lambda(\widetilde{u}, \widetilde{v})_f\, \mathbf{f} \\ &= -\lambda([\widetilde{u}, \widetilde{v}])_f\, \mathbf{f} \\ &= -d\pi([\widetilde{u}, \widetilde{v}])_x \mod D_x,\ x = \pi(f) \\ &= -\omega_x(d\pi(u^H), d\pi(v^H)) \\ &= -\sigma(u, v)\, \mathbf{f}. \end{aligned}$$

□

Lemma 3 *As in Lemma 2, let $H \subset \mathcal{D}_f$ be a subspace such that $\mathcal{D}_f = T^{\pi}_f \oplus H$. Define*

$$\text{Null}^{d\lambda}_H := \{\mathbf{w} \in T_{\mathcal{P},f},\ d\lambda(\mathbf{w}, \mathbf{v}) = 0 \textit{ for all } \mathbf{v} \in H\}.$$

Then there exists a unique 1-dimensional subspace $\ell_H \subset T_{M,x}$, $x = \pi(f)$, such that

$$(d_f\pi)^{-1}(\ell_H) = \text{Null}^{d\lambda}_H \textit{ and } \text{Null}^{d\lambda}_H / T^{\pi}_f \cong \ell_H.$$

Proof Lemma 1 implies $T^{\pi}_{f} \subset \mathrm{Null}^{\mathrm{d}\lambda}_{H}$. Lemma 2 shows that $\mathrm{d}\lambda|_{H}$ is nondegenerate. Thus

$$H \cap \mathrm{Null}^{\mathrm{d}\lambda}_{H} = 0.$$

Choose a subspace $H^{+} \subset T_{\mathcal{P},f}$ such that

$$H \subset H^{+} \text{ and } T_{\mathcal{P},f} = T^{\pi}_{f} \oplus H^{+}.$$

Then $\mathrm{d}\lambda|_{H^{+}}$ cannot be nondegenerate because $\dim H^{+} = \dim D_{x} + 1$ is odd. It follows that

$$\mathrm{Null}^{\mathrm{d}\lambda}_{H} = T^{\pi}_{f} \oplus (H^{+} \cap \mathrm{Null}^{\mathrm{d}\lambda}_{H}) \text{ and } \dim H^{+} \cap \mathrm{Null}^{\mathrm{d}\lambda}_{H} = 1.$$

Thus $\ell_{H} := \mathrm{d}\pi(H^{+} \cap \mathrm{Null}^{\mathrm{d}\lambda}_{H})$ satisfies the requirement. It is clear that ℓ_{H} is uniquely determined by H. □

Lemma 4 *In the setting of Lemma 3, define* $\Pi_{H} \in \mathrm{Hom}(\wedge^{2} V, V)$ *as follows. Extend* $\ell_{H} \subset T_{M,x}$ *in Lemma 3 to a line subbundle* $\mathcal{L}_{H} \subset T_{M}$ *complementary to* D *in a neighborhood of* x*. By Definition 5 (2), we have the* V*-valued 1-form* $\theta^{\mathcal{L}_{H}}$ *in a neighborhood of* $\pi^{-1}(x) \subset \mathcal{P}$*. For* $u, v \in V$*, let* $u^{H}, v^{H} \in H$ *be as in Lemma 2 and define*

$$\Pi_{H}(u, v) := \mathrm{d}\theta^{\mathcal{L}_{H}}(u^{H}, v^{H}).$$

Then

(i) Π_{H} *is determined by* H*, independent of the choice of* $\mathcal{L}_{H}$*; and*
(ii) $\mathrm{d}\lambda(v^{H}, \mathbf{w}) = \sigma(\theta^{\mathcal{L}_{H}}(\mathbf{w}), v)$ *for any* $v \in V$ *and* $\mathbf{w} \in T_{\mathcal{P},f}$.

Proof To check (i), let $\mathcal{L}'_{H} \subset T_{M}$ be another line subbundle in a neighborhood of x complementary to D such that

$$(\mathcal{L}'_{H})_{x} = (\mathcal{L}_{H})_{x} = \ell_{H}.$$

Then $\beta := \theta^{\mathcal{L}'_{H}} - \theta^{\mathcal{L}_{H}}$ is a V-valued 1-form in a neighborhood of $\pi^{-1}(x)$ which vanishes on $\mathcal{D}$. Thus we can write $\beta = h \cdot \lambda$ for some V-valued function h. We have $h|_{\pi^{-1}(x)} = 0$ from $(\mathcal{L}'_{H})_{x} = (\mathcal{L}_{H})_{x}$. Then

$$\mathrm{d}\theta^{\mathcal{L}'_{H}} - \mathrm{d}\theta^{\mathcal{L}_{H}} = \mathrm{d}\beta = \mathrm{d}h \wedge \lambda + h\, \mathrm{d}\lambda$$

satisfies $\mathrm{d}\beta(\mathbf{u}, \mathbf{v}) = 0$ for any $\mathbf{u}, \mathbf{v} \in \mathcal{D}_{f}$, because $\lambda(\mathbf{u}) = \lambda(\mathbf{v}) = 0$ and $h(f) = 0$. Thus Π_{H} does not depend on the choice of $\mathcal{L}_{H}$.

To check (ii), first assume that $\mathbf{w} \in \mathcal{D}_{f}$. Then $\theta^{\mathcal{L}_{H}}(\mathbf{w}) = f^{-1}(\mathrm{d}\pi(\mathbf{w}))$. As in the proof of Lemma 2, choose local sections $\widetilde{v}$ and $\widetilde{w}$ of $\mathcal{D}$ near $f \in \mathcal{P}$ whose values at

f are v^H and $\mathbf{w}$, respectively. Then the same argument as in the proof of Lemma 2 shows

$$\begin{aligned} \mathrm{d}\lambda(v^H, \mathbf{w})\, \mathbf{f} &= -\omega_x(\mathrm{d}\pi(v^H), \mathrm{d}\pi(\mathbf{w})) \\ &= -\sigma(v, \theta^{\mathcal{L}_H}(\mathbf{w}))\, \mathbf{f}. \end{aligned}$$

This proves (ii) when $\mathbf{w} \in \mathcal{D}_{(f,\mathbf{f})}$. Since $\mathrm{Null}_H^{\mathrm{d}\lambda} \cap \mathcal{D}_f$ has codimension 1 in $\mathrm{Null}_H^{\mathrm{d}\lambda}$, it remains to check (ii) when $\mathbf{w} \in \mathrm{Null}_H^{\mathrm{d}\lambda}$. But in this case, the left hand side of (ii) vanishes by the definition of $\mathrm{Null}_H^{\mathrm{d}\lambda}$, while the right hand side of (ii) vanishes by the definitions of ℓ_H and $\theta^{\mathcal{L}_H}$. This completes the proof of (ii). □

Definition 6 In the setting of Lemma 4, let H and H' be two subspaces of $\mathcal{D}_f$ complementary to T_f^π. Then we have $s_{H,H'} \in \mathrm{Hom}(V, \mathfrak{g})$ defined in the following way. For $u \in V$, we have vectors $u^H \in H$ and $u^{H'} \in H'$ defined as in Lemma 2. Then $u^H - u^{H'} \in T_f^\pi$. We define $s_{H,H'}(u) \in \mathfrak{g}$ as the unique element such that

$$\widetilde{s_{H,H'}(u)}_f = u^H - u^{H'},$$

namely, the fundamental vector field on $\mathcal{P}$ generated by $s_{H,H'}(u)$ has value $u^H - u^{H'}$ at f.

The following is immediate.

Lemma 5 *In Definition 6, fix a choice of $H \subset \mathcal{D}_f$ complementary to T_f^π. Then the map $H' \mapsto s_{H,H'} \in \mathrm{Hom}(V, \mathfrak{g})$ gives an identification of the set of complementary subspaces of $T_f^\pi \subset \mathcal{D}_f$ and $\mathrm{Hom}(V, \mathfrak{g})$.*

Lemma 6 *In Definition 6, let $\mathbf{s}_{H,H'} \in V$ be the vector associated to $s_{H,H'} \in \mathrm{Hom}(V, \mathfrak{g})$ as in Definition 3 (2). Then*

$$-\lambda \cdot \mathbf{s}_{H,H'} = \theta^{\mathcal{L}_H} - \theta^{\mathcal{L}_{H'}}$$

on $T_{\mathcal{P},f}$. Consequently,

$$-\mathrm{d}\lambda(\mathbf{u}, \mathbf{v})\, \mathbf{s}_{H,H'} = \mathrm{d}\theta^{\mathcal{L}_H}(\mathbf{u}, \mathbf{v}) - \mathrm{d}\theta^{\mathcal{L}_{H'}}(\mathbf{u}, \mathbf{v})$$

for any $\mathbf{u}, \mathbf{v} \in \mathcal{D}_f$.

Proof From Definition 3 (2), it suffices to check

$$\sigma(\theta^{\mathcal{L}_H}(\mathbf{w}) - \theta^{\mathcal{L}_{H'}}(\mathbf{w}), v) = -\frac{2}{\dim V} \mathrm{tr}(s_{H,H'}(v))\, \lambda(\mathbf{w})$$

for any $v \in V$ and $\mathbf{w} \in T_{\mathcal{P},f}$. But Lemma 4 (ii) implies

$$\sigma(\theta^{\mathcal{L}_H}(\mathbf{w}) - \theta^{\mathcal{L}_{H'}}(\mathbf{w}), v) = \mathrm{d}\lambda(v^H - v^{H'}, \mathbf{w}) = \mathrm{d}\lambda(\widetilde{s_{H,H'}(v)}, \mathbf{w}),$$

which is equal to $-\frac{2}{\dim V}\mathrm{tr}(s_{H,H'}(v))\, \lambda(\mathbf{w})$ by Lemma 1. This completes the proof. □

Lemma 7 *In Definition 6, we have*

$$\Pi_{H'} - \Pi_H = \delta s_{H,H'}$$

where δ *is as in Definition 3 (3).*

Proof For $u, v \in V$, let $\widetilde{s_u}$ (resp. $\widetilde{s_v}$) be the fundamental vector field on $\mathcal{P}$ corresponding to $s_u := s_{H,H'}(u) \in \mathfrak{g}$ (resp. $s_v := s_{H,H'}(v) \in \mathfrak{g}$). Then

$$\begin{aligned}
\Pi_{H'}(u, v) - \Pi_H(u, v) &= \mathrm{d}\theta^{\mathcal{L}_{H'}}(u^{H'}, v^{H'}) - \mathrm{d}\theta^{\mathcal{L}_H}(u^H, v^H) \\
&= \mathrm{d}\theta^{\mathcal{L}_{H'}}((u^{H'} - u^H), v^{H'}) + \mathrm{d}\theta^{\mathcal{L}_{H'}}(u^H, v^{H'}) \\
&\quad + \mathrm{d}\theta^{\mathcal{L}_H}(u^H, (v^{H'} - v^H)) - \mathrm{d}\theta^{\mathcal{L}_H}(u^H, v^{H'}) \\
&= -\mathrm{d}\theta^{\mathcal{L}_{H'}}(\widetilde{s_u}, v^{H'}) - \mathrm{d}\theta^{\mathcal{L}_H}(u^H, \widetilde{s_v}) \\
&\quad + \mathrm{d}\lambda(u^H, v^{H'})\, \mathbf{s}_{H,H'} \\
&= s_{H,H'}(u) \cdot v - s_{H,H'}(v) \cdot u - \sigma(u, v)\, \mathbf{s}_{H,H'} \\
&= \delta s_{H,H'}(u, v),
\end{aligned}$$

where the third equality is from Lemma 6 and the fourth equality is from Lemma 1 and Lemma 2. This finishes the proof. □

Proof of Theorem 4 We use the terminology in Lemma 7. Fix a subspace $W \subset \mathrm{Hom}(\wedge^2 V, V)$ complementary to $\mathrm{Im}(\delta)$ once and for all. (This W may not be stable under the natural G-action on $\mathrm{Hom}(\wedge^2 V, V)$.)

Let $f \in \mathcal{P}$ be a given point. We claim that there exists a unique subspace $\mathcal{H}_f \subset \mathcal{D}_f$ complementary to T^π_f such that $\Pi_{\mathcal{H}_f} \in W$.

To see the existence, fix a complement $H \subset \mathcal{D}_f$ to T^π_f. Then there exists an element $s \in \mathrm{Hom}(V, \mathfrak{g})$ such that $\Pi_H + \delta s \in W$. By Lemma 5, we have $s = s_{H,H'}$ for some $H' \subset \mathcal{D}_f$ complementary to T^π_f. Then Lemma 7 gives $\Pi_{H'} = \Pi_H + \delta s_{H,H'} \in W$. So we can put $\mathcal{H}_f = H'$.

To see the uniqueness, suppose that H, H' are two subspaces of $\mathcal{D}_f$ complementary to T^π_f satisfying $\Pi_H, \Pi_{H'} \in W$. Then $\Pi_H - \Pi_{H'} \in W$, but $\Pi_H - \Pi_{H'} \in \mathrm{Im}(\delta)$ by Lemma 7. Thus we have $\Pi_H = \Pi_{H'}$ and $\delta s_{H,H'} = 0$. By the assumption that $\mathfrak{g}$ has no nonzero contact prolongation, the homomorphism δ is injective. Thus we have $s_{H,H'} = 0$, which implies $H = H'$ by Lemma 5.

By the claim, at each $f \in \mathcal{P}$, we have a unique horizontal subspace $\mathcal{H}_f \subset \mathcal{D}_f$. This defines a subbundle $\mathcal{H} \subset \mathcal{D}$ complementary to T^π. □

3 Contact Prolongation and Legendrian Submanifolds

Let (V, σ) be a symplectic vector space. For a Legendrian submanifold $Z \subset \mathbb{P}V$, let $G^Z \subset \mathrm{CSp}(V)$ be the subgroup in Definition 1 and let $\mathfrak{g}^Z \subset \mathfrak{csp}(V)$ be its Lie algebra. The main result of this section is the following.

Theorem 5 *A nondegenerate Legendrian submanifold $Z \subset \mathbb{P}V$ is a subadjoint variety if $\mathfrak{g}^Z \subset \mathfrak{csp}(V)$ has a nonzero contact prolongation.*

The converse of Theorem 5 holds, although we do not need it. One can see it from the fact that Z-isotrivial Legendrian cone structures are parabolic contact structures and the latter has prolongations (Section 4.2 of [3]). To prove Theorem 5, it is convenient to use the following notion from [5].

Definition 7 Let V be a vector space. A subvariety $Z \subset \mathbb{P}V$ is *Euler-symmetric* if for a general point $z \in Z$, there exists a $\mathbb{C}^\times$-subgroup $E_z \subset \mathrm{GL}(V)$ such that E_z preserves Z with an isolated fixed point at z and acts trivially on $\mathbb{P}T_{Z,z}$.

The following result is Proposition 2.7 in [5]. We refer the reader to [5] for the definition of fundamental forms of projective varieties.

Proposition 1 *Let Z_1 and Z_2 be two Euler-symmetric subvarieties of $\mathbb{P}V$ of equal dimension. Let $z_1 \in Z_1$ and $z_2 \in Z_2$ be general points. If the systems of fundamental forms of Z_1 at z_1 and Z_2 at z_2 are isomorphic, then Z_1 and Z_2 are isomorphic by a projective transformation of $\mathbb{P}V$.*

To apply this to Legendrian submanifolds, we recall some basic properties of fundamental forms of Legendrian submanifolds. The following is proved in pp. 348-349 of [11]. Recall that for a cubic form $P \in \mathrm{Sym}^3 W^\vee$ on a vector space W, its Hessian is the homomorphism $\mathrm{Sym}^2 W \to W^\vee$ defined by

$$w_1 \odot w_2 \in \mathrm{Sym}^2 W \mapsto P(w_1, w_2, \cdot) \in W^\vee.$$

We say that a cubic form P is nondegenerate if its Hessian is surjective to $W^\vee$.

Proposition 2 *Let (V, σ) be a symplectic vector space. for a point $z \in \mathbb{P}V$, denote by $\widehat{z} \subset V$ the corresponding 1-dimensional subspace and by $z^{\perp\sigma} \subset V$ the subspace of codimension 1 defined by $\sigma(\widehat{z}, \cdot) = 0$. Let $Z \subset \mathbb{P}V$ be a Legendrian submanifold.*

(1) If Z is not a linear subspace, then Z is nondegenerate, i.e, the affine cone $\widehat{Z}$ spans V.
(2) If Z is nondegenerate and $z \in Z$ is a general point, then the image of the second fundamental form

$$\mathrm{II}_{Z,z} : \mathrm{Sym}^2 T_{Z,z} \to N_{Z,z} = \mathrm{Hom}(\widehat{z}, V/T_{\widehat{Z},z})$$

is $\mathrm{Hom}(\widehat{z}, z^{\perp\sigma}/T_{\widehat{Z},z})$.
(3) In (2), the third fundamental form $\mathrm{III}_{Z,z}$ is given by a nondegenerate cubic form on $T_{Z,z}$ and

$$\mathrm{II}_{Z,z} : \mathrm{Sym}^2 T_{Z,z} \to \mathrm{Hom}(\widehat{z}, z^{\perp\sigma}/T_{\widehat{Z},z}) \cong (\widehat{z}^\vee)^{\otimes 2} \otimes T^\vee_{Z,z}$$

is isomorphic to the Hessian of this cubic form. In particular, the second fundamental form is determined by the third fundamental form.

The following is Theorem 2 of [1]. As the proof in [1] is somewhat complicated, we will give a simple proof using Proposition 2.

Proposition 3 *Let (V, σ) be a symplectic vector space. Let $Z \subset \mathbb{P}V$ be a nondegenerate Legendrian submanifold. Then an element $A \in \mathrm{GL}(V)$ preserving $\widehat{Z} \subset V$ belongs to $\mathrm{CSp}(V)$.*

Proof To prove the proposition, it suffices to show that if σ' is a symplectic form on V with respect to which Z is Legendrian, then $\sigma' \in \mathbb{C} \cdot \sigma$.

Suppose that $\sigma' \notin \mathbb{C}\sigma$. Then there exists a linear combination $\tau := \sigma' + c \cdot \sigma$ for some $c \in \mathbb{C}$ such that its null space

$$\mathrm{Null}_\tau := \{v \in V,\ \tau(v, u) = 0 \text{ for all } u \in V\}$$

is nonzero. By Proposition 2 (2), for a general $z \in Z$, we have $z^{\perp\sigma} = z^{\perp\sigma'}$. Thus $\mathrm{Null}_\tau \subset z^{\perp\sigma} = z^{\perp\sigma'}$ for all $z \in Z$. This implies that the isomorphism $\iota_\sigma : \mathbb{P}V \to \mathbb{P}V^\vee$ given by $z \mapsto \mathbb{P}z^{\perp\sigma}$ sends Z into the linear subspace annihilating Null_τ. This is a contradiction to the assumption that Z is nondegenerate. □

Proposition 3 says that when we apply results like Proposition 1 to nondegenerate Legendrian submanifolds, we do not need to check whether isomorphisms as projective subvarieties respect the underlying symplectic form on the vector space: they always do.

Proposition 4 *For a symplectic vector space (V, σ), if a nondegenerate Legendrian submanifold $Z \subset \mathbb{P}V$ is Euler-symmetric, then it is a subadjoint variety.*

Proof When $\dim V = 2n + 2$, for each nondegenerate cubic polynomial P in n variables, a special Legendrian subvariety $\widetilde{Z}_P \subset \mathbb{P}^{2n+1}$ whose third fundamental form at a general point is isomorphic to P is constructed in p. 357 of [11]. Their construction implies that $\widetilde{Z}_P$ is Euler-symmetric. (In fact, this is a special case of a more general construction of Euler symmetric varieties in Definition 3.6 of [5], as mentioned in Example 3.10 of [5].) By Proposition 2 (3), the system of fundamental forms of a nondegenerate Legendrian submanifold is determined by its third fundamental form. Thus by Proposition 1, an Euler-symmetric Legendrian submanifold $Z \subset \mathbb{P}V$ is projectively isomorphic to $\widetilde{Z}_P \subset \mathbb{P}^{2n+1}$. Then by Corollary 26 of [11], it is a subadjoint variety. □

Proposition 5 *Let $Z \subset \mathbb{P}V$ be a Legendrian submanifold and let $\mathfrak{g}^Z \subset \mathfrak{csp}(V)$ be as before. Let A be a contact prolongation of $\mathfrak{g}^Z$ with the corresponding vector $\mathbf{A} \in V$ as in Definition 3 (2). Then for a nonzero point $v \in \widehat{Z}$ and $w \in T_{\widehat{Z},v}$,*

(i) $A_v(v) = 2\sigma(\mathbf{A}, v)v$; *and*
(ii) $A_v(w) = \sigma(\mathbf{A}, v)w + \sigma(\mathbf{A}, w)v$.

Proof By the description of $\mathfrak{csp}(V)$ in Definition 3, we have

$$\sigma(A_v(u), y) + \sigma(u, A_v(y)) = \frac{2}{\dim V} \mathrm{tr}(A_v)\, \sigma(u, y)$$

for any $u, v, y \in V$. Putting $v = y$ and using $\frac{2}{\dim V}\mathrm{tr}(A_v) = \sigma(\mathbf{A}, v)$, we obtain

$$\sigma(A_v(u), v) - \sigma(A_v(v), u) = \sigma(\mathbf{A}, v)\, \sigma(u, v).$$

On the other hand, the condition $A_v(u) - A_u(v) = \sigma(v, u)\mathbf{A}$ for a contact prolongation from Definition 3 (3) implies

$$\sigma(A_v(u), v) - \sigma(A_u(v), v) = \sigma(v, u)\, \sigma(\mathbf{A}, v).$$

Combining the two equalities above, we obtain

$$\sigma(A_v(v), u) = \sigma(A_u(v), v) + 2\sigma(v, u)\, \sigma(\mathbf{A}, v).$$

Now, if $v \in \widehat{Z}$, then $A_u \in \mathfrak{g}$ implies that $A_u(v) \in T_{\widehat{Z},v}$. Since $T_{\widehat{Z},v}$ is isotropic with respect to σ, this implies that $\sigma(A_u(v), v) = 0$. Thus the above equality gives

$$\sigma(A_v(v), u) = 2\sigma(\sigma(\mathbf{A}, v)v, u)$$

for all $u \in V$. We conclude that

$$A_v(v) = 2\sigma(\mathbf{A}, v)v$$

for all nonzero $v \in \widehat{Z}$, proving (i).

From (i), if we choose an arc $v(t) := v + tw + t^2(\cdots)$ on $\widehat{Z}$ through v in the direction of $w \in T_{\widehat{Z},v}$, we have

$$A_{v(t)}(v(t)) = 2\sigma(\mathbf{A}, v(t))\, v(t).$$

Expanding in t, the linear terms of both sides read

$$A_v(w) + A_w(v) = 2\sigma(\mathbf{A}, v)w + 2\sigma(\mathbf{A}, w)v.$$

From Definition 3 (3),

$$A_v(w) - A_w(v) = \sigma(v, w)\mathbf{A},$$

which is equal to zero, because v and w are two vectors in the subspace $T_{\widehat{Z},v}$ isotropic with respect to σ. We obtain $A_v(w) = A_w(v)$. It follows that

$$A_v(w) = \sigma(\mathbf{A}, v)w + \sigma(\mathbf{A}, w)v,$$

proving (ii). □

Theorem 5 is a direct consequence of Proposition 4 and the following.

Proposition 6 *Let $Z \subset \mathbb{P}V$ be a nondegenerate Legendrian submanifold. If $\mathfrak{g}^Z \subset \mathfrak{csp}(V)$ has a nonzero contact prolongation, then Z is Euler-symmetric.*

Proof Pick a nonzero contact prolongation A of $\mathfrak{g}^Z$. First, assume that $\mathbf{A} = 0$. Then we have $A_u(v) = A_v(u)$ for all $u, v \in V$. This means A is the usual prolongation of the Lie algebra $\mathfrak{g}^Z \subset \mathfrak{gl}(V)$, in the sense of the section (1.1) of [9] or Definition 7.9 of [4]. Moreover, Proposition 5 shows $A_v(v) = 0$ for all $v \in \widehat{Z}$. This implies $A = 0$ by Lemma 2.2.1 of [9], a contradiction.

Thus we assume that $\mathbf{A} \neq 0$. Pick $v \in \widehat{Z}$ with $\omega(\mathbf{A}, v) = 1$. The endomorphism $A_v \in \mathfrak{g}^Z$ has v as an eigenvector by Proposition 5 (i) and acts on the tangent space of Z at the point $z \in Z$ corresponding to v,

$$T_{Z,z} = \mathrm{Hom}(\mathbb{C}v, T_{\widehat{Z},v}/\mathbb{C}v),$$

as the identity by Proposition 5 (ii). In particular, the semi-simple part of A_v under the Jordan decomposition, to be denoted by S_v, is a nonzero element of $\mathfrak{g}^Z$ and acts on $T_{Z,z}$ as the identity. Thus the $\mathbb{C}^\times$-subgroup $\{\exp(tS_v), t \in \mathbb{C}\}$ acts on Z with an isolated fixed point at z and acts trivially on $\mathbb{P}T_{Z,z}$. Since this is true for any $z \in Z$ outside the hyperplane $\sigma(\mathbf{A}, \cdot) = 0$, we see that Z is Euler-symmetric. □

References

1. J. Buczyński, Toric Legendrian subvarieties. Transform. Groups **12**, 631–646 (2007)
2. J. Buczyński, Hyperplane sections of Legendrian subvarieties. Math. Res. Lett. **15**, 623–629 (2008)
3. A. Čap, J. Slovák, in *Parabolic Geometries I. Background and General Theory.* Mathematical Surveys and Monographs, vol. 154 (Amer. Math. Soc., Providence, 2009)
4. B. Fu, J.-M. Hwang, Special birational transformations of type (2, 1). J. Alg. Geom. **27**, 55–89 (2018)
5. B. Fu, J.-M. Hwang, Euler-symmetric projective varieties. Algebraic Geom. **7**, 377–389 (2020)
6. J. Hong, Fano manifolds with geometric structures modeled after homogeneous contact manifolds. Internat. J. Math. **11**, 1203–1230 (2000)
7. J.-M. Hwang, Geometry of minimal rational curves on Fano manifolds, in *School on Vanishing Theorems and Effective Results in Algebraic Geometry (Trieste, 2000).* ICTP Lect. Notes, vol. 6 (Abdus Salam Int. Cent. Theoret. Phys., Trieste, 2001), pp. 335–393
8. J.-M. Hwang, N. Mok, Varieties of minimal rational tangents on uniruled projective manifolds, in *Several Complex Variables, MSRI Publications*, vol. 37 (Cambridge University Press, Cambridge, 1999), pp. 351–389
9. J.-M. Hwang, N. Mok, Prolongations of infinitesimal linear automorphisms of projective varieties and rigidity of rational homogeneous spaces of Picard number 1 under Kähler deformation. Invent. Math. **160**, 591–645 (2005)
10. S. Kebekus, Lines on complex contact manifolds. II. Compos. Math. **141**, 227–252 (2005)
11. J. Landsberg, L. Manivel, Legendrian varieties. Asian J. Math. **11**, 341–359 (2007)
12. C. LeBrun, Fano manifolds, contact structures, and quaternionic geometry. Int. J. Math. **6**, 419–437 (1995)

13. N. Mok, Recognizing certain rational homogeneous manifolds of Picard number 1 from their varieties of minimal rational tangents, in *Third International Congress of Chinese Mathematicians. Part 1, 2*. AMS/IP Stud. Adv. Math., vol. 42 (Amer. Math. Soc., Providence, 2008), pp. 41–61
14. T. Morimoto, Geometric structures on filtered manifolds. Hokkaido Math. J. **22**, 263–347 (1993)
15. S. Sternberg, *Lectures on Differential Geometry*, 2nd edn. (Amer. Math. Soc. Chelsea Publ., Providence, 1983)

The Search for Solitons on Homogeneous Spaces

Jorge Lauret

Abstract The concept of soliton, in its most general version, allows us to find canonical or distinguished elements on any set provided with an equivalence relation and an 'optimal' tangent direction at each point. We study in this paper solitons on homogeneous spaces, which are a quite useful tool to find soliton geometric structures in Riemannian, pseudo-Riemannian, complex, symplectic and G_2 geometries.

1 Introduction

We aim to continue here the study of geometric flows and their solitons on homogeneous spaces, with emphasis in a unified approach regardless of the type of geometry, as it has previously done in [31, 37, 38, 41]. The present paper is intended to be an invitation to explore the class of homogeneous spaces beyond Lie groups, with regard to the existence, uniqueness and structure of solitons in pseudo-Riemannian, Hermitian, almost-Kähler and exceptional holonomy geometries.

A *homogeneous space* is a quotient $M = G/K$ of a Lie group G by a closed subgroup $K \subset G$. They are differentiable manifolds and their algebraic provenance promises a rich interplay. The presentation of a differentiable manifold as a homogeneous space is far from being unique, and attached to each $M = G/K$ we have the subgroup $\mathrm{Aut}(G/K) \subset \mathrm{Diff}(M)$ of *equivariant diffeomorphisms* defined by automorphisms of G taking K onto K. On the other hand, the natural left action of G on M determines G-actions by pull-back on any space of tensor fields on M, giving rise to the concept of *G-invariant* geometric structure on M.

The author gratefully acknowledges support from FONCyT (ANPCyT) and SECyT (UNC).

J. Lauret (✉)
Universidad Nacional de Córdoba and CIEM, CONICET, Córdoba, Argentina
e-mail: lauret@famaf.unc.edu.ar

S. Hervik et al. (eds.), *Geometry, Lie Theory and Applications*, Abel Symposia 16,
https://doi.org/10.1007/978-3-030-81296-6_8

Two G-invariant geometric structures (or tensor fields) are said to be *equivariantly equivalent* when they belong to the same $\mathrm{Aut}(G/K)$-orbit.

Given a homogeneous space $M = G/K$ and some kind of geometry, we ask ourselves the following question borrowed from the first page of Besse's book [6]:

Are there any best (or nicest, or distinguished) G-invariant structures on M?

The meaning of these adjectives are, of course, part of the problem. Classical possibilities to approach this question include special curvature properties, critical points of geometric functionals, parallelism relative to some connection, etc. The property is expected to be weak enough to allow existence results for large classes of homogeneous spaces, but also sufficiently strong to produce uniqueness or finiteness results up to equivariant equivalence.

In this paper, the following dynamical approach is considered. Let Γ be a space of G-invariant geometric structures on a fixed homogeneous space $M = G/K$. Assume that at each $\gamma \in \Gamma$, there is a G-invariant optimal or preferred direction $q(\gamma) \in T_\gamma\Gamma$, viewed as a 'direction of improvement' in some sense. The corresponding geometric flow on Γ, defined by the ODE

$$\tfrac{d}{dt}\gamma(t) = q(\gamma(t)), \qquad \gamma(0) = \gamma, \tag{1}$$

is therefore supposed to 'improve' the structures while they are flowing. In this light, it is natural to consider a structure γ distinguished in the case when the solution to (1) is given by

$$\gamma(t) = c(t) f(t)^*\gamma, \qquad \text{for some} \quad c(t) \in \mathbb{R}^*, \quad f(t) \in \mathrm{Aut}(G/K),$$

i.e. $\gamma(t)$ is self-similar relative to equivariant equivalence. In that case, $(G/K, \gamma)$ is called a *semi-algebraic soliton* in the literature. The condition is equivalent to the following nice blend of geometric and algebraic aspects of the homogeneous structure $(G/K, \gamma)$:

$$q(\gamma) = c\gamma + \theta(D_{\mathfrak{p}})\gamma, \qquad \text{for some} \quad c \in \mathbb{R}, \quad D = \begin{bmatrix} * & * \\ 0 & D_{\mathfrak{p}} \end{bmatrix} \in \mathrm{Der}(\mathfrak{g}), \tag{2}$$

where θ is the usual $\mathfrak{gl}(\mathfrak{p})$-representation on tensors. Here we are using a *reductive decomposition* $\mathfrak{g} = \mathfrak{k} \oplus \mathfrak{p}$ (i.e. $\mathrm{Ad}(K)\mathfrak{p} \subset \mathfrak{p}$) for the homogeneous space G/K and the fact that any G-invariant tensor field on $M = G/K$ can be identified with an $\mathrm{Ad}(K)$-invariant tensor on $T_oM \equiv \mathfrak{p}$.

If the derivation D in (2) satisfies in addition that $D_{\mathfrak{p}}$ is orthogonal to the stabilizer $\mathfrak{gl}(\mathfrak{p})_\gamma$, then $(G/K, \gamma)$ is said to be an *algebraic soliton*. They also distinguish themselves by being precisely the fixed points of the bracket flow as well as by their 'diagonal' evolution. The *bracket flow* is a dynamical system defined on the variety of Lie algebras which is equivalent to the geometric flow (1) in a very precise sense.

The concepts of algebraic and semi-algebraic solitons have a long and fruitful history in the Ricci flow case. More recently, they have also provided a quite useful tool to study the existence problem of homogeneous soliton geometric structures in many different geometries, including pseudo-Riemannian, complex, symplectic and G_2 (see Sect. 4 for references).

After surveying the topics of general solitons, homogeneous geometric structures and semi-algebraic solitons in Sects. 2, 3, and 4, respectively, the following new results are obtained.

- Provided that a certain mild condition on the flow holds, we prove that any soliton (M, γ) in the general sense (i.e. $q(\gamma) = c\gamma + \mathcal{L}_X\gamma$ for some $c \in \mathbb{R}$ and $X \in \mathfrak{X}(M)$), which is homogeneous, is a semi-algebraic soliton when presented as a homogeneous space $(G/K, \gamma)$ with $G = \mathrm{Aut}(M, \gamma)$ (see Sect. 5).
- In Sect. 6, we first give overviews on the moving-bracket approach (to study homogeneous geometric structures) and on the bracket flow. Secondly, we compute the evolution of semi-algebraic solitons relative to the bracket flow.
- We geometrically characterize algebraic solitons among homogeneous solitons as those developing a 'simultaneously diagonalizable' solution $\gamma(t)$ to the ODE (1) (see Sect. 7).

2 Solitons in Differential Geometry

From a very general point of view, the necessary ingredients to define a *soliton* are just the following (see [41]):

- A set Γ with a notion of tangent space $T_\gamma\Gamma$ at each $\gamma \in \Gamma$.
- An equivalence relation $\simeq$ on Γ.
- An optimal or preferred direction at each point, $q(\gamma) \in T_\gamma\Gamma$, viewed as a 'direction of improvement' in some sense.

In that case, $\gamma \in \Gamma$ is called a *soliton* if

$$q(\gamma) \in T_\gamma[\gamma], \tag{3}$$

where $[\gamma]$ is the equivalence class of γ, that is, γ is in a way 'nice' enough that it does not need to be 'improved'.

In the context of differential geometry, we consider a space Γ of geometric structures on a fixed differentiable manifold M and identify Γ with a subset of the vector space $\mathcal{T}^{r,s}M$ of all tensor fields of some type (r, s) (or tuples of tensors). As usual, the equivalence relation is scaling and pulling-back by diffeomorphisms, i.e. the equivalence class of $\gamma \in \Gamma$ is given by

$$[\gamma] = \{ch^*\gamma : c \in \mathbb{R},\ h \in \mathrm{Diff}(M)\} \cap \Gamma.$$

Typically, a preferred direction,

$$\gamma \mapsto q(\gamma) \in T_\gamma \Gamma \subset \mathcal{T}^{r,s} M,$$

is given by a curvature tensor associated to some affine connection attached to γ, or the gradient of a natural geometric functional, or the Hodge-Laplacian on differential forms, etc.

Once the space Γ and the preferred direction q have been specified, it follows from (3) that $\gamma \in \Gamma$ is a soliton if and only if

$$q(\gamma) = c\gamma + \mathcal{L}_X \gamma, \qquad \text{for some} \quad c \in \mathbb{R}, \quad X \in \mathfrak{X}(M), \tag{4}$$

where $\mathcal{L}_X$ denotes Lie derivative with respect to a vector field X of M (recall that $T_\gamma(\mathrm{Diff}(M) \cdot \gamma) = \mathcal{L}_{\mathfrak{X}(M)}\gamma$). In the case when q is *diffeomorphism equivariant* (i.e. $q(f^*\gamma) = f^*q(\gamma)$ for any $f \in \mathrm{Diff}(M)$ and $\gamma \in \Gamma$), which will be assumed in this paper, any geometric structure in Γ which is equivalent to a soliton is also a soliton.

Example 2.1 (Ricci Solitons) The space Γ of all Riemannian metrics on M is open in $\mathcal{S}^2 M$, the vector space of all symmetric 2-tensors, and a natural preferred direction is

$$q(g) := -2\,\mathrm{Ric}_g \in T_g\Gamma = \mathcal{S}^2 M \subset \mathcal{T}^{2,0} M,$$

where Ric_g is the Ricci tensor of the metric $g \in \Gamma$. This gives rise to the well-known Ricci solitons.

Example 2.2 If a complex manifold (M, J) is fixed and a space Γ of hermitian metrics on (M, J) are to be considered, then the equivalence is determined by the action of the group $\mathrm{Aut}(M, J)$ of bi-holomorphic maps rather than $\mathrm{Diff}(M)$, so X has to be a holomorphic field. In the symplectic case, the group of symplectomorphisms $\mathrm{Aut}(M, \omega)$ of a fixed symplectic manifold (M, ω) must be considered.

Concerning the associated geometric flow,

$$\tfrac{\partial}{\partial t}\gamma(t) = q(\gamma(t)), \qquad \gamma(0) = \gamma, \tag{5}$$

one easily obtains that γ is a soliton if and only if

$$\gamma(t) = c(t) f(t)^* \gamma, \qquad \text{for some} \quad c(t) \in \mathbb{R}, \quad f(t) \in \mathrm{Diff}(M), \tag{6}$$

i.e. $\gamma(t)$ is a *self-similar* solution.

Example 2.3 (Ricci Flow) The evolution equation determined by Example 2.1 is precisely the famous Ricci flow $\frac{\partial}{\partial t} g(t) = -2\,\mathrm{Ric}_{g(t)}$ introduced in the 80s by Hamilton and used as a primary tool by Perelman to prove the Poincaré and Geometrization conjectures.

Note that solitons are not necessarily fixed points of the flow (i.e. zeroes of q). Nevertheless, they are not either improved by the flow and may attract or stop other solutions in its way to a fixed point. Thus their existence is undesirable if one is hoping to use the flow to find zeroes of q in Γ; on the other hand, the existence of solitons is great news for the search of canonical or distinguished structures in Γ beyond zeroes of q.

Remark 2.4 A natural preferred direction q may or may not produce a flow, as the existence of solutions to the PDE (5) is not guaranteed. So possibly, a study of solitons can be worked out without any reference to a flow, as in the case of Ricci solitons in pseudo-Riemannian geometry (see [9]).

Let us assume for the rest of the paper that the scaling behavior of the preferred direction q is given by

$$q(c\gamma) = c^{\alpha}\gamma, \qquad \forall c \in \mathbb{R}^*, \quad \gamma \in \Gamma, \tag{7}$$

for some fixed $\alpha < 1$. In that case, the scaling in (6) is given by

$$c(t) = ((1-\alpha)ct + 1)^{\frac{1}{1-\alpha}},$$

where c is the constant appearing in the soliton equation (4) (see [38]). The soliton γ is therefore called *expanding*, *steady* or *shrinking* depending on whether $c > 0$, $c = 0$ or $c < 0$, and the maximal time interval of the corresponding self-similar solution is respectively given by,

$$(-T_\alpha, \infty), \quad (-\infty, \infty), \quad (-\infty, T_\alpha), \qquad \text{where} \quad T_\alpha := \frac{1}{(1-\alpha)|c|} > 0,$$

often called *immortal*, *eternal* and *ancient* solutions, respectively. For instance, $\alpha = 0$ if q is the Ricci tensor or form of any connection associated to a metric or to an almost-hermitian structure, and $\alpha = \frac{1}{3}$ for most of the flows for G_2-structures in the literature.

We refer to [38, 41] and Sect. 4 for overviews on solitons for different kinds of geometric flows in complex, symplectic and G_2 geometries, including:

- Ricci flow [11].
- Chern-Ricci flow [59].
- Pluriclosed flow [56].
- Hermitian curvature flow [57].
- Anti-complexified Ricci flow [46].
- Symplectic curvature flow [58].
- G_2-Laplacian flow [26, 47, 48].

Remark 2.5 The word *soliton* has been used in PDE theory since nineteenth century in the context of Korteweg-de Vries equation to name certain solutions resembling solitary water waves. More generally, in the study of geometric flows, solitons refer to geometric structures which evolve along symmetries of the flow (i.e. self-similar solutions) and the use of the word soliton was initiated by Hamilton in the 1980s (see [20]) in the context of Ricci flow to name *Ricci solitons*. Nowadays, solitons are spread over the fields of differential geometry and geometric analysis.

3 Homogeneous Geometric Structures

A differentiable manifold endowed with a geometric structure, say (M,γ), is said to be *homogeneous* if its automorphism group,

$$\mathrm{Aut}(M,\gamma) := \{f \in \mathrm{Diff}(M) : f^*\gamma = \gamma\},$$

acts transitively on M. Note that this property makes the points of M geometrically indistinguishable.

Recall that a *homogeneous space* is a quotient G/K of a Lie group G over a closed subgroup $K \subset G$. If (M,γ) is homogeneous, then each Lie group $G \subset \mathrm{Aut}(M,\gamma)$ which is still transitive on M gives rise to a presentation of M as a homogeneous space G/K, where K is the isotropy subgroup of G at some *origin* point $o \in M$. In this way, γ becomes a G-invariant geometric structure on the homogeneous space $M = G/K$. On the other hand, one can also start with a homogeneous space $M = G/K$ endowed with a G*-invariant* geometric structure γ (i.e. $\tau_a^*\gamma = \gamma$ for any $a \in G$, where $\tau_a \in \mathrm{Diff}(M)$ is defined by $\tau_a(bK) = abK$ for all $b \in G$), giving rise to a homogeneous (M,γ).

Any homogeneous space G/K will be assumed in this paper to be only *almost-effective* (i.e. the subgroup $\{a \in K : \tau_a = id\}$ is discrete) rather than *effective* (i.e. $\tau_a = id$ if and only if $a = e$). Note that G/K is effective if $G \subset \mathrm{Aut}(M,\gamma)$.

If $\mathfrak{g} = \mathfrak{k} \oplus \mathfrak{p}$ is a *reductive decomposition* (i.e. $\mathrm{Ad}(K)\mathfrak{p} \subset \mathfrak{p}$) for the homogeneous space G/K, where $\mathfrak{g}$ and $\mathfrak{k}$ respectively denote the Lie algebras of G and K, then the tangent space at the origin of M is identified with $\mathfrak{p}$,

$$T_oM \equiv \mathfrak{p},$$

via the isomorphism

$$\mathfrak{p} \longrightarrow T_oM, \qquad X \mapsto d\pi|_e X = X_o, \tag{8}$$

since $\mathrm{Ker}\, d\pi|_e = \mathfrak{k}$. Here $\pi : G \to G/K$ is the usual projection and each $X \in \mathfrak{g}$ is also viewed as the vector field on M defined by $X_p := \frac{d}{dt}\big|_{t=0} \tau_{\exp tX}(p)$.

Remark 3.1 In the case when a Riemannian metric g_γ is involved in the geometric structure γ, i.e. contained in or uniquely determined by γ, one obtains the following:

- $\mathrm{Aut}(M,\gamma)$ is a Lie group, as it is a closed subgroup of the Lie group $\mathrm{Iso}(M,g_\gamma)$ of all isometries of the Riemannian manifold (M,g_γ). In particular, G is closed in $\mathrm{Aut}(M,\gamma)$ if and only if K is compact just as in the Riemannian case.

- The existence of a reductive decomposition for G/K is guaranteed since $\overline{\mathrm{Ad}(K)} \subset \mathrm{GL}(\mathfrak{g})$ turns out to be compact. Anyway, one can just work with the identification $T_o M \equiv \mathfrak{g}/\mathfrak{k}$ if a reductive decomposition is preferred not to be chosen.

3.1 Tensors

Since a G-invariant tensor field on a homogeneous space $M = G/K$ is determined by its value at the origin o, it can always be identified with a tensor γ on $\mathfrak{p}$ which is $\mathrm{Ad}(K)$-*invariant*, i.e. $\mathrm{Ad}(z)|_{\mathfrak{p}} \cdot \gamma = \gamma$ for any $z \in K$, or equivalently if K is connected, $\theta(\mathrm{ad}\, Z|_{\mathfrak{p}})\gamma = 0$ for all $Z \in \mathfrak{k}$. We are considering here the usual left $\mathrm{GL}(\mathfrak{p})$-action and corresponding $\mathfrak{gl}(\mathfrak{p})$-representation θ on the finite-dimensional vector space

$$T^{r,s}\mathfrak{p} := \{\gamma : \underbrace{\mathfrak{p} \times \cdots \times \mathfrak{p}}_{r} \longrightarrow \underbrace{\mathfrak{p} \otimes \cdots \otimes \mathfrak{p}}_{s} : \gamma \text{ is multi-linear}\}$$

of all tensors of type (r, s), given by

$$h \cdot \gamma := h \cdot_s \gamma(h^{-1}\cdot, \dots, h^{-1}\cdot), \quad \forall h \in \mathrm{GL}(\mathfrak{p}), \tag{9}$$

$$\theta(A)\gamma := \theta_s(A)\gamma - \gamma(A\cdot, \dots, \cdot) - \cdots - \gamma(\cdot, \dots, A\cdot), \quad \forall A \in \mathfrak{gl}(\mathfrak{p}), \tag{10}$$

where $\cdot_s$ and θ_s denote the $\mathrm{GL}(\mathfrak{p})$-action and $\mathfrak{gl}(\mathfrak{p})$-representation on $\mathfrak{p} \otimes \cdots \otimes \mathfrak{p}$ (s times), respectively. Note that $h \cdot \gamma = (h^{-1})^*\gamma$.

Conversely, it is easy to see that any $\mathrm{Ad}(K)$-invariant tensor on $\mathfrak{p}$ can be walked around M via the G-action to become a G-invariant tensor field on M. To sum up, the linear map

$$(\mathcal{T}^{r,s}M)^G \longrightarrow (T^{r,s}\mathfrak{p})^K, \qquad \gamma \longmapsto \gamma_o,$$

is an isomorphism between the vector space $(\mathcal{T}^{r,s}M)^G$ of all G-invariant tensors fields on M and the finite-dimensional vector space $(T^{r,s}\mathfrak{p})^K$ of all $\mathrm{Ad}(K)$-invariant tensors on $\mathfrak{p}$.

An alternative identification of $(\mathcal{T}^{r,s}M)^G$ is given by the isomorphism

$$(\mathcal{T}^{r,s}M)^G \longrightarrow \mathcal{T}_0^{r,s}G, \qquad \gamma \longmapsto \pi^*\gamma,$$

where $\mathcal{T}_0^{r,s}G \subset \mathcal{T}^{r,s}G$ is the vector subspace of all (r, s)-tensors γ on the Lie group G such that the following conditions hold:

(i) γ is left-invariant.
(ii) γ is invariant under the action of K on G by conjugation.
(iii) $\gamma(\dots, Z, \dots) = 0$ for any $Z \in \mathfrak{k}$.

Note that, algebraically, $\mathcal{T}_0^{r,s}G$ is identified with

$$T_0^{r,s}\mathfrak{g} := \left\{\gamma \in T^{r,s}\mathfrak{g} : \gamma \text{ is } \operatorname{Ad}(K)\text{-invariant and (iii) holds}\right\},$$

and the linear map

$$(T^{r,s}\mathfrak{p})^K \longrightarrow T_0^{r,s}\mathfrak{g}, \qquad \gamma \longmapsto d\pi|_e^*\gamma,$$

is an isomorphism. In the context of differential forms, the differentials of forms d_G and d_M on G and M, respectively, satisfy that

$$d_G\,(d\pi|_e)^*\,\gamma = (d\pi|_e)^*\,d_M\gamma, \qquad \forall\gamma \in (\Lambda^k\mathfrak{p}^*)^K \subset (T^{k,0}\mathfrak{p})^K,$$

and thus

$$d_M\gamma(X_1,\dots,X_{k+1}) = \sum_{i<j}(-1)^{i+j}\gamma([X_i,X_j]_\mathfrak{p}, X_1,\dots,\widehat{X}_i,\dots,\widehat{X}_j,\dots,X_k),$$

for any $X_1,\dots,X_{k+1} \in \mathfrak{p}$, where $[X_i,X_j]_\mathfrak{p}$ is the projection of the Lie bracket $[X_i,X_j]$ on $\mathfrak{p}$ relative to $\mathfrak{g} = \mathfrak{k}\oplus\mathfrak{p}$ and $\widehat{X}_i$ means that X_i has to be deleted.

3.2 *Equivalence*

A diffeomorphism of a homogeneous space $M = G/K$ is called *equivariant* if it is given by an automorphism of G taking K onto K. Let $\operatorname{Aut}(G/K)$ denote the group of all equivariant diffeomorphisms of $M = G/K$. It is easy to check that $\operatorname{Aut}(G/K)$ acts on $(\mathcal{T}^{r,s}M)^G$ and two G-invariant geometric structures (or tensor fields) are said to be *equivariantly equivalent* when they belong to the same $\operatorname{Aut}(G/K)$-orbit.

In the case when G is simply connected and K connected (in particular, M simply connected), which will be assumed in the rest of this subsection, $\operatorname{Aut}(G/K)$ can be identified with the Lie group

$$\operatorname{Aut}(\mathfrak{g}/\mathfrak{k}) := \left\{\overline{h} \in \operatorname{Aut}(\mathfrak{g}) : \overline{h}(\mathfrak{k}) = \mathfrak{k}\right\},$$

with Lie algebra $\operatorname{Der}(\mathfrak{g}/\mathfrak{k}) := \{D \in \operatorname{Der}(\mathfrak{g}) : D(\mathfrak{k}) \subset \mathfrak{k}\}$. At the Lie algebra level, $\operatorname{Aut}(\mathfrak{g}/\mathfrak{k})$ acts on $(T^{r,s}\mathfrak{p})^K$ by

$$\gamma \mapsto h\cdot\gamma = (h^{-1})^*\gamma, \qquad \text{where} \quad \overline{h} = \begin{bmatrix} * & * \\ 0 & h \end{bmatrix} \in \operatorname{Aut}(\mathfrak{g}/\mathfrak{k}),$$

giving rise to the equivariant equivalence in the simply connected case. Note that $\operatorname{Ad} K \subset \operatorname{Aut}(\mathfrak{g}/\mathfrak{k})$.

Remark 3.2 Two non-equivariantly equivalent G-invariant geometric structures γ and γ' on a homogeneous space $M = G/K$ may be still equivalent in the general sense as in Sect. 2, that is, $\gamma' = f^*\gamma$ for some $f \in \mathrm{Diff}(M)$.

The computation of $(T^{r,s}\mathfrak{p})^K$ is a technical problem in representation theory whose difficulty strongly depends on the decomposition of $\mathfrak{p}$ into irreducible and isotypic components and the type of each irreducible component $\mathfrak{p}_i$. Recall that $\mathfrak{p}_i$ is said to be of *real*, *complex* or *quaternionic type*, depending on whether the space of intertwining operators $\mathrm{End}_K(\mathfrak{p}_i)$ is $\mathbb{R}$, $\mathbb{C}$ or $\mathbb{H}$, respectively. However, the existence problem of a structure $\gamma \in (T^{r,s}\mathfrak{p})^K$ satisfying in addition certain non-degeneracy condition can be very tricky. On the other hand, the moduli space

$$\Gamma/\,\mathrm{Aut}(\mathfrak{g}/\mathfrak{k}) \subset (T^{r,s}\mathfrak{p})^K/\,\mathrm{Aut}(\mathfrak{g}/\mathfrak{k})$$

of a space Γ of G-invariant geometric structures on $M = G/K$, up to equivariant equivalence, is an object much harder to compute or understand.

A larger group acting on $(T^{r,s}\mathfrak{p})^K$ is the normalizer

$$N_{\mathrm{GL}(\mathfrak{p})}(\operatorname{ad}\mathfrak{k}|_{\mathfrak{p}}) := \{h \in \mathrm{GL}(\mathfrak{p}) : h \operatorname{ad}\mathfrak{k}|_{\mathfrak{p}} h^{-1} \subset \operatorname{ad}\mathfrak{k}|_{\mathfrak{p}}\}.$$

This follows from the fact that

$$\theta(\operatorname{ad}\mathfrak{k}|_{\mathfrak{p}})h\cdot\gamma = h\cdot\theta(h^{-1}\operatorname{ad}\mathfrak{k}|_{\mathfrak{p}}h)\gamma = 0, \qquad \forall\, h \in N_{\mathrm{GL}(\mathfrak{p})}(\operatorname{ad}\mathfrak{k}|_{\mathfrak{p}}), \quad \gamma \in (T^{r,s}\mathfrak{p})^K.$$

Note that $\mathrm{Aut}(\mathfrak{g}/\mathfrak{k})|_{\mathfrak{p}}$ and the centralizer $C_{\mathrm{GL}(\mathfrak{p})}(\operatorname{ad}\mathfrak{k}|_{\mathfrak{p}})$ are both subgroups of $N_{\mathrm{GL}(\mathfrak{p})}(\operatorname{ad}\mathfrak{k}|_{\mathfrak{p}})$. This action is many times a useful tool to obtain non-equivariantly equivalent G-invariant geometric structures from a given one.

3.3 *Operators Versus Tensors*

A property that is satisfied by many classes of geometric structures is that the orbit $\mathrm{GL}(\mathfrak{p})\cdot\gamma$ is open in a certain vector subspace $T \subset T^{r,s}\mathfrak{p}$ and consists precisely of those tensors which are non-degenerate in some sense. In that case, $\theta(\mathfrak{gl}(\mathfrak{p}))\gamma = T$, and thus if we consider the $\mathrm{GL}(\mathfrak{p})_\gamma$-invariant decomposition,

$$\mathfrak{gl}(\mathfrak{p}) = \mathfrak{gl}(\mathfrak{p})_\gamma \oplus \mathfrak{q}_\gamma, \tag{11}$$

then $\theta(\mathfrak{q}_\gamma)\gamma = T$, where $\mathfrak{gl}(\mathfrak{p})_\gamma := \{A \in \mathfrak{gl}(\mathfrak{p}) : \theta(A)\gamma = 0\}$ is the Lie algebra of the stabilizer subgroup $\mathrm{GL}(\mathfrak{p})_\gamma$. Moreover, for every tensor $q \in T$, there exists a unique operator $Q \in \mathfrak{q}_\gamma$ such that

$$q = \theta(Q)\gamma. \tag{12}$$

Note that $\theta(I)\gamma = (s-r)\gamma$ for any $\gamma \in T^{r,s}\mathfrak{p}$ (see (10)).

Some examples of geometric structures for which this translation to operators works follow (see [38, 41] for further information).

- A Riemannian metric $\gamma = g$:

$$T = \mathcal{S}^2\mathfrak{p}^* \subset T^{2,0}\mathfrak{p}, \qquad \mathfrak{gl}(\mathfrak{p})_\gamma = \mathfrak{so}(\mathfrak{p}), \qquad \mathfrak{q}_\gamma = \operatorname{sym}(\mathfrak{p}) := \{A \in \mathfrak{gl}(\mathfrak{p}) : A^t = A\}.$$

For example, if $q = -2\operatorname{Rc}(g)$, where $\operatorname{Rc}(g)$ is the Ricci tensor of g, then $Q = \operatorname{Ric}(g)$, the Ricci operator of g.
- A pseudo-Riemannian metric $\gamma = g$ of signature (p, q) ($\dim\mathfrak{p} = p + q$): $T = \mathcal{S}^2\mathfrak{p}^* \subset T^{2,0}\mathfrak{p}$,

$$\begin{aligned} \mathfrak{gl}(\mathfrak{p})_\gamma =& \{A \in \mathfrak{gl}(\mathfrak{p}) : g(A\cdot, \cdot) = -g(\cdot, A\cdot)\} \simeq \mathfrak{so}(p, q), \\ \mathfrak{q}_\gamma =& \{A \in \mathfrak{gl}(\mathfrak{p}) : g(A\cdot, \cdot) = g(\cdot, A\cdot)\}. \end{aligned}$$

- An almost-symplectic structure $\gamma = \omega$ ($\dim\mathfrak{p} = 2n$): $T = \Lambda^2\mathfrak{p}^* \subset T^{2,0}\mathfrak{p}$,

$$\begin{aligned} \mathfrak{gl}(\mathfrak{p})_\gamma =& \{A \in \mathfrak{gl}(\mathfrak{p}) : \omega(A\cdot, \cdot) = -\omega(\cdot, A\cdot)\} \simeq \mathfrak{sp}(n, \mathbb{R}), \\ \mathfrak{q}_\gamma =& \{A \in \mathfrak{gl}(\mathfrak{p}) : \omega(A\cdot, \cdot) = \omega(\cdot, A\cdot)\}. \end{aligned}$$

- An almost-hermitian structure $\gamma = (\omega, g, J)$, $\omega = g(J\cdot, \cdot)$ ($\dim\mathfrak{p} = 2n$):

$$T = \left\{(\overline{\omega}, \overline{g}) \in \Lambda^2\mathfrak{p}^* \times \mathcal{S}^2\mathfrak{p}^* : \overline{g}^{1,1} = \overline{\omega}^{1,1}(\cdot, J\cdot)\right\} \subset T^{2,0}\mathfrak{p} \times T^{2,0}\mathfrak{p},$$

$$\mathfrak{gl}(\mathfrak{p})_\gamma = \{A \in \mathfrak{so}(\mathfrak{p}) : AJ = JA\} \simeq \mathfrak{u}(n), \qquad \mathfrak{q}_\gamma = \mathfrak{q}_1 \oplus \mathfrak{q}_2 \oplus \mathfrak{q}_3,$$

where,

$$\begin{aligned} \mathfrak{q}_1 =& \{A \in \mathfrak{gl}(\mathfrak{p}) : A^t = A, \quad AJ = -JA\}, \\ \mathfrak{q}_2 =& \{A \in \mathfrak{gl}(\mathfrak{p}) : A^t = -A, \quad AJ = -JA\}, \\ \mathfrak{q}_3 =& \{A \in \mathfrak{gl}(\mathfrak{p}) : A^t = A, \quad AJ = JA\}. \end{aligned}$$

Note that $\mathfrak{sp}(n, \mathbb{R}) = \mathfrak{u}(n) \oplus \mathfrak{q}_1$, $\mathfrak{so}(2n) = \mathfrak{u}(n) \oplus \mathfrak{q}_2$ and $\mathfrak{gl}_n(\mathbb{C}) = \mathfrak{u}(n) \oplus \mathfrak{q}_3$.
- A G_2-structure $\gamma = \varphi$ ($\dim\mathfrak{p} = 7$):

$$T = \Lambda^3\mathfrak{p}^* \subset T^{3,0}\mathfrak{p}, \qquad \mathfrak{gl}(\mathfrak{p})_\gamma = \mathfrak{g}_2, \qquad \mathfrak{q}_\gamma = \mathfrak{q}_1 \oplus \mathfrak{q}_7 \oplus \mathfrak{q}_{27},$$

where $\mathfrak{q}_1 = \mathbb{R}I$, $\mathfrak{so}(\mathfrak{p}) = \mathfrak{g}_2 \oplus \mathfrak{q}_7$ and $\mathfrak{q}_{27} = \operatorname{sym}_0(\mathfrak{p}) := \{A \in \operatorname{sym}(\mathfrak{p}) : \operatorname{tr} A = 0\}$. Here, an example of preferred direction is $q(\varphi) = \Delta_\varphi\varphi$, where Δ_φ is the Hodge-Laplacian on 3-forms (giving rise to the G_2-Laplacian flow), and Q_φ can be written in terms of the Ricci operator and the torsion (see [40, Section 3]).

4 Solitons on Homogeneous Spaces

In this section, we study solitons on a fixed homogeneous space $M = G/K$. As in Sect. 2, we consider a space of geometric structures Γ and assume that every $\gamma \in \Gamma$ is G-invariant, that is,

$$\Gamma \subset (T^{r,s}\mathfrak{p})^K,$$

where $\mathfrak{g} = \mathfrak{k} \oplus \mathfrak{p}$ is any reductive decomposition as in Sect. 3.1. Accordingly, the equivalence between G-invariant structures on M is defined by scalings and equivariant diffeomorphisms of M (see Sect. 3.2). Let us assume that Γ is $\mathrm{Aut}(G/K)$-invariant.

As a preferred direction q we can choose any direction from the general case (see Sect. 2), which produces a G-invariant preferred direction determined by a function

$$\gamma \mapsto q(\gamma) \in T_\gamma\Gamma \subset (T^{r,s}\mathfrak{p})^K.$$

In this light, once the space Γ and the preferred direction q have been chosen, we obtain that if $\gamma \in \Gamma$ is a soliton in the sense of (3), called a *semi-algebraic soliton* in the literature, then

$$q(\gamma) = c\gamma + \theta(D_\mathfrak{p})\gamma, \qquad \text{for some} \quad c \in \mathbb{R}, \quad D = \begin{bmatrix} * & * \\ 0 & D_\mathfrak{p} \end{bmatrix} \in \mathrm{Der}(\mathfrak{g}/\mathfrak{k}), \tag{13}$$

where θ denotes the usual $\mathfrak{gl}(\mathfrak{p})$-representation on tensors (see (10)). Indeed, we have that

$$\theta(D_\mathfrak{p})\gamma = -\mathcal{L}_{X_D}\gamma = -\tfrac{d}{dt}\big|_{t=0} f(t)^*\gamma \in T_\gamma\, \mathrm{Aut}(G/K)\cdot\gamma, \tag{14}$$

where $f(t) \in \mathrm{Aut}(G/K)$, $D = \frac{d}{dt}\big|_{t=0} df(t)|_e \in \mathrm{Der}(\mathfrak{g}/\mathfrak{k})$ and $\mathcal{L}_{X_D}$ denotes the Lie derivative with respect to the vector field $X_D \in \mathfrak{X}(M)$ defined by

$$X_D(p) = \tfrac{d}{dt}\big|_{t=0} f(t)(p), \qquad \forall p \in M.$$

In particular, $(G/K, \gamma)$ is of course a soliton from the general point of view considered in (4).

Conversely, in the simply connected case, any $(G/K, \gamma)$ satisfying (13) is in fact a semi-algebraic soliton. Indeed, the existence of the automorphisms $f(t) \in \mathrm{Aut}(G/K)$ such that $df(t)|_e = e^{tD} \in \mathrm{Aut}(\mathfrak{g}/\mathfrak{k})$ is guaranteed for any $D \in \mathrm{Der}(\mathfrak{g}/\mathfrak{k})$. Note that the fact that $\theta(D_\mathfrak{p})\gamma \in T_\gamma\Gamma$ automatically follows from (13).

Remark 4.1 The vector fields X_D's described above can be viewed as generalizations of *linear vector fields* on $\mathbb{R}^n$ (i.e. $X_v = Av$, $A \in \mathfrak{gl}_n(\mathbb{R})$) and have been strongly used in control theory on Lie groups after the pioneering article [4]. Since $X_D(o) = 0$, none of these vector fields can belong to $\mathfrak{p}$ (see (8)).

Unlike the concept of soliton, which is invariant under pull-back by diffeomorphisms, the concept of semi-algebraic soliton is not a geometric invariant, as it may depend on the presentation of the homogeneous geometric structure (M, γ) as a homogeneous space $(G/K, \gamma)$.

On a given homogeneous space $M = G/K$, the corresponding geometric flow

$$\tfrac{d}{dt}\gamma(t) = q(\gamma(t)), \qquad \gamma(0) = \gamma, \tag{15}$$

is actually an ODE rather than a PDE. Indeed, the solutions to (15) are precisely the integral curves of the vector field q on the finite-dimensional vector space $(T^{r,s}\mathfrak{p})^K$. In particular, short time existence (forward and backward) of G-invariant solutions and their uniqueness (among G-invariant solutions) are always guaranteed.

We assume from now on that G is simply connected and K is connected. Notice that γ is a semi-algebraic soliton if and only if the solution $\gamma(t)$ to (15) is given by

$$\gamma(t) = c(t)f(t)^*\gamma, \qquad \text{for some} \quad c(t) \in \mathbb{R}^*, \quad f(t) \in \operatorname{Aut}(G/K), \tag{16}$$

that is, $\gamma(t)$ is self-similar relative to equivariant equivalence (cf. (6)). More precisely, it is easy to check that condition (13) holds for $(G/K, \gamma)$ if and only if

$$\gamma(t) = c(t)e^{s(t)D_{\mathfrak{p}}} \cdot \gamma, \tag{17}$$

is a solution to (15), where

$$c(t) := ((1-\alpha)ct + 1)^{\frac{1}{1-\alpha}}, \qquad s(t) := \frac{\log((1-\alpha)ct + 1)}{(1-\alpha)c} \qquad (\text{set } s(t) := t \text{ if } c = 0).$$

Recall that $\alpha < 1$ (see (7)) and note that $c(0) = 1$, $c'(t) = c\,c(t)^\alpha$ and $s(0) = 0$, $s'(t) = c(t)^{\alpha-1}$.

The following result is essentially contained in [23, Theorem 3.1].

Proposition 4.2 *Let $(G/K, \gamma)$ be a semi-algebraic soliton and assume that K is compact. Then there exist $f(t) \in \operatorname{Aut}(G/K)$, $f(0) = id$, such that,*

(i) the solution to (15) is given by $\gamma(t) = c(t)f(t)^*\gamma$, for some $c(t) \in \mathbb{R}^*$;
(ii) $f(t)|_K = id$ for all t.

Proof According to (16), $\gamma(t) = c(t)g(t)^*\gamma$, for some $c(t) \in \mathbb{R}^*$ and $g(t) \in \operatorname{Aut}(G/K)$. We can assume that $g(0) = id$ by just composing with $g(0)^{-1}$, which belongs to $\operatorname{Aut}(M, \gamma)$ as $\gamma(0) = \gamma$. Since K is connected and compact, the identity component $\operatorname{Aut}(K)_0$ of $\operatorname{Aut}(K)$ consists of inner automorphisms, and thus there exist $a(t) \in K$, $a(0) = e$, such that $g(t)|_K = I_{a(t)}$, the conjugation map by $a(t)$. But $I_{a(t)} \in \operatorname{Aut}(G/K)$ and is an automorphism of γ (recall that $dI_{a(t)}|_o = \operatorname{Ad}(a(t)) \in$

$\mathrm{Ad}(K)$), so $f(t) := I_{a(t)}^{-1} \circ g(t)$ satisfies that $f(t)^*\gamma = g(t)^*\gamma$ and $f(t)|_K = id$ for all t, concluding the proof. □

In the case when the reductive decomposition $\mathfrak{g} = \mathfrak{k} \oplus \mathfrak{p}$ for G/K with $B(\mathfrak{k}, \mathfrak{p}) = 0$ is considered, where B is the Killing form of $\mathfrak{g}$, it follows that $D\mathfrak{p} \subset \mathfrak{p}$ for any $D \in \mathrm{Der}(\mathfrak{g}/\mathfrak{k})$ (see [28, Lemma 3.10]). If in addition K is compact, then by Proposition 4.2 the derivation in the definition (13) of a semi-algebraic soliton $(G/K, \gamma)$ takes the simpler form

$$D = \begin{bmatrix} 0 & 0 \\ 0 & D_{\mathfrak{p}} \end{bmatrix} \in \mathrm{Der}(\mathfrak{g}/\mathfrak{k}). \tag{18}$$

Recall that Γ is typically contained in a single $\mathrm{GL}(\mathfrak{p})$-orbit. We assume for the rest of this section that $\mathrm{GL}(\mathfrak{p}) \cdot \gamma$ is in addition open in a suitable subspace $T \subset T^{r,s}\mathfrak{p}$, as in Sect. 3.3. In that case, in terms of the operator viewpoint proposed in Sect. 3.3, the semi-algebraic soliton condition (13) for $(G/K, \gamma)$ can be rewritten as

$$Q_\gamma = \tfrac{c}{s-r} I + \mathrm{pr}_\gamma(D_{\mathfrak{p}}), \tag{19}$$

where $q(\gamma) = \theta(Q_\gamma)\gamma$, $Q_\gamma \in \mathfrak{q}_\gamma$ (see (12)) and $\mathrm{pr}_\gamma : \mathfrak{gl}(\mathfrak{p}) \to \mathfrak{q}_\gamma$ is the projection relative to decomposition (11). The matrix equation (19) has been extremely useful to explore the existence and study the structure of semi-algebraic solitons in many different contexts.

In the special case when (19) holds for a derivation $D \in \mathrm{Der}(\mathfrak{g}/\mathfrak{k})$ such that

$$\mathrm{pr}_\gamma(D_{\mathfrak{p}}) = D_{\mathfrak{p}}, \tag{20}$$

$(G/K, \gamma)$ is called an *algebraic soliton*. It is worth noticing that being an algebraic soliton may a priori not only depend on the homogeneous space presentation, but also on the reductive decomposition $\mathfrak{g} = \mathfrak{k} \oplus \mathfrak{p}$ chosen. Algebraic solitons are also distinguished from other points of view, as bracket flow evolution (see Sect. 6) and the flow diagonal property (see Sect. 7).

Any homogeneous Ricci soliton is isometric to an algebraic soliton (see [22]). On the other hand, examples of G_2-Laplacian and pluriclosed semi-algebraic solitons which are not isometric to any algebraic soliton were respectively found in [39, 51] and [3].

The strong interplay between geometric and algebraic aspects of the homogeneous structure $(G/K, \gamma)$ was very fruitful in the study of homogeneous Ricci solitons. On the other hand, the role of algebraic solitons has also been crucial to obtain explicit examples of homogeneous soliton geometric structures in the case of pseudo-Riemannian, complex, symplectic and G_2 geometries. A far from complete list of references follows.

- Ricci solitons [2, 13, 22–25, 28, 29, 33, 52, 60] (see the survey [32] for references previous to 2008).

- Pseudo-Riemannian Ricci solitons [9, 10, 53].
- Chern-Ricci solitons [37, 43].
- Pluriclosed solitons [3].
- Hermitian curvature flow solitons [30, 55].
- Anti-complexified Ricci solitons [14, 31].
- Symplectic curvature flow solitons [14, 37, 44].
- G_2-Laplacian solitons [5, 16, 17, 39, 40, 42, 50, 51].

Remark 4.2 Given a semi-algebraic soliton $(G/K, \gamma)$, if G admits a cocompact discrete subgroup Λ, then Λ acts freely and properly discontinuous on G/K and the solution $\gamma(t)$ is also a solution to (15) on the compact manifold $M' = \Lambda\backslash G/K$, which is locally diffeomorphic to $M = G/K$. However, the locally homogeneous manifold (M', γ) is not longer a soliton in general since the field X_D may not descend to M'. The solution $(M, \gamma(t))$ is very peculiar though, it is 'locally self-similar' in the sense that $\gamma(t)$ is locally equivalent to γ up to scaling for all t.

5 On the Evolution of Homogeneous Geometric Structures

We assume in this section that $\mathrm{Aut}(M, \gamma)$ is a Lie group. Recall that this holds, for example, when there is a Riemannian metric attached to γ (see Remark 3.1).

Proposition 5.1 *For any homogeneous geometric structure (M, γ), there exists a unique solution $\gamma(t)$ to (5) which is $\mathrm{Aut}(M, \gamma)$-invariant. Furthermore, $\gamma(t)$ has the following properties:*

(i) *$\gamma(t)$ is defined in a maximal interval of time (T_-, T_+), where $T_- < 0 < T_+$.*
(ii) $\mathrm{Aut}(M, \gamma(t)) = \mathrm{Aut}(M, \gamma)$ *for all $t \in (T_-, T_+)$.*
(iii) *For any transitive Lie subgroup of automorphisms $G \subset \mathrm{Aut}(M, \gamma)$, $\gamma(t)$ is the unique G-invariant solution to (15) on the corresponding homogeneous space $M = G/K$.*

Proof The existence and uniqueness of the solution $\gamma(t)$ and part (i) follow from the paragraph below (15) applied to $G = \mathrm{Aut}(M, \gamma)$. On the other hand, part (iii) follows from uniqueness and the fact that $\gamma(t)$ is also G-invariant for any $G \subset \mathrm{Aut}(M, \gamma)$. This implies that given t_0, since $\mathrm{Aut}(M, \gamma) \subset \mathrm{Aut}(M, \gamma(t))$ for all t, we have that $\gamma(t + t_0)$ is the $\mathrm{Aut}(M, \gamma)$-invariant solution starting at $\gamma(t_0)$. Thus $\gamma(t + t_0)$ is also $\mathrm{Aut}(M, \gamma(t_0))$-invariant and so $\mathrm{Aut}(M, \gamma(t_0)) \subset \mathrm{Aut}(M, \gamma)$, by evaluating at $t = -t_0$, and part (ii) holds, concluding the proof. □

Since the uniqueness of geometric flow solutions is in most cases established only for compact manifolds, we do not know a priori if there are homogeneous solutions other than $\gamma(t)$ starting at a noncompact homogeneous (M, γ). This seems to be very unlikely though.

The following result was proved in [23, Theorem 3.1] for Ricci solitons; we essentially follow the lines of that proof.

Theorem 5.2 *Let (M, γ) be a soliton as in (4) which is homogeneous. If the solution $\gamma(t)$ (see Proposition 5.1) is self-similar as in (6), then $(G/K, \gamma)$ is a semi-algebraic soliton for the presentation $M = G/K$ with $G = \mathrm{Aut}(M, \gamma)$.*

Remark 5.3 In particular, $\gamma(t) = c(t)f(t)^*\gamma$ for some $c(t) \in \mathbb{R}^*$ and $f(t) \in \mathrm{Aut}(G/K)$ such that $f(0) = id$ and $f(t)|_K = id$ for all t by Proposition 4.2, since the isotropy K is compact.

Proof Consider the presentation $M = G/K$, where $G = \mathrm{Aut}(M, \gamma)$ and K is the isotropy subgroup of G at some $p \in M$. It follows from (6) that $\gamma(t) = c(t)g(t)^*\gamma$ for some $c(t) \in \mathbb{R}^*$ and $g(t) \in \mathrm{Diff}(M)$. It can be assumed that $g(0) = id$, and also that $g(t)(p) = p$ for all t, by composing with $h(t) \in G$ such that $h(t)^{-1}(p) = g(t)(p)$. Since $G = \mathrm{Aut}(M, \gamma(t)) = g(t)^{-1}Gg(t)$ for all t (see Proposition 5.1, (ii)), we can define an isomorphism $\tilde{f}(t) : G \longrightarrow G$ by $\tilde{f}(t)(a) := g(t)ag(t)^{-1}$ for all $a \in G$, which in turns determines $f(t) \in \mathrm{Aut}(G/K)$ as $\tilde{f}(t)(K) \subset K$. But $f(t) = g(t)$ on G/K for all t:

$$f(t)(aK) = g(t)ag(t)^{-1}K = g(t)ag(t)^{-1}(p) = g(t)(a(p)) = g(t)(aK), \qquad \forall a \in G,$$

which concludes the proof. □

Remark 5.4 It follows from the proof of the above proposition that if $\gamma(t) = c(t)f(t)^*\gamma$ for some $c(t) \in \mathbb{R}^*$ and $f(t) \in \mathrm{Diff}(M)$ such that $f(0) = id$, $f(t)(o) = o$, then $f(t) \in \mathrm{Aut}(G/K)$ for all t for $G = \mathrm{Aut}(M, \gamma)$.

The uniqueness of solutions has been proved in some cases (e.g. for the Ricci flow) among bounded curvature solutions on non-compact manifolds. We note that this applies to homogeneous solutions.

Corollary 5.5 *Assume that for each homogeneous (M, γ), there exists a unique solution $\gamma(t)$ to (5) such that $(M, \gamma(t))$ is homogeneous for all t. Then any homogeneous soliton (M, γ) is a semi-algebraic soliton when presented as a homogeneous space $(G/K, \gamma)$ with $G = \mathrm{Aut}(M, \gamma)$.*

6 The Moving-Bracket Approach

Let $(G/K, \gamma)$ be a homogeneous space endowed with a G-invariant geometric structure. We note that all the geometry of $(G/K, \gamma)$ is essentially encoded only in the tensor $\gamma \in (T^{r,s}\mathfrak{p})^K$ and the Lie bracket μ of $\mathfrak{g}$. Therefore, in order to vary homogeneous geometric structures, it is natural to wonder about what if we vary μ rather than γ. To do so, we consider a fixed $(k+n)$-dimensional real vector space $\mathfrak{g}$ together with a direct sum decomposition

$$\mathfrak{g} = \mathfrak{k} \oplus \mathfrak{p}, \qquad \dim \mathfrak{k} = k, \qquad \dim \mathfrak{p} = n, \tag{21}$$

a fixed suitable tensor $\gamma \in T^{r,s}\mathfrak{p}$ (or a tuple of tensors) and the *variety of Lie algebras* $\mathcal{L} \subset \Lambda^2\mathfrak{g}^* \otimes \mathfrak{g} \subset T^{2,1}\mathfrak{g}$ (i.e. the algebraic subset of all Lie brackets on the vector space $\mathfrak{g}$). In this way, each $\mu \in \mathcal{L}$ determines a simply connected almost-effective homogeneous space endowed with an invariant geometric structure with reductive decomposition $\mathfrak{g} = \mathfrak{k} \oplus \mathfrak{p}$, say $(G_\mu/K_\mu, \gamma)$, provided that the following conditions hold:

(i) $\mu(\mathfrak{k}, \mathfrak{k}) \subset \mathfrak{k}$ and $\mu(\mathfrak{k}, \mathfrak{p}) \subset \mathfrak{p}$.
(ii) K_μ is closed in G_μ, where G_μ denotes the simply connected Lie group with Lie algebra $(\mathfrak{g}, \mu)$ and K_μ is the connected Lie subgroup of G_μ with Lie algebra $\mathfrak{k}$.
(iii) $\{Z \in \mathfrak{k} : \mu(Z, \mathfrak{p}) = 0\} = 0$.
(iv) $\theta(\mathrm{ad}_\mu Z|_\mathfrak{p})\gamma = 0$ for all $Z \in \mathfrak{k}$.

If $\mathcal{H}^\gamma_{k,n}$ denotes the subset of $\mathcal{L}$, consisting of the elements satisfying (i)-(iv), then via the identification

$$\mathcal{H}^\gamma_{k,n} \ni \mu \longleftrightarrow (G_\mu/K_\mu, \gamma), \tag{22}$$

the space $\mathcal{H}^\gamma_{k,n}$ parametrizes the set of all n-dimensional simply connected homogeneous spaces with k-dimensional isotropy endowed with an invariant geometric structure of the same type as the fixed γ (see [35, 38] for more detailed treatments).

Recall that any Lie group isomorphism $G \to G'$ taking K onto K' defines a diffeomorphism $G/K \to G'/K'$ called an *equivariant diffeomorphism*. In that case, the homogeneous spaces are said to be *equivariantly equivalent*.

The natural actions on tensors provide the following key equivariant equivalence between geometric structures:

$$(G_{\overline{h}\cdot\mu}/K_{\overline{h}\cdot\mu}, \gamma) \longleftarrow (G_\mu/K_\mu, h^*\gamma), \qquad \forall\, \overline{h} = \begin{bmatrix} h_\mathfrak{k} & 0 \\ 0 & h \end{bmatrix} \in \mathrm{GL}(\mathfrak{g}), \tag{23}$$

given by the Lie group isomorphism $G_\mu \to G_{\overline{h}\cdot\mu}$ with derivative $\overline{h}$. In the case when $\Gamma \subset \mathrm{GL}(\mathfrak{p}) \cdot \gamma$, it follows from (22) and (23) that the orbit $(\mathrm{GL}(\mathfrak{k}) \times \mathrm{GL}(\mathfrak{p})) \cdot \mu$ contains all the geometric structures of the same type of γ (up to equivalence) on the homogeneous space G_μ/K_μ, for each $\mu \in \mathcal{H}^\gamma_{k,n}$. Note that all the homogeneous spaces involved share the same reductive decomposition $\mathfrak{g} = \mathfrak{k} \oplus \mathfrak{p}$.

The scaling of tensors can be viewed on the space $\mathcal{H}^\gamma_{k,n}$ as follows. Since $(cI)^*\gamma = c^{r-s}\gamma$ for any $c \in \mathbb{R}^*$ and $\gamma \in T^{r,s}\mathfrak{p}$ (see (9)), if we set $\overline{h} := \left(I, c^{-1}I\right) \in \mathrm{GL}(\mathfrak{k}) \times \mathrm{GL}(\mathfrak{p})$, then we obtain from (23) that

$$(G_{c\cdot\mu}/K_{c\cdot\mu}, \gamma) \simeq (G_\mu/K_\mu, c^{s-r}\gamma),$$

where $c \cdot \mu := \overline{h} \cdot \mu$ is given by

$$c \cdot \mu|_{\mathfrak{k}\times\mathfrak{k}} = \mu, \qquad c \cdot \mu|_{\mathfrak{k}\times\mathfrak{p}} = \mu, \qquad c \cdot \mu|_{\mathfrak{p}\times\mathfrak{p}} = c^2\mu_\mathfrak{k} + c\mu_\mathfrak{p}, \tag{24}$$

and the subscripts denote the $\mathfrak{k}$- and $\mathfrak{p}$-components of $\mu|_{\mathfrak{p}\times\mathfrak{p}}$.

Remark 6.1 The usual convergence of a sequence of brackets produces convergence of the corresponding geometric structures in well-known senses, like pointed (or Cheeger-Gromov) and smooth up to pull-back by diffeomorphisms, under suitable conditions (see [35, 38]). Remarkably, a degeneration (i.e. $\lambda \in \overline{\mathrm{GL}(\mathfrak{g}) \cdot \mu} \backslash \mathrm{GL}(\mathfrak{g}) \cdot \mu$) may give rise to the convergence of a sequence of geometric structures on a given homogeneous space (see (23)) toward a structure on a different homogeneous space, which may be topologically quite different.

The moving-bracket approach has actually been used for decades in homogeneous geometry (see the overviews given in [38, Section 5] and [45, Section 3]). In most applications, concepts and results from geometric invariant theory, including moment maps and their convexity properties, closed orbits, stability, categorical quotients and Kirwan stratification, have been exploited in one way or another.

6.1 The Bracket Flow

We assume for the rest of the paper that $\mathrm{GL}(\mathfrak{p}) \cdot \gamma$ is open in $T \subset T^{r,s}\mathfrak{p}$ as in Sect. 3.3.

Motivated by equivalence (23), a main tool to study the geometric flow (15) is a dynamical system defined on the variety of Lie algebras $\mathcal{L}$ called the *bracket flow*, defined by

$$\tfrac{d}{dt}\mu(t) = -\theta\left(\begin{bmatrix} 0 & 0 \\ 0 & Q_{\mu(t)} \end{bmatrix}\right)\mu(t), \qquad \mu(0) = \mu, \tag{25}$$

where $Q_\mu \in \mathfrak{q}_\gamma \subset \mathfrak{gl}(\mathfrak{p})$ is the unique operator such that $\theta(Q_\mu)\gamma = q(G_\mu/K_\mu, \gamma)$ (see (12)).

The bracket flow is equivalent to the geometric flow (15) in the following precise sense. Given $\mu \in \mathcal{H}^\gamma_{k,n}$, we consider the one-parameter families

$$(G_\mu/K_\mu, \gamma(t)) \qquad \text{and} \qquad \left(G_{\mu(t)}/K_{\mu(t)}, \gamma\right),$$

where $\gamma(t)$ is the solution to the geometric flow (15) and $\mu(t)$ is the solution to the bracket flow (25). Recall that $\mathfrak{g} = \mathfrak{k} \oplus \mathfrak{p}$ is a reductive decomposition for each of the homogeneous spaces involved.

Theorem 6.2 ([38, Theorem 5]) *There exist equivariant diffeomorphisms*

$$\tilde{h}(t) : G_\mu/K_\mu \longrightarrow G_{\mu(t)}/K_{\mu(t)} \qquad \textit{such that} \qquad \gamma(t) = \tilde{h}(t)^*\gamma, \qquad \forall t.$$

Moreover, each $\tilde{h}(t)$ can be chosen to be the Lie group isomorphism $G_\mu \longrightarrow G_{\mu(t)}$ with derivative

$$\overline{h}(t) := \begin{bmatrix} I & 0 \\ 0 & h(t) \end{bmatrix} : \mathfrak{g} \longrightarrow \mathfrak{g},$$

where $h(t) = d\tilde{h}(t)|_o : \mathfrak{p} \longrightarrow \mathfrak{p}$ is the solution to any of the following ODE's:

(i) $\frac{d}{dt}h(t) = -h(t)Q_{\gamma(t)}$, $h(0) = I$, *where* $Q_{\gamma(t)} \in \mathfrak{q}_{\gamma(t)} \subset \mathfrak{gl}(\mathfrak{p})$ *is defined by*

$$\theta(Q_{\gamma(t)})\gamma(t) = q(G/K, \gamma(t)).$$

(ii) $\frac{d}{dt}h(t) = -Q_{\mu(t)}h(t)$, $h(0) = I$, *where* $Q_{\mu(t)} \in \mathfrak{q}_{\gamma} \subset \mathfrak{gl}(\mathfrak{p})$ *is defined by*

$$\theta(Q_{\mu(t)})\gamma = q(G_{\mu(t)}/K_{\mu(t)}, \gamma).$$

Remark 6.3 The following conditions also hold:

(iii) $\gamma(t) = h(t)^*\gamma$ for all t.
(iv) $\mu(t) = \overline{h}(t) \cdot \mu$ for all t.

The bracket flow is useful to better visualize the possible limits of solutions, under diverse rescalings, as well as to address regularity issues. Since the flows differ only by pullback by time-dependent diffeomorphisms, the behavior of any geometric quantity can be studied along the bracket flow solutions. We now list some of its properties and applications.

- The flows (15) and (25) are also equivalent in the following strong sense: each one can be obtained from the other by solving the corresponding ODE in part (i) or (ii) and applying parts (iv) or (iii), accordingly. In particular, the maximal interval of time (T_-, T_+) where a solution exists is exactly the same for both flows.
- The velocity of the flow $q(\gamma(t))$ must blow up at any finite-time singularity. More precisely, if $r > s$ and $T_+ < \infty$, then

$$|q(\gamma(t))|_t \geq \frac{C}{T_+ - t}, \qquad |\mu(t)| \geq \frac{C}{(T_+ - t)^{\frac{1}{(r-s)(1-\alpha)}}}, \qquad \forall t \in [0, T_+),$$

  for some $C > 0$ depending only on the dimension $k + n$ and the geometric flow (see [38, Corollary 5] and [38, Proposition 5], respectively).
- $\mu(t)$ may converge to a geometric structure on a different homogeneous space, as explained in Remark 6.1 (see [38, Corollary 4]). This occurs already in dimension 3 for the Ricci flow (see [36]).
- Many geometric flows have been studied in the homogeneous case using the bracket flow, including the Ricci flow [1, 18, 19, 27, 28, 34, 36, 54], the Chern-Ricci flow [37, 43], the pluriclosed flow [3, 12], the hermitian curvature flow [30], the symplectic curvature flow [44] and the G_2-Laplacian flow [5, 15, 39, 50].
- The bracket flow was recently applied by Böhm-Lafuente in [7, 8] to prove the convergence of any immortal homogeneous Ricci flow solution to a homogeneous expanding Ricci soliton.

6.2 Bracket Flow Evolution of Solitons

In the light of the equivalence between the flows (15) and (25), a natural question arises: how do solitons evolve according to the bracket flow? It is natural to expect an evolution of a very special kind.

If μ is a fixed point (up to scaling) of the bracket flow (25), i.e. $\mu(t) = c(t) \cdot \mu$ for some $c(t) \in \mathbb{R}^*$ (see (24)), then by just evaluating the equation at $t = 0$ one obtains

$$Q_\mu = cI + D_{\mathfrak{p}}, \qquad \text{for some} \quad c \in \mathbb{R}, \quad D = \begin{bmatrix} 0 & 0 \\ 0 & D_{\mathfrak{p}} \end{bmatrix} \in \operatorname{Der}(\mathfrak{g}/\mathfrak{k}). \tag{26}$$

This implies that $(G_\mu/K_\mu, \gamma)$ is an algebraic soliton (see (19) and (20)) satisfying (18). The converse assertion also holds, see Corollary 6.5 below.

We now describe the bracket flow evolution of any semi-algebraic soliton.

Proposition 6.4 *Let $(G_\mu/K_\mu, \gamma)$ be a semi-algebraic soliton as in (13), say $q(\gamma) = c\gamma + \theta(D_{\mathfrak{p}})$, and assume that (18) holds. Then the bracket flow solution starting at μ is given by*

$$\mu(t) = c(t)^{\frac{1}{s-r}} \cdot \left(\begin{bmatrix} I & 0 \\ 0 & e^{s(t)A} \end{bmatrix} \cdot \mu \right), \qquad A := \operatorname{pr}_{\mathfrak{gl}(\mathfrak{p})_\gamma}(D_{\mathfrak{p}}), \tag{27}$$

where $c(t)$ and $s(t)$ are as in (17) and $\operatorname{pr}_{\mathfrak{gl}(\mathfrak{p})_\gamma} : \mathfrak{gl}(\mathfrak{p}) \to \mathfrak{gl}(\mathfrak{p})_\gamma$ is the projection relative to the decomposition $\mathfrak{gl}(\mathfrak{p}) = \mathfrak{gl}(\mathfrak{p})_\gamma \oplus \mathfrak{q}_\gamma$ given in (11).

Proof It follows from (17) that $\gamma(t) = c(t)e^{s(t)D_{\mathfrak{p}}} \cdot \gamma$ and thus

$$\begin{aligned} q(\gamma(t)) =& c(t)^\alpha e^{s(t)D_{\mathfrak{p}}} \cdot q(\gamma) = cc(t)^\alpha e^{s(t)D_{\mathfrak{p}}} \cdot \gamma + c(t)^\alpha \theta(D_{\mathfrak{p}}) e^{s(t)D_{\mathfrak{p}}} \cdot \gamma \\ =& cc(t)^{\alpha-1}\gamma(t) + c(t)^{\alpha-1}\theta(D_{\mathfrak{p}})\gamma(t). \end{aligned}$$

This implies that

$$\begin{aligned} Q_{\gamma(t)} =& \tfrac{cc(t)^{\alpha-1}}{s-r} I + c(t)^{\alpha-1} \operatorname{pr}_{\gamma(t)}(D_{\mathfrak{p}}) \\ =& \tfrac{cc(t)^{\alpha-1}}{s-r} I + c(t)^{\alpha-1} e^{s(t)D_{\mathfrak{p}}} \operatorname{pr}_\gamma(D_{\mathfrak{p}}) e^{-s(t)D_{\mathfrak{p}}} \\ =& e^{s(t)D_{\mathfrak{p}}} \left(\tfrac{cc(t)^{\alpha-1}}{s-r} I + c(t)^{\alpha-1}(D_{\mathfrak{p}} - A) \right) e^{-s(t)D_{\mathfrak{p}}} \\ =& c(t)^{\alpha-1} e^{s(t)D_{\mathfrak{p}}} Q_\gamma e^{-s(t)D_{\mathfrak{p}}}, \end{aligned} \tag{28}$$

where $A := \operatorname{pr}_{\mathfrak{gl}(\mathfrak{p})_\gamma}(D_{\mathfrak{p}})$. According to Theorem 6.2, one can compute the bracket flow solution $\mu(t)$ by solving the differential equation given in its part (i). It is straightforward to check that the solution is given by

$$h(t) := c(t)^{\frac{1}{r-s}} e^{s(t)A} e^{-s(t)D_{\mathfrak{p}}},$$

where $c(t)$ and $s(t)$ are as in (17), and therefore formula (27) follows from (18), concluding the proof. □

Corollary 6.5 *$(G_\mu/K_\mu, \gamma)$ is an algebraic soliton satisfying (18) if and only if μ is a fixed point (up to scaling) of the bracket flow.*

In particular, algebraic solitons are precisely the possible limits of any normalized bracket flow (cf. Remark 6.1). Furthermore, given a starting point, one can obtain at most one non-flat algebraic soliton as a limit by running all possible normalized bracket flow solutions.

We note that for expanding (resp. shrinking) solitons, the function s is strictly increasing and $s(t) \to \infty$ as $t \to \infty$ (resp. $s(t) \to -\infty$ as $t \to -\infty$). Recall also that $s(t) = t$ for steady solitons.

In the case when A is skew-symmetric (e.g. if a metric is attached to γ, since in that case $\mathfrak{gl}(\mathfrak{p})_\gamma \subset \mathfrak{so}(\mathfrak{p})$), its eigenvalues are either purely imaginary numbers or zero, say $\pm i a_1, \dots, \pm i a_m, 0, \dots, 0$ $(a_j > 0)$. Thus exactly one of the following two behaviors occur:

(i) If the set $\{a_1, \dots, a_m\}$ is linearly dependent over $\mathbb{Q}$, then there exists a sequence t_k, with $t_k \to \pm\infty$ (depending on the type of soliton), such that $e^{s(t_k)A} = I$ for all k, and thus the bracket flow solution projected on the sphere is periodic.
(ii) If on the contrary, the set $\{a_1, \dots, a_m\}$ is linearly independent over $\mathbb{Q}$, then by Kronecker's theorem, for each $t_0 \in (T_-, T_+)$, there exists a sequence t_k, with $t_k \to \pm\infty$, such that $e^{s(t_k)A} \to e^{s(t_0)A}$. This implies that the solution projected on the sphere is not periodic but develop the following chaotic behavior: for each t_0 there exists a sequence $t_k \to \infty$ such that $\mu(t_k)/|\mu(t_k)|$ converges to $\mu(t_0)/|\mu(t_0)|$, that is, each point of the solution on the sphere is contained in the ω-limit.

There are examples of G_2-Laplacian (see [39, 51]) and pluriclosed (see [3]) semi-algebraic solitons satisfying (i). However, the existence of a semi-algebraic soliton as in part (ii) is still an open problem.

7 Simultaneous Diagonalization of a Homogeneous Geometric Flow

In this section, we aim to characterize algebraic solitons among homogeneous solitons in a geometric way. It will be assumed here that γ comes with a Riemannian metric g_γ (see Remark 3.1).

Definition 7.1 A homogeneous geometric structure (M, γ) is said to be *flow diagonal* if the $\mathrm{Aut}(M, \gamma)$-invariant solution $\gamma(t)$ starting at γ (see Proposition 5.1) satisfies the following property: at some point $p \in M$, there exists an orthonormal basis β of T_pM such that the matrix $[Q_{\gamma(t)}]_\beta$ is diagonal for all t.

Note that the point p can be replaced by any other point of M by homogeneity. The property of being flow diagonal is a geometric invariant since given $f \in \mathrm{Diff}(M)$, the operators corresponding to $(M, f^*\gamma)$ are simultaneously conjugate, via $df|_{f^{-1}(p)}$, to those of (M, γ).

If (M, γ) is flow diagonal, then the velocity of the geometric flow $q(\gamma(t)) = \theta(Q_{\gamma(t)})\gamma(t)$ takes a very simple form in terms of the fixed basis β, making the study of any aspect of the solutions to the ODE (15) much more workable, including its qualitative behavior and the search for exact solutions. Indeed, once we consider a presentation $(M, \gamma) = (G/K, \gamma)$ and a reductive decomposition $\mathfrak{g} = \mathfrak{k} \oplus \mathfrak{p}$, recall from Theorem 6.2 that $\gamma(t) = h(t)^*\gamma$ for the solution $h(t) \in \mathrm{GL}(\mathfrak{p})$ to the ODE $\frac{d}{dt}h(t) = -h(t)Q_{\gamma(t)}$, $h(0) = I$. It follows that the following conditions are equivalent:

(i) $(G/K, \gamma)$ is flow diagonal.
(ii) The family of operators $\{h(t) : t \in (T_-, T_+)\}$ is simultaneously diagonalizable with respect to an orthonormal basis of $\mathfrak{p}$.

For instance, the flow diagonal condition was assumed in the study of the Ricci flow on 4-dimensional homogeneous manifolds in [21, 49] and it was shown to hold in most of the exact Laplacian flow solutions on nilpotent Lie groups given in [15, Section 4].

Example 7.2 Let $(G/K, \gamma)$ be a simply connected semi-algebraic soliton such that (18) holds. It follows from (28) that

$$Q_{\gamma(t)} = c(t)^{\alpha-1} e^{s(t)D_{\mathfrak{p}}} Q_\gamma e^{-s(t)D_{\mathfrak{p}}}, \qquad \forall t. \tag{29}$$

In particular, if $(G/K, \gamma)$ is actually an algebraic soliton, then $[Q_\gamma, D_{\mathfrak{p}}] = 0$ and so $Q_{\gamma(t)} = c(t)^{\alpha-1}Q_\gamma$. Thus $(G/K, \gamma)$ is flow diagonal as soon as only Q_γ is diagonalizable with respect to an orthonormal basis, i.e. Q_γ symmetric.

We now show that the flow diagonal condition characterizes algebraic solitons among homogeneous solitons. Recall that according to Theorem 5.2, any homogeneous soliton can be presented as a semi-algebraic soliton.

Theorem 7.3 *A semi-algebraic soliton $(G/K, \gamma)$ with K compact and Q_γ symmetric is flow diagonal if and only if it is an algebraic soliton.*

Remark 7.4 In particular, any semi-algebraic soliton which is equivalent (via a diffeomorphism) to an algebraic soliton must be an algebraic soliton itself.

Proof We have that $Q_\gamma = cI + S$, where $D_{\mathfrak{p}} = A + S$, $A \in \mathfrak{gl}(\mathfrak{p})_\gamma$, $A^t = -A$ (since there is a metric attached to γ), and $S \in \mathfrak{q}_\gamma$, $S^t = S$ (since $Q_\gamma^t = Q_\gamma$). According to Proposition 4.2, we can assume that D satisfies (18). If $(G/K, \gamma)$ is flow diagonal, then it follows from (29) that

$$[e^{sD_{\mathfrak{p}}} Q_\gamma e^{-sD_{\mathfrak{p}}}, Q_\gamma] = 0, \qquad \forall s \in (-\epsilon, \epsilon).$$

Hence $[[D_\mathfrak{p}, Q_\gamma], Q_\gamma] = 0$ and

$$0 = \operatorname{tr} D_\mathfrak{p}[[D_\mathfrak{p}, Q_\gamma], Q_\gamma] = -\operatorname{tr}[D_\mathfrak{p}, Q_\gamma]^2 = -\operatorname{tr}[A, S]^2,$$

from which follows that $[A, S] = 0$ as it is symmetric, and thus $D_\mathfrak{p}$ is normal. This implies that D is normal as well relative to any extension of the inner product to $\mathfrak{g}$ and so $D^t \in \operatorname{Der}(\mathfrak{g})$. Thus $Q_\gamma = cI + \frac{1}{2}(D_\mathfrak{p} + D_\mathfrak{p}^t)$, with $\frac{1}{2}(D + D^t) \in \operatorname{Der}(\mathfrak{g})$, showing that $(G/K, \gamma)$ is an algebraic soliton. □

Acknowledgement The author gratefully acknowledges Valeria Gutiérrez and Marina Nicolini for helpful comments during the preparation of this paper.

References

1. R. Arroyo, The Ricci flow in a class of solvmanifolds. Differ. Geom. Appl. **31**, 472–485 (2013)
2. R. Arroyo, R. Lafuente, Homogeneous Ricci solitons in low dimensions. Int. Math. Res. Notices **2015**, 4901–4932 (2015)
3. R. Arroyo, R. Lafuente, The long-time behavior of the homogeneous pluriclosed flow. Proc. Lond. Math. Soc. **119**, 266–289 (2019)
4. V. Ayala, J. Tirao, Linear control systems on Lie groups and controllability. Proc. Symp. Pure Math. **64**, 47–64 (1999)
5. L. Bagaglini, A. Fino, The Laplacian coflow on almost-abelian Lie groups. Ann. Mat. Pura Appl. **197**, 1855–1873 (2018)
6. A. Besse, Einstein manifolds, in *Ergeb. Math.*, vol. 10 (Springer, Berlin-Heidelberg, 1987)
7. C. Böhm, R. Lafuente, Immortal homogeneous Ricci flows. Invent. Math. **212**, 461–529 (2018)
8. C. Böhm, R. Lafuente, The Ricci flow on solvmanifolds of real type. Adv. Math. **352**, 516–540 (2019)
9. M. Brozos-Vázquez, G. Calvaruso, E. García-Río, S. Gavino-Fernández, Three- dimensional Lorentzian homogeneous Ricci solitons. Israel J. Math. **188**, 385–403 (2012)
10. G. Calvaruso, A. Fino, Four-dimensional pseudo-Riemannian homogeneous Ricci solitons. Int. J. Geom. Methods Mod. Phys. **12**(5), 1550056 (2015)
11. B. Chow, S.-C. Chu, D. Glickenstein, C. Guenther, J. Isenberg, T. Ivey, D. Knopf, P. Lu, F. Luo, L. Ni, *The Ricci flow: Techniques and Applications, Part I: Geometric Aspects.* AMS Math. Surv. Mon., vol. 135 (Amer. Math. Soc., Providence, 2007)
12. N. Enrietti, A. Fino, L. Vezzoni, The pluriclosed flow on nilmanifolds and tamed symplectic flow. J. Geom. Anal. **25**, 883–909 (2015)
13. E. Fernández-Culma, Classification of 7-dimensional Einstein Nilradicals. Transf. Groups **17**, 639–656 (2012)
14. E. Fernández-Culma, Soliton almost Kähler structures on 6-dimensional nilmanifolds for the symplectic curvature flow. J. Geom. Anal. **25**, 2736–2758 (2015)
15. M. Fernández, A. Fino, V. Manero, Laplacian flow of closed G_2-structures inducing nilsolitons. J. Geom. Anal. **26**, 1808–1837 (2016)
16. A. Fino, A. Raffero, Closed warped G_2-structures evolving under the Laplacian flow. Ann. Scuola Sup. Pisa **20**, 315–348 (2020)
17. A. Fino, A. Raffero, Remarks on homogeneous solitons of the G_2-Laplacian flow. Comp. Rend. Math. (in press)
18. D. Glickenstein, T.L. Payne, Ricci flow on three-dimensional, unimodular metric Lie algebras. Commun. Anal. Geom. **18**, 927–962 (2010)

19. G. Guzhvina, The action of the Ricci flow on almost flat manifolds, Ph.D. Thesis, Universität Münster (2007)
20. R.S. Hamilton, The Ricci flow on surfaces. Contemp. Math. **71**, 237–261 (1988)
21. J. Isenberg, M. Jackson, P. Lu, Ricci flow on locally homogeneous closed 4-manifolds. Commun. Anal. Geom. **14**, 345–386 (2006)
22. M. Jablonski, Homogeneous Ricci solitons are algebraic. Geom. Topol. **18**, 2477–2486 (2014)
23. M. Jablonski, Homogeneous Ricci solitons. J. Reine Angew. Math. **699**, 159–182 (2015)
24. M. Jablonski, P. Petersen, M.B. Williams, Linear stability of algebraic Ricci solitons. J. Reine Angew. Math. **713**, 181–224 (2016)
25. H. Kadioglu, T.L. Payne, Computational methods for nilsoliton metric Lie algebras I. J. Symb. Comp. **50**, 350–373 (2013)
26. S. Karigiannis, B. McKay, M.-P. Tsui, Soliton solutions for the Laplacian co-flow of some G_2-structures with symmetry. Differ. Geom. Appl. **30**, 318–333 (2012)
27. R. Lafuente, Scalar curvature behavior of homogeneous Ricci flows. J. Geom. Anal. **25**, 2313–2322 (2015)
28. R. Lafuente, J. Lauret, On homogeneous Ricci solitons. Q. J. Math. **65**, 399–419 (2014)
29. R. Lafuente, J. Lauret, Structure of homogeneous Ricci solitons and the Alekseevskii conjecture. J. Differ. Geom. **98**, 315–347 (2014)
30. R. Lafuente, M. Pujia, L. Vezzoni, Hermitian Curvature flow on Lie groups and static invariant metrics. Trans. Am. Math. Soc. **373**, 3967–3993 (2020)
31. J. Lauret, A canonical compatible metric for geometric structures on nilmanifolds. Ann. Global Anal. Geom. **30**, 107–138 (2006)
32. J. Lauret, Einstein solvmanifolds and nilsolitons. Contemp. Math. **491**, 1–35 (2009)
33. J. Lauret, Ricci soliton solvmanifolds. J. Reine Angew. Math. **650**, 1–21 (2011)
34. J. Lauret, The Ricci flow for simply connected nilmanifolds. Commun. Anal. Geom. **19**, 831–854 (2011)
35. J. Lauret, Convergence of homogeneous manifolds. J. Lond. Math. Soc. **86**, 701–727 (2012)
36. J. Lauret, Ricci flow of homogeneous manifolds. Math. Z. **274**, 373–403 (2013)
37. J. Lauret, Curvature flows for almost-hermitian Lie groups. Trans. Am. Math. Soc. **367**, 7453–7480 (2015)
38. J. Lauret, Geometric flows and their solitons on homogeneous spaces (Workshop in memory of Sergio Console). Rend. Sem. Mat. Torino **74**, 55–93 (2016)
39. J. Lauret, Laplacian flow of homogeneous G_2-structures and its solitons. Proc. Lond. Math. Soc. **114**, 527–560 (2017)
40. J. Lauret, Laplacian solitons: questions and homogeneous examples. Differ. Geom. Appl. **54**, 345–360 (2017)
41. J. Lauret, Finding solitons, in *Notices Amer. Math. Soc.* (2020)
42. J. Lauret, M. Nicolini, Extremally Ricci pinched G_2-structures on Lie groups. Commun. Anal. Geom. (in press)
43. J. Lauret, E. Rodríguez-Valencia, On the Chern-Ricci flow and its solitons for Lie groups. Math. Nachr. **288**, 1512–1526 (2015)
44. J. Lauret, C.E. Will, On the symplectic curvature flow for locally homogeneous manifolds. J. Symp. Geom. **15**, 1–49 (2017)
45. J. Lauret, C.E. Will, On Ricci negative Lie groups, in *Abel Symposia* (Springer, Berlin, 2019)
46. H-V. Le, G. Wang, Anti-complexified Ricci flow on compact symplectic manifolds. J. Reine Angew. Math. **530**, 17–31 (2001)
47. C. Lin, Laplacian solitons and symmetry in G_2-geometry. J. Geom. Phys. **64**, 111–119 (2013)
48. J. Lotay, Geometric flows of G_2 structures, in *Fields Institute Communications* (Springer, Berlin, in press)
49. J. Lott, On the long-time behavior of type-III Ricci flow solutions. Math. Ann. **339**, 627–666 (2007)
50. A. Moreno, H. Sá Earp, Explicit soliton for the Laplacian co-flow on a solvmanifold. São Paulo J. Math. Sci. (in press)

51. M. Nicolini, Laplacian solitons on nilpotent Lie groups. Bull. Belgian Math. Soc. **25**, 183–196 (2018)
52. Y. Nikolayevsky, Einstein solvmanifolds and the pre-Einstein derivation. Trans. Am. Math. Soc. **363**, 3935–3958 (2011)
53. K. Onda, Examples of algebraic Ricci Solitons in the Pseudo-Riemannian case. Acta Math. Hung. (in press)
54. T. Payne, The Ricci flow for nilmanifolds. J. Modern Dyn. **4**, 65–90 (2010)
55. M. Pujia, Expanding solitons to the Hermitian curvature flow on complex Lie groups. Differ. Geom. App. (in press)
56. J. Streets, Classification of solitons for pluriclosed flow on complex surfaces. Math. Ann. **375**, 1555–1595 (2019)
57. J. Streets, G. Tian, Hermitian curvature flow. J. Eur. Math. Soc. **13**, 601–634 (2011)
58. J. Streets, G. Tian, Symplectic curvature flow. J. Reine Angew. Math. **696**, 143–185 (2014)
59. V. Tosatti, B. Weinkove, On the evolution of a hermitian metric by its Chern-Ricci form. J. Differ. Geom. **99**, 125–163 (2015)
60. C.E. Will, The space of solvsolitons in low dimensions. Ann. Global Anal. Geom. **40**, 291–309 (2011)

On Ricci Negative Lie Groups

Jorge Lauret and Cynthia E. Will

Abstract We give an overview of what is known on Lie groups admitting a left-invariant metric of negative Ricci curvature, including many natural questions and conjectures in the solvable case. We also introduce an open and convex cone $\mathcal{C}(\mathfrak{n})$ of derivations attached to each nilpotent Lie algebra $\mathfrak{n}$, which is defined as the image of a moment map and parametrizes a set of solvable Lie algebras with nilradical $\mathfrak{n}$ admitting Ricci negative metrics.

1 Introduction

The main objects of study in this paper can be introduced as generalizations of the Euclidean space as follows: Riemannian manifolds which are also groups and left translations are all isometries. This somewhat naive presentation of Lie groups endowed with left-invariant metrics could be hiding the rich variety of possible curvature behaviors this class of metrics exhibits, as shown by Milnor in his seminal paper [31].

Given a Lie group G, it is natural to expect a nice and strong interplay between any prescribed curvature condition of left-invariant Riemannian metrics on G and not only the topology of G but also the algebraic structure of its Lie algebra $\mathfrak{g}$. Structurally, the only strict pinching curvature behavior which is still not understood in the homogeneous case is $\operatorname{Ric} < 0$ (see e.g. the overview given in [33, Introduction]). Indeed, the following natural problem is wide open and represents our leitmotiv in this paper.

This research was partially supported by grants from FONCyT and SeCyT (UNC).

J. Lauret (✉) · C. E. Will
FaMAF, Universidad Nacional de Córdoba and CIEM, CONICET, Córdoba, Argentina
e-mail: lauret@famaf.unc.edu.ar; cwill@famaf.unc.edu.ar

S. Hervik et al. (eds.), *Geometry, Lie Theory and Applications*, Abel Symposia 16,
https://doi.org/10.1007/978-3-030-81296-6_9

Question 1.1 Which Lie groups admit a left-invariant metric of negative Ricci curvature?

We recall that in the general case there are no topological obstructions on a differentiable manifold M to the existence of a complete Riemannian metric with $\mathrm{Ric} < 0$ (see [30]).

After surveying the subject in Sects. 2 and 3, we propose several natural questions and conjectures in Sect. 4 in the context of solvable Lie groups. Finally, some new results are proved in Sect. 5, including the construction of an open and convex cone $\mathcal{C}(\mathfrak{n})$ of derivations, attached to each nilpotent Lie algebra $\mathfrak{n}$, parametrizing a subset (conjectured to be essentially the full set) of one-dimensional solvable extensions of $\mathfrak{n}$ admitting $\mathrm{Ric} < 0$ metrics.

2 Some History

In what follows, we overview the rich history of the existence problem for negatively curved metrics on Lie groups (from now on, any metric on a Lie group is assumed to be left-invariant).

2.1 Negative Sectional

We start with four theorems concerning scalar and sectional curvature.

Theorem 2.1 ([31]) *Any non-abelian Lie group admits* $\mathrm{scal} < 0$ *metrics. Moreover, any non-flat metric on a solvable Lie group has* $\mathrm{scal} < 0$.

Theorem 2.2 ([2, 15]) *A Lie group admits* $\mathrm{Sec} < 0$ *metrics if and only if,*

(i) $\dim \mathfrak{n} = \dim \mathfrak{s} - 1$, *where* $\mathfrak{s}$ *denotes the corresponding Lie algebra and* $\mathfrak{n}$ *is the nilradical of* $\mathfrak{s}$ *(i.e. the maximal nilpotent ideal); in particular,* $\mathfrak{s}$ *is solvable.*
(ii) *There is an element* $Y \in \mathfrak{s}$ *such that* $\operatorname{ad} Y|_{\mathfrak{n}} > 0$ *(i.e. its eigenvalues have all positive real parts).*

Thus the structure of a $\mathrm{Sec} < 0$ Lie algebra is simply given by a nilpotent Lie algebra together with a positive derivation. Unfortunately, the question of which nilpotent Lie algebras admit a positive derivation seems to be hopeless.

Theorem 2.3 ([12]) *A (solvable) Lie group admits a metric such that* $-4 \le \mathrm{Sec} \le -1$ *if and only if,*

(i) $\mathfrak{n}$ *is 2-step nilpotent or abelian.*
(ii) *There is an element* $Y \in \mathfrak{s}$ *such that the real part of any eigenvalue of* $\operatorname{ad} Y|_{\mathfrak{n}}$ *belongs to the interval* $[1, 2]$.
(iii) $\operatorname{ad} Y|_{\mathfrak{n}}$ *is given by* αI *when restricted to the generalized eigenspace of any eigenvalue* α *such that* $\mathrm{Re}(\alpha) = 1$ *or* 2.

It is further obtained in [12] that if $\mathfrak{n}$ is non-abelian, then $\operatorname{ad} Y|_{\mathfrak{n}}$ must actually be, up to amalgamated products, the canonical derivation of a 2-step algebra given by $2I$ on the center and I on its orthogonal complement.

Theorem 2.4 *A Lie group admits a* $\mathrm{Sec} \leq 0$ *metric if and only if*

(i) $[\mathfrak{a}, \mathfrak{a}] = 0$*, where* $\mathfrak{s} = \mathfrak{a} \oplus \mathfrak{n}$ *is the orthogonal decomposition; in particular,* $\mathfrak{s}$ *is solvable.*
(ii) *There is an element* $Y \in \mathfrak{s}$ *such that* $\operatorname{ad} Y|_{\mathfrak{n}} \geq 0$.
(iii) *For any* $Y' \in \mathfrak{a}$*,* $\operatorname{ad} Y'|_{\mathfrak{n}}$ *coincides with* αI *when restricted to the generalized eigenspace of any eigenvalue* α *such that* $\mathrm{Re}(\alpha) = 0$.
(iv) *Some extra technical conditions.*

Moreover, any homogeneous Riemannian manifold with $\mathrm{Sec} \leq 0$ *is isometric to a metric on a simply connected solvable Lie group.*

2.2 Negative Ricci, Non-solvable Case

Let us now consider the curvature behavior addressed in Question 1.1: $\mathrm{Ric} < 0$. Unlike negative sectional curvature, non-solvable Lie groups come into play in the Ricci negative case.

Theorem 2.5 ([9]) *No unimodular solvable Lie group can admit* $\mathrm{Ric} < 0$ *metrics.*

Theorem 2.6 ([11]) *The only unimodular Lie groups that can admit* $\mathrm{Ric} < 0$ *metrics are the non-compact semisimple ones.*

The following is a list of some of the known results, constructions and examples in the non-solvable case.

- Dotti and Leite [10] $\mathrm{SL}_n(\mathbb{R})$ admits $\mathrm{Ric} < 0$ metrics for any $n \geq 3$. The corresponding compact quotients $\mathrm{SL}_n(\mathbb{R})/\Gamma$ by any lattice Γ gave the first examples of compact manifolds admitting both a $\mathrm{scal} > 0$ metric and a $\mathrm{Ric} < 0$ metric, answering a question in [35, pp. 242].
- Dotti et al. [11] Most non-compact simple Lie groups do admit $\mathrm{Ric} < 0$ metrics, with some low dimensional exceptions, including $\mathrm{SL}_2(\mathbb{C})$, $\mathrm{Sp}(2, \mathbb{R})$ and G_2^*. The existence of $\mathrm{Ric} < 0$ metrics on these simple Lie groups is still open.
- Milnor [31] $\mathrm{SL}_2(\mathbb{R})$ does not admit $\mathrm{Ric} < 0$ metrics. This is actually the only simple Lie group for which the non-existence of $\mathrm{Ric} < 0$ metrics has been proved. Moreover, we do not know of any example of a non-unimodular non-solvable Lie group G such that G admits no $\mathrm{Ric} < 0$ metric.
- Jablonski and Petersen A semisimple Lie group admitting a metric with $\mathrm{Ric} < 0$ can not have compact factors, i.e. it is of non-compact type. On the other hand, we do not know whether the existence of $\mathrm{Ric} < 0$ metrics on a semisimple Lie group implies the existence of such metrics on each of its factors.

- Will [36, 37] Unexpected examples of Lie groups admitting Ric < 0 metrics which are neither semisimple nor solvable. Moreover, the Levi factors of some of these examples are compact, including SU(n) ($n \geq 2$), SO(n) ($n \geq 3$) and Sp(n) ($n \geq 2$).
- Will [37] Any non-compact semisimple Lie group admitting a Ric < 0 metric can be the Levi factor of a non-semisimple Lie group with a Ric < 0 metric. Non-abelian nilradicals are possible in most of these constructions.
- Will [36] For any (non-trivial) real representation W of $\mathfrak{su}(2)$, the Lie algebra

$$\mathfrak{g} = (\mathfrak{su}(2) \oplus \mathbb{R}) \ltimes W$$

admits Ric < 0 metrics, where $\mathbb{R}$ is acting on W by possibly different multiples of the identity on each irreducible component. This produces the example of lowest possible dimension of a Ric < 0 non-semisimple and non-solvable Lie algebra: $\mathfrak{g} = (\mathfrak{su}(2) \oplus \mathbb{R}) \ltimes \mathbb{R}^3$, $\dim \mathfrak{g} = 7$.
- Lauret and Will [20] Any compact semisimple Lie group can be the Levi factor of a Lie group admitting a Ric < 0 metric. Moreover, given a compact simple Lie algebra $\mathfrak{u}$, all but finitely many irreducible representations of its complexification $\mathfrak{u} \otimes \mathbb{C}$ (including the adjoint representation) can be the abelian nilradical of a Lie algebra $\mathfrak{g}$ with Levi factor $\mathfrak{u}$ admitting Ric < 0 metrics. We have listed in Table 1, for each $\mathfrak{u}$, the example of lowest dimension known of such a $\mathfrak{g} = (\mathfrak{u} \oplus \mathbb{R}) \ltimes V$, where (V, π) is the (complex) irreducible representation of $\mathfrak{u} \otimes \mathbb{C}$ viewed as a real representation of $\mathfrak{u}$; in particular

$$\dim \mathfrak{g} = \dim \mathfrak{u} + 1 + 2 \dim_{\mathbb{C}} V.$$

In Table 1, the fundamental representations of $\mathfrak{u} \otimes \mathbb{C}$ are denoted by $\omega_1, \omega_2, \ldots$ and so on in the order used by the software package LiE.

Table 1 Smallest dimension known of a Ric < 0 Lie algebra $\mathfrak{g} = (\mathfrak{u} \oplus \mathbb{R}) \ltimes V$ with a given compact simple Levi factor $\mathfrak{u}$

Type	$\mathfrak{u}$	$\dim \mathfrak{u}$	π	$\dim \mathfrak{g}$
A_n, $n \geq 2$	$\mathfrak{su}(n+1)$	$n(n+2)$	$2\omega_1 = Sym^2(\mathbb{C}^{n+1})$	$2n^2 + 5n + 3$
B_n, $n \geq 2$	$\mathfrak{so}(2n+1)$	$n(2n+1)$	$\omega_2 = \mathrm{ad}$	$6n^2 + 3n + 1$
C_n, $n \geq 3$	$\mathfrak{sp}(n)$	$n(2n+1)$	$2\omega_1 = \mathrm{ad}$	$6n^2 + 3n + 1$
D_n, $n \geq 4$	$\mathfrak{so}(2n)$	$n(2n-1)$	$\omega_2 = \mathrm{ad}$	$6n^2 - 3n + 1$
E_6	E_6	78	$\omega_2 = \mathrm{ad}$	235
E_7	E_7	133	$\omega_1 = \mathrm{ad}$	400
E_8	E_8	248	$\omega_8 = \mathrm{ad}$	745
F_4	F_4	52	$\omega_1 = \mathrm{ad}$	157
G_2	G_2	14	$\omega_2 = \mathrm{ad}$	43

2.3 Negative Ricci, Solvable Case

We now focus on the existence of Ricci negative metrics in the solvable case. Note that unimodular solvable (in particular nilpotent) Lie groups are not allowed in the Ricci negative case by Theorem 2.5. Recall the orthogonal decomposition $\mathfrak{s} = \mathfrak{a} \oplus \mathfrak{n}$, where $\mathfrak{n}$ is the nilradical of the solvable Lie algebra $\mathfrak{s}$.

- Nikolayevsky and Nikonorov [33] *Sufficient condition*: There exists $Y \in \mathfrak{s}$ such that $\operatorname{ad} Y|_{\mathfrak{n}} > 0$ (recall that this means $\operatorname{Re}\operatorname{Spec}(\operatorname{ad} Y|_{\mathfrak{n}}) > 0$). Note that the nilradicals involved in these examples are the same as those needed for Sec < 0 (cf. Theorem 2.2), although the condition $[\mathfrak{a}, \mathfrak{a}] = 0$ is not mandatory here as in the case of Sec ≤ 0 (cf. Theorem 2.4).
- Nikolayevsky and Nikonorov [33] *Necessary condition*: There exists $Y \in \mathfrak{s}$ such that $\operatorname{tr}\operatorname{ad} Y > 0$ and $\operatorname{ad} Y|_{\mathfrak{z}(\mathfrak{n})} > 0$, where $\mathfrak{z}(\mathfrak{n})$ is the center of $\mathfrak{n}$. Surprisingly (or not), this is the only known general obstruction so far.

For a given $D \in \operatorname{Der}(\mathfrak{n})$, we consider its additive Jordan decomposition,

$$D = D^{\mathbb{R}} + D^{i\mathbb{R}} + D^{n}, \qquad [D^{\mathbb{R}}, D^{i\mathbb{R}}] = 0, \quad [D^{\mathbb{R}}, D^{n}] = 0, \quad [D^{i\mathbb{R}}, D^{n}] = 0,$$

where $D^{\mathbb{R}}$ is diagonalizable (over $\mathbb{R}$), $D^{i\mathbb{R}}$ is diagonalizable over $\mathbb{C}$ with only imaginary eigenvalues and D^{n} is nilpotent. It is well known that $D^{\mathbb{R}}, D^{i\mathbb{R}}, D^{n} \in \operatorname{Der}(\mathfrak{n})$. A maximal abelian subspace $\mathfrak{t}(\mathfrak{n})$ of diagonalizable (over $\mathbb{R}$) derivations is called a *maximal torus* of $\mathfrak{n}$, which is known to be unique up to $\operatorname{Aut}(\mathfrak{n})$-conjugation.

In the light of the above and some other results in the particular cases of Heisenberg and filiform Lie algebras as nilradicals obtained in [32, 33], a complete characterization of solvable Lie algebras admitting Ric < 0 metrics is expected to take the following form.

Conjecture 2.7 ([33]) *For each nilpotent Lie algebra* $\mathfrak{n}$*, there is an open and convex cone* $C \subset \mathfrak{t}(\mathfrak{n})$ *such that a solvable Lie algebra* $\mathfrak{s}$ *with nilradical* $\mathfrak{n}$ *admits a* Ric < 0 *metric if and only if there exists* $Y \in \mathfrak{s}$ *such that* $\operatorname{ad} Y|_{\mathfrak{n}}^{\mathbb{R}} \in C$ *(up to automorphism conjugation).*

The following rank-one reduction was proved to hold in the Einstein case in [14].

Conjecture 2.8 ([33]) *A solvable Lie algebra* $\mathfrak{s}$ *with nilradical* $\mathfrak{n}$ *admits a* Ric < 0 *metric if and only if there exists* $Y \in \mathfrak{s}$ *such that the Lie subalgebra* $\mathbb{R}Y \oplus \mathfrak{n}$ *admits a* Ric < 0 *metric.*

On the other hand, the structural conditions on a solvable Lie algebra imposed by the negativity of sectional or Ricci curvature described above motivates the following natural question.

Question 2.9 Which nilpotent Lie algebras can be the nilradical of a solvable Lie algebra admitting Ric < 0 metrics?

Such a Lie algebra will be called a *Ricci negative nilradical* (RN-nilradical for short). Since the existence of a positive derivation is sufficient, even for the existence of a Sec < 0 solvable extension by Theorem 2.2, any nilpotent Lie algebra which is 2-step or has dimension ≤ 6 is a RN-nilradical. It is also known that any 7-dimensional nilpotent Lie algebra (admitting a non-nilpotent derivation, or equivalently, a non-nilpotent solvable extension) is a RN-nilradical (see [8]). However, the following examples suggest that, as in the case of Einstein nilradicals (see Remark 2.10 below), a structural characterization of RN-nilradicals may be hopeless.

- Deré and Lauret [8] Explicit examples of nilpotent Lie algebras with the following properties (cf. the necessary condition above):

  (i) There are no diagonalizable derivations but not every derivation is nilpotent.
  (ii) There is a diagonalizable derivation but any derivation is traceless.

- Deré and Lauret [8] Explicit examples of nilpotent Lie algebras which are not RN-nilradicals. They all have a derivation of positive trace and one example for each of the following features is provided (for the first three examples any diagonalizable derivation has a zero eigenvalue on the center):

  (i) $\dim \mathfrak{n} = 8$ (5-step).
  (ii) $\mathfrak{n}$ is 3-step nilpotent ($\dim \mathfrak{n} = 10$).
  (iii) A continuous family of pairwise non-isomorphic 6-step nilpotent Lie algebras of dimension 13.
  (iv) $\mathfrak{n}$ has a non-singular derivation but any diagonalizable derivation has a negative eigenvalue on the center ($\dim \mathfrak{n} = 13, 17$, 5-step).

- Deré and Lauret [8] A 10-dimensional 5-step example of a RN-nilradical such that any diagonalizable derivation has a negative eigenvalue.

Remark 2.10 If in Question 2.9 we replace Ric < 0 by *Einstein* (i.e. $\mathrm{Ric}_g = cg$), then what it is analogously obtained are the so called *Einstein nilradicals* (see the survey [24] for further information). To admit a derivation with natural numbers as eigenvalues (so defining an $\mathbb{N}$-gradation) is the only known general obstruction for Einstein nilradicals and no sufficient structural condition is available. Several classification results within special classes of algebras are known in the literature though. Remarkably, Einstein nilradicals are precisely the nilpotent Lie algebras admitting a Ricci soliton metric, which is known to be unique up to isometry and scaling.

Remark 2.11 There are also in the literature fine results on the existence of negatively curved G-invariant metrics on a homogeneous space G/K. For instance, the homogeneous space $\mathrm{SO}^+(n,2)/\mathrm{SO}(n)$ ($n \geq 2$), which is homeomorphic to $S^1 \times \mathbb{R}^k$, does admit a Ric < 0 invariant metric (see [34, Example 1]). On the other hand, it is proved in [3] that any invariant metric on a given G/K has scal < 0 if and only if the universal cover of G/K is the Euclidean space.

3 Degeneration Principle

It is evident that all the geometric information on a Lie group endowed with a left-invariant metric, say (G, g), is encoded just in the inner product $\langle\cdot,\cdot\rangle := g_e$ and the Lie bracket of the Lie algebra $\mathfrak{g}$. This suggests the following point of view, called the *moving-bracket approach*, which proposes to vary Lie brackets rather than inner products.

We fix a real vector space $\mathfrak{g}$ endowed with a fixed inner product $\langle\cdot,\cdot\rangle$. Let $\mathcal{L} \subset \Lambda^2\mathfrak{g}^* \otimes \mathfrak{g}$ denote the algebraic subset of all Lie brackets on $\mathfrak{g}$. Each $\mu \in \mathcal{L}$ is identified with the (left-invariant) metric determined by $\langle\cdot,\cdot\rangle$ on the simply connected Lie group G_μ with Lie algebra $(\mathfrak{g}, \mu)$:

$$\mu \longleftrightarrow (G_\mu, \langle\cdot,\cdot\rangle).$$

In this way, the isomorphism class $\mathrm{GL}(\mathfrak{g}) \cdot \mu$ is identified with the set of all metrics on G_μ as follows:

$$(G_{h\cdot\mu}, \langle\cdot,\cdot\rangle) \longleftrightarrow (G_\mu, \langle h\cdot, h\cdot\rangle), \qquad \forall h \in \mathrm{GL}(\mathfrak{g}),$$

where $h \cdot \mu := h\mu(h^{-1}\cdot, h^{-1}\cdot)$. Note that h^{-1} is an isometric isomorphism that determines an isometry between these two Riemannian manifolds. Thus any two Lie brackets in the same $\mathrm{O}(\mathfrak{g}, \langle\cdot,\cdot\rangle)$-orbit are isometric as Riemannian metrics; the converse assertion is known to hold (see [1]) for *completely solvable* Lie brackets (i.e. $\mathrm{Spec}(\mathrm{ad}_\mu X) \subset \mathbb{R}$ for any $X \in \mathfrak{g}$).

A key observation is that any kind of geometric quantity associated to the Riemannian manifold $(G_\mu, \langle\cdot,\cdot\rangle)$ depends continuously on the Lie bracket $\mu \in \mathcal{L}$, which can be used to study pinching curvature properties as follows.

Definition 3.1 Given $\mu, \lambda \in \mathcal{L}$, we say that μ *degenerates to* λ, denoted by $\mu \to \lambda$, if $\lambda \in \overline{\mathrm{GL}(\mathfrak{g}) \cdot \mu}$, where the closure relative to the usual vector space topology of $\Lambda^2\mathfrak{g}^* \otimes \mathfrak{g}$ is considered.

Recall the geometric role of the orbit $\mathrm{GL}(\mathfrak{g}) \cdot \mu$ as the set of all metrics on G_μ.

Proposition 3.2 (Degeneration Principle, See Fig. 1) *If $\mu \to \lambda$ and G_λ admits a metric satisfying a strict pinching curvature condition (e.g.* Ric < 0*), then there is also a metric on G_μ for which the same pinching curvature condition holds.*

Proof Without any lost of generality, we can assume that $(G_\lambda, \langle\cdot,\cdot\rangle)$ has Ric < 0. If $\mu \to \lambda$, then there is a sequence $h_k \in \mathrm{GL}(\mathfrak{g})$ such that $h_k \cdot \mu \to \lambda$, as $k \to \infty$. The pinching curvature condition will therefore hold for $(G_{h_k\cdot\mu}, \langle\cdot,\cdot\rangle)$ for sufficiently large k by continuity, which implies that it holds for the corresponding isometric metrics $\langle h_k\cdot, h_k\cdot\rangle$ on G_μ for sufficiently large k, concluding the proof. □

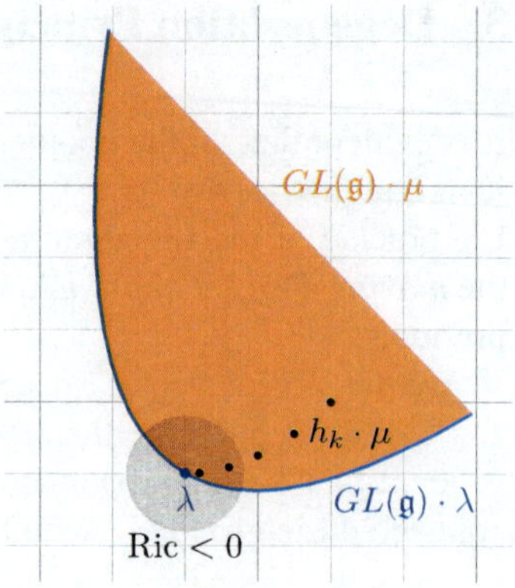

Fig. 1 Degeneration principle

We now overview many applications, starting back in the 70s, of the moving-bracket approach and the above principle in the study of negatively curved Lie groups.

- Milnor proved the first part of Theorem 2.1 by showing that any non-abelian Lie algebra of dimension n degenerates to exactly one of the following:

$$\mu_{heis}(e_1, e_2) = e_3; \qquad \mu_{hyp}(e_n, e_i) = e_i, \quad i = 1, \dots, n-1,$$

  which are easily seen to admit scal < 0 metrics. These two are actually the only non-abelian Lie groups admitting a unique metric up to isometry and scaling (see [22]), and the metric on μ_{hyp} is isometric to the real hyperbolic space $\mathbb{R}H^n$.
- To show that conditions (i) and (ii) of Theorem 2.2 are sufficient, Heintze considered the curve of Lie brackets μ_t on $\mathfrak{s}$ defined by $\mathrm{ad}_{\mu_t} Y|_{\mathfrak{n}} = t \,\mathrm{ad}\, Y|_{\mathfrak{n}}$, $\mu_t|_{\mathfrak{n}\times\mathfrak{n}} = [\cdot,\cdot]_{\mathfrak{n}}$ and proved that $\mathrm{Sec}_{\mu_t} < 0$ for sufficiently large t (see [15, Theorem 2]).
- In order to prove the above mentioned sufficient condition, Nikolayevsky-Nikonorov first showed that any solvable Lie algebra $\mu \in \mathcal{L}$ degenerates to a λ such that $\lambda(\mathfrak{a},\mathfrak{a}) = 0$, $\lambda(\mathfrak{n},\mathfrak{n}) = 0$, $\lambda(\mathfrak{a},\mathfrak{n}) \subset \mathfrak{n}$ and $\mathrm{ad}_\lambda X$ is a semisimple operator with the same spectrum as $\mathrm{ad}_\mu X$ for any $X \in \mathfrak{a}$, where $\mathfrak{g} = \mathfrak{a} \oplus \mathfrak{n}$ is any decomposition such that $\mathfrak{n}$ is the nilradical of μ. Then they also applied the Degeneration Principle to prove the sufficient conditions in [33, Theorems 3 and 4] and Nikolayevsky did it in the proof of [32, Theorem 1.3].
- A main part of most constructions of a non-solvable Lie algebra $\mathfrak{g}$ admitting Ric < 0 metrics in [36, 37] is to show that $\mathfrak{g}$ degenerates to a solvable Lie algebra $\mathfrak{s} = \mathfrak{a} \oplus \mathfrak{n}$ such that $[\mathfrak{a},\mathfrak{a}] = 0$. Interestingly, $\mathfrak{n}$ is non-abelian in spite of the nilradical of $\mathfrak{g}$ is abelian.
- In [8, Theorem 3.2], the Degeneration Principle is applied to obtain a characterization of Ricci negative one-dimensional solvable extensions of a given nilpotent Lie algebra in terms of the set of all Ricci operators at the points in the closure X of an orbit of nilpotent Lie brackets. Since X is irreducible (relative to the

real algebraic Zariski topology), this allowed the application of nice convexity properties of a certain moment map (see Theorem 4.10 below).

Remark 3.3 The moving-bracket approach was considered by J. Heber in his seminal article [14] on Einstein solvmanifolds and it has been further developed in [21, 23, 26]. We refer to [27, Section 5] and [6, 7, 28] for more applications. In most of the applications, concepts and results from geometric invariant theory, including moment maps, closed orbits, stability, categorical quotients and Kirwan stratification, have been exploited in one way or another.

4 Ricci Negative Derivations

As explained in Sect. 2, the diverse kinds of examples provided in [8] seems to indicate that Question 2.9 is too ambitious. In this section, we consider instead the following also very natural problem.

Question 4.1 Given a nilpotent Lie algebra $\mathfrak{n}$, which are the solvable Lie algebras with nilradical $\mathfrak{n}$ admitting a $\operatorname{Ric} < 0$ metric?

With the aim of keeping it simple, we first focus on one-dimensional solvable extensions. Let $\mathfrak{n}$ be a nilpotent Lie algebra. Each $D \in \operatorname{Der}(\mathfrak{n})$ defines a solvable Lie algebra

$$\mathfrak{s}_D = \mathbb{R}f \oplus \mathfrak{n},$$

with Lie bracket defined as the semi-direct product such that $\operatorname{ad} f|_{\mathfrak{n}} = D$.

Definition 4.2 A derivation D of a nilpotent Lie algebra $\mathfrak{n}$ with $\operatorname{tr} D > 0$ is said to be *Ricci negative* (or rn-derivation for short) if the solvable Lie algebra $\mathfrak{s}_D$ admits an inner product of negative Ricci curvature. We denote by $\operatorname{Der}(\mathfrak{n})_{rn}$ the set of all Ricci negative derivations of $\mathfrak{n}$.

It follows from Theorem 2.2 that any positive derivation D (i.e. $\operatorname{Spec}\left(D^{\mathbb{R}}\right) > 0$) belongs to $\operatorname{Der}(\mathfrak{n})_{rn}$. Note also that any $\mathfrak{n}$ admitting a rn-derivation is a RN-nilradical; the validity of the converse assertion, which is strongly related to Conjecture 2.8, is not known.

Question 4.3 Does any RN-nilradical admit a rn-derivation?

The following particular instance of Question 4.1 arises.

Question 4.4 What kind of set is $\operatorname{Der}(\mathfrak{n})_{rn}$?

Note that $\operatorname{Der}(\mathfrak{n})_{rn}$ parametrizes the set of all one-dimensional solvable extensions of $\mathfrak{n}$ admitting $\operatorname{Ric} < 0$ metrics.

4.1 Ricci Operator

Before going through more questions and conjectures, it is worth pointing out at this point some of the technical difficulties involved.

We fix an inner product $\langle\cdot,\cdot\rangle$ on $\mathfrak{s} = \mathbb{R}f \oplus \mathfrak{n}$ such that $|f| = 1$ and $f \perp \mathfrak{n}$, in order to consider the moving-bracket approach described in Sect. 3. Let $[\cdot,\cdot]_{\mathfrak{n}}$ denote the Lie bracket of $\mathfrak{n}$. The Lie bracket of $\mathfrak{s}_D$ is therefore determined by the pair $(D, [\cdot,\cdot]_{\mathfrak{n}})$, for any $D \in \operatorname{Der}(\mathfrak{n})$. It is easy to see (using e.g. [25] or [19, (11)]) that the Ricci operator of $(\mathfrak{s}_D, \langle\cdot,\cdot\rangle)$ is given by

$$\operatorname{Ric}_{(D,[\cdot,\cdot]_{\mathfrak{n}})} = \left[\begin{array}{c|c} -\operatorname{tr} S(D)^2 & * \\ \hline * & \operatorname{Ric}_{[\cdot,\cdot]_{\mathfrak{n}}} + \frac{1}{2}[D, D^t] - \operatorname{tr}(D)S(D) \end{array}\right], \tag{1}$$

where $\operatorname{Ric}_{[\cdot,\cdot]_{\mathfrak{n}}} = \frac{|[\cdot,\cdot]_{\mathfrak{n}}|^2}{4}\,\mathrm{m}([\cdot,\cdot]_{\mathfrak{n}})$ is the Ricci operator of $(\mathfrak{n}, \langle\cdot,\cdot\rangle|_{\mathfrak{n}\times\mathfrak{n}})$ (see (6) for the definition of the moment map m), $S(D) := \frac{1}{2}(D + D^t)$ and

$$\langle \operatorname{Ric} f, X\rangle = -\operatorname{tr} S(D)\operatorname{ad}_{\mathfrak{n}} X, \qquad \forall X \in \mathfrak{n}.$$

It is known that $* = 0$ if D is normal (see e.g. the proof of [25, Proposition 4.3]).

If with respect to a fixed orthonormal basis $\{f, e_1, \ldots, e_n\}$ of $(\mathfrak{s}, \langle\cdot,\cdot\rangle)$, where $e_i \in \mathfrak{n}$ for all i,

$$\overline{h} = \left[\begin{array}{c|c} c^{-1} & 0 \\ \hline X & h \end{array}\right], \qquad c \in \mathbb{R}^*, \quad X \in \mathfrak{n}, \quad h \in \operatorname{GL}(\mathfrak{n}), \tag{2}$$

then $\left(\mathfrak{s}_D, \langle\overline{h}\cdot, \overline{h}\cdot\rangle\right)$ is isometric to the pair

$$\left(ch(D - \operatorname{ad}_{\mathfrak{n}} h^{-1}X)h^{-1}, h\cdot[\cdot,\cdot]_{\mathfrak{n}}\right), \tag{3}$$

since $\overline{h}$ is an isometric isomorphism between the corresponding metric Lie algebras. The fact that any metric on $\mathfrak{s}_D$ is of the form $\langle\overline{h}\cdot, \overline{h}\cdot\rangle$ for some $\overline{h}$ as above implies the following characterization of Ricci negative derivations of $\mathfrak{n}$.

Lemma 4.5 *$D \in \operatorname{Der}(\mathfrak{n})_{rn}$ if and only if there exist $X \in \mathfrak{n}$ and $h \in \operatorname{GL}(\mathfrak{n})$ such that $\operatorname{Ric}_{(D_{X,h}, h\cdot[\cdot,\cdot]_{\mathfrak{n}})} < 0$, where $D_{X,h} := (h(A - \operatorname{ad}_{\mathfrak{n}} h^{-1}X)h^{-1} \in \operatorname{Der}(\mathfrak{n}, h\cdot[\cdot,\cdot]_{\mathfrak{n}})$.*

The terms $[D_{X,h}, (D_{X,h})^t]$ and $S(D_{X,h})$ in formula (1) for $\operatorname{Ric}_{(D_{X,h}, h\cdot[\cdot,\cdot]_{\mathfrak{n}})}$ are extremely difficult to control, even in the case when D is diagonalizable.

4.2 Diagonalizable Ricci Negative Derivations

We now list some immediate though very useful consequences of Definition 4.2.

- $\operatorname{Der}(\mathfrak{n})_{rn}$ is open in $\operatorname{Der}(\mathfrak{n})$ and it is a *cone* (i.e. $\mathbb{R}_{>0}$-invariant).
- $\operatorname{Der}(\mathfrak{n})_{rn}$ is $\operatorname{Aut}(\mathfrak{n})$-invariant. Moreover, if $D \in \operatorname{Der}(\mathfrak{n})_{rn}$, then

$$D' := c\left(hDh^{-1} - \operatorname{ad}_{\mathfrak{n}} X\right) \in \operatorname{Der}(\mathfrak{n})_{rn},$$

  for any $c > 0$, $X \in \mathfrak{n}$ and $h \in \operatorname{Aut}(\mathfrak{n})$ (see (3)). Indeed, it is easy to check that $\mathfrak{s}_{D'}$ is isomorphic to $\mathfrak{s}_D$.
- $\operatorname{Der}(\mathfrak{n})^{\mathbb{R}}_{rn} := \{D^{\mathbb{R}} : D \in \operatorname{Der}(\mathfrak{n})_{rn}\}$ is an open cone in the $\operatorname{Aut}(\mathfrak{n})$-invariant subspace $\operatorname{Der}(\mathfrak{n})^{\mathbb{R}} := \{D^{\mathbb{R}} : D \in \operatorname{Der}(\mathfrak{n})\}$. We do not known a priori if $D^{\mathbb{R}}$ is also Ricci negative for any $D \in \operatorname{Der}(\mathfrak{n})_{rn}$.
- $\operatorname{Der}(\mathfrak{n})^{\mathbb{R}}_{rn} = \operatorname{Aut}(\mathfrak{n}) \cdot \mathfrak{t}(\mathfrak{n})_{rn}$, where

$$\mathfrak{t}(\mathfrak{n})_{rn} := \operatorname{Der}(\mathfrak{n})^{\mathbb{R}}_{rn} \cap \mathfrak{t}(\mathfrak{n}), \tag{4}$$

  and $\mathfrak{t}(\mathfrak{n})$ is any maximal torus of diagonalizable derivations of $\mathfrak{n}$ (cf. Conjecture 2.7). Note that $\operatorname{Der}(\mathfrak{n})^{\mathbb{R}} = \operatorname{Aut}(\mathfrak{n}) \cdot \mathfrak{t}(\mathfrak{n})$ since all the maximal tori are $\operatorname{Aut}(\mathfrak{n})$-conjugate and any diagonalizable derivation belongs to some maximal torus.
- $\mathfrak{t}(\mathfrak{n})_{rn}$ is an open cone in $\mathfrak{t}(\mathfrak{n})$.
- The *Weyl group*

$$W(\mathfrak{n}) := N_{\operatorname{Aut}(\mathfrak{n})}(\mathfrak{t}(\mathfrak{n}))/C_{\operatorname{Aut}(\mathfrak{n})}(\mathfrak{t}(\mathfrak{n})) \tag{5}$$

  acts on $\mathfrak{t}(\mathfrak{n})_{rn}$ by conjugation, where N and C denote normalizer and centralizer, respectively.

It is not too hard to see that if Conjecture 2.7 turns out to be true, then the following must hold.

Conjecture 4.6 $D \in \operatorname{Der}(\mathfrak{n})_{rn}$ *if and only if* $D^{\mathbb{R}} \in \operatorname{Der}(\mathfrak{n})_{rn}$.

This would imply that $\mathfrak{t}(\mathfrak{n})_{rn} = \operatorname{Der}(\mathfrak{n})_{rn} \cap \mathfrak{t}(\mathfrak{n})$ and so $\mathfrak{t}(\mathfrak{n})_{rn}$ stands out from the sets to be understood in order to give a satisfactory answer to Question 4.4. The relationship with the image of certain moment map and its convexity properties described below lead us to propose the following conjecture.

Conjecture 4.7 *The open cone* $\mathfrak{t}(\mathfrak{n})_{rn}$ *is convex.*

It is easy to see that $\mathfrak{t}(\mathfrak{n})_{rn}$ must contain the cone $\mathcal{C}$ in Conjecture 2.7, provided Conjecture 2.8 holds true.

4.3 Strongly Ricci Negative Derivations and the Moment Map

The following stronger condition on a derivation was studied in [8].

Definition 4.8 A derivation D of a nilpotent Lie algebra $\mathfrak{n}$ with $\operatorname{tr} D > 0$ is said to be *strongly Ricci negative* (srn-derivation for short) if the solvable Lie algebra $\mathfrak{s}_D$ admits an inner product of negative Ricci curvature such that $D^t = D$ and $f \perp \mathfrak{n}$. We denote by $\operatorname{Der}(\mathfrak{n})_{srn}$ the cone of all strongly Ricci negative derivations of $\mathfrak{n}$.

Note that $\operatorname{Der}(\mathfrak{n})_{srn} \subset \operatorname{Der}(\mathfrak{n})^{\mathbb{R}}$ and it is also $\operatorname{Aut}(\mathfrak{n})$-invariant. Indeed, if $(\mathfrak{s}_D, \langle\cdot,\cdot\rangle)$ has $\operatorname{Ric} < 0$, then the isometric pair $\left(\mathfrak{s}_{hDh^{-1}}, \left\langle \overline{h}^{-1}\cdot, \overline{h}^{-1}\cdot \right\rangle\right)$ also has $\operatorname{Ric} < 0$, where $c = 1$ and $X = 0$ (see (2)), and hDh^{-1} is symmetric with respect to the metric $\left\langle \overline{h}^{-1}\cdot, \overline{h}^{-1}\cdot \right\rangle$.

Question 4.9 What kind of set is the cone $\operatorname{Der}(\mathfrak{n})_{srn}$? Is $\operatorname{Der}(\mathfrak{n})_{srn}$ open in $\operatorname{Der}(\mathfrak{n})^{\mathbb{R}}$?

The following result characterizes strongly Ricci negative derivations in terms of the moment map $\mathrm{m} : \Lambda^2\mathfrak{n}^* \otimes \mathfrak{n} \to \operatorname{sym}(\mathfrak{n})$ for the $\operatorname{GL}(\mathfrak{n})$-representation $\Lambda^2\mathfrak{n}^* \otimes \mathfrak{n}$, defined by

$$\langle \mathrm{m}(\mu), E \rangle = \tfrac{1}{|\mu|^2} \langle E \cdot \mu, \mu \rangle, \qquad \mu \in \Lambda^2\mathfrak{n}^* \otimes \mathfrak{n}, \quad E \in \operatorname{sym}(\mathfrak{n}). \tag{6}$$

See Sect. 7 for more details on the moment map. We fix a basis $\{e_i\}$ of $\mathfrak{n}$ such that

$$\mathfrak{t}(\mathfrak{n}) \subset \operatorname{Dg}(\mathfrak{n}),$$

where $\operatorname{Dg}(\mathfrak{n})$ denotes the vector space of all operators of $\mathfrak{n}$ whose matrix in terms of $\{e_i\}$ is diagonal. Note that the vector space $\mathfrak{n}$ is identified with $\mathbb{R}^n$ using the basis $\{e_i\}$ and so the whole setting described in Sect. 7 can be used. In particular, the inner products appearing in (6) are the canonical ones, i.e. $\{e_i \otimes e^j\}$ and $\{(e^i \wedge e^j) \otimes e_k\}$ are respectively orthonormal bases.

Let G_D denote the connected component of the centralizer subgroup of D in $\operatorname{GL}(\mathfrak{n})$ and let $\mathfrak{g}_D$ be its Lie algebra.

Theorem 4.10 ([8, Corollary 3.4]) *Let $\mathfrak{n}$ be a nilpotent Lie algebra with Lie bracket $[\cdot,\cdot]$ and consider $D \in \mathfrak{t}(\mathfrak{n})$ such that $\operatorname{tr} D > 0$. Then the following conditions are equivalent:*

(i) *D is strongly Ricci negative.*
(ii) *$D \in \mathbb{R}_{>0}\, \mathrm{m}\left(G_D \cdot [\cdot,\cdot]\right) \cap \operatorname{Dg}(\mathfrak{n}) + \operatorname{Dg}(\mathfrak{n})_{>0}$.*
(iii) *$D \in \mathbb{R}_{>0}\, \mathrm{m}\left(\overline{G_D \cdot [\cdot,\cdot]}\right) \cap \operatorname{Dg}(\mathfrak{n}) + \operatorname{Dg}(\mathfrak{n})_{>0}$.*
(iv) *$D \in \mathbb{R}_{>0}\, \mathrm{m}\left(\overline{G_D \cdot [\cdot,\cdot]}\right) \cap \mathfrak{a}_+^D + \operatorname{Dg}(\mathfrak{n})_{>0}$, where $\mathfrak{a}_+^D \subset \operatorname{Dg}(\mathfrak{n})$ is any Weyl chamber of G_D.*

It would be very helpful to know the answer to the following question.

Question 4.11 Does condition (iii) imply that $\operatorname{tr} D > 0$?

Note that all the cones in the above theorem depend on D and are open in $\mathrm{Dg}(\mathfrak{n})$ since the space $\mathrm{Dg}(\mathfrak{n})_{>0}$ of positive diagonal matrices is so. On the other hand, the cone in part (iv) is in addition convex by Theorem 7.1.

Since we do not know if $\mathrm{Der}(\mathfrak{n})_{srn}$ is open in $\mathrm{Der}(\mathfrak{n})^{\mathbb{R}}$ (see Question 4.9), we do not know whether the corresponding cone

$$\mathfrak{t}(\mathfrak{n})_{srn} := \mathrm{Der}(\mathfrak{n})_{srn} \cap \mathfrak{t}(\mathfrak{n}) \tag{7}$$

is open in $\mathfrak{t}(\mathfrak{n})$ either. Note that the Weyl group $W(\mathfrak{n})$ (see (5)) is also acting on $\mathfrak{t}(\mathfrak{n})_{srn}$ since $\mathrm{Der}(\mathfrak{n})_{srn}$ is $\mathrm{Aut}(\mathfrak{n})$-invariant.

The cone $\mathfrak{t}(\mathfrak{n})_{srn}$ was proved to be open in $\mathfrak{t}(\mathfrak{n})$ and convex for Heisenberg and filiform Lie algebras in [32, 33] (see [13]). Furthermore, it is shown in [13] that the same holds for every $\mathfrak{n}$ of dimension ≤ 5. All this and the results given in the next section suggest that the following may be true.

Conjecture 4.12 *The cone* $\mathfrak{t}(\mathfrak{n})_{srn}$ *is open in* $\mathfrak{t}(\mathfrak{n})$ *and convex.*

5 An Open and Convex Cone of Ricci Negative Derivations

In this section, we use Theorem 4.10 to construct, for any nilpotent Lie algebra $\mathfrak{n}$, an open and convex cone $\mathcal{C}(\mathfrak{n})$ of diagonalizable Ricci negative derivations, containing all the generic ones.

We keep the basis $\{e_1, \ldots, e_n\}$ of $\mathfrak{n}$ such that $\mathfrak{t}(\mathfrak{n}) \subset \mathrm{Dg}(\mathfrak{n})$ fixed. Note that $\mathfrak{t}(\mathfrak{n})$ is precisely the orthogonal complement in $\mathrm{Dg}(\mathfrak{n})$ of the set $\{F_{ij}^k : c_{ij}^k \neq 0\}$, where c_{ij}^k are the structural constants of the Lie bracket $[\cdot,\cdot]$ of $\mathfrak{n}$ and F_{ij}^k is the diagonal matrix with nonzero entries -1 at i, j and 1 at k. Let $\alpha_1, \ldots, \alpha_r \in \mathfrak{t}(\mathfrak{n})^*$ be the weights for $\mathfrak{t}(\mathfrak{n})$ and let $\mathfrak{n} = \mathfrak{n}_1 \oplus \cdots \oplus \mathfrak{n}_r$ be the decomposition of $\mathfrak{n}$ in weight subspaces, that is,

$$DX = \alpha_i(D)X, \qquad \forall X \in \mathfrak{n}_i, \quad D \in \mathfrak{t}(\mathfrak{n}).$$

We call a derivation $D \in \mathfrak{t}(\mathfrak{n})$ *generic* when $\alpha_i(D) \neq \alpha_j(D)$ for any $i \neq j$. Let $\mathfrak{t}(\mathfrak{n})_{gen}$ denote the subset of all generic derivations, which is open and dense in $\mathfrak{t}(\mathfrak{n})$.

If $G_{\mathfrak{t}(\mathfrak{n})}$ denotes the connected component of the centralizer of $\mathfrak{t}(\mathfrak{n})$ in $\mathrm{GL}(\mathfrak{n})$, then $G_{\mathfrak{t}(\mathfrak{n})} \subset G_D$ for any $D \in \mathfrak{t}(\mathfrak{n})$ and equality holds if and only if $D \in \mathfrak{t}(\mathfrak{n})_{gen}$. In the *multiplicity-free* case, i.e. $\dim \mathfrak{n}_i = 1$ for all i, $G_{\mathfrak{t}(\mathfrak{n})}$ is given by the connected torus $\mathrm{Dg}(\mathfrak{n})_{>0}$ with Lie algebra $\mathrm{Dg}(\mathfrak{n})$ and $\mathfrak{t}(\mathfrak{n})$ is the only Weyl chamber. Note that the space $\mathfrak{t}(\mathfrak{n})$ is multiplicity-free if and only if there is at least one $D \in \mathfrak{t}(\mathfrak{n})$ with pairwise different n eigenvalues; in that case, the basis is necessarily nice (see Sect. 5.1).

We consider the smallest of the open cones appearing in Theorem 4.10, (iii) to define

$$\mathcal{C}(\mathfrak{n}) := \left(\mathbb{R}_{>0}\, m\left(\overline{G_{\mathfrak{t}(\mathfrak{n})} \cdot [\cdot,\cdot]}\right) \cap \mathrm{Dg}(\mathfrak{n}) + \mathrm{Dg}(\mathfrak{n})_{>0}\right) \cap \mathfrak{t}(\mathfrak{n})_{\mathrm{tr}>0}, \tag{8}$$

where $\mathfrak{t}(\mathfrak{n})_{\mathrm{tr}>0} := \{D \in \mathfrak{t}(\mathfrak{n}) : \mathrm{tr}\, D > 0\}$. A positive answer to Question 4.11 would allow us to take the intersection with $\mathfrak{t}(\mathfrak{n})_{\mathrm{tr}>0}$ out of the definition of the open cone $\mathcal{C}(\mathfrak{n})$. By Theorem 4.10, (iv) one can use any Weyl chamber $\mathfrak{a}_+ \subset \mathrm{Dg}(\mathfrak{n})$ of $G_{\mathfrak{t}(\mathfrak{n})}$ instead of the whole $\mathrm{Dg}(\mathfrak{n})$ to define $\mathcal{C}(\mathfrak{n})$, that is,

$$\mathcal{C}(\mathfrak{n}) = \left(\mathbb{R}_{>0}\, m\left(\overline{G_{\mathfrak{t}(\mathfrak{n})} \cdot [\cdot,\cdot]}\right) \cap \mathfrak{a}_+ + \mathrm{Dg}(\mathfrak{n})_{>0}\right) \cap \mathfrak{t}(\mathfrak{n})_{\mathrm{tr}>0}. \tag{9}$$

Let $\mathrm{O}(\mathfrak{n})$ denote the orthogonal group relative to the inner product making $\{e_i\}$ an orthonormal basis. In order to have a group acting on $\mathcal{C}(\mathfrak{n})$, which may be helpful for computing it, we consider the subgroup of the Weyl group $W(\mathfrak{n})$ (see (5)) given by

$$W_{ort}(\mathfrak{n}) := N_{\mathrm{Aut}(\mathfrak{n})\cap \mathrm{O}(\mathfrak{n})}(\mathfrak{t}(\mathfrak{n}))/C_{\mathrm{Aut}(\mathfrak{n})\cap \mathrm{O}(\mathfrak{n})}(\mathfrak{t}(\mathfrak{n})) \tag{10}$$

and call it the *orthogonal Weyl group*.

Proposition 5.1 *$\mathcal{C}(\mathfrak{n})$ is an open and convex cone in $\mathfrak{t}(\mathfrak{n})$ such that*

$$\mathfrak{t}(\mathfrak{n})_{srn} \cap \mathfrak{t}(\mathfrak{n})_{gen} \subset \mathcal{C}(\mathfrak{n}) \subset \mathfrak{t}(\mathfrak{n})_{srn}. \tag{11}$$

Moreover, $W_{ort}(\mathfrak{n})$ acts on $\mathcal{C}(\mathfrak{n})$ by conjugation and $\mathcal{C}_t(\mathfrak{n}) := \{D \in \mathcal{C}(\mathfrak{n}) : \mathrm{tr}\, D = t\}$ is a $W_{ort}(\mathfrak{n})$-invariant convex polytope (i.e. the convex hull of finitely many points) for any $t > 0$.

Proof The open cone $\mathcal{C}(\mathfrak{n})$ is convex and $\mathcal{C}_t(\mathfrak{n})$ is a polytope by Theorem 7.1 and (9). Condition (11) follows from Theorem 4.10 and the fact that $G_{\mathfrak{t}(\mathfrak{n})} = G_D$ for any $D \in \mathfrak{t}(\mathfrak{n})_{gen}$.

It only remains to prove that $\mathcal{C}(\mathfrak{n})$ is $W_{ort}(\mathfrak{n})$-invariant. If $f \in W_{ort}(\mathfrak{n})$ and $D \in \mathcal{C}(\mathfrak{n})$, say $D = r\,\mathrm{m}(h \cdot [\cdot,\cdot]) + E$, for some $r > 0$, $h \in G_{\mathfrak{t}(\mathfrak{n})}$ and $E \in \mathrm{Dg}(\mathfrak{n})_{>0}$, then

$$fDf^{-1} = rf\,\mathrm{m}(h \cdot [\cdot,\cdot])f^{-1} + fEf^{-1} = r\,\mathrm{m}((fhf^{-1}) \cdot [\cdot,\cdot]) + fEf^{-1},$$

with $fhf^{-1} \in G_{\mathfrak{t}(\mathfrak{n})}$ and thus $fDf^{-1} \in \mathcal{C}(\mathfrak{n})$, concluding the proof. □

In view of the suggestive property (11) satisfied by $\mathcal{C}(\mathfrak{n})$, it is natural to expect that the following must hold.

Conjecture 5.2 $\mathcal{C}(\mathfrak{n}) = \mathfrak{t}(\mathfrak{n})_{srn}$.

This has recently been proved for any nilpotent Lie algebra of dimension ≤ 5 in [13]. Note that the conjecture holds in the rank one case (i.e. $\dim \mathfrak{t}(\mathfrak{n}) = 1$). Furthermore, in [13], the cone $\mathcal{C}(\mathfrak{n})$ is computed for some classical families of nilpotent Lie algebras, obtaining for example that the convex polytope $\mathcal{C}_1(\mathfrak{h}_{2n+1})$ is the n-dimensional hypercube for the Heisenberg Lie algebra $\mathfrak{h}_{2n+1}$.

The following result shows that we only need $\mathfrak{t}(\mathfrak{n})_{srn}$ to be open in $\mathfrak{t}(\mathfrak{n})$ to prove Conjectures 4.12 and 5.2.

Proposition 5.3 *If* $\mathfrak{t}(\mathfrak{n})_{srn}$ *is open in* $\mathfrak{t}(\mathfrak{n})$*, then* $\mathcal{C}(\mathfrak{n}) = \mathfrak{t}(\mathfrak{n})_{srn}$.

Proof For a given $D \in \mathfrak{t}(\mathfrak{n})_{srn}$, consider an open neighborhood $U \subset \mathfrak{t}(\mathfrak{n})_{srn}$. It is easy to show that D belongs to a segment contained in U with extremes in $\mathfrak{t}(\mathfrak{n})_{gen}$, and since such extremes belong to the convex cone $\mathcal{C}(\mathfrak{n})$ by (11) we obtain that $D \in \mathcal{C}(\mathfrak{n})$, concluding the proof. □

Let us summarize the different cones in the maximal torus $\mathfrak{t}(\mathfrak{n}) \subset \mathrm{Der}(\mathfrak{n})$ that we have considered in this and the previous sections, we refer to (4), Definition 4.2, (7) and (8) for their definition, respectively:

$$\mathfrak{t}(\mathfrak{n}) \supseteq \mathfrak{t}(\mathfrak{n})_{rn} \supseteq \mathrm{Der}(\mathfrak{n})_{rn} \cap \mathfrak{t}(\mathfrak{n}) \supseteq \mathfrak{t}(\mathfrak{n})_{srn} \supseteq \mathcal{C}(\mathfrak{n}). \tag{12}$$

They are all known to be open except $\mathfrak{t}(\mathfrak{n})_{srn}$; on the other hand, $\mathcal{C}(\mathfrak{n})$ is the only one known to be convex (see Conjectures 4.7, 4.6 and 5.2).

Remark 5.4 It is important to notice that throughout all this section and Sect. 4.3, $\mathfrak{t}(\mathfrak{n})$ can be replaced by $\mathrm{Der}(\mathfrak{n}) \cap \mathrm{Dg}(\mathfrak{n})$, where $\mathrm{Dg}(\mathfrak{n})$ is defined with respect to any basis of $\mathfrak{n}$. It is not necessary for $\mathrm{Der}(\mathfrak{n}) \cap \mathrm{Dg}(\mathfrak{n})$ to be maximal.

5.1 Nice Bases

The following type of bases is very friendly with the computation of Ricci curvature.

Definition 5.5 A basis $\{e_1, \dots, e_n\}$ of a Lie algebra is said to be *nice* if every bracket $[e_i, e_j]$ is a scalar multiple of some element e_k in the basis and two different brackets $[e_i, e_j]$, $[e_r, e_s]$ can be a nonzero multiple of the same e_k only if $\{i, j\}$ and $\{r, s\}$ are either equal or disjoint.

It is proved in [29] (see also [8, Lemma 3.12]) that a basis $\{e_i\}$ is a nice basis if and only if

$$\mathrm{m}\left(\overline{\mathrm{Dg}(\mathfrak{n})_{>0} \cdot [\cdot,\cdot]}\right) = \mathrm{CH}_{[\cdot,\cdot]} := \mathrm{CH}(F_{ij}^k : c_{ij}^k \neq 0), \tag{13}$$

where c_{ij}^k are the structure coefficients and CH denotes convex hull. If $\{e_i\}$ is a nice basis, then by (13),

$$\mathrm{CH}_{[\cdot,\cdot]} \subset \mathrm{m}\left(\overline{G_{\mathfrak{t}(\mathfrak{n})} \cdot [\cdot,\cdot]}\right) \cap \mathrm{Dg}(\mathfrak{n}),$$

and equality holds if in addition $\mathfrak{t}(\mathfrak{n})$ is multiplicity-free since in that case, $G_{\mathfrak{t}(\mathfrak{n})} = \mathrm{Dg}(\mathfrak{n})_{>0}$ and $\mathfrak{a}_+ = \mathrm{Dg}(\mathfrak{n})$ is the only Weyl chamber.

The following example shows that both $\mathrm{m}\left(\overline{\mathrm{Dg}(\mathfrak{n})_{>0} \cdot [\cdot,\cdot]}\right) \cap \mathrm{Dg}(\mathfrak{n})$ and $\mathrm{m}\left(\overline{G_{\mathfrak{t}(\mathfrak{n})} \cdot [\cdot,\cdot]}\right) \cap \mathrm{Dg}(\mathfrak{n})$ can be tricky subsets if the basis is not nice.

Example 5.6 (See [8, Example 3.17] for a Detailed Treatment) Let $\mathfrak{n}$ be the 5-dimensional nilpotent Lie algebra with basis $\{e_1, \dots, e_5\}$ and Lie bracket

$$[e_1, e_2] = e_3 + e_4, \quad [e_1, e_3] = e_5, \quad [e_1, e_4] = e_5,$$

for which $\mathrm{CH}_{[\cdot,\cdot]} = \mathrm{CH}(F_{12}^3, F_{12}^4, F_{13}^5, F_{14}^5)$ is a rectangle centered at $\mathrm{m}([\cdot,\cdot])$ (see Fig. 2) and

$$G_{\mathfrak{t}(\mathfrak{n})} = \left\{ \begin{bmatrix} h_1 & & & \\ & h_2 & & \\ & & H & \\ & & & h_5 \end{bmatrix} : H \in \mathrm{GL}_2^+(\mathbb{R}), \quad h_i > 0 \right\}.$$

We consider $G_{\mathfrak{t}(\mathfrak{n})}$ acting on the cone $\overline{G_{\mathfrak{t}(\mathfrak{n})} \cdot [\cdot,\cdot]}$ of nilpotent Lie brackets $\mu = \mu(x, y, z, w)$ defined by

$$\mu(e_1, e_2) = xe_3 + ye_4, \quad \mu(e_1, e_3) = ze_5, \quad \mu(e_1, e_4) = we_5, \qquad x, y, z, w \geq 0.$$

The moment map $\mathrm{m} : \overline{G_{\mathfrak{t}(\mathfrak{n})} \cdot [\cdot,\cdot]} \longrightarrow \mathrm{sym}(\mathfrak{n})$ is given by

$$\mathrm{m}(\mu) = \tfrac{1}{x^2+y^2+z^2+w^2} \left(x^2 F_{12}^3 + y^2 F_{12}^4 + z^2 F_{13}^5 + w^2 F_{14}^5 + (xy - zw)(E_{34} + E_{43}) \right),$$

where E_{ij} denotes as usual the matrix whose only nonzero entry is a 1 at ij, and so

$$\begin{aligned} \mathrm{m}\left(\overline{G_{\mathfrak{t}(\mathfrak{n})} \cdot [\cdot,\cdot]}\right) \cap \mathrm{Dg}(\mathfrak{n}) = & \{aF_{12}^3 + bF_{12}^4 + cF_{13}^5 + dF_{14}^5 : a, b, c, d \geq 0, \\ & a + b + c + d = 1, \quad ab = cd\}. \end{aligned} \tag{14}$$

It is not so hard to see by using (14) that $\mathrm{m}\left(\overline{G_{\mathfrak{t}(\mathfrak{n})} \cdot [\cdot,\cdot]}\right) \cap \mathrm{Dg}(\mathfrak{n})$ is the union of two triangles as in Fig. 2, each of which being the intersection with one of the Weyl chambers

$$(\mathfrak{a}_+)_1 = \{\mathrm{Dg}(a_1, \dots, a_5) : a_3 \leq a_4\}, \qquad (\mathfrak{a}_+)_2 = \{\mathrm{Dg}(a_1, \dots, a_5) : a_3 \geq a_4\}.$$

On the other hand, the closure $\overline{\mathrm{Dg}(\mathfrak{n})_{>0} \cdot [\cdot,\cdot]}$ is the union of the open orbit $\mathrm{Dg}(\mathfrak{n})_{>0} \cdot [\cdot,\cdot] = \{\mu : xz = yw \neq 0\}$ and other eight nonzero T-orbits corresponding to the edges and vertexes of the rectangle $\mathrm{CH}_{[\cdot,\cdot]}$ (see (19)). One therefore obtains from (14) that

$$\begin{aligned} \mathrm{m}\left(\overline{\mathrm{Dg}(\mathfrak{n})_{>0} \cdot [\cdot,\cdot]}\right) \cap \mathrm{Dg}(\mathfrak{n}) = & \{aF_{12}^3 + bF_{12}^4 + cF_{13}^5 + dF_{14}^5 : a, b, c, d \geq 0, \\ & a + b + c + d = 1, \quad ab = cd, \quad ac = bd\}, \end{aligned} \tag{15}$$

given by the union of three segments as in Fig. 2.

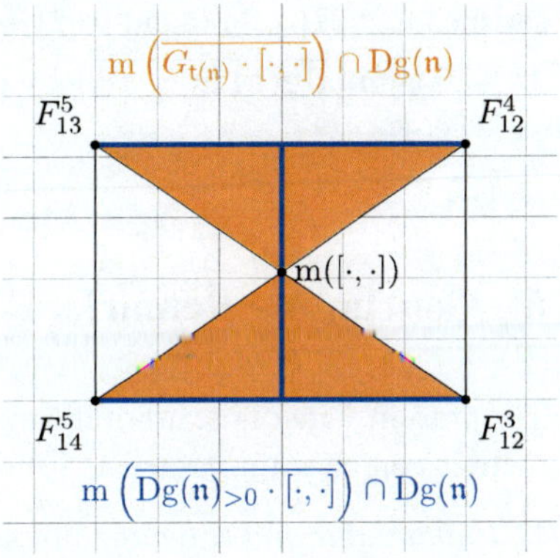

Fig. 2 Moment map images

The following fact from Lie theory will be useful. Recall that any ideal of a nilpotent Lie algebra meets the center non-trivially by Engel's Theorem.

Lemma 5.7 *Let $\mathfrak{n}$ be a nilpotent Lie algebra and assume that $\mathfrak{n} = \mathfrak{n}_0 \oplus \mathfrak{n}_1$, where $\mathfrak{n}_0$ is a subalgebra, $\mathfrak{n}_1$ an ideal and $\mathfrak{z}(\mathfrak{n}) \subset \mathfrak{n}_1$. Then for any nonzero $X_0 \in \mathfrak{n}_0$, there exists $X_1 \in \mathfrak{n}_1$ such that $[X_0, X_1] \neq 0$.*

Proof Assume there is a nonzero $X_0 \in \mathfrak{n}_0$ such that $[X_0, \mathfrak{n}_1] = 0$. Since $\mathfrak{n}_0$ is nilpotent and $X_0 \notin \mathfrak{z}(\mathfrak{n})$, there exists $k \in \mathbb{N}$ such that

$$\mathfrak{b}_k := [\dots[X_0, \mathfrak{n}_0], \dots, \mathfrak{n}_0] \quad (k\text{-times})$$

is nonzero and $\mathfrak{b}_{k+1} = 0$. It follows by induction that $[\mathfrak{b}_k, \mathfrak{n}_1] = 0$ for all k; indeed, for any $Z \in \mathfrak{b}_k$, $X \in \mathfrak{n}_0$ (i.e. $[Z, X] \in \mathfrak{b}_{k+1}$) and $Y \in \mathfrak{n}_1$,

$$[[Z, X], Y] = [[Z, Y], X] + [Z, [X, Y]] = 0,$$

by inductive hypothesis. Thus $\mathfrak{b}_k \subset \mathfrak{n}_0 \cap \mathfrak{z}(\mathfrak{n}) = \{0\}$, which is a contradiction. □

As an application, we next show that any nilpotent Lie algebra admitting a non-negative diagonalizable (relative to a nice basis) derivation which is positive on the center is a RN-nilradical.

Proposition 5.8 *Let $\mathfrak{n}$ be a nilpotent Lie algebra endowed with a nice basis $\{e_i\}$. If $D \in \mathrm{Der}(\mathfrak{n}) \cap \mathrm{Dg}(\mathfrak{n})$ satisfies that $D \geq 0$ and $D|_{\mathfrak{z}(\mathfrak{n})} > 0$, then $D \in \mathrm{Der}(\mathfrak{n})_{srn}$.*

Proof We consider the decomposition $\mathfrak{n} = \mathfrak{n}_0 \oplus \mathfrak{n}_1$, where $\mathfrak{n}_0$ is the kernel of D and $\mathfrak{n}_1$ the sum of all eigenspaces of D with positive eigenvalues. Note that such decomposition satisfies the hypothesis of Lemma 5.7. We can assume that $\{e_1, \dots, e_r\}$ is a basis for $\mathfrak{n}_0$. It follows from Lemma 5.7 that for each $i = 1, \dots, r$, there exist j_i, k_i such that $e_{j_i}, e_{k_i} \in \mathfrak{n}_1$ and $c_{ij_i}^{k_i} \neq 0$. Thus

$$M := F_{1j_1}^{k_1} + \dots + F_{rj_r}^{k_r},$$

satisfies that $M|_{\mathfrak{n}_0} < 0$ and so $D - \epsilon M > 0$ for some $\epsilon > 0$ since $D|_{\mathfrak{n}_1} > 0$. Since $M \in \mathbb{R}_{>0}m\left(\overline{\mathrm{Dg}(\mathfrak{n})_{>0} \cdot [\cdot,\cdot]}\right)$ by (13), we conclude from Theorem 4.10, (iii) that $D \in \mathrm{Der}(\mathfrak{n})_{srn}$. □

6 Concluding Remarks

The present state of the art of the main question of which Lie algebras admit Ric < 0 metrics can be summarized as follows:

1. *Unimodular and solvable*: not allowed.
2. *Unimodular and non-solvable*: only semisimple Lie algebras are allowed and most non-compact simple Lie algebras are known to admit.
3. *Non-unimodular and non-solvable*: there are several large classes of examples, including all compact and most non-compact semisimple Lie algebras as possible Levi factors as well as many of their representations as possible nilradicals. On the other hand, no obstruction is known and not a single example where the non-existence of Ric < 0 metrics can be proved is available.
4. *Non-unimodular and solvable*: only one obstruction is available and if all the conjectures in this paper turn out to be true, then $\mathfrak{s}$ admits Ric < 0 metrics if and only if the diagonalizable part of some ad Y restricted to the nilradical $\mathfrak{n}$ belongs (up to automorphism conjugation) to the open and convex cone of derivations $\mathcal{C}(\mathfrak{n}) \subset \mathfrak{t}(\mathfrak{n})$ introduced in Sect. 5.

Independently of the truth of the characterization given in 4), it is important to highlight the role of the open and convex cone $\mathcal{C}(\mathfrak{n})$ defined via a moment map, in actually producing Ric < 0 metrics on different solvable Lie algebras with the same nilpotent Lie algebra $\mathfrak{n}$ as nilradical.

7 Appendix: The Moment Map

We consider the vector space $V := \Lambda^2(\mathbb{R}^n)^* \otimes \mathbb{R}^n$ of all skew-symmetric algebras of dimension n, on which $\mathrm{GL}_n(\mathbb{R})$ is acting by $g \cdot \mu := g\mu(g^{-1}\cdot, g^{-1}\cdot)$ and $\mathfrak{gl}_n(\mathbb{R})$ by

$$E \cdot \mu = E\mu(\cdot,\cdot) - \mu(E\cdot,\cdot) - \mu(\cdot, E\cdot), \qquad E \in \mathfrak{gl}_n(\mathbb{R}), \quad \mu \in V.$$

The weight vectors for the above representation are $\mu_{ijk} := (e^i \wedge e^j) \otimes e_k$ with corresponding weights

$$F_{ij}^k := E_{kk} - E_{ii} - E_{jj} \in \mathfrak{t}^n, \qquad i < j,$$

where $\mathfrak{t}^n$ denotes the set of all diagonal $n \times n$ matrices. The structural constants $c(\mu)_{ij}^k$ of an algebra $\mu \in V$ are therefore given by

$$\mu(e_i, e_j) = \sum_k c(\mu)_{ij}^k e_k, \qquad \mu = \sum_{i<j,k} c(\mu)_{ij}^k \mu_{ijk}.$$

All these vector spaces will be endowed with their canonical inner products.

The moment map (or $\mathrm{GL}_n(\mathbb{R})$-gradient map) from real geometric invariant theory (see [5, 17] for further information) for the above representation is the $\mathrm{O}(n)$-equivariant map

$$\mathrm{m} : V \setminus \{0\} \longrightarrow \mathrm{sym}(n),$$

defined by

$$\langle \mathrm{m}(\mu), E \rangle = \frac{1}{|\mu|^2} \langle E \cdot \mu, \mu \rangle, \qquad \mu \in V \setminus \{0\}, \quad E \in \mathrm{sym}(n), \tag{16}$$

or equivalently, for any $X, Y \in \mathbb{R}^n$,

$$\langle \mathrm{m}(\mu)X, Y \rangle = -\tfrac{1}{2} \sum \langle \mu(X, e_i), e_j \rangle \langle \mu(Y, e_i), e_j \rangle + \tfrac{1}{4} \sum \langle \mu(e_i, e_j), X \rangle \langle \mu(e_i, e_j), Y \rangle. \tag{17}$$

We are using $\mathfrak{gl}_n(\mathbb{R}) = \mathfrak{so}(n) \oplus \mathrm{sym}(n)$ as a Cartan decomposition, where $\mathfrak{so}(n)$ and $\mathrm{sym}(n)$ denote the subspaces of skew-symmetric and symmetric matrices, respectively. Note that m is also well defined on the projective space $\mathbb{P}(V)$ and that $\operatorname{tr} \mathrm{m}(\mu) = -1$ for any $\mu \in V$. In [4, 16] many nice and useful results on the convexity of the image of the moment map have been obtained, which were used in [8] to study Ricci negative solvmanifolds (see Sect. 4).

For any compatible subgroup $G \subset \mathrm{GL}_n(\mathbb{R})$, one has that $K := G \cap \mathrm{O}(n)$ is a maximal compact subgroup of G, $\mathfrak{g} = \mathfrak{k} \oplus \mathfrak{p}$ is a Cartan decomposition, where $\mathfrak{p} := \mathfrak{g} \cap \mathrm{sym}(n)$, and $G = K \exp \mathfrak{p}$. Consider $\mathfrak{a} \subset \mathfrak{p}$, a maximal abelian subalgebra; thus the corresponding torus $A = \exp \mathfrak{a} \subset G$ is also a compatible subgroup. Let m and $\mathrm{m}_{\mathfrak{a}}$ denote the moment maps for the G-action and for the A-action, respectively, and let $\mathfrak{a}_+ \subset \mathfrak{a}$ be any Weyl chamber of G. Consider also $W := N_K(\mathfrak{a})/C_K(\mathfrak{a})$, the corresponding Weyl group.

Theorem 7.1 ([16]) *For any closed G-invariant subset $X \subset \mathbb{P}(V)$, $\mathrm{m}(X) \cap \mathfrak{a}$ is the union of finitely many convex polytopes. If in addition X is irreducible, then $\mathrm{m}(X) \cap \mathfrak{a}_+$ is a convex polytope and*

$$\mathrm{m}(X) \cap \mathfrak{a} = \bigcup_{k \in W} k \cdot (\mathrm{m}(X) \cap \mathfrak{a}_+).$$

Remark 7.2 In particular, $\mathrm{m}_{\mathfrak{a}}(X)$ is always a convex polytope if X is irreducible since $\mathfrak{a}$ is the only Weyl chamber of A.

Therefore $\mathrm{m}(\overline{G\cdot\mu})\cap\mathfrak{a}_+$ and $\mathrm{m}_{\mathfrak{a}}(\overline{G\cdot\mu})$ are both convex polytopes for any $\mu\in V$ if G is connected, but $\mathrm{m}(\overline{G\cdot\mu})\cap\mathfrak{a}$ and $\mathrm{m}(\overline{A\cdot\mu})\cap\mathfrak{a}$ are not convex in general (see Example 5.6).

For each $\mu\in V$, we define the following convex polytope in $\mathrm{Dg}(\mathfrak{n})$,

$$\mathrm{CH}_\mu := \mathrm{CH}(F_{ij}^k : c(\mu)_{ij}^k \neq 0).$$

Note that $F_{ij}^k = \mathrm{m}(\mu_{ijk})$. It is proved in [8, Section 3.2] that

$$\mathrm{Diag}\left(\mathrm{m}\left(\overline{\mathrm{Dg}(\mathfrak{n})_{>0}\cdot\mu}\right)\right) = \mathrm{CH}_\mu, \qquad \mathrm{Diag}\left(\mathrm{m}\left(\mathrm{Dg}(\mathfrak{n})_{>0}\cdot\mu\right)\right) = \mathrm{CH}_\mu^\circ, \tag{18}$$

where $\mathrm{Diag}(A)$ denotes the diagonal part of a matrix A. Conversely, $\overline{\mathrm{Dg}(\mathfrak{n})_{>0}\cdot\mu}$ is combinatorically determined by CH_μ in the following way (see [8, Lemma 3.10]):

$$\overline{\mathrm{Dg}(\mathfrak{n})_{>0}\cdot\mu} = \left\{\lambda_J : \mathrm{CH}(F_{ij}^k : (i,j,k)\in J) \text{ is a face of } \mathrm{CH}_\mu\right\}, \tag{19}$$

where for each subset $J\subset I_\mu := \{(i,j,k) : c(\mu)_{ij}^k \neq 0\}$,

$$\lambda_J := \sum_{(i,j,k)\in J} c(\mu)_{ij}^k\, \mu_{ijk}.$$

Moreover, if $c(\mu)_{ij}^k \neq 0$, then $\mu_{ij}^k \in \overline{\mathrm{Dg}(\mathfrak{n})_{>0}\cdot\mu}$ and F_{ij}^k is an extreme point of CH_μ. In particular,

$$F_{ij}^k \in \mathrm{m}(\overline{\mathrm{Dg}(\mathfrak{n})_{>0}\cdot\mu})\cap\mathfrak{t}^n, \qquad \forall\, c(\mu)_{ij}^k \neq 0. \tag{20}$$

Acknowledgement The authors are very grateful to Romina Arroyo and Emilio Lauret for very helpful comments.

References

1. D. Alekseevskii, Homogeneous Riemannian spaces of negative curvature. Math. Sb. (N.S.) **96**, 93–117 (1975) (Russian). English translation. In: Math. USSR-Sb. **25**, 87–109 (1976)
2. R. Azencott, E. Wilson, Homogeneous manifolds with negative curvature. I. Trans. Am. Math. Soc. **215**, 323–362 (1976)
3. L. Bérard-Bergery, Sur la courbure des métriques riemanniennes invariantes des groupes de Lie et des espaces homogènes. Ann. Sci. E.N.S. **11**, 543–576 (1978)
4. L. Billiotti, A. Ghigi, P. Heinzner, Invariant convex sets in polar representations. Israel J. Math. **213**, 423–441 (2016)
5. C. Böhm, R. Lafuente, Real geometric invariant theory. Differ. Geom. Large (2020). Cambridge Univ. Press, in press
6. C. Böhm, R. Lafuente, Immortal homogeneous Ricci flows. Invent. Math. **212**, 461–529 (2018)

7. C. Böhm, R. Lafuente, The Ricci flow on solvmanifolds of real type. Adv. Math. **352**, 516–540 (2019)
8. J. Deré, J. Lauret, On Ricci negative solvmanifolds and their nilradicals. Math. Nach. **292**, 1462–1481 (2019)
9. I. Dotti, Ricci curvature of left-invariant metrics on solvable unimodular Lie groups. Math. Z. **180**, 257–263 (1982)
10. I. Dotti, M.L. Leite, Metrics of negative Ricci curvature on SL$(n, \mathbb{R})$, $n \geq 3$. J. Differ. Geom. **17**, 635–641 (1982)
11. I. Dotti, M.L. Leite, R. Miatello, Negative Ricci curvature on complex semisimple Lie groups. Geom. Dedicata **17**, 207–218 (1984)
12. P. Eberlein, J. Heber, Quarter pinched homogeneous spaces of negative curvature. Int. J. Math. **7**, 441–500 (1996)
13. V. Gutiérrez, On Ricci negative derivations, in preparation
14. J. Heber, Noncompact homogeneous Einstein spaces. Invent. Math. **133**, 279–352 (1998)
15. E. Heintze, On homogeneous manifolds of negative curvature. Math. Ann. **211**, 23–34 (1974)
16. P. Heinzner, P. Schützdeller, Convexity properties of gradient maps. Adv. Math. **225**, 1119–1133 (2010)
17. P. Heinzner, G.W. Schwarz, H. Stötzel, Stratifications with respect to actions of real reductive groups. Comp. Math. **144**, 163–185 (2008)
18. M. Jablonski, P. Petersen, A step towards the Alekseevskii conjecture. Math. Ann. **368**, 197–212 (2017)
19. R. Lafuente, J. Lauret, Structure of homogeneous Ricci solitons and the Alekseevskii conjecture. J. Differ. Geom. **98**, 315–347 (2014)
20. E.A. Lauret, C.E. Will, Non-solvable Lie groups with negative Ricci curvature. Transf. Groups, in Press (arXiv)
21. J. Lauret, Ricci soliton homogeneous nilmanifolds. Math. Ann. **319**, 715–733 (2001)
22. J. Lauret, Degenerations of Lie algebras and geometry of Lie groups. Differ. Geom. Appl. **18**, 177–194 (2003)
23. J. Lauret, A canonical compatible metric for geometric structures on nilmanifolds. Ann. Global Anal. Geom. **30**, 107–138 (2006)
24. J. Lauret, Einstein solvmanifolds and nilsolitons. Contemp. Math. **491**, 1–35 (2009)
25. J. Lauret, Ricci soliton solvmanifolds. J. Reine Angew. Math. **650**, 1–21 (2011)
26. J. Lauret, Convergence of homogeneous manifolds. J. Lond. Math. Soc. **86**, 701–727 (2012)
27. J. Lauret, Geometric flows and their solitons on homogeneous spaces (Workshop for Sergio Console). Rend. Sem. Mat. Torino **74**, 55–93 (2016)
28. J. Lauret, The search for solitons on homogeneous spaces, in *Abel Symposia* (Springer, Berlin, 2019)
29. J. Lauret, C.E. Will, On the diagonalization of the Ricci flow on Lie groups. Proc. Am. Math. Soc. **141**(10), 3651–3663 (2013)
30. J. Lohkamp, Metrics of negative Ricci curvature. Ann. Math. **140**, 655–683 (1994)
31. J. Milnor, Curvature of left-invariant metrics on Lie groups. Adv. Math. **21**, 293–329 (1976)
32. Y. Nikolayevsky, Solvable extensions of negative Ricci curvature of filliform Lie Groups. Math. Nach. **289**, 321–331 (2016)
33. Y. Nikolayevsky, Yu.G. Nikonorov, On solvable Lie groups of negative Ricci curvature. Math. Z. **280**, 1–16 (2015)
34. Yu.G. Nikonorov, On the Ricci curvature of homogeneous metrics on noncompact homogeneous spaces. Siberian Math. J. **41**, 349–356 (2000)
35. S.-T. Yau, The role of partial differential equations in differential geometry, in *Proc. ICM Helsinki* (1978), pp. 235–250
36. C.E. Will, Negative Ricci Curvature on some non-solvable Lie groups. Geom. Dedicata **186**, 181–195 (2017)
37. C.E. Will, Negative Ricci Curvature on some non-solvable Lie groups II. Math. Z. **294**, 1085–1105 (2020)
38. J.A. Wolf, Homogeneity and bounded isometrics in manifolds of negative curvature. Illinois J. Math. **8**, 14–18 (1964)

Semi-Riemannian Cones

Thomas Leistner

Abstract Due to a result by Gallot a Riemannian cone over a complete Riemannian manifold is either flat or has an irreducible holonomy representation. This is false in general for indefinite cones but the structures induced on the cone by holonomy invariant subspaces can be used to study the geometry on the base of the cone. The purpose of this paper is twofold: first we will give a survey of general results about semi-Riemannian cones with non irreducible holonomy representation and then, as the main result, we will derive improved versions of these general statements in the case when the cone admits a parallel vector field. We will show that if the base manifold is complete and the fibre of the cone and the parallel vector field have the same causal character, then the cone is flat, and that otherwise, the base manifold admits a certain global warped product structure. We will use these results to give a new proof of the classification results for Riemannian manifolds with imaginary Killing spinors and Lorentzian manifolds with real Killing spinors which are due to Baum and Bohle.

Keywords Lorentzian manifolds · Pseudo-Riemannian manifolds · Metric cones · Special holonomy · Geodesic completeness · Killing spinors

---

This work was supported by the Niels Henrik Abel Memorial Fund in relation to the 2019 Abel Symposium "Geometry, Lie Theory and Applications" and by the Australian Research Council (Discovery Program DP190102360).

T. Leistner (✉)
School of Mathematical Sciences, University of Adelaide, Adelaide, SA, Australia
e-mail: thomas.leistner@adelaide.edu.au

S. Hervik et al. (eds.), *Geometry, Lie Theory and Applications*, Abel Symposia 16,
https://doi.org/10.1007/978-3-030-81296-6_10

1 Introduction

Given a semi-Riemannian manifold (M, g), the *(space-like or time-like) semi-Riemannian cone over* (M, g) is the manifold $\widehat{M} = \mathbb{R}_{>0} \times M$ together with the metric

$$\widehat{g}_\epsilon = \epsilon\, \mathrm{d}r^2 + r^2 g, \tag{1.1}$$

where $\epsilon = 1$ in case of a space-like cone and $\epsilon = -1$ in case of a time-like cone. The original manifold (M, g) is then called the base of the cone. One reason for considering semi-Riemannian cones is that certain systems of PDE on the base manifold simplify when considered on the cone. The key example is the equation for a *Killing spinor field*, which is an overdetermined system of PDE. A solution to this PDE corresponds to a spinor field on the cone that is parallel for the Levi-Civita connection of the cone metric, which is a closed system of PDE that can be understood as the prolongation of the original PDE and that is easier to analyse, for example, by using tools from holonomy theory. Another example is the existence of a Sasaki structure on the base, which correspond to a Kähler structure on the cone and hence to a holonomy reduction to the unitary group. Semi-Riemannian cones play also an important role in conformal geometry where they appear as conformal ambient metrics for conformal structures containing an Einstein metric.

As mentioned, the most prominent application is the classification of complete Riemannian manifolds with real Killing spinors by C. Bär in [3]. He showed that the cone over such manifold admits a parallel spinor. By a fundamental theorem of Gallot, the cone is either irreducible or flat. With Gallot's result, the holonomy of the cone is one of the irreducible holonomy groups from Berger's list [8] that admit invariant spinors [14]. This leads to a short list of structures on the cone which correspond to certain structures on the base, all of which had been shown to admit Killing spinors [7].

In an attempt to apply this method to Killing spinor (and related) equations on manifolds with *indefinite* metrics, in [1, 2] possible generalisations of Gallot's theorem in the semi-Riemannian context were studied, yielding a comprehensive analysis of the case when the cone admits an invariant subspace under its holonomy representation. In the first part of this paper, in Sect. 3 we will give a brief survey of these results, including a theorem from [12]. However instead of giving the details of these general results, we will then focus on the special case when the cone admits a parallel vector field. The focus on this case enables us to explain the essential steps in the proofs of the general result without too much technical detail and at the same time give self-contained proofs for the special case. More importantly, we will be able to improve some of the general results in this special case, in particular in regards to their global character. In Sect. 4 we will prove the main result of the paper:

Theorem 1.1 *Let (M, g) be a geodesically complete semi-Riemannian manifold and let $(\widehat{M}, \widehat{g}_\epsilon)$ be the (time-like or space-like) cone over (M, g). Assume that $(\widehat{M}, \widehat{g}_\epsilon)$ admits a parallel vector field V.*

(1) If $\widehat{g}(V, V) = \epsilon$, then the cone is flat and (M, g) is of constant curvature ϵ.
(2) If $\widehat{g}(V, V) = -\epsilon$, then (M, g) is globally isometric to

$$(\mathbb{R} \times N, -\epsilon ds^2 + \cosh^2(s)\, g_N),$$

where (N, g_N) is a complete semi-Riemannian manifold.
(3) If $\widehat{g}(V, V) = 0$, then M is a disjoint union $M = M_- \cup M_0 \cup M_+$ with $M_\pm$ open and such that M_0 is either empty (in which case one of $M_\pm$ is also empty) or a smooth totally geodesic hypersurface and $(M_\pm, g)$ is globally isometric to

$$(\mathbb{R} \times N_\pm, -\epsilon ds^2 + e^{2s} g_{N_\pm}),$$

where $(N_\pm, g_{N_\pm})$ are complete semi-Riemannian manifolds. Moreover, $M_0 = \emptyset$ if and only if (M, g) is Riemannian or negative definite.

Note that the cases (2) and (3) also include the possibility that (M, g) has constant curvature: in (2) g_N has constant curvature ϵ if and only if g also has constant curvature ϵ, whereas in (3), g_N is flat, if and only if g has constant curvature ϵ (see [1, Section 2]).

The improvements in this theorem over the of the general result will allow us to give an alternative proof of the classification of complete Riemannian manifolds with imaginary Killing spinors by Baum [5] and of Lorentzian manifolds with real Killing spinors [9]. In both cases, the parallel spinor on the cone induces a parallel vector field on the Lorentzian cone. In fact, the results in this paper will be applicable to the classification of Killing spinors in the case when the parallel spinor on the cone induces a parallel vector field. Working out the detail of this is however beyond the scope of this paper.

2 Preliminaries

2.1 Curvature and Geodesics of Semi-Riemannian Cones

Let $\widehat{g}_\epsilon = \epsilon dr^2 + r^2 g$ be the cone metric on $\widehat{M} = \mathbb{R}_{>0} \times M$, where $\epsilon = \pm 1$ and (M, g) is a pseudo-Riemannian manifold. The cone is called space-like if $\epsilon = 1$ and time-like if $\epsilon = -1$. We denote by $\partial_r = \frac{\partial}{\partial r}$ the radial unit vector field. The Levi-Civita connection of the cone $(\widehat{M}, \widehat{g}_\epsilon)$ is given by

$$\widehat{\nabla}_{\partial_r} \partial_r = 0, \qquad \widehat{\nabla}_X \partial_r = \frac{1}{r} X, \qquad \widehat{\nabla}_X Y = \nabla_X Y - \epsilon g(X, Y)\partial_r, \tag{2.1}$$

for all vector fields $X, Y \in \Gamma(T\widehat{M})$ orthogonal to ∂_r. The curvature $\widehat{R}$ of the cone is given by the following formulas including the curvature R of the base metric g:

$$\partial_r \lrcorner \widehat{R} = 0, \qquad \widehat{R}(X, Y)Z = R(X, Y)Z - \epsilon\,(g(Y, Z)X - g(X, Z)Y)\,, \tag{2.2}$$

for $X, Y, Z, U \in TM$. This implies that if (M, g) is a space of constant curvature κ, i.e.,

$$R(X, Y, Z, U) = \kappa\,(g(X, U)g(Y, Z) - g(X, Z)g(Y, U))\,,$$

then the cone has the curvature $r^2\,(\kappa - \epsilon)\,(g(X, U)g(Y, Z) - g(X, Z)g(Y, U))$. In particular, if $\kappa = \epsilon$, then the cone is flat, as it is the case for the $\epsilon = 1$ cone over the standard sphere of radius 1 or the $\epsilon = -1$ cone over the hyperbolic space.

Let I be an interval containing 0 and $\widehat{\gamma} = (\rho, \gamma) : I \to \widehat{M} = \mathbb{R}_{>0} \times M$ be a geodesic of $(\widehat{M}, \widehat{g}_\epsilon)$ with $\widehat{\gamma}(0) = \hat{p}$ and $\widehat{\gamma}'(0) = a\partial_r + X$. The geodesic equations are easily checked to be

$$0 = \rho''(t) - \epsilon r(t) g\left(\gamma'(t), \gamma'(t)\right), \qquad 0 = 2\,\rho'(t)\gamma'(t) + \rho(t)\nabla_{\gamma'(t)}\gamma'(t). \tag{2.3}$$

Let γ be a reparametrisation of a geodesic β of (M, g),

$$\gamma(t) = \beta(f(t)), \quad \text{with } \beta(0) = p \text{ and } \beta'(0) = X, \tag{2.4}$$

implying the initial conditions $f(0) = 0$ and $f'(0) = 1$ for f.

Now let $g(X, X) = cL^2$ with $c \in \{0, \pm 1\}$ and $L > 0$. Hence, from (2.3) we get

$$0 = \rho''(t) - c\epsilon\rho(t) f'(t)^2 L^2, \qquad 0 = 2\,\rho'(t) f'(t) + \rho(t) f''(t) \tag{2.5}$$

with initial conditions

$$\rho(0) = r, \quad f(0) = 0, \quad \rho'(0) = a, \quad f'(0) = 1.$$

If the initial speed X satisfies $cL^2 = g(X, X) = 0$, i.e., if it is zero or light-like, then the equations become

$$0 = \rho''(t), \qquad 0 = 2\,\rho f'(t) + (\rho t + \rho) f''(t),$$

i.e., with solutions

$$\rho(t) = at + r, \qquad f(t) = \frac{rt}{at + r}. \tag{2.6}$$

This implies that f and thus $\hat{\gamma}$ is defined for $t \in [0, -\frac{r}{a})$ if $a < 0$, and for $t \geq 0$ otherwise.

If $cL^2 \neq 0$, the solutions to Eqs. (2.5) are then given by

$$\rho(t) = \sqrt{(at+r)^2 + c\epsilon L^2 r^2 t^2},$$
$$f(t) = \begin{cases} \frac{1}{L}\operatorname{artan}\left(\frac{Lrt}{at+r}\right), & \text{if } c\epsilon = 1, \\ \frac{1}{L}\operatorname{artanh}\left(\frac{Lrt}{at+r}\right), & \text{if } c\epsilon = -1, \end{cases} \tag{2.7}$$

This gives us the maximal domain of the cone geodesics under the assumption that (M, g) is complete: in case of $c\epsilon = 1$, in particular if the cone is Riemannian, all geodesics are defined on $\mathbb{R}$ if $a \geq 0$ and on $t \in [0, -\frac{r}{a})$ if $a < 0$. Otherwise, if the functions ρ and f are defined on an interval $[0, T)$, where T is the first positive zero of the polynomial

$$\left(\frac{Lrt}{at+r} - 1\right)\left(\frac{Lrt}{at+r} + 1\right)(at+r)^2 = L^2r^2t^2 - (at+r)^2 = ((Lr-a)t - r)((Lr+a)t + r),$$

or $T = \infty$ if the polynomial has no positive zero. More explicitly, $T = \frac{r}{Lr-a}$ if $a < Lr$ and $T = \infty$ if $a \geq Lr$. We summarise this:

Proposition 2.1 *Let (M, g) be a complete semi-Riemannian manifold and $(\widehat{M}, \widehat{g}_\epsilon)$ be the cone. Let $\hat{p} = (r, p) \in \widehat{M}$ and $\widehat{X} = a\partial_r|_{\hat{p}} + X \in T_{\hat{p}}\widehat{M}$ with $g(X, X) = cL^2$ with $c \in \{0, \pm 1\}$ and $L > 0$. Then there is a geodesic $\widehat{\gamma} : [0, T) \to \widehat{M}$ of $(\widehat{M}, \widehat{g})$ starting at $\hat{p}$ with $\widehat{\gamma}'(0) = \widehat{X}$ and where*

$$T = \begin{cases} \infty, & \text{if } c\epsilon \in \{0, 1\} \text{ and } a \geq 0, \text{ or if } c\epsilon = -1 \text{ and } a \geq Lr, \\ -\frac{r}{a}, & \text{if } c\epsilon \in \{0, 1\} \text{ and } a < 0, \\ \frac{r}{Lr-a}, & \text{if } c\epsilon = -1 \text{ and } a < Lr. \end{cases} \tag{2.8}$$

This geodesic is given by (2.4) together with (2.6) or (2.7).

2.2 Completeness of Certain Warped Products

In this section we are going to study the completeness of warped products of the form

$$(M = \mathbb{R} \times N, g = -\epsilon \mathrm{d}s^2 + f^2(s) g_N),$$

where (N, g_N) is a semi-Riemannian manifold, f is a positive function on N and $\epsilon = \pm 1$. We will need these results in Sect. 4. The Levi-Civita connection of such metrics is given by

$$\begin{aligned}
\nabla_{\partial_s}\partial_s &= 0,\\
\nabla_X \partial_s &= \tfrac{f'(s)}{f(s)} X, \qquad (2.9)\\
\nabla_X Y &= \nabla^N_X Y + \epsilon f'(s) f(s) g_N(X,Y)\partial_s,
\end{aligned}$$

Proposition 2.2 *Let $f : \mathbb{R} \to \mathbb{R}_{>0}$ be a smooth function and (N, g_N) be a semi-Riemannian manifold and define $(M = \mathbb{R} \times N, g = -\epsilon \mathrm{d}s^2 + f^2(s) g_N)$.*

(1) If all geodesics of (M, g) with initial velocity tangent to N are defined on $\mathbb{R}$, then (N, g_N) is complete. In particular, if (M, g) is complete, then (N, g_N) is complete.
(2) If $f = \cosh$, then (M, g) is complete if (N, g_N) is complete.
(3) If $f(s) = \mathrm{e}^s$, then (M, g) is complete if and only if (N, g_N) is complete and (M, g) is definite, i.e., if $-\epsilon g_N$ is a complete Riemannian metric.

Proof

(1) Let $(\sigma, \gamma) : \mathbb{R} \to$ be a geodesic of (M, g) with $\sigma'(0) = 0$. Then the geodesic equations are

$$\sigma'' + \epsilon f'(\sigma) f(\sigma) g(\gamma', \gamma') = 0, \qquad \nabla^N_{\gamma'}\gamma' + 2\frac{f'(\sigma)}{f(\sigma)}\sigma'\gamma' = 0, \qquad (2.10)$$

in particular, γ is a pre-geodesic for g_N. The first equation shows that, if $\gamma'(t_0) = 0$ for some t_0, then $\sigma(t) = at + b$ and $\gamma(t) \equiv \gamma(t_0)$ constant. Hence, if $\gamma'(0) \neq 0$, then $\gamma'(t) \neq 0$ for all t, and so we can parametrise γ by arc-length. The second geodesic equation shows that the reparametrised curve is a geodesic for g_N. Hence, (N, g_N) is complete.

(2) Assume that (M, g) is incomplete. Hence there is a maximal geodesic $(\sigma, \gamma) : (a, b) \to M$ with $b \in \mathbb{R}$. Then the first geodesic equation in (2.10) and the equation that the geodesic is of constant length,

$$-\epsilon(\sigma')^2 + f^2(\sigma) g(\gamma', \gamma') = c,$$

for a constant c, imply that

$$\cosh(\sigma)\sigma'' + \sinh(\sigma)(\sigma')^2 + \epsilon c \sinh(\sigma) = 0.$$

Then, with substituting $\xi = \sinh(\sigma)$, this equation becomes

$$\xi'' + \epsilon c \xi = 0.$$

This is a linear ODE for ξ and hence we can extend ξ and also σ beyond b and in fact to $\mathbb{R}$. Moreover, the second geodesic equation in (2.10) implies that

$\gamma = \beta \circ \tau$, where β is a geodesic equation and σ and τ satisfy the equations

$$\tau'' + 2\sigma'\tau' \tanh(\sigma) = 0.$$

With $\sigma : \mathbb{R} \to \mathbb{R}$, this is a linear ODE for τ and hence can be extended beyond b. This yields a contradiction to the incompleteness of (M, g).

(3) Assume that g is complete but indefinite. With g indefinite we can consider a light-like geodesic (σ, γ), i.e., with

$$0 = -\epsilon(\sigma')^2 + e^{2\sigma} g_N(\gamma', \gamma').$$

Moreover, from the first geodesic equation we obtain

$$0 = \left((\sigma')^2 + \sigma''\right) = \xi^{-1}\xi'',$$

where we substitute $\xi = e^{\sigma} > 0$. This however yields the equation $\xi'' = 0$, so ξ is affine and, since (M, g) is complete, defined on $\mathbb{R}$. This contradicts $\xi = e^{\sigma} > 0$, so (M, g) cannot have light-like geodesics and hence g is definite.

Conversely, assume that (N, g_N) is a complete Riemannian manifold. If (M, g) is not complete, there is a maximal geodesic $\gamma = (\sigma, \beta) : [0, b) \to M$ that leaves every compact set in M. For such a geodesic we have

$$1 = (\sigma')^2 + e^{2\sigma} g_N(\beta', \beta').$$

Hence, with g_N Riemannian, we have $0 \leq (\sigma')^2 \leq 1$ and hence that σ is bounded on $[0, b)$. This implies that σ remains in a compact set, which implies that β leaves every compact set in N. It also implies that e^{σ} is bounded away from zero and so $g_N(\beta', \beta')$ is bounded on $[0, b)$ say by c^2. Then we have that $\beta(t)$ is contained in the geodesic ball around $\beta(0)$ of radius bc since

$$\text{dist}_{g_N}(\beta(0), \beta(t)) \leq \text{length}_{g_N}(\beta|_{[0,t]}) \leq bc.$$

Since (N, g_N) is complete, its geodesic balls are compact, which gives a contradiction. Hence (M, g) is complete.

□

3 Survey of General Results

3.1 Holonomy Groups and Gallot's Theorem

Let (M, g) be a connected semi-Riemannian manifold. The *holonomy group* $\text{Hol}_p(M, g)$ of (M, g) at $p \in M$ is defined as the group of parallel transports, with respect to the Levi-Civita connection of g, along piecewise smooth loops that are closed at p. Since the Levi-Civita connection preserves the metric, the holonomy group is a subgroup of the orthogonal groupp $O(T_pM, g|_p)$ acting on T_pM. By fixing a basis of T_pM, it can be identified with a subgroup of $O(r, s)$, where (r, s) is the signature of (M, g). The holonomy groups at different points in M are conjugated within $O(r, s)$. Hence the holonomy group as a subgroup in $O(r, s)$ is well defined up to conjugation and we refer to this as the holonomy group $\mathrm{Hol}(M, g)$. If $G \subset O(r, s)$ is a subgroup and $\mathrm{Hol}(r, s) \subset G$ we say that the holonomy reduces to G.

The holonomy group is a Lie group. Its connected component is given by parallel transport along *contractible* loops. Its Lie algebra is denoted by $\mathfrak{hol}_p(M, g)$, the *holonomy algebra*. One can show that the holonomy algebra contains all curvature endomorphisms $R|_p(X, Y)$ at p, with $X, Y \in T_pM$ and all derivatives of curvature endomorphisms. Moreover, the Ambrose-Singer Holonomy Theorem states that the holonomy algebra at p is spanned as a vector space by the following linear maps,

$$P_\gamma^{-1} \circ R|_q(X, Y) \circ P_\gamma,$$

where $q \in M$, γ is a path from p to q, P_γ the parallel transport along γ and $X, Y \in T_qM$.

The importance of the holonomy group arises from the well-known holonomy principles. First, parallel sections (with respect to the Levi-Civita connection of g) of TM or of any tensor bundle over M are in one-to-one correspondence with vectors (or tensors) that are fixed under the holonomy representation. For example, the existence of parallel vector field reduces the holonomy to a the stabiliser in $O(r, s)$ of a vector in $\mathbb{R}^{r,s}$. Another example is the existence of a parallel complex structure, which reduces the holonomy to the unitary group $U(r/2, s/2)$. The other principle is that subspaces in T_pM, or in $\mathbb{R}^{r,s}$, that are invariant under the holonomy group are in one-to-one correspondence with vector distributions that are invariant under parallel transport, or for short, a *parallel distribution*. The parallel distribution $\mathbb{V} \subset TM$ is obtained from an holonomy invariant subspace $E \subset T_pM$ by parallel transport: the fibre $\mathbb{V}|_q$ is defined as by the parallel transport of $E \subset T_pM$ by any curve from p to q. Because of the holonomy invariance of E, this is a well defined procedure and $\mathbb{V}|_q$ does not depend on the chosen loop. A parallel distribution $\mathbb{V}$ is involutive and defines a foliation of M into totally geodesic leaves of $\mathbb{V}$.

The holonomy group acts *irreducibly* if it does not admit any invariant subspace. In this case we also say that (M, g) is irreducible. If $\mathrm{Hol}(M, g)$ does admit an invariant subspace E, since $\mathrm{Hol}(M, g) \subset O(r, s)$, the orthogonal space $E^\perp$ is also invariant under $\mathrm{Hol}(M, g)$. Hence, every holonomy invariant subspace defines two parallel distributions $\mathbb{V}$ and $\mathbb{V}^\perp$. If g is indefinite and E is a degenerate subspace, i.e., $E \cap E^\perp \neq \{0\}$, there is a totally light-like distribution $\mathbb{V} \cap \mathbb{V}^\perp$ with totally geodesic leaves. If E is non-degenerate, i.e., if $T_pM = E \oplus E^\perp$, then we also have $TM = \mathbb{V} \oplus \mathbb{V}^\perp$. In this case we say that the holonomy group acts *decomposably*, or for short that (M, g) is decomposable. If there is no non-degenerate subspace that is invariant under $\mathrm{Hol}(M, g)$ we say that the holonomy acts *indecomposably*, or that (M, g) is indecomposable. If g is indefinite, the holonomy group may act indecomposably without acting irreducibly. This is the case if the holonomy group admits a totally light-like invariant subspace, but no non-degenerate invariant subspace.

If the holonomy group acts decomposably, not just the tangent space decomposes into holonomy invariant subspaces, but under certain global assumptions also the manifold decomposes into a semi-Riemannian product. This is due to the splitting theorems of de Rham [10] and Wu [13]: If (M, g) is complete and simply connected and the holonomy group acts decomposably, then (M, g) is isometric to a global semi-Riemannian product $(M_1, g_1) \times (M_2, g_2)$ and the holonomy representation of (M, g) is isomorphic to the product of the holonomy representations of (M_i, g_i). The manifolds M_i correspond to the totally geodesic foliations of M into the leaves of the parallel complementary distributions $\mathbb{V}$ and $\mathbb{V}^{\perp}$.

The notions of irreducibility and (in-)decomposability can also be formulated for the holonomy algebras $\mathfrak{hol}(M, g)$, depending on wether the holonomy algebra admits a (non-degenerate) invariant subspace. Note that if M is not simply connected, the holonomy algebra acting decomposably does not imply that the holonomy group does act decomposably. In particular, the existence of a non-degenerate subspace that is invariant under the holonomy algebra does not necessarily imply the existence of a globally defined parallel distribution.

The following result about the holonomy algebra of a Riemannian cone is due to Gallot.

Theorem 3.1 (S. Gallot, [11]) *Let (M, g) be a complete Riemannian manifold of dimension ≥ 2 such that the holonomy algebra of the cone $(\widehat{M}, \widehat{g}_+)$ does not act irreducibly. Then $(\widehat{M}, \widehat{g}_+)$ is flat and hence (M, g) has constant curvature* 1. *If, in addition, (M, g) is simply connected, then (M, g) is isometric to the standard sphere.*

We will present Gallot's proof of this theorem in Sect. 4. Here we will only explain its first step, which is needed in order to understand possible generalisations and which is based on the aforementioned *holonomy principle*: since the aim is to show that the cone is flat, we can pass to the universal cover, which is the cone over the universal cover of M, and assume that the holonomy *group* of this cone admits an invariant subspace $E \subset T_p\widehat{M}$. Hence, this invariant subspace defines a vector distribution $\mathbb{V} \subset T\widehat{M}$ that is invariant under parallel transport. With E holonomy invariant, its orthogonal space $E^{\perp}$ is also holonomy invariant and defines a parallel distribution $\mathbb{V}^{\perp}$. Since the cone is Riemannian, $\mathbb{V}$ and $\mathbb{V}^{\perp}$ are non-degenerate and hence the tangent space splits into a direct sum of parallel vector distributions $TM = \mathbb{V} \oplus \mathbb{V}^{\perp}$. Both distributions are parallel and hence involutive and define totally geodesic leaves. This splitting and the induced foliation is then used in Gallot's proof.

If one considers indefinite cone metrics, for example by considering time-like cones over Riemannian manifolds or by considering an indefinite base (M, g), a holonomy invariant subspace may be degenerate, i.e., $E \cap E^{\perp} \neq 0$, and hence

$\mathbb{V} \cap \mathbb{V}^{\perp} \neq \{0\}$, so that the resulting parallel distributions are not complementary. The following example shows that Gallot's Theorem is false in this case.

Example 3.2 Consider the semi-Riemannian manifold

$$\left(M = \mathbb{R} \times N, g = \mathrm{d}s^2 + e^{-2s} g_N\right), \tag{3.1}$$

where (N, g_N) is a semi-Riemannian manifold. Then the light-like vector field

$$V = e^{-s}(\partial_r + \tfrac{1}{r}\partial_s)$$

on the time-like cone $(\widehat{M}, \widehat{g}_-)$ is parallel. The manifold (M, g) has constant negative curvature only if g_N is flat. If we now assume that (N, g_N) is a complete Riemannian manifold, then, by Proposition 2.2, (M, g) is a complete Riemannian manifold whose time-like cone $(\widehat{M}, \widehat{g}_-)$ admits a parallel light-like vector field and hence has a non irreducible holonomy group. However, unless g_N is flat, the cone $\widehat{g}_-$ is not flat. This shows that Gallot's Theorem cannot hold when the cone has a parallel light-like vector field.

This example suggests that one has to strengthen the assumptions in Gallot's Theorem in the indefinite setting. To get Gallot's proof started, instead of assuming the existence of *some* holonomy invariant subspace, on should require the existence of a *non-degenerate* invariant subspace, that gives complementary parallel distributions $T\widehat{M} = \mathbb{V} \oplus \mathbb{V}^{\perp}$. Unfortunately, the following example shows that such a modification of Gallot's Theorem also fails.

Example 3.3 Let (N, g_N) be a complete semi-Riemannian manifold of dimension at least 2 and which is not of constant curvature 1. Then the semi-Riemannian manifold

$$(M = \mathbb{R} \times N, g = ds^2 + \cosh(s) g_N) \tag{3.2}$$

is complete by Proposition 2.2. Using Eq. (2.9) it is easily established that the spacelike vector field

$$V = -\sinh(s)\partial_r + \frac{\cosh(s)}{r}\partial_s$$

on the time-like cone $(\widehat{M}, \widehat{g}_-)$ is parallel. For the curvature tensor R of (M, g) we have

$$R(X, Y)Z = R_N(X, Y)Z + \tanh^2(s)\left(g_N(Y, Z)X - g_N(X, Z)Y\right),$$

where $X, Y, Z, \; U \in TF$ and R_N is the curvature tensor of (N, g_N). This shows that (M, g) cannot have constant sectional curvature, unless N has constant curvature 1. Thus, in general the cone $(\widehat{M}, \widehat{g})$ over the complete manifolds (M, g) is decomposable but not flat.

3.2 Decomposable Cones Over Complete and Over Compact Manifolds

In this section we will review a few results that show to which extent Gallot's Theorem does generalise to the semi-Riemannian context, having in mind the counter examples of the previous section. We will mainly focus on space-like cones, as the corresponding results for time-like can be obtained by multiplying the cone metric by -1. First we consider cones over complete semi-Riemannian manifolds.

Theorem 3.4 ([1]) *Let (M, g) be a complete semi-Riemannian manifold of dimension ≥ 2 and assume that the holonomy algebra of the cone $(\widehat{M}, \widehat{g}_+)$ acts decomposably. Then there exists an open dense submanifold $M' \subset M$ such that each connected component of M' is isometric to a pseudo-Riemannian manifold of the form:*

(1) *a pseudo-Riemannian manifold M_1 of constant sectional curvature 1, or*
(2) *a pseudo-Riemannian manifold $M_2 = \mathbb{R}_{>0} \times N_1 \times N_2$ with the metric*

$$-\mathrm{d}s^2 + \cosh^2(s) g_1 + \sinh^2(s) g_2,$$

where (N_1, g_1) and (N_2, g_2) are semi-Riemannian manifolds and (N_2, g_2) has constant sectional curvature -1 or $\dim N_2 \leq 1$.

Moreover, the cone $\widehat{M}_2$ is isometric to the open subset $\{r_1 > r_2\}$ in the product of the space-like cone $(\mathbb{R}_{>0} \times N_1, \mathrm{d}r^2 + r^2 g_1)$ over (N_1, g_1) and the time-like cone $(\mathbb{R}_{>0} \times N_2, -\mathrm{d}r^2 + r^2 g_2)$ over (N_2, g_2).

Note that Example 3.3 shows that this theorem is sharp.

Next we consider cones over closed semi-Riemannian manifolds (M, g), i.e., when M compact without boundary. Recall that for indefinite metrics compactness of M does *not* imply the geodesic completeness of (M, g), so we have to assume it, in order to get a version of Gallot's Theorem under these strengthened assumptions.

Theorem 3.5 ([1]) *Let (M, g) be a closed and geodesically complete semi-Riemannian manifold of dimension ≥ 2. If the cone $(\widehat{M}, \widehat{g}_+)$ is decomposable, then it is flat and hence (M, g) has constant curvature 1.*

Since there is no simply connected compact indefinite pseudo-Riemannian manifold of constant curvature 1, we obtain the following corollary.

Corollary 3.6 ([1]) *If (M, g) is a simply connected closed and geodesically complete indefinite pseudo-Riemannian manifold, then the holonomy group of the cone $(\widehat{M}, \widehat{g})$ is indecomposable.*

Theorem 3.5 was strengthened by Matveev in [12], see also [13].

Theorem 3.7 (V. Matveev [12]) *Let M be a closed manifold.*

(1) If g is a light-like complete indefinite semi-Riemannian metric on M, then the cone $(\widehat{M}, \widehat{g}_+)$ is indecomposable.
(2) If g is a Riemannian metric on M, then the cone $(\widehat{M}, \widehat{g}_-)$ is indecomposable.

Note that (2) in Theorem 3.7 implies that even though the time-like cone over a compact quotient $M = \mathbb{H}^n/\Gamma$ of hyperbolic space $\mathbb{H}^n$ is flat, its holonomy group acts indecomposably, which is rather surprising.

3.3 Local Structure of Non Irreducible Cones

In this section we will review some results about the local structure of non irreducible cones. We start with decomposable cones.

Theorem 3.8 ([1]) *Let (M, g) be a semi-Riemannian manifold such that the holonomy algebra of the cone $(\widehat{M}, \widehat{g}_+)$ acts decomposably. Then there exists an open dense submanifold $M' \subset M$ such that any point $p \in M'$ has a neighborhood U that is isometric to a semi-Riemannian manifold of the form $(a, b) \times N_1 \times N_2$ with the metric given either by*

$$g_+ = \mathrm{d}s^2 + \cos^2(s) g_1 + \sin^2(s) g_2 \quad or \qquad g_- = -\mathrm{d}s^2 + \cosh^2(s) g_1 + \sinh^2(s) g_2, \tag{3.3}$$

where g_1 and g_2 are metrics on N_1 and N_2 respectively.

Moreover, $\mathbb{R}_{>0} \times U \subset \widehat{M}$ with the metric $g_\pm$ in (3.3) is locally isometric to the product of cone metrics

$$(\mathrm{d}r_1^2 + r_1^2 g_1) + (\pm \mathrm{d}r_2^2 + r_2^2 g_2).$$

Note that this theorem also applies to the Riemannian context. The cone over the incomplete Riemannian metric g_+ in (3.3), with g_1 and g_2 Riemannian, is decomposable without being flat.

Next we consider the case when the holonomy of the cone admits an invariant degenerate subspace E. This implies the existence of an invariant subspace $E \cap E^\perp$ that is totally light-like. We restrict ourselves to the case when the dimension of $E \cap E^\perp$ is 1 or 2. In this case we have a parallel distribution of totally light-like lines or planes.

Theorem 3.9 ([2]) *Let $(\widehat{M}, \widehat{g}_-)$ be the time-like cone over a semi-Riemannian manifold (M, g). If the cone admits a parallel light-like line field* **L**, *then locally there is a parallel trivializing section of* **L**. *Moreover, on a dense open subset $\widehat{M}_{\mathrm{reg}} \subset \widehat{M}$, the metric $\widehat{g}$ is locally isometric to a warped product of the form*

$$\widetilde{g}_0 = 2\,\mathrm{d}u\,\mathrm{d}v + u^2 g_0, \tag{3.4}$$

with a semi-Riemannian metric g_0, and the metric g is locally of the form

$$g = \mathrm{d}s^2 + \mathrm{e}^{2s} g_0.$$

The results in the case when $E \cap E^{\perp}$ is of dimenion 2 are more technical and related to the existence of s shearfree, geodesic, light-like congruence on the base:

Theorem 3.10 ([2]) *The time-like cone $(\widehat{M}, \widehat{g})$ over a semi-Riemannian manifold (M, g) admits a parallel, totally light-like 2-plane field if and only if, locally over an open dense subset, the base (M, g) admits two vector fields V and Z satisfying*

$$g(V, V) = 0, \quad g(Z, Z) = 1, \quad g(V, Z) = 0, \tag{3.5}$$

and such that

$$\nabla_X V = \alpha(X)V + g(X, V)Z, \quad \nabla_X Z = -X + \beta(X)V + g(X, Z)Z, \tag{3.6}$$

with 1-forms α and β on M. In particular, the base (M, g) admits a geodesic, shearfree light-like congruence defined by V.

Note that the first equation in Eq. (3.6) implies that $V^{\perp}$ is integrable. This allows us to determine the local form of the metrics with vector fields V and Z satisfying Eqs. (3.5) and (3.6):

Proposition 3.11 ([2]) *A semi-Riemannian metric (M, g) admits vector fields V and Z with (3.5) and (3.6) if and only if (M, g) is locally of the form $M = M_0 \times \mathbb{R}^3$ and*

$$g = \mathrm{d}s^2 + e^{-2s} g_0(u) + 2\,\mathrm{d}u\,\eta,$$

for a family of metrics $g_0(u)$ on M_0 depending on u and a 1-form η on M such that $\eta(\partial_t)$ is nowhere vanishing satisfying the following system of first order PDEs:

$$\begin{aligned} \partial_t \eta_t = \partial_s \eta_t \;=\; X\eta_t \;=\; \partial_t(\eta(X)) &= 0, \\ \partial_t \eta_s &= 2\eta_t, \\ \partial_s\, \eta(X) - X\, \eta_s &= -2\eta(X) \end{aligned} \tag{3.7}$$

for all $X \in \Gamma(TM_0)$ and where we denote $\eta_t = \eta(\partial_t)$ and $\eta_s = \eta(\partial_s)$.

One can solve explicitly the system (3.7) in the following way: Let $f_1 = f_1(u)$ be an arbitrary nowhere vanishing smooth function on the real line equipped with the coordinate u and $f_2 = f_2(x, s, u)$ an arbitrary smooth function on M which does not depend on t. Let $h_i = h_i(x, s, u)$ be a (t-independent) solution of the ordinary differential equation

$$\partial_s h_i + 2h_i = \partial_i f_2$$

for all $i = 1, \ldots, n_0$, where $\partial_i = \partial/\partial x^i$. Then

$$\eta_t := f_1(u), \quad \eta_s := 2t f_1(u) + f_2(x, s, u), \quad \eta(\partial_i) := h_i(x, s, u)$$

solves (3.7) and every solution is of this form.

This provides us with a construction method of metrics whose cone admits a totally light-like 2-plane.

Remark 3.12 For completeness we should mention further results in [1] for the case when the cone admits a holonomy invariant maximal isotropic subspace $\mathbb{V} = \mathbb{V}^\perp$ and an invariant maximally isotropic complement. This is equivalent to the existence of a para-Kähler structure on the cone. In [1, Section 8] we have shown that the existence of a para-Kähler structure on the cone over (M, g) is equivalent to the existence of a para-Sasaki structure on (M, g) and a similar correspondence for para-hyper-Kähler structures on the cone and para-3-Sasakian structures on (M, g).

3.4 Holonomy of Cones

In the last part of this survey section we are going to review results about the possible holonomy groups of cones. We will consider the fundamental cases when the holonomy group acts irreducibly or not irreducibly but indecomposably.

3.4.1 Irreducible Cone Holonomies

If the holonomy algebra of the cone is irreducible, we can use Berger's list and single those out that can be cone holonomies. They key here is to observe that $\partial_r \lrcorner \widehat{R} = 0$ prevents cones from being Einstein with non zero Einstein constant. This shows, for example, that cones cannot be irreducible locally symmetric spaces. So by ruling out all holonomy groups of Einstein cones, we obtain:

Theorem 3.13 ([2]) *If $(\widehat{M}, \widehat{g})$ is a cone with irreducible holonomy algebra $\mathfrak{g}$, then $\mathfrak{g}$ is isomorphic to one of the following Lie algebras*

$$\begin{array}{rrr} \mathfrak{so}(t,s),\ \mathfrak{u}(p,q),\ \mathfrak{su}(p,q) \subset \mathfrak{so}(2p,2q), & & \mathfrak{sp}(p,q) \subset \mathfrak{so}(4p,4q), \\ \mathfrak{so}(n,\mathbb{C}) \subset \mathfrak{so}(n,n), & \mathfrak{g}_2^{\mathbb{C}} \subset \mathfrak{so}(7,7), & \mathfrak{spin}(7,\mathbb{C}) \subset \mathfrak{so}(8,8), \\ & \mathfrak{g}_2 \subset \mathfrak{so}(7), & \mathfrak{spin}(7) \subset \mathfrak{so}(8), \\ & \mathfrak{g}_{2(2)} \subset \mathfrak{so}(3,4), & \mathfrak{spin}(3,4) \subset \mathfrak{so}(4,4). \end{array} \tag{3.8}$$

3.4.2 Holonomy of Non Irreducible, Indecomposable Cones

In general the classification of non irreducible, indecomposable holonomy groups is widely open and only solved in Lorentzian and in some special cases in signature $(2, n)$ and (n, n). We will focus here in the case where the invariant totally light-like subspace has dimension 1. The key here is the result in Theorem 3.9, where it was shown that that a cone that admits a parallel light-like line distribution is locally isometric to a metric of the form (3.4).

Theorem 3.14 ([2]) *Let (N, g_0) be a semi-Riemannian manifold in dimension n and $\widetilde{g}_0$ the metric defined in (3.4). If the holonomy of $\widetilde{g}$ acts indecomposably, then*

$$\mathfrak{hol}(\widetilde{g}_0) \subset \mathfrak{hol}(g_0) \ltimes \mathbb{R}^{t,s} \tag{3.9}$$

where $\mathfrak{hol}(g_0) \ltimes \mathbb{R}^{t,s}$ is a subalgebra of the stabiliser algebra of ∂_v in $\mathfrak{so}(t+1, s+1)$, i.e., in $\mathfrak{so}(t, s) \ltimes \mathbb{R}^{t,s} = \mathfrak{so}(t + 1, s + 1)_{\partial_v}$, and

$$\mathrm{pr}_{\mathfrak{so}(t,s)}(\mathfrak{hol}(\widetilde{g}_0)) = \mathfrak{hol}(g_0).$$

There is an equality in (3.9) whenever (N, g_0) is one of the following:

(1) an irreducible locally symmetric space, or a product thereof;
(2) a Riemannian manifold;
(3) a Lorentzian manifold without a parallel light-like vector field.

4 Semi-Riemannian Cones with Parallel Vector Fields

In this section we will consider the special case when the invariant subspace under the holonomy group of the cone is given by a parallel vector field, that is, the rank of the invariant distribution is one and the distribution admits a global parallel section. For this special case we will prove versions of the theorems in the previous section that are slightly stronger and more specific, and we will prove Theorem 1.1. Before we do this we will review Gallot's original proof of his theorem in order to see when we can generalise it to the case of a parallel vector field. We will see that this can be done when the radial vector field and the parallel vector field have the same causal character.

4.1 The Proof of Gallot's Theorem

Gallot's proof of Theorem 3.1 uses the following fundamental observation, which holds not only for Riemannian cones.

Lemma 4.1 *Let $(\widehat{M}, \widehat{g}_\epsilon)$ be semi-Riemannian cone and let $\mathbb{V} \subset T\widehat{M}$ be a non degenerate, parallel distribution. Let $p \in M$ such that $\partial_r|_p \in \mathbb{V}|_p$ and $N_p^\perp$ the leaf of $\mathbb{V}^\perp$ through p. Then the image in $N_p^\perp$ under the exponential map restricted to $\mathbb{V}_p^\perp \in T_p\widehat{M}$ is flat.*

Proof Let $\widehat{\gamma} = (\rho, \gamma) : I \to \widehat{M} = \mathbb{R}_{>0} \times M$ be a geodesic in $(\widehat{M}, \widehat{g})$ with $\widehat{\gamma}(0) = p$ and $\widehat{\gamma}'(0) \in \mathbb{V}^{\perp}|_p$. It is easy to check using (2.1) that the vector field

$$F(t) = \rho(t)\partial_r - t\widehat{\gamma}'(t) \tag{4.1}$$

is parallel transported along $\hat{\gamma}$. Then with $F(0) = r(p)\partial_r|_{\hat{p}} \in \mathbb{V}|_{\hat{p}}$ we have that $F(t) \in \mathbb{V}|_{\widehat{\gamma}(t)}$ for all t. Since the curvature tensor leave parallel distributions invariant and because of $\partial_r \lrcorner \widehat{R} = 0$, we have that

$$\widehat{R}(X, Y)F(t) = \widehat{R}(X, Y)\widehat{\gamma}'(t) \in \mathbb{R} \cdot \mathbb{V}|_{\widehat{\gamma}(t)}$$

for all t. On the other hand we have that $\widehat{\gamma}'(t) \in \mathbb{V}^{\perp}_{\hat{\gamma}(t)}$ for all t. Hence, with $\mathbb{V} \cap \mathbb{V}^{\perp} = \{0\}$ this implies that

$$\widehat{R}(X, Y)\widehat{\gamma}'(t) = 0,$$

for all vector fields X and Y along $\widehat{\gamma}$ and all $t \in I$. From this we see that the Jacobi fields along $\widehat{\gamma}$ are those of a flat manifold, which implies that N is flat. □

Using this lemma, we can now proceed with the proof of Gallot's Theorem.

Proof of Theorem 3.1 By passing to the universal cover of the cone, which is the cone over the universal cover of M, we can assume that $\widehat{M}$ is simply connected. Let $\mathbb{V}$ be a parallel distribution in $T\widehat{M}$ and $\mathbb{V}^{\perp}$ the orthogonal distribution that are induced by the subspace that is invariant under the holonomy group. If we assume that $\mathbb{V}$ is non degenerate, as we can in the case of a Riemannian manifold (M, g), we have $T\widehat{M} = \mathbb{V} \oplus \mathbb{V}^{\perp}$. For a given point $p \in \widehat{M}$ denote by N_p and $N_p^{\perp}$ the totally geodesic leaves of $\mathbb{V}$ and $\mathbb{V}^{\perp}$. Moreover, denote

$$C = \{p \in \widehat{M} \mid \partial_r|_p \in \mathbb{V}|_p\}, \quad C_{\perp} = \{q \in \widehat{M} \mid \partial_r|_q \in \mathbb{V}^{\perp}|_q\}.$$

Note that $\widehat{\nabla}_X \partial_r = \frac{1}{r} X$ for $X \in TM$ implies that neither C nor $C_{\perp}$ can contain an open set and hence that $\widehat{M}_0 = \widehat{M} \setminus (C \cup C_{\perp})$ is dense in $\widehat{M}$. □

Lemma 4.2 *Let $(\widehat{M}, \widehat{g})$ be a Riemannian cone over a complete Riemannian manifold (M, g). Then for each point $x \in \widehat{M} \setminus (C \cup C_{\perp})$ there is a $p \in C$ and a $q \in C_{\perp}$ such that x lies in the image of the exponential map $\exp_p$ restricted to $\mathbb{V}^{\perp}|_p$ and in the image of $\exp_q$ restricted to $\mathbb{V}|_q$.*

Proof Let $x \in M$ and assume that $x \notin C \cup C_{\perp}$. Let $\partial_r|_x = V + W$ with $V = \mathrm{pr}_{\mathbb{V}|_x}(\partial_r|_x) \in \mathbb{V}|_x$ and $W = \mathrm{pr}_{\mathbb{V}^{\perp}|_x}(\partial_r|_x) \in \mathbb{V}^{\perp}|_x \neq 0$. Then

$$\widehat{g}(\partial_r, W) = \widehat{g}(V + W, W) = \widehat{g}(W, W)$$

and

$$\widehat{g}(V, V) = \widehat{g}(\partial_r - W, \partial_r - W) = 1 - \widehat{g}(W, W),$$

which implies that

$$0 < \widehat{g}(\partial_r, W) < 1. \tag{4.2}$$

Let $\hat{\gamma} = (\rho, \gamma)$ be the maximal geodesic starting at x with $\rho(x) = r$, satisfying the initial condition

$$\widehat{\gamma}'(0) = -rW = -r\mathrm{pr}_{\mathbb{V}^{\perp}|_x}(\partial_r|_x).$$

Now we have $a = \rho'(0) = -r\widehat{g}(\partial_r, W)$, and hence, by the Sect. 2.1, the maximal geodesic is defined for $t < T$ with

$$T = -\frac{r}{a} = \frac{1}{\widehat{g}(\partial_r, W)} > 1,$$

by (4.2). Let $F(t)$ be the parallel transported vector field defined in (4.1) along $\hat{\gamma}$. Then

$$F(0) + \widehat{\gamma}'(0) = r\partial_r|_x - r\mathrm{pr}_{\mathbb{V}^{\perp}|_x}(\partial_r|_x) \in \mathbb{V}|_x.$$

The parallel transport of this vector up to $t = 1$ is

$$F(1) + \hat{\gamma}(1) = r(\hat{\gamma}(1))\partial_r|_{\hat{\gamma}(1)},$$

which is in $\mathbb{V}|_{\hat{\gamma}(1)}$ as $\mathbb{V}$ is a parallel distribution. This implies that $\hat{\gamma}(1) \in C$.

The argument for $C_{\perp}$ works completely analogously. □

Both lemmas imply that each point in $\widehat{M} \setminus (C \cup C_{\perp})$ lies in the intersection of two flat leaves of $\mathbb{V}$ and $\mathbb{V}^{\perp}$ and hence has a flat neighbourhood. This implies that $\widehat{g}$ on $\widehat{M} \setminus (C \cup C_{\perp})$ is flat. Since $\widehat{M} \setminus (C \cup C_{\perp})$ is dense in $\widehat{M}$, this implies that $(\widehat{M}, \widehat{g})$ is flat. This finishes the proof of Theorem 3.1.

4.2 A Generalisation of Gallot's Theorem

Let $(\widehat{M}, \widehat{g}_{\epsilon})$ be a time-like or space-like cone over a semi-Riemannian (M, g). From now on we restrict to the case when $\mathbb{V} = \mathbb{R} \cdot V$, where V is a parallel vector field, normalised such that

$$\widehat{g}(V, V) = \nu \in \{-1, 0, 1\}.$$

Since V is assumed to be parallel, the leaves of $\mathbb{R}\cdot V$ are flat, so in order to generalise Gallot's Theorem we would need to show that the leaves of $V^\perp$ are also flat. In order show this using Gallot's method, we need that the set

$$C = \{p \in \widehat{M} \mid \partial_r|_p \in \mathbb{R}V|_p\}$$

is not empty. This however can *only* be the case when ∂_r and V have the same causal character, i.e., only when $\epsilon = \nu$, i.e.,

$$C \neq \emptyset \text{ implies } \epsilon = \nu.$$

We have already seen Examples 3.2 and 3.3, which show that Gallot's Theorem does not generalise when this condition is not satisfied, i.e., when $\epsilon \neq \nu$. We will deal with this case in the next section. Here we consider the case when $\nu = \epsilon$. In this special case we obtain a generalisation of Gallots Theorem as a stronger version of Theorem 3.4.

Theorem 4.3 *Let (M, g) be a complete semi-Riemannian manifold and let $(\widehat{M}, \widehat{g}_\epsilon)$ be the cone over (M, g). If $(\widehat{M}, \widehat{g})$ admits a parallel vector field V with $\widehat{g}(V, V) = \epsilon$, then the cone is flat and (M, g) is of constant curvature ϵ.*

Proof Let V be the parallel vector field on $(\widehat{M}, \widehat{g}_\epsilon)$ with $\widehat{g}(V, V) = \epsilon$. As in the proof of Theorem 3.1 we consider the set $C = \{p \in \widehat{M} \mid \partial_r|_p = \mathbb{R}\cdot V|_p\}$ and show that each $q \in \widehat{M} \setminus C$ admits a flat neighbourhood. Let $\partial_r|_q = \alpha V + W$ with $W \in V^\perp$ and, since $\widehat{g}(V, V) = \epsilon$, with $\alpha = \epsilon\widehat{g}(V, \partial_r)$. Again we have

$$\widehat{g}(\partial_r, W) = \widehat{g}(W, W) = w \neq 0,$$

and

$$\alpha^2\epsilon = \widehat{g}(\partial_r - W, \partial_r - W) = \epsilon - \widehat{g}(W, W) = \epsilon - w.$$

Hence we obtain

$$0 < \alpha^2 = 1 - \epsilon w. \tag{4.3}$$

On the other hand we write

$$W = \epsilon\widehat{g}(\partial_r, W)\partial_r + W_0 = \epsilon w \partial_r + W_0,$$

with a $W_0 \in T_qM$. Hence,

$$\widehat{g}(W, W) = \widehat{g}(W, W)^2\epsilon + \widehat{g}(W_0, W_0),$$

and hence

$$\widehat{g}(W_0, W_0) = w(1 - \epsilon w). \tag{4.4}$$

Now let $\hat{\gamma}$ be a geodesic starting at q with $r(q) = r$ and with $\hat{\gamma}'(0) = -rW$. We will show that $\hat{\gamma}$ is defined on $[0, 1]$. We have $\hat{\gamma}'(0) = a\partial_r - rW_0$ with (as in Lemma 2.1)

$$a = -\epsilon wr, \qquad cL^2 = r^2 g(W_0, W_0) = \hat{g}(W_0, W_0) = w(1 - \epsilon w),$$

with $c = \pm 1$. We now consider the cases $c\epsilon = 1$, $c\epsilon = -1$ and $c = 0$.

If $c\epsilon = 1$, then

$$0 < L^2 = \epsilon w(1 - \epsilon w),$$

which, together with (4.3) implies that $\epsilon w > 0$ and $a = -\epsilon wr < 0$. By Lemma 2.1, $\hat{\gamma}$ is defined for $t < T$ with

$$T = -\frac{r}{a} = \frac{1}{\epsilon w} > 1,$$

because of (4.3).

If $c\epsilon = -1$ we get

$$0 < L^2 = -\epsilon w(1 - \epsilon w),$$

and hence that $a = -\epsilon wr > 0$ and moreover

$$r^2 L^2 = a(a + r) = a^2 + ar > a^2.$$

Hence, we are in the case $a < rL$ in Lemma 2.1, and $\hat{\gamma}$ is defined for $t < T$ with with $T = \frac{r}{Lr - a}$. We show now that $T > 1$. For this note that by the previous displayed equation we have

$$L^2 r^2 - (r + a)^2 = a^2 + ra - (r + a)^2 = -r(r + a) < 0$$

since $a > 0$. This shows that $Lr < r + a$ and therefore $T = \frac{r}{Lr - a} > 1$.

Finally, in the case $c = 0$ we must have $w = 0$ and hence $a = 0$, so $\hat{\gamma}$ is defined on $[0, \infty)$.

Now we proceed as in the proof of Theorem 3.1: the vector field $F(t)$ along $\hat{\gamma}$ satisfies $F(0) + \hat{\gamma}(0) = r\alpha V|_q$ whose parallel transport is given by $F(1) - \hat{\gamma}(1) = r(\hat{\gamma}(1))\partial_r|_{\hat{\gamma}(1)}$. This implies that $\hat{\gamma}(1) \in C$ and by Lemma 4.1 the leaf of $V^{\perp}$ though q is flat. Since V is a parallel vector field, this implies that q has a flat neighbourhood and hence, since $\widehat{M} \setminus C$ is dense, that $(\widehat{M}, \widehat{g})$ is flat. □

4.3 Non-Flat Cones with Parallel Vector Field

Recall the two Examples 3.2 and 3.3. We will now show that cone with parallel vector fields satisfying the condition $\widehat{g}(V, V) \neq \epsilon$ are always of the form of these examples and thus obtain a stronger version of Theorem 3.4 in the case of a parallel vector field on the cone.

First we define the function $u = \widehat{g}(V, \partial_r)$ and observe:

Lemma 4.4 *Let V be a parallel vector field on the cone $(\widehat{M}, \widehat{g}_\epsilon)$ over a (not necessarily complete) semi-Riemannian manifold (M, g). Then $u = \widehat{g}(V, \partial_r)$ is a smooth function on M, $u \in C^\infty(M)$, that satisfies*

$$V = \epsilon u \partial_r + \frac{1}{r}\nabla u, \tag{4.5}$$

where ∇u is the gradient of u with respect to g. Moreover, for the Hessian of u we have

$$\nabla \mathrm{d}u = -\epsilon u g. \tag{4.6}$$

Proof With $u = \widehat{g}(V, \partial_r)$, we split V as $V = -u\partial_r + W$ where W is a section of $TM \to \widehat{M}$. Since V is parallel, we use Eqs. (2.1) to get $0 = \hat{\nabla}_{\partial_r} V$ which implies that $\partial_r(u) = 0$ and $[\partial_r, W] + \frac{1}{r}W = 0$. The latter implies that $W = \frac{1}{r}U$ with $U \in \Gamma(TM)$ is a vector field on M. The equation $\nabla V|_{TM} = 0$ implies that $\nabla u = U$, where ∇u denotes the gradient of u with respect to g, and $\nabla\nabla u = -\epsilon u \mathrm{Id}$, i.e., that $\nabla \mathrm{d}u = \epsilon u g$. □

Recall that in the case when $\nu = \widehat{g}(V, V) = 0$ or $\nu = -\epsilon$ we have that

$$C = \{p \in \widehat{M} \mid \partial_r|p \in \mathbb{R} \cdot V|_p\} = \emptyset$$

and also that the set of critical points of u is empty,

$$C_0 = \{p \in M \mid \nabla u|_p = 0\} = \emptyset. \tag{4.7}$$

Moreover we have

$$g(\nabla u, \nabla u) = \begin{cases} -\epsilon u^2, & \text{if } \nu = 0, \\ -\epsilon(1 + u^2), & \text{if } \nu = -\epsilon. \end{cases} \tag{4.8}$$

This yields the following result.

Theorem 4.5 *Let (M, g) be a complete semi-Riemannian manifold and $(\widehat{M}, \widehat{g}_\epsilon)$ be the cone over (M, g). If $(\widehat{M}, \widehat{g})$ admits a parallel vector field V with $\widehat{g}(V, V) = -\epsilon$, then (M, g) is globally isometric to*

$$(\mathbb{R} \times N, -\epsilon \mathrm{d}s^2 + \cosh^2(s) g_N),$$

where (N, g_N) is a complete semi-Riemannian manifold.

Proof The idea of the proof is to rescale the gradient ∇u in a way that the rescaled vector field is a geodesic gradient vector field. To this end consider the function $s = -\epsilon \operatorname{arcsinh} \circ u$ on M, i.e., $u(p) = \sinh(-\epsilon s(p))$, for which we write $u = \sinh(-\epsilon s)$. Then we have

$$\nabla u|_p = -\epsilon \cosh(s(p)) \nabla s|_p,$$

and hence

$$g(\nabla u, \nabla u) = \cosh^2(s) g(\nabla s, \nabla s) = (1+\sinh^2(s)) g(\nabla s, \nabla s) = (1+u^2) g(\nabla s, \nabla s).$$

Hence, from (4.8) we get $g(\nabla s, \nabla s) = -\epsilon$, so $S = \nabla s$ is a unit gradient vector field. Moreover, from (4.6) we get

$$-\epsilon \sinh(-\epsilon s) X = \nabla_X \nabla u = \sinh(s) g(X, S) S - \epsilon \cosh(s) \nabla_X S,$$

and therefore

$$\nabla_X S = \tanh(-\epsilon s) \left(X + \epsilon g(X, S) S\right). \tag{4.9}$$

This implies that S is a geodesic vector field. Since (M, g) is assumed to be complete, the flow ϕ of S is defined on $\mathbb{R} \times M$. By the above observation (4.7) we have $\nabla u \neq 0$ and hence all level sets are smooth hypersurfaces. Moreover, the Lie derivative of ds in direction of S vanishes,

$$\mathcal{L}_S \mathrm{d}s(X) = \mathrm{d}^2 s(X) + X(\mathrm{d}s(S)) = X(g(S, S)) = 0.$$

This implies that the flow of S maps each level set of s to a level set of s.

For a fixed $p \in M$ we define the function $\sigma(t) = s(\phi_t(p))$. Since $S = \nabla s$ is complete, σ is defined on $\mathbb{R}$ and satisfies the differential equation

$$\sigma'(t) = \mathrm{d}s|_{\phi_t(p)}(S) = g_{\phi_t(p)}(S, S) \equiv -\epsilon.$$

Hence $\sigma(t) = -\epsilon t + s(p)$, which shows that

$$\phi_t(N_c) = N_{-\epsilon t + c}, \tag{4.10}$$

where $N_c = s^{-1}(c)$ denotes the level set of s, Now set $N = N_0 = \{u = 0\} = \{s = 0\}$, which is a smooth hypersurface and denote by g_N the restriction of g to N. We define a smooth map

$$\Phi : \mathbb{R} \times N \ni (t, p) \mapsto \phi_t(p) \in M,$$

which, because of (4.10), has the inverse

$$\Phi^{-1}(q) = \big(s(q), \phi_{-s(q)}(q)\big) \in \mathbb{R} \times N.$$

This shows that ϕ is a diffeomorphism.

Finally, Eq. (4.9) implies that

$$\mathcal{L}_S g(X, Y) = 2\tanh(-\epsilon s) g(X, Y),$$

for all $X, Y \in S^\perp$, i.e., al X, Y tangent to the level sets of s. This shows that

$$\Phi^* g = -\epsilon \mathrm{d}s^2 + (\cosh(s))^2 g_N.$$

Since (M, g) was assumed to be complete, (N, g_N) has to be complete by Proposition 2.2. □

Now, let (M, g) be a semi-Riemannian manifold and $(\widehat{M}, \widehat{g}_\epsilon)$ be the cone over (M, g). We consider the case that V is a parallel *light-like* vector field. Recall that in this case we have, in addition to Lemma 4.4, that $g(\nabla u, \nabla u) = -\epsilon u^2$. In this situation we observe:

Lemma 4.6 *If $\gamma : I \to M$ is a geodesic on (M, g) with $g(\gamma'(0), \gamma'(0)) = -\epsilon$ and $f = u \circ \gamma$, then $f'' = f$, i.e.,*

$$f(t) = u(\gamma(0)) \cosh(t) + g(\nabla u|_{\gamma(0)}, \gamma'(0)) \sinh(t). \tag{4.11}$$

In particular, if (M, g) is complete, then the image of u contains $(0, \infty)$ if $\{u > 0\} \neq \emptyset$ and $(-\infty, 0)$ if $\{u < 0\} \neq \emptyset$.

Proof With $f = u \circ \gamma$ we have $f' = g|_\gamma(\nabla u|_\gamma, \gamma')$ and hence by Lemma 4.4,

$$f'' = g|_\gamma(\nabla_{\gamma'} \nabla u, \gamma') = -\epsilon f g|_\gamma(\gamma', \gamma') = f.$$

The general solution to this equation is given by (4.11). If (M, g) is complete, the maximal geodesics through a point with $u(p) \neq 0$ are defined on $\mathbb{R}$ and hence, by choosing a geodesic with $\gamma'(0) = \frac{1}{u(p)} \nabla u|_p$, i.e., with $g(\nabla u, \gamma'(0)) = -\epsilon u(p)$, we get

$$f(t) = \frac{u(p)}{2} \left((1 - \epsilon)\mathrm{e}^t + (1 + \epsilon)\mathrm{e}^{-t}\right) = u(p)\mathrm{e}^{-\epsilon t}. \tag{4.12}$$

This implies the statement about the image of u. □

Theorem 4.7 *Let (M, g) be a complete semi-Riemannian manifold and $(\widehat{M}, \widehat{g}_\epsilon)$ be the cone over (M, g). If $(\widehat{M}, \widehat{g})$ admits a parallel light-like vector field V, then M is a disjoint union $M = M_- \cup M_0 \cup M_+$ with $M_\pm$ open and such that M_0 is* *either empty (in which case one of $M_\pm$ is also empty) or a smooth totally geodesic hypersurface and $(M_\pm, g)$ is globally isometric to*

$$(\mathbb{R} \times N_\pm, -\epsilon \mathrm{d}s^2 + \mathrm{e}^{2s} g_{N_\pm}),$$

where $(N_\pm, g_{N_\pm})$ are complete semi-Riemannian manifolds. Moreover, $M_0 = \emptyset$ if and only if (M, g) is Riemannian.

Proof Recall that for in the case of V being light-like we have that $g(\nabla u, \nabla u) = -\epsilon u^2$. The proof is analogous to the previous proof, with a difficulty arising from the possibility that the set

$$M_0 = \{p \in M \mid g(\nabla u|_p, \nabla u|_p) = 0\} = \{p \in M \mid u(p) = 0\}$$

may be non empty, so that the geodesic gradient vector field S from the previous proof may not be defined on all of M. However, since V is light-like, we have $\nabla u \neq 0$, and so M_0 is either empty or a smooth hypersurface. In fact, if $M_0 \neq \emptyset$, it is totally geodesic: if $X \in TM_0 = \nabla u^\perp|_{M_0}$, then Lemma 4.6 shows that $f(t) \equiv 0$, so the geodesics starting in direction of M_0 remain in M_0.

We set $M_\pm = \{\pm u > 0\}$. Without loss of generality, we assume that $M_+ \neq \emptyset$, in which case we get that $N_+ = \{u = 1\} \neq \emptyset$ by the previous lemma.

We consider the function $s = -\epsilon \ln \circ (\pm u)$ on $M_\pm$, i.e., $u = \pm \mathrm{e}^{-\epsilon s}$. Then we have

$$\nabla u = \mp \epsilon \mathrm{e}^{-\epsilon s} \nabla s,$$

and hence, for $S = \nabla s$,

$$g(\nabla u, \nabla u) = \mathrm{e}^{-2\epsilon s} g(S, S) = u^2 g(S, S),$$

and so $g(\nabla s, \nabla s) = -\epsilon$ by (4.8). Next we get from (4.6) that

$$\mp \epsilon \mathrm{e}^{-\epsilon s} X = \nabla_X \nabla u = \mp \epsilon \mathrm{e}^{-\epsilon s} \left(-\epsilon g(X, S) S + \nabla_X S\right),$$

and hence

$$\nabla_X S = (X + \epsilon g(X, S) S) . \tag{4.13}$$

Again, this shows that S is a geodesic vector field on $M_\pm$. Equation 4.12 in the proof of Lemma 4.6 then shows that the geodesics with initial speed given by $S|_p$ for $p \in M_\pm$ remain in $M_\pm$ for all $t \in \mathbb{R}$. Hence S is a complete vector field on $M_\pm$ with its flow defined on $\mathbb{R} \times M_\pm$, so we can continue with the proof as for the previous theorem yielding a diffeomorphism

$$\Phi_\pm : \mathbb{R} \times N_\pm \ni (t, p) \mapsto \phi_t(p) \in M,$$

where $N_\pm = \{p \in M \mid u(p) = \pm 1\} = \{p \in M_\pm \mid s(p) = 0\}$, with the inverse

$$\Phi_\pm^{-1}(q) = \big(s(q), \phi_{-s(q)}(q)\big) \in \mathbb{R} \times N_\pm.$$

Now Eq. (4.13) implies that

$$\mathcal{L}_S g(X, Y) = 2g(X, Y),$$

for all $X, Y \in S^\perp$, i.e., al X, Y tangent to the level sets of s. This shows that

$$\Phi_\pm^* g = -\epsilon \mathrm{d}s^2 + \mathrm{e}^{2s} g_{N_\pm}$$

with a semi-Riemannian manifold $(N_\pm, g_{N_\pm})$. In order to conclude that $(N_\pm, g_{N_\pm})$ are complete, we observe that (4.11) in Lemma 4.6 shows that geodesics of (M, g) with initial speed tangent to $N_\pm$, i.e., with initial speed orthogonal to $\nabla u|_{N_\pm}$, remain in $M_\pm$ and hence, because (M, g) is complete, are defined on $\mathbb{R}$. With this, Proposition 2.2 implies that $(N_\pm, g_{N_\pm})$ are complete.

For the last statement, first note that if (M, g) is Riemannian, then, since $\nabla u \neq 0$, we get that $M_0 = \emptyset$. On the other hand assume that $M_0 = \emptyset$ and without loss of generality that $M_+ = M$, so that globally $(M = \mathbb{R} \times N, g = -\epsilon \mathrm{d}s^2 + \mathrm{e}^{2s} g_N))$. By Proposition 2.2, the metric g is only complete if it is definite and g_N is complete.

□

As a corollary we obtain a global version of Theorem 3.9.

Corollary 4.8 *Let (M, g) be a complete semi-Riemannian manifold and $(\widehat{M}, \widehat{g}_\epsilon)$ be the cone over (M, g). If $(\widehat{M}, \widehat{g})$ admits a parallel light-like vector field V, then $\widehat{M}$ is a disjoint union $\widehat{M} = \widehat{M}_- \cup \widehat{M}_0 \cup \widehat{M}_+$ with*

$$\widehat{M}_\pm = \{p \in \widehat{M} \mid \pm\widehat{g}(V, \partial_r) > 0\}, \quad \widehat{M}_0 = \{p \in \widehat{M} \mid \widehat{g}(V, \partial_r) = 0\} = \mathbb{R}_{>0} \times M_0,$$

and such that $(\widehat{M}_\pm, \widehat{g})$ is globally isometric to

$$(\mathbb{R}_+ \times \mathbb{R}_\epsilon \times N_\pm, \widetilde{g} = 2\mathrm{d}u\mathrm{d}v + u^2 g_{N_\pm}),$$

where $(N_\pm, g_{N_\pm})$ are a complete semi-Riemannian manifolds and where $\mathbb{R}_\pm = \{x \in \mathbb{R} \mid \pm x > 0\}$. The isometry is given by

$$\Psi_\pm : \widehat{M} \ni (r, s, p) \mapsto (u = r\mathrm{e}^s, v = \frac{\epsilon}{2} r\mathrm{e}^{-s}, p) \in (\mathbb{R}_+ \times \mathbb{R}_\epsilon \times N_\pm).$$

Proof We have $u^2 = r^2 \mathrm{e}^{2s}$ and

$$2\mathrm{d}u\mathrm{d}v = \epsilon \left(\mathrm{e}^s \mathrm{d}r \pm r\mathrm{e}^s ds\right)\left(\mathrm{e}^{-s}\mathrm{d}r \mp r\mathrm{e}^{-s}\mathrm{d}s\right) = \epsilon(\mathrm{d}r^2 - r^2\mathrm{d}s^2).$$

Hence, by the previous theorem, $\Psi_\pm^* \widetilde{g} = \widehat{g}$. □

5 Lorentzian Cones and Applications to Killing Spinors

5.1 Parallel Spinors and Killing Spinors

Let (M, g) be a semi-Riemannian spin manifold, i.e., a space and time oriented semi-Riemannian manifold with a spin structure, and let Σ its complex spinor bundle. This is a complex vector bundle that is equipped with the following structures.

1. the Clifford multiplication

$$TM \otimes \Sigma \ni X \otimes \varphi \mapsto X \cdot \varphi \in \Sigma,$$

2. a Hermitian bundle metric $\langle .,. \rangle \in \Gamma(\Sigma^* \otimes \overline{\Sigma}^*)$ on Σ, conjugate-linear in the second component, that is positive definite if g is Riemannian and of neutral signature if g is indefinite,
3. the lift ∇^Σ of the Levi-Civita connection to Σ,

that satisfy the following properties, where r is the number of negative eigenvalues of g,

$$\begin{aligned} (X \cdot Y + Y \cdot X) \cdot \varphi &= -2\, g(X, Y)\, \varphi, \\ \langle X \cdot \varphi, \psi \rangle_\Sigma &= (-1)^{r+1} \langle \varphi, X \cdot \psi \rangle_\Sigma, \\ \nabla^\Sigma_Y (X \cdot \varphi) &= (\nabla_Y X) \cdot \varphi + X \cdot \nabla^\Sigma_Y \varphi, \\ X(\langle \varphi, \psi \rangle_\Sigma) &= \langle \nabla^\Sigma_X \varphi, \psi \rangle_\Sigma + \langle \varphi, \nabla^\Sigma_X \psi \rangle_\Sigma. \end{aligned} \tag{5.1}$$

The second of these relations together with $\langle .,. \rangle$ being Hermitian shows that to each spinor field φ one can assign a (real) vector field $V_\varphi \in \Gamma(TM)$

$$g(V_\varphi, X) := i^{r+1} \langle \varphi, X \cdot \varphi \rangle_\Sigma \quad \text{for all } X \in TM.$$

This vector field is sometimes called the *Dirac current of* φ. The above relations also show that $\nabla V_\varphi = 0$ if φ is a parallel a parallel spinor field, i.e., if $\nabla^\Sigma \varphi = 0$. However, V_φ may be identically zero even if φ is not. This happens for example for parallel spinors on Riemannian manifolds.

Moreover, the Ricci tensor of a semi-Riemannian manifold with parallel spinor satisfies $g(Ric(X), Ric(X)) = 0$. In particular, Riemannian manifolds with parallel spinors are Ricci-flat.

A *Killing spinor with Killing number* $z \in \mathbb{C}$ is a spinor field $\varphi \in \Gamma(\Sigma)$ that satisfies the equation

$$\nabla^\Sigma_X \varphi = z\, X \cdot \varphi.$$

Using the above formula one can show that the scalar curvature of a semi-Riemannian manifold with a Killing spinor is equal to $4n(n-1)z^2$. This implies that z is either real or imaginary and hence the scalar curvature is a positive or negative constant. A Killing spinor φ with Killing number $z = \pm\frac{1}{2}$ is called *real Killing spinor*, and φ is called *imaginary Killing spinor* if the Killing number is $z = \pm\frac{i}{2}$. Moreover, Riemannian manifolds with Killing spinors are Einstein, so Riemannian manifolds with real/imaginary Killing spinor provide examples of Einstein manifolds with positive/negative scalar curvature. The question which Einstein manifolds (or constant scalar curvature manifolds) can be constructed in this way lead to the problem of classifying manifolds with Killing spinors. The fundamental observation for solving this problem is the relation to semi-Riemannian cones:

Theorem 5.1 ([3, 9]) *Let (M, g) be a semi-Riemannian spin manifold that admits a Killing spinor with Killing number $\pm\frac{\sqrt{\epsilon}}{2}$ if and only if the semi-Riemannian cone $(\widehat{M}, \widehat{g}_\epsilon)$ admits a parallel spinor field.*

Remark 5.2 ([9]) In [9] Bohle proved a more general result: Let (M, g) be a semi-Riemannian spin manifold and $f : I \to \mathbb{R}$ be a smooth function. Then the warped product metric

$$g_{\epsilon,f} = \epsilon \mathrm{d}s^2 + f^2(s)g$$

on $I \times M$ admits a Killing spinor with Killing number $\hat{\lambda} \in \{0, \pm\frac{1}{2}, \pm\frac{i}{2}\}$ if and only if

(1) The warping function satisfies the ODE $f'' = -4\epsilon\hat{\lambda}^2 f$, and
(2) (M, g) admits a Killing spinor with Killing number $\pm\lambda$, where $\lambda^2 = \hat{\lambda}^2 f^2 + \frac{\epsilon}{4}(f')^2$.

Theorem 5.1 together with Gallot's Theorem 3.1 was used by Bär [3] to derive a classification of complete Riemannian manifolds with real Killing spinors: if (M, g) admits a real Killing spinor, the cone admits a parallel spinor and under the assumption of completeness, by Gallot's theorem, the cone is irreducible. Then by Berger's classification of irreducible holonomy groups [8], Wangs classification of those admitting an invariant spinor [14] under their spin representation, and the correspondence between holonomy groups and geometric structures, Bär arrived at the following classification:

Theorem 5.3 (C. Bär [3]) *Let M be a complete, simply connected Riemannian spin manifold with a real Killing spinor. Then M is isometric to round sphere, or to a compact Einstein space with one of the following structures: Sasaki, 3-Sasaki, 6-dimensional nearly-Kähler, or nearly parallel G_2.*

Baum gave a classification of Riemannian manifolds with imaginary Killing spinors [5]. Baum's proof does not use the cone construction of Theorem 5.1 explicitly. In other signatures the classification of semi-Riemannian manifolds with Killing spinors is only known in special cases: for example, Bohle and Baum classified Lorentzian manifolds with real Killing spinors [9, Section 5], with an addition made in [6, Proposition 7.1], again without using the cone construction explicitly. In the next section we will use our results from the previous section to obtain Baum's and Bohle's classification results.

5.2 *Lorentzian Cones and Killing Spinors*

In this section we will use our results of Sect. 4 to derive the classification of complete Riemannian manifolds with imaginary Killing spinors and of complete Lorentzian manifolds with real Killing spinors. In both cases Theorem 5.1 yields a parallel spinor on a Lorentzian cone and hence a parallel Dirac current by the observations in Sect. 5.1. In Lorentzian signature one can show that the Dirac current is a causal vector field:

Lemma 5.4 *Let φ be a parallel spinor field on a spin Lorentzian manifold (M, g). Then V_φ is a causal parallel vector field, i.e, $V_\varphi \neq 0$, $\nabla V_\varphi = 0$ and $g(V_\varphi, V_\varphi) \leq 0$.*

Proof We have already seen that V_φ is parallel, so it is either identically zero or non vanishing and we have to verify its causal character. Since (M, g) is time orientable we fix a time-like unit vector field T and split $V_\varphi \neq 0$ as

$$V_\varphi = -g(T, V_\varphi)T + g(N, V_\varphi)N,$$

where N is a spacelike unit normal field orthogonal to T. Then we have by (5.1) that

$$g(V_\varphi, V_\varphi) = -g(T, V_\varphi)^2 + g(N, V_\varphi)^2 = -\langle T \cdot \varphi, \varphi\rangle^2 + \langle N \cdot \varphi, \varphi\rangle^2,$$

and we have to show that this is not positive. For this observe that the endomorphism $T \cdot N$ on Σ squares to the identity by the defining relation for the Clifford algebra in (5.1),

$$T \cdot N \cdot T \cdot N = -N \cdot T \cdot T \cdot N = -N \cdot N = 1.$$

Hence $T \cdot N$ has eigenvalues ± 1 and we can split $\varphi = \varphi_+ + \varphi_-$ into its components in the corresponding eigenspaces. Note that

$$T \cdot \varphi_\pm = \pm N \cdot \varphi_\pm,$$

which, together with (5.1), implies that

$$\langle T \cdot \varphi_+, \varphi_-\rangle = \langle \varphi_+, T \cdot \varphi_-\rangle = -\langle \varphi_+, N \cdot \varphi_-\rangle = -\langle N \cdot \varphi_+, \varphi_-\rangle = -\langle T \cdot \varphi_+, \varphi_-\rangle,$$

so that $\langle T \cdot \varphi_+, \varphi_- \rangle = 0$. Then we use the fact (see [4] for a proof) that the Hermitian form $(\phi, \psi)_T = \langle T \cdot \phi, \psi \rangle$ on Σ is positive definite. The last equation then shows that $(\varphi_+, \varphi_-)_T = 0$ and we get

$$g(V_\varphi, V_\varphi) = -(\varphi, \varphi)_T^2 + (T \cdot N \cdot \varphi, \varphi)_T^2 = -4(\varphi_+, \varphi_+)_T(\varphi_-, \varphi_-)_T \leq 0.$$

This shows that V_φ is either time-like or light-like. □

In fact, on a Lorentzian manifold the Dirac current of spinor field is always causal even if the spinor is not parallel, but it may change its causal character from light-like to time-like. The proof of this has to take into account that V_φ may have zeros so that N may not be well defined.

The following theorem gives a classification of Riemannian manifolds with imaginary Killing spinors.

Theorem 5.5 ([5]) *Let (M, g) be a complete Riemannian manifold with an imaginary Killing spinor. Then (M, g) is globally isometric to hyperbolic space or to a warped product of the form*

$$\left(\mathbb{R} \times N, \mathrm{d}s^2 + \mathrm{e}^{2s} g_N\right), \tag{5.2}$$

where (N, g_N) is a complete Riemannian manifold with a parallel spinor field.

Proof Let (M, g) be a complete Riemannian manifold with an imaginary Killing spinor field. Then, by Theorem 5.1, the Lorentzian cone $(\widehat{M}, g_-)$ admits a parallel spinor field φ, which by Lemma 5.4 provides us with a parallel vector field V_φ that is either light-like or time-like. In case it is time-like, Theorem 4.3 yields that (M, g) has constant sectional curvature -1 and hence is isometric to hyperbolic space. If V_φ is light-like, we can apply Theorem 4.7 to get the desired warped product in (5.2) with a complete Riemannian manifold (N, g_N). To get that (N, g_N) admits a parallel spinor field we can either use the result in Remark 5.2 or recall Corollary 4.8 and Theorem 3.14 to obtain that the holonomy algebra of $(\widehat{M}, \widehat{g}_-)$ is equal to $\mathfrak{hol}(N, g_N) \ltimes \mathbb{R}^{\dim(N)}$. This is an indecomposable holonomy algebra that admits an invariant spinor under its spin representation if and only if $\mathfrak{hol}(N, g_N)$ admits an invariant spinor. □

The next theorem provides a classification of Lorentzian manifolds with real Killing spinors.

Theorem 5.6 ([6, 9]) *Let (M, g) be a complete Lorentzian manifold with a real Killing spinor. Then*

(1) either (M, g) is globally isometric to de Sitter space or space or to a warped product of the form

$$\left(\mathbb{R} \times N, -\mathrm{d}s^2 + \cosh^2(s) g_N\right), \tag{5.3}$$

where (N, g_N) *is a complete Riemannian manifold with a real Killing spinor (i.e., with one of the structures in Theorem 5.3), or*

(2) M *is a disjoint union* $M = M_- \cup M_0 \cup M_+$ *with* M_0 *a smooth totally geodesic hypersurface and* $M_\pm$ *and such that* $(M_\pm, g)$ *are globally isometric to*

$$(\mathbb{R} \times N_\pm, -\mathrm{d}s^2 + \mathrm{e}^{2s} g_{N_\pm}),$$

where $(N_\pm, g_{N_\pm})$ *are complete Riemannian manifolds with parallel spinors.*

Proof If (M, g) admits a real Killing spinor, then the cone $(\widehat{M}, \widehat{g}_+)$ admits a parallel spinor and hence a parallel causal vector field V.

If V is time-like, then we apply Theorem 4.5, to get that (M, g) is isometric to the Lorentzian manifolds in (5.3) with a complete Riemannian manifold (N, g_N). If (N, g_N) is the round metric on the sphere then (M, g) is de Sitter space. The result in Remark 5.2 shows that (M, g) admits a real Killing spinor if and only if (N, g_N) does.

If V is light-like, Theorem 4.7 shows that (2) holds with complete Riemannian manifolds $(N_\pm, g_{N_\pm})$. To obtain that (N, g_N) admits a parallel spinor, we use again Remark 5.2 or recall Corollary 4.8 and Theorem 3.14, as for the proof of Theorem 5.5. □

Acknowledgments This paper originated from a talk given at the Abel Symposium "Geometry, Lie Theory and Applications" in June 2019. The author would like to thank the organisers for their hospitality and the Niels Henrik Abel Memorial Fund for financial support.

References

1. D. Alekseevsky, V. Cortés, A. Galaev, T. Leistner, Cones over pseudo-Riemannian manifolds and their holonomy. J. Reine Angew. Math. **635**, 23–69 (2009)
2. D. Alekseevsky, V. Cortés, T. Leistner, Geometry and holonomy of indecomposable cones (2019). Preprint arXiv:1902.02493
3. C. Bär, Real Killing spinors and holonomy. Commun. Math. Phys. **154**(3), 509–521 (1993)
4. H. Baum, in *Spin-Strukturen und Dirac-Operatoren über pseudoriemannschen Mannigfaltigkeiten*, vol. 41. Teubner-Texte zur Mathematik (Teubner-Verlagsgesellschaft, Leipzig, 1981)
5. H. Baum, Complete Riemannian manifolds with imaginary Killing spinors. Ann. Global Anal. Geom. **7**(3), 205–226 (1989)
6. H. Baum, Twistor and Killing spinors in Lorentzian geometry, in *Global Analysis and Harmonic Analysis (Marseille-Luminy, 1999)*, vol. 4. Sémin. Congr. (Soc. Math. France, Paris, 2000), pp. 35–52
7. H. Baum, T. Friedrich, R. Grunewald, I. Kath, in *Twistors and Killing spinors on Riemannian manifolds*, vol. 124. Teubner-Texte zur Mathematik [Teubner Texts in Mathematics]. (B. G. Teubner Verlagsgesellschaft mbH, Stuttgart, 1991)
8. M. Berger, Sur les groupes d'holonomie homogène des variétés à connexion affine et des variétés riemanniennes. Bull. Soc. Math. France **83**, 279–330 (1955)
9. C. Bohle, Killing spinors on Lorentzian manifolds. J. Geom. Phys. **45**(3–4), 285–308 (2003)

10. G. de Rham, Sur la reductibilité d'un espace de Riemann. Comment. Math. Helv. **26**, 328–344 (1952)
11. S. Gallot, Équations différentielles caractéristiques de la sphère. Ann. Sci. École Norm. Sup. (4) **12**(2), 235–267 (1979)
12. V.S. Matveev, Gallot-Tanno theorem for pseudo-Riemannian metrics and a proof that decomposable cones over closed complete pseudo-Riemannian manifolds do not exist. Differ. Geom. Appl. **28**(2), 236–240 (2010)
13. V.S. Matveev, P. Mounoud, Gallot-Tanno theorem for closed incomplete pseudo-Riemannian manifolds and applications. Ann. Global Anal. Geom. **38**(3), 259–271 (2010)
14. M.Y. Wang, Parallel spinors and parallel forms. Ann. Global Anal. Geom. **7**(1), 59–68 (1989)
15. H. Wu, On the de Rham decomposition theorem. Illinois J. Math. **8**, 291–311 (1964)

Building New Einstein Spaces by Deforming Symmetric Einstein Spaces

David McNutt, Sigbjørn Hervik, and Eivind E. Dahl

Abstract We introduce an approach to determine new pseudo-Riemannian Einstein spaces by deforming symmetric pseudo-Riemannian Einstein spaces. The metrics of the spaces we will deform are associated with complex hyperbolic spaces and are (para-)Kähler manifolds. That is, they admit a parallel field of skew-symmetric endomorphisms, called a (para-)complex structure K, such that $K^2 = -Id$ ($K^2 = Id$), where Id denotes the identity endomorphism. These metrics also belong to the class of $\mathcal{I}$-degenerate metrics and can be deformed to produce new metrics with the same set of scalar polynomial curvature invariants. Using the metric deformations of $\mathcal{I}$-degenerate metrics which preserve the set of scalar polynomial curvature invariants we explicitly construct examples of new (para-)Kähler or almost-(para-)Kähler Einstein spaces.

1 Introduction

A pseudo-Riemannian space is a d-dimensional differentiable manifold, M, equipped with a metric acting on the tangent space, $T_x M$, at each point x of the manifold, $\mathbf{g}$, of signature (p, q) with $d = p + q$, i.e., a metric admitting p eigenvectors with eigenvalue -1 and q eigenvectors with eigenvalue $+1$. This implies there is an anholonomic orthonormal coframe, $\{e^1, \ldots, e^d\}$ of the cotangent space, $T_x^* M$, such that the metric is of the form

$$\mathbf{g} = -\sum_{a=1}^{p} e^a \otimes e^a + \sum_{b=1}^{q} e^b \otimes e^b. \tag{1.1}$$

A space is *Riemannian* if $p = 0$, *Lorentzian* if $p = 1$ and *neutral* if $q = p$. Hereafter, we will omit the tensor product symbol, $\otimes$, when displaying the metric

D. McNutt (✉) · S. Hervik · E. E. Dahl
Faculty of Science and Technology, University of Stavanger, Stavanger, Norway

S. Hervik et al. (eds.), *Geometry, Lie Theory and Applications*, Abel Symposia 16,
https://doi.org/10.1007/978-3-030-81296-6_11

relative to a coframe basis. In particular, we will write symmetric tensor products as $2\mathbf{vw} = \mathbf{v} \otimes \mathbf{w} + \mathbf{w} \otimes \mathbf{v}$ to display the metric relative to a null coframe basis.

Unlike the Riemannian case, where the metric is positive-definite, pseudo-Riemannian metrics admit non-zero vector fields with vanishing norm, called null vector fields. This is reflected in the isotropy group of the metric (1.1), which is non-compact if $p, q > 0$. The non-compactness of this Lie group has implications for the classification of pseudo-Riemannian spaces by the set of all scalar polynomial curvature invariants (SPIs):

$$\mathcal{I} = \{R, R_{abcd}R^{abcd}, \dots, R_{abcd;e}R^{abcd;e}, \dots\}.$$

The condition that the metric, say $\mathbf{g}_0$, of a pseudo-Riemannian space is not uniquely characterized locally by its set of SPIs, $\mathcal{I}_0$, implies that there exists a smooth (one-parameter) deformation of the metric, $\mathbf{g}_t$ $t \geq 0$, with $\mathbf{g}_0 = \mathbf{g}$, and $\mathbf{g}_\infty := \lim_{t\to\infty} \mathbf{g}_t$ not diffeomorphic to $\mathbf{g}_0$ with $\mathcal{I}_t = \mathcal{I}_0$. Any metric with this property is called $\mathcal{I}$-*degenerate* [1, 2].

As a Riemannian metric can always be uniquely characterized by the associated set of curvature invariants $\mathcal{I}$, there are no $\mathcal{I}$-degenerate Riemannian spaces [3]. In the Lorentzian case, the class of Kundt metrics which admit a geodesic, shear-free, twist-free, non-expanding null congruence, $\boldsymbol{\ell}$ [4], contains a subclass of metrics known as the *degenerate Kundt metrics* which are $\mathcal{I}$-degenerate [1, 5]. It is conjectured that all $\mathcal{I}$-degenerate Lorentzian manifolds are contained within the class of degenerate Kundt metrics and this has been confirmed in three and four dimensions [5]. For other pseudo-Riemannian spaces, with $p > 1$, there are additional possibilities for $\mathcal{I}$-degenerate spaces. For example, in the class of neutral metrics which admit an invariant null plane [6, 7], known as Walker type metrics, there are $\mathcal{I}$-degenerate spaces, in addition to the neutral Kundt type metrics [2].

As a generalization of the Kundt and Walker type metrics we can consider the class of pseudo-Riemmanian spaces admitting a k-form constructed from k mutually orthogonal null vector fields, satisfying two surface-forming conditions [2]. The form of these metrics and the subcases relevant to the current work will be reviewed in Sect. 3. The underlying assumption of the analysis was inspired by invariant theory which states that a certain (pointwise) boost limit should exist which produces a new non-diffeomorphic metric with the same set $\mathcal{I}$. By assuming that the boost limit extends to a neighbourhood the authors were able to construct a rich class of metrics containing all known examples of $\mathcal{I}$-degenerate metrics. Notably, this class contains examples of locally homogeneous spaces, and more specifically, symmetric spaces.

In this article, we will consider symmetric pseudo-Riemannian spaces of even dimension which are generated by Wick-rotating [8, 9] symmetric Riemannian spaces associated with the complex hyperbolic plane. As an example, in dimension four, the complex hyperbolic plane [10] can be treated as a four-dimensional (4D) neutral metric by Wick-rotating the Fubini-Study metric [11, 12]. However, unlike the Riemannian case where the manifold is Kähler and there exists a parallel complex structure in terms of a skew-symmetric endomorphism $K : T_pM \to T_pM$ such that

$$\mathbf{g}(KX, KY) = -\mathbf{g}(X, Y), \tag{1.2}$$

for arbitrary vector fields X and Y at p, the resulting Wick-rotated metric, $\mathbf{g}'$, is different. It is no longer Kähler, as the Wick-rotation K' of the parallel complex structure K satisfies

$$\mathbf{g}'(K'X, K'Y) = \mathbf{g}'(X, Y). \tag{1.3}$$

This structure is called a *para-complex structure*, and such manifolds are called *para-Kähler*. Their properties will be briefly discussed in Sect. 5. We can also Wick-rotate the metrics of complex hyperbolic spaces into metrics of signature $(2p, 2p + 2q + 2)$. For any such Wick-rotated metric $\mathbf{g}'$, the Wick-rotation K' of the parallel complex structure K satisfies (1.2), implying that the pseudo-Riemannian manifold is Kähler. The form of the Wick-rotated Kähler metrics of signature $(2p, 2p+2q+2)$ and their associated symplectic forms will be displayed in Sect. 6.

For any skew-symmetric endomorphism, K, satisfying $K^2 = -1$ ($K^2 = 1$) there is an associated two-form, $\boldsymbol{\omega}$, with $\boldsymbol{\omega}(X, Y) = \mathbf{g}(X, KY)$. A manifold equipped with a closed $\boldsymbol{\omega}$ is called an *almost (para-)Kähler manifold*. Relative to the metric compatible Levi-Civita connection, ∇, if the covariant derivative $\nabla\boldsymbol{\omega}$ is a 3-form, then it is called a *nearly-(para-)Kähler manifold*. If a manifold is both nearly and almost (para-)Kähler, then it satisfies the integrability condition $\nabla\boldsymbol{\omega} = 0$, and consequently it is called a *(para-)Kähler manifold*.

The metrics of the symmetric pseudo-Riemannian spaces discussed in this paper will be either Kähler or para-Kähler, unless explicitly stated otherwise. However, they are all $\mathcal{I}$-degenerate spaces, so that distinct pseudo-Riemannian spaces can be constructed with the same invariants as the original metric, $\mathcal{I}_0$, using metric deformations that leave SPIs unchanged. The form of the deformations which preserve SPIs is known for these metrics [2] and we can restrict these deformations to subclasses that preserve the almost-(para-)Kähler ($d\boldsymbol{\omega} = 0$), the nearly-(para-)Kähler ($\nabla\boldsymbol{\omega}$ is a 3-form) condition or the (para-)Kähler ($\nabla\boldsymbol{\omega} = 0$) condition. We will always consider the deformations that preserve the Einstein condition, $\mathbf{R} = k\mathbf{g}$, in addition to the almost-(para-)Kähler or (para-)Kähler conditions.

2 $\mathcal{I}$-Degeneracy and the Boost Weight Decomposition of Tensors

Setting $p = n$ and $q = n+m$, we will consider a $(2n+m)$-dimensional manifold of signature $(n, n+m)$. The original metric in (1.1), defined relative to an orthonormal coframe basis, e^a, can be reformulated as a null coframe basis by taking n pairs of timelike and spacelike coframe elements: $t^i = e^i$ and $s^i = e^{n+i}$, with $i = 1, \ldots, n$, and constructing null directions:

$$\theta^{2i-1} = \boldsymbol{\ell}^i = \frac{1}{\sqrt{2}}(t^i - s^i),\ \theta^{2i} = \mathbf{n}^i = \frac{1}{\sqrt{2}}(t^i + s^i). \tag{2.1}$$

Completing the coframe with $\theta^{2n+a} = \mathbf{m}^a = e^{2n+a}$, where $a = 1, \ldots, m$, a null coframe can be chosen to express the metric in the form

$$ds^2 = 2(\boldsymbol{\ell}^1\mathbf{n}^1 + \boldsymbol{\ell}^2\mathbf{n}^2 + \ldots + \boldsymbol{\ell}^n\mathbf{n}^n) + \delta_{ab}\mathbf{m}^a\mathbf{m}^b, \tag{2.2}$$

where a and b range from 1 to m. Relative to the coframe $\{\boldsymbol{\ell}^i, \mathbf{m}^a, \mathbf{n}^{\hat{i}}\}$, with the indices i and $\hat{i}$ ranging from 1 to n, the abelian subgroup of the group $SO(n, n+m)$ consists of boosts in each of the n planes spanned by the null directions,

$$(\boldsymbol{\ell}^i, \mathbf{n}^i) \to (e^{\lambda_i}\boldsymbol{\ell}^i, e^{-\lambda_i}\mathbf{n}^i), \tag{2.3}$$

where λ_i are real-valued. In analogy with the Lorentzian case [13], we have the concept of *boost weights* $\mathbf{b} \in \mathbb{Z}^n$ such that for an arbitrary component of a rank r tensor $\mathbf{T}$ with respect to the coframe (2.2), a boost in each of the n planes, $(\boldsymbol{\ell}^i, \mathbf{n}^i)$, gives the transformation

$$T_{\mu_1 \ldots \mu_r} \to e^{(b_1\lambda_1 + b_2\lambda_2 + \ldots + b_n\lambda_n)} T_{\mu_1 \ldots \mu_r}, \tag{2.4}$$

where the indices $\mu_i, \ldots \mu_r$ range from 1 to $2n + m$, $b_1, \ldots, b_n$ are integers and $\mathbf{b} = (b_1, b_2, \ldots, b_n)$ is the boost weight vector of the component $T_{\mu_1 \ldots \mu_r}$. Each boost weight b_i records the difference in the number of appearances of $\boldsymbol{\ell}^i$ and $\mathbf{n}^i$, respectively, in the associated tensor product of a given component $T_{\mu_1 \ldots \mu_r}$. We can write the tensor $\mathbf{T}$ in the following decomposition:

$$\mathbf{T} = \sum_{\mathbf{b} \in \mathbb{Z}^n} (\mathbf{T})_{\mathbf{b}}. \tag{2.5}$$

Here $(\mathbf{T})_{\mathbf{b}}$ denotes the projection onto the subspace of components of boost weight $\mathbf{b}$.

In the Lorentzian case, with $n = 1$, we can define the maximum boost weight of a tensor, $\mathbf{T}$, for a null direction $\boldsymbol{\ell}$ as the boost order, and denote it as $\mathcal{B}_{\mathbf{T}}(\boldsymbol{\ell})$. For a given null direction $\boldsymbol{\ell}$, $\mathcal{B}_{\mathbf{T}}(\boldsymbol{\ell})$ remains unchanged under boosts, spatial rotations and null rotations about $\boldsymbol{\ell}$, implying that the choice of $\mathbf{n}$ will not affect the integer value of $\mathcal{B}_{\mathbf{T}}(\boldsymbol{\ell})$.

The Weyl tensor and any rank two tensor, $\mathbf{T}$, can be broadly classified into five *alignment types*: if for all null directions $\boldsymbol{\ell}$, $\mathcal{B}_{\mathbf{T}}(\boldsymbol{\ell}) = 2$ then $\mathbf{T}$ is of alignment type G, if there exists an $\boldsymbol{\ell}$ such that $\mathcal{B}_{\mathbf{T}}(\boldsymbol{\ell}) = 1, 0, -1, -2$ then $\mathbf{T}$ is of alignment type I, II, III, or N, respectively, while if $\mathbf{T}$ vanishes then it belongs to alignment type

O. While this definition of type focusses on a given null direction, $\boldsymbol{\ell}$, the existence of such a null direction with $\mathcal{B}_{\mathbf{T}}(\boldsymbol{\ell}) < 2$ is an invariant property of the tensor $\mathbf{T}$. We will say $\boldsymbol{\ell}$ is $\mathbf{T}$-aligned if $\mathcal{B}_{\mathbf{T}}(\boldsymbol{\ell}) < 2$.

As an illustration we will consider the alignment types for a rank 2 symmetric tensor in a Lorentzian manifold.

$$\mathbf{R} = R_{11}\mathbf{nn} + R_{1i}\mathbf{nm}^i + R_{12}\boldsymbol{\ell\ell} + R_{ab}\mathbf{m}^a\mathbf{m}^b + R_{2a}\boldsymbol{\ell}\mathbf{m}^a + R_{22}\boldsymbol{\ell\ell}$$

The alignment types for $\mathbf{R}$ correspond to the existence of a null direction $\boldsymbol{\ell}$ where particular components vanish:

- Type I: $R_{11} = 0$.
- Type II: $R_{11} = R_{1a} = 0$.
- Type III: $R_{11} = R_{1a} = R_{12} = R_{ab} = 0$.
- Type N: $R_{11} = R_{1a} = R_{12} = R_{ab} = R_{2a} = 0$.
- Type O: $R_{11} = R_{1a} = R_{12} = R_{ab} = R_{2a} = R_{22} = 0$.

If no such $\boldsymbol{\ell}$ exists where both R_{11} and R_{22} vanish then $\mathbf{R}$ is of alignment type $\mathbf{G}$.

For higher rank tensors, like the covariant derivatives of the curvature tensor, the alignment types are still applicable despite the possibility that $|\mathcal{B}_{\mathbf{T}}(\boldsymbol{\ell})|$ may be greater than two. Any $\mathcal{I}$-degenerate spacetime admits a null coframe such that all of the positive boost weight terms of the curvature tensor and its covariant derivatives are zero in this common coframe; that is, they are all of alignment type II.

With the boost weight decomposition, we can introduce properties to classify tensors in a similar manner to the alignment classification for more general pseudo-Riemannian spaces [13]. Denoting the maximum value of each boost weight b_i of a tensor $\mathbf{T}$ with respect to a null direction $\boldsymbol{\ell}^i$ as $\mathcal{B}_T(\boldsymbol{\ell}^i)$, this quantity is invariant under the subgroup of $SO(n, n+m)$ which leaves the direction of $\boldsymbol{\ell}^i$ unchanged. As this subgroup affects the form of the other coframe elements in the coframe, it is possible to vary the choice of $\mathbf{n}^i$ while leaving $\mathcal{B}_T(\boldsymbol{\ell}^i)$ unchanged. It follows that $\mathcal{B}_T(\boldsymbol{\ell}^i)$ depends on the null direction $\boldsymbol{\ell}^i$ alone. For a given tensor $\mathbf{T}$, the existence of a null directions $\boldsymbol{\ell}^i$ with $\mathcal{B}_T(\boldsymbol{\ell}^i) = 0$ leads to the following properties:

Definition 1 Consider the conditions

$$(B1) \qquad (\mathbf{T})_{\mathbf{b}} = 0, \text{ for all } \mathbf{b} = (b_1, b_2, \ldots, b_n), b_1 > 0,$$

$$(B2) \qquad (\mathbf{T})_{\mathbf{b}} = 0, \text{ for all } \mathbf{b} = (0, b_2, \ldots, b_n), b_2 > 0,$$

$$\ldots$$

$$(Bn) \qquad (\mathbf{T})_{\mathbf{b}} = 0, \text{ for all } \mathbf{b} = (0, 0, \ldots, 0, b_n), b_n > 0.$$

A tensor $\mathbf{T}$ possesses the $\mathbf{S}_i$ property, $1 \leq i \leq n$, if there exists a null coframe such that the conditions $B1) - Bi)$ hold.

Definition 2 A tensor **T** posses the N property if a null coframe exists such that $B1) - Bn)$ are satisfied and

$$(\mathbf{T})_{\mathbf{b}} = 0, \text{ for } \mathbf{b} = (0, 0, \ldots 0, 0).$$

When $n > 1$ there are additional properties that must be introduced. A tensor which does not have the S_i property can still have a degenerate structure, since the boost weights form a lattice $\mathbf{b} \in \mathbb{Z}^n \subset \mathbb{R}^n$ and we can use a linear transformation $G \in GL(n)$ to map the boost weights onto another lattice, say Γ, in $\mathbb{R}^n$. If a $G \in GL(n)$ exists so that the image of the boost weights $G\mathbf{b}$ of the tensor **T** now satisfies some of the properties above, we say the tensor **T** possesses the S_i^G or N^G property.

With these properties the following theorem has an important implication for $\mathcal{I}$-degenerate spaces [2].

Theorem 1 *A tensor is not characterized by its invariants if and only if it possesses (at least) the S_1^G property.*

Using this theorem we can state a definition for $\mathcal{I}$-degenerate pseudo-Riemannian spaces using the boost weight decomposition.

Corollary 1 *A pseudo-Riemannian space is $\mathcal{I}$-degenerate if and only if the curvature tensor and its covariant derivatives satisfy (at least) the S_1^G property relative to a common null coframe.*

2.1 Classification of Pseudo-Riemannian Spaces

The set $\mathcal{I}$ may not be sufficient to uniquely characterize a given pseudo-Riemannian space locally. However, there is another set of curvature invariants that will always characterize pseudo-Riemannian spaces, called *Cartan invariants*, arising from the components of the curvature tensor and its covariant derivatives, relative to a fixed coframe which has been determined in a coordinate independent manner. Such a coframe is a geometrically preferred coframe and is called an *invariantly defined coframe* [26]. The process of determining an invariantly defined coframe can be formalized into an algorithm, and this has been implemented for the Lorentzian case [4, 25] and the neutral case [27, 28]. If two spaces have identical sets of Cartan invariants, the spaces are related by a coordinate transformation and we will say they are *equivalent*.

The algorithm relies on canonical forms of the irreducible parts of the curvature tensor and its covariant derivatives at each iteration, say $\tilde{q}$, in order to record the number of functionally independent components and the dimension of the linear isotropy group of the $\tilde{q}$-th covariant derivative of the curvature tensor, which determine the stopping condition for the algorithm. In the Lorentzian case the alignment types give clear canonical forms. However, for more general pseudo-Riemannian spaces the S_i^G property introduces considerable difficulty in the choice of canonical forms. In addition to this problem, the size of the isotropy group in higher dimensions can make an invariantly defined coframe difficult to determine in practice.

For the (para-)Kähler spaces considered in this paper, the coframe used to present the (para-)complex structure K coincides with the invariantly defined coframe since K must necessarily admit the same linear isotropy group shared by the curvature tensor and all of its curvature tensors relative to the invariantly defined coframe. Thus, any new (para-)Kähler space constructed from metric deformations can be readily determined to be distinct from the original (para-)Kähler by comparing the components of the curvature tensor and its covariant derivatives relative to the initial coframe. We note that for any metric deformation, the choice of the deformed coframe will affect the form of K. We will see in Sect. 5 that the choice of coframe yields different forms for the para-Kähler one-form.

3 $\mathcal{I}$-Degenerate Metrics and Their Deformations

In the Lorentzian case, a significant subset of the $\mathcal{I}$-degenerate spacetimes are contained in the subclass of degenerate Kundt spacetimes, where corollary 1 implies that the curvature tensors and its covariant derivatives are of alignment type II or more special. It is conjectured that any d-dimensional $\mathcal{I}$-degenerate spacetime is a degenerate Kundt spacetime [1]. For other pseudo-Riemannian spaces, it has been shown that the Walker and Kundt type metrics admit $\mathcal{I}$-degenerate metrics but it is possible that there are other $\mathcal{I}$-degenerate metrics which are neither Kundt or Walker.

In order to discuss the class of deformations that leave the set $\mathcal{I}$ invariant, we must review an approach to generate $\mathcal{I}$-degenerate metrics introduced in [2]. For a metric of signature $(n, n+m)$, relative to the canonical coframe $\{\boldsymbol{\ell}^i, \mathbf{m}^a, \mathbf{n}^{\hat{i}}\}$, we define the n-form constructed from the n null and mutually orthogonal null directions, $\boldsymbol{\ell}^i$:

$$\mathbf{F} = \boldsymbol{\ell}^1 \wedge \boldsymbol{\ell}^2 \wedge \ldots \wedge \boldsymbol{\ell}^n.$$

If this n-form satisfies the two surface-forming conditions:

$$\begin{aligned} d\mathbf{F} &= \boldsymbol{\Sigma} \wedge \mathbf{F}, \quad \boldsymbol{\Sigma} \in T_p^* M, \\ \nabla_{\hat{i}} F_{a\mu_1 \ldots \mu_{n-1}} &= 0, \end{aligned} \tag{3.1}$$

then there exists coordinates, (u^i, x^a, v^j), such that the line-element is of the form

$$ds^2 = 2du^i(a_{ij}dv^j + A_{ij}du^j + B_{ia}dx^a) + g_{ab}dx^a dx^b. \tag{3.2}$$

There are several important subclasses of these metrics and in order to enumerate the subclasses we will use the null forms, $\boldsymbol{\ell}^i = du^i$ that constitute $\mathbf{F}$ as part of a coframe basis:

$$\boldsymbol{\ell}^i = du^i,\ \ \mathbf{n}^i = a_{ij}dv^j + A_{ij}du^j + B_{ia}dx^a,\ \ \mathbf{m}^a = e^a{}_b dx^b,$$

with $g_{ab} = \eta_{cd}e^c{}_a e^d{}_b$. The corresponding frame basis $\{\mathbf{n}_i, \mathbf{m}_a, \boldsymbol{\ell}_{\hat{j}}\}$ is constructed by contracting the elements of the coframe basis with the inverse of the metric. To clarify the notation between the frame and coframe basis, we have used a hat on the index i when it is necessary to distinguish between the appearance of the coordinates u^i or v^j. For example, $\boldsymbol{\ell}_{\hat{j}} = \partial_{v^j}$, while $\boldsymbol{\ell}^i = du^i$.

- **Class I: Shear-free and expansion-free.**
  This occurs when $\mathbf{F}$ satisfies

$$\nabla_a F_{b\mu_1\ldots\mu_{n-1}} = 0, \tag{3.3}$$

  relative to the canonical coframe, and implies there is a coordinate system where g_{ab} is independent of the v^j's. If $n = 1$, this corresponds to the Kundt class of metrics.
- **Class II:** $a_{ij} = \delta_{ij}$.
  If relative to the canonical coframe, $\mathbf{F}$ satisfies

$$\nabla_{\hat{j}} F_{i\mu_1\ldots\mu_{k-1}} = 0, \tag{3.4}$$

  where the covariant derivative is taken in the direction of $\boldsymbol{\ell}_{\hat{j}}$, then a coordinate system exists where $a_{ij} = \delta_{ij}$.
- **Class III: The Walker case.**
  If $\mathbf{F}$ is invariant so that

$$\nabla_\mu \mathbf{F} = k_\mu \mathbf{F}, \tag{3.5}$$

  for a vector field $\mathbf{k}$. This class contains class I and II as special cases and further implies that B_{ia} is independent of the v^i coordinates.

We note that the pseudo-Riemannian metrics we consider in this paper are all of class II.

To generate $\mathcal{I}$-degenerate metrics, we recall from [2] that a necessary component of the proof of theorem 1 is the existence of a boost B_t of the form (2.3) for which the components of a tensor $\mathbf{T}$ have well-defined limits under the action of the boost as $t \to \infty$. That is, there exists an element of the Lie algebra of boosts, X, such that the limit

$$\lim_{t\to\infty} exp(tX)(\mathbf{T}) = \mathbf{T}_\infty, \tag{3.6}$$

is finite. Writing $\tilde{\mathbf{b}}$ as the boost represented by X, if $(\mathbf{T})_{\mathbf{b}} \neq 0$ then it is necessary that $\tilde{\mathbf{b}} \cdot \mathbf{b} \leq 0$ since

$$exp(t\tilde{\mathbf{b}} \cdot \mathbf{b}) \to \begin{cases} (\mathbf{T})_{\mathbf{b}}, & \tilde{\mathbf{b}} \cdot \mathbf{b} = 0 \\ 0, & \tilde{\mathbf{b}} \cdot \mathbf{b} < 0 \end{cases}, \tag{3.7}$$

with all other components $(\mathbf{T})_{\mathbf{b}} = 0$, $\tilde{\mathbf{b}} \cdot \mathbf{b} > 0$ vanishing to ensure the limit exists.

In the current coordinate system, there is a helpful lemma that relates vector fields to the boost relative to the coframe basis [2].

Lemma 1 *Consider a null coframe $\{\boldsymbol{\ell}^i, \mathbf{m}^a, \mathbf{n}^i\}$ at a point x. Let B be a boost of the coframe at x:*

$$\{\boldsymbol{\ell}^i, \mathbf{m}^a, \mathbf{n}^i\} \to \{e^{\lambda_i t}\boldsymbol{\ell}^i, \mathbf{m}^a, e^{-\lambda_i t}\mathbf{n}^i\},\ t \in \mathbb{R}, (\lambda_i) \in \mathbb{R}^n. \tag{3.8}$$

Then there exist neighbourhoods U and $\tilde{U}$ of x and coordinate systems (u^i, x^a, v^i) of U and $(\tilde{u}^i, \tilde{x}^a, \tilde{v}^i)$ of $\tilde{U}$ where x is the origin of each coordinate system such that the diffeomorphism $\phi_t : U \to \tilde{U}$ given by

$$(\tilde{u}^i, \tilde{x}^a, \tilde{v}^i) = (e^{-\lambda_i t}u^i, x^a, e^{\lambda_i t}v^i), \tag{3.9}$$

induces the boost B (3.8) at x. Furthermore the diffeomorphism ϕ_t can be considered as a 1-parameter family of diffeomorphisms generated by the vector field

$$X = \sum_i -\lambda_i \left(u^i \frac{\partial}{\partial u^i} - v^i \frac{\partial}{\partial v^i}\right). \tag{3.10}$$

With this boost, $\mathcal{I}$-degenerate metrics will be constructed by considering the limit $t \to \infty$ of the pullback

$$\mathbf{g}_t := \phi_t^* \mathbf{g}. \tag{3.11}$$

Given U and defining $\mathbb{M}$ as the space of smooth metrics on U, we will assume the following.

Assumption *There exists a neighbourhood U, so that the metric, $\mathbf{g}$, relative to the chosen coordinates has a finite limit $\lim_{t\to\infty} \phi_t^* \mathbf{g} \in \mathbb{M}$ with a boost with respect to any $p \in U$.* □

This assumption imposes strong constraints on the potential metric. To see why, consider a neighbourhood, U, that is sufficiently small to be covered by one coordinate chart. As the boost above is with respect to the origin of the coordinate system, we can choose an arbitrary point $x \in U$, which is given by (u_0^i, v_0^i, x_0^a) and shift this point to the origin, by introducing $(\bar{u}^i, \bar{v}^i, \bar{x}^a) = (u^i - u_0^i, v^i -$

$v^i_0, x^a - x^a_0)$ to then apply the above boost. The above assumption now implies that the corresponding limit should be finite for all constant values (u^i_0, v^i_0, x^i_0) in the neighbourhood U.

The metric consists of symmetric components $g_{\mu\nu}(u^i, v^j, x^a)dx^\mu dx^\nu$, which after a translation of the point p to the origin of the coordinate system is $g_{\mu\nu}(u^i + \bar{u}^i, v^j + \bar{v}^j, x^a + \bar{x}^a)d\bar{x}^\mu d\bar{x}^\nu$. Following the analysis in [2], we may summarize the effect of this assumption on the components of the metric. The components involving $dv^i dv^j$ and $dv^i dx^a$ must vanish on U. The metric for the transversal space is independent of the v^i coordinates, i.e.,

$$g_{ab}(u^i, x^a)dx^a dx^b.$$

Finally, the components containing one or two copies of $d\bar{u}^i$ must be polynomials in the coordinates $\bar{v}^i$ but are arbitrary smooth functions in $(\bar{u}^i, \bar{x}^a)$. The order of each polynomial (in terms of the $\bar{v}^i$ coordinates) will depend on the actual boost, as well as which component is considered.

In the following, we will display the indices of the v^i-coordinates as subscripts instead of superscripts for less awkward typesetting. We define $P(v_1, v_2, \ldots, v_n)$ as a polynomial in the v_i-coordinates with smooth functions in u_i, x_a as coefficients. The ring of all such polynomials is

$$\mathcal{P} = \left\{ P(v_1, v_2, \ldots, v_n) \mid P \text{ a polynomial with coefficients } c(u^i, x^a) \right\}.$$

We denote certain subsets of polynomials in $\mathcal{P}$ by a bracket $[-, \ldots, -]$ enclosing a list of monomials in the v_i-coordinates, indicating the highest allowable possible power of the v_i-coordinates in a polynomial in the set.

Returning to the form of the metric (3.2), in light of the above assumption, in order for the limit of the metric to be well-defined, the metric components, g_{ab}, must be independent of the v_i-coordinates, $g_{ab} = g_{ab}(u^i, x^a)$, and the metric components a_{ij}, A_{ij} and B_{ia} are polynomials belonging to some subset, $[v_1^{d_1}, v_2^{d_2}, \ldots, v_n^{d_n}]$, of $\mathcal{P}$. The values $[d_1, \ldots, d_n]$ for each of the metric components a_{ij}, A_{ij} and B_{ia} are determined by the choice of boost $\mathbf{b}$ used and the behaviour of the differentials $du^i du^i$, $du^i dv^j$ and $du^i dx^a$, respectively, under the boost limit.

This motivates the definition of a n-dimensional integer-valued vector which encodes the boost weights of the differentials $du^i du^i$, $du^i dv^j$ and $du^i dx^a$ in addition to the highest powers of the polynomials in a_{ij}, A_{ij} and B_{ia}, $[d_1, \ldots d_n]$,

$$\mathbf{V} = (d_1, \ldots, d_n) + {}_a\mathbf{v} + {}_A\mathbf{v} + {}_B\mathbf{v} \in \mathbb{Z}^n$$

and the components of the integer n-vectors, $({}_a v)_i$, $({}_A v)_i$ and $({}_B v)_i$ for i ranging from 1 to n are defined in the following manner, starting with zero vectors for ${}_a\mathbf{v}$, ${}_A\mathbf{v}$ and ${}_B\mathbf{v}$:

- For each non-zero component, a_{ij}, subtract 1 to $({}_a v)_i$ and add 1 to $({}_a v)_j$. If $i = j$ then leave $({}_a v)_i$ unchanged.

- For each non-zero component, A_{ij}, subtract 1 from $({}_{A}v)_i$ and subtract 1 from $({}_{A}v)_j$. If $i = j$ then subtract 2 from $({}_{A}v)_i$.
- For each non-zero component, B_{ia}: $({}_{B}v)_i = -1$.

Then, by requiring that the limit of the metric is well-defined it follows that $\mathbf{V} \cdot \mathbf{b} \leq 0$. We note that the lower degree terms in the v_i-polynomials with degree $0 < d_i' < d_i$ for i from 1 to n must satisfy a similar boost-vector calculation with $\mathbf{V}' = (d_1', \ldots, d_n') + {}_{a}\mathbf{v} + {}_{A}\mathbf{v} + {}_{B}\mathbf{v}$ where ${}_{a}\mathbf{V}$, ${}_{A}\mathbf{V}$ and ${}_{B}\mathbf{V}$ are defined above. However, if $\mathbf{V} \cdot \mathbf{b} \leq 0$ holds for the highest powers, then $\mathbf{V}' \cdot \mathbf{b} \leq 0$ automatically.

For a subclass of metrics of the form (3.2) where the limit of the boost (3.8) is well-defined, so that $g_{ab} = g_{ab}(u^i, x^a)$, and a_{ij}, A_{ij} and $B_{ia} \in [v_1^{d_1}, v_2^{d_2}, \ldots, v_n^{d_n}]$, the limiting process yields a new metric where all lower order terms in the polynomials are eliminated. The following theorem ensures that the SPIs of this subclass of metrics will only depend on the coefficients of the highest degree in the v_i-coordinates [2].

Theorem 2 *Consider a point $x \in U$ and assume that there is a one-parameter family of diffeomorphisms ϕ_t such that $\phi_t(x) = x$ and $\lim_{t\to\infty} \phi_t^* g = g_0 \in \mathbb{M}$. Then any SPI of g evaluated at x is identical to the corresponding invariant at g_0 at x.*

This result provides a clear approach to deforming $\mathcal{I}$-degenerate metrics: we will modify the lower degree coefficients of the polynomials in a_{ij}, A_{ij} and $B_{ia} \in [v_1^{d_1}, v_2^{d_2}, \ldots, v_n^{d_n}]$ to produce non-diffeomorphic metrics with the same set $\mathcal{I}$.

4 Pseudo-Riemannian Spaces with Constant Scalar Curvature Invariants

We have reviewed a general approach to generate a large class of $\mathcal{I}$-degenerate spaces. Of particular interest within this set are spaces for which all elements of $\mathcal{I}$ vanish or are constant, such spaces are known as *vanishing scalar invariant* (VSI) or *constant scalar invariant* (CSI) spaces, respectively [2]. The class of CSI Lorentzian spaces are applicable to many theories of gravity, as they contain a subset of spacetimes that are universal, and hence solve the vacuum field equations of all gravitational theories with a Lagrangian constructed from SPIs [14–16].

In the Riemannian case, a space is CSI if and only if the space is locally homogeneous. For Lorentzian spaces, it is conjectured that a CSI spacetime will either be locally homogeneous or belong to the degenerate Kundt class [14] and this conjecture has been proven for dimension less than or equal to four [17, 18]. It has been shown that the VSI Lorentzian spaces belong to the degenerate Kundt class in all dimensions [19]. The subset of CSI spacetimes belonging to the Kundt class are called *Kundt-CSI*. For Kundt-CSI metrics, the transverse space is a locally homogeneous Riemannian manifold and the metric functions must satisfy particular differential equations [17, 18, 20, 21].

For any Lorentzian Kundt-CSI spacetime, taking the limit of the boost transformation (2.3) yields a locally homogeneous Kundt-CSI spacetime with the same set $\mathcal{I}$ which are necessarily constant [2, 21]. The pseudo-Riemannian case admits CSI metrics, with two known classes of metrics containing CSI solutions, namely the Kundt and Walker pseudo-Riemannian metrics [2]. Unlike the Lorentzian case, there exist CSI pseudo-Riemannian spaces which are mapped to simpler CSI spaces which are not locally homogeneous and lack the sufficient number of Killing vector fields required to prove they are CSI. In such cases, all possible SPIs up to an appropriate order must be checked explicitly to prove the CSI property. For such pseudo-Riemannian spaces it is conjectured that there exist $\mathcal{I}$-preserving nil-Killing vector fields in addition to the Killing vector fields that lead to the CSI property [22, 23]. However, it is still possible to take locally homogeneous or symmetric $\mathcal{I}$-degenerate pseudo-Riemannian spaces and deform the metrics to construct new pseudo-Riemannian CSI spaces. All of the examples we will construct in the following sections are CSI, although the isometry group of each distinct space will be smaller than the original space.

5 Para-Kähler Metrics

To define the para-Kähler metrics in a more rigorous manner, let $K : V \to V$ be an involution on a $2n$-dimensional vector space V. Assume also that K is a para-complex structure on V so that $K^2 = 1$, and $V = V_+ \oplus V_-$ is the eigenspace decomposition of V where $\dim(V_+) = \dim(V_-) = n$.

Given a $2n$-dimensional manifold M admitting a pseudo-Riemannian metric g, we will assume also that the manifold possesses a symplectic form $\boldsymbol{\omega}$, and a K given above, acting on the tangent space. Then the triple $(\mathbf{g}, K, \boldsymbol{\omega})$ is said to be *compatible* if for all $X, Y \in TM$:

1. $\mathbf{g}(X, Y) = \boldsymbol{\omega}(KX, Y)$,
2. $\boldsymbol{\omega}(X, Y) = -\mathbf{g}(X, KY)$, and
3. $K(X) = \phi_{\mathbf{g}}^{-1}(\phi_{\boldsymbol{\omega}} X)$,

where $\phi_{\boldsymbol{\alpha}} : TM \to T^*M$ is, for a non-degenerate bi-linear form $\boldsymbol{\alpha}$, the bundle isomorphism $X \mapsto \boldsymbol{\alpha}(X, -)$. We note that the three conditions for $(\mathbf{g}, K, \boldsymbol{\omega})$ to be compatible are equivalent.

Any metric admitting such a structure must be neutral, because if X is an eigenvector of K, so that $KX = \lambda X$, then

$$\mathbf{g}(X, X) = \boldsymbol{\omega}(KX, X) = \lambda\boldsymbol{\omega}(X, X) = 0.$$

Hence, V_+ and V_- are both n-dimensional null planes and the metric $\mathbf{g}$ is of neutral signature.

In this section we first consider the conditions on $\mathcal{I}$-preserving metric deformations of neutral metrics that conserve the para-Kähler condition. We show that the four dimensional example is rigid under two different coframe deformations that yield the same metric deformation and argue that the higher dimensional examples will also be rigid with respect to a particular deformed coframe. We conclude the section by showing that Einstein almost-para-Kähler metrics can be readily constructed.

5.1 Symmetric Metrics of Neutral Signature

Using a null-coframe we can write the metric and the symplectic form, respectively, as

$$\mathbf{g} = 2\left(\theta^1\theta^2 + \theta^3\theta^4 + \cdots + \theta^{2n-1}\theta^{2n}\right) \tag{5.1}$$

and

$$\boldsymbol{\omega} = \theta^1 \wedge \theta^2 + \theta^3 \wedge \theta^4 + \cdots + \theta^{2n-1} \wedge \theta^{2n}. \tag{5.2}$$

The structure group, which defines the isotropy group for the metric, can then be seen to be $GL(n) \subset O(n,n)$, given by the following action on $T_pM = V_+ \oplus V_-$: For $A \in GL(n)$, and $X_\pm \in V_\pm$: $(X_+, X_-) \longmapsto (AX_+, (A^t)^{-1}X_-)$.

Examples of symmetric spaces admitting this structure group can be generated by considering such a space as the quotient $M = SL(n+1,\mathbb{R})/GL(n)$, where $GL(n) \subset SL(n+1,\mathbb{R})$ by the natural embedding

$$A \in GL(n), \qquad \begin{bmatrix} \det(A^{-1}) & 0 \\ 0 & A \end{bmatrix} \in SL(n+1,\mathbb{R}).$$

Using coordinates (x^i, y^i) we can now write the null coframe as

$$\theta^{2i-1} = dx^i, \quad \theta^{2i} = dy^i - ay^i\boldsymbol{\alpha}, \text{ with } \boldsymbol{\alpha} = y^1dx^1 + \ldots + y^ndx^n, a \in \mathbb{R}. \tag{5.3}$$

We note that $d\boldsymbol{\alpha} = -\boldsymbol{\omega}$ and consequently $\boldsymbol{\omega}$ is closed, as required. The constant a cannot be eliminated by coordinate transformations as it is proportional to the Ricci scalar, $\mathbf{Ric} = -2g = -12a$ which is a coordinate independent scalar.

5.2 $\mathcal{I}$-Preserving Metric Deformations

Let the metric and symplectic form be as in Eqs. (5.1) and (5.2), but with the null coframe

$$\theta^{2i-1} = dx^i, \qquad \theta^{2i} = dy^i - ay^i\alpha + \left[y^kH_k{}^{ij}(x^l) + B^{ij}(x^l)\right]dx^j, \tag{5.4}$$

where a is a constant, $\alpha = y^1dx^1 + \ldots + y^ndx^n$, and $H_l^{ij}(x^k)$ and $B^{ij}(x^l)$ are functions of x^l and symmetric in the indices i and j. The constant a cannot be rescaled since $\mathbf{Ric} = -12a$ which is coordinate independent.

5.3 Example: Four-Dimensional Para-Kähler Einstein Neutral Metrics

We will consider the neutral signature metric associated to the complex two-dimensional hyperbolic plane. Explicitly, the metric and sympletic form relative to the coordinate basis are, respectively,

$$\mathbf{g} = 2dx^1(-a(y^1)^2dx^1 + dy^1 - ay^1y^2dx^2) + 2dx^2(a(y^2)^2dx^2 + dy^2) \quad (5.5)$$

and

$$\boldsymbol{\omega} = dx^1 \wedge dy^2 + dx^2 \wedge dy^2. \quad (5.6)$$

This space is Einstein with $\mathbf{Ric} = -2\mathbf{g}$.

After deforming the metric by introducing the functions $H_k{}^{ij}(x^l)$ and $B^{ij}(x^l)$ into the coframes, given in Eq. (5.4), the para-Kähler form is still

$$\boldsymbol{\omega} = dx^1 \wedge dy^2 + dx^2 \wedge dy^2. \quad (5.7)$$

Without loss of generality, we can choose coordinates so that two functions are zero,

$$H_1{}^{11} = H_2{}^{22} = 0, \quad (5.8)$$

by translating the y^1 and y^2 appropriately. With this choice of metric deformation and coframe, the almost-para-Kähler condition $d\boldsymbol{\omega} = 0$ will be automatically satisfied. At this stage, the para-Kähler condition $\nabla\boldsymbol{\omega} = 0$ is difficult to solve, and we will use the Einstein condition to find further algebraic conditions on the deformation functions:

$$\begin{aligned} aB^{11} &= H_2{}^{11}H_1{}^{12} + (H_2{}^{12})^2 - \partial_{x^1}H_2{}^{12} + \partial_{x^2}H_2{}^{11}, \\ aB^{22} &= H_2{}^{12}H_1{}^{22} + (H_1^{12})^2 + \partial_{x^1}H_1{}^{22} - \partial_{x^2}H_1{}^{12}, \\ 2aB^{12} &= 2H_2{}^{11}H_1{}^{22} - 2H_2{}^{12}H_1{}^{12} + \partial_{x^2}H_2{}^{12} + \partial_{x^1}H_1{}^{12}. \end{aligned} \quad (5.9)$$

This drastically simplifies the para-Kähler condition, giving three equations:

$$\begin{aligned}
0 =& 4\partial_{x^2} H_1{}^{12} H_2{}^{11} - 4\partial_{x^1} H_1{}^{22} H_2{}^{11} + 4\partial_{x^2} H_2{}^{11} H_1{}^{12} + 5\partial_{x^2} H_2{}^{12} H_2{}^{12} \\
&+ 3\partial_{x^1} H_1{}^{12} H_2{}^{12} - 2\partial_{x^1} H_2{}^{11} H_1{}^{22} + 2\partial^2_{x^2} H_2{}^{11} - 3\partial_{x^1}\partial_{x^2} H_2{}^{12} - \partial^2_{x^1} H_1{}^{12}, \\
0 =& \partial_{x^2} H_2{}^{12} - \partial_{x^1} H_1{}^{12}, \\
0 =& 4\partial_{x^2} H_2{}^{11} H_1{}^{22} - 4\partial_{x^1} H_1{}^{22} H_2{}^{12} - 4\partial_{x^1} H_2{}^{12} H_1{}^{22} - 5\partial_{x^1} H_1{}^{12} H_1{}^{12} \\
&- 3\partial_{x^2} H_2{}^{12} H_1{}^{12} + 2\partial_{x^2} H_1{}^{22} H_2{}^{11} + \partial^2_{x^2} H_2{}^{12} + 3\partial_{x^1}\partial_{x^2} H_1{}^{12} - 2\partial^2_{x^1} H_1{}^{22}.
\end{aligned} \tag{5.10}$$

The second equation is satisfied if

$$H_1{}^{12} = \partial_{x^2} f(x^1, x^2), \quad H_2{}^{12} = \partial_{x^1} f(x^1, x^2). \tag{5.11}$$

The remaining two equations become

$$\begin{aligned}
0 =& 4\partial_{x^1} f \partial_{x^1}\partial_{x^2} f + 2H_2{}^{11}\partial^2_{x^2} f - 2\partial_{x^1} H_1{}^{22} H_2{}^{11} - \partial_{x^1} H_2{}^{11} H_1{}^{22} \\
&+ 2\partial_{x^2} H_2{}^{11}\partial_{x^2} f + \partial^2_{x^2} H_2{}^{11} - 2\partial^2_{x^1}\partial_{x^2} f, \\
0 =& 4\partial_{x^2} f \partial_{x^1}\partial_{x^2} f + 2H_1{}^{22}\partial^2_{x^1} f - 2\partial_{x^2} H_2{}^{11} H_1{}^{22} - \partial_{x^2} H_1{}^{22} H_2{}^{11} \\
&+ 2\partial_{x^1} H_1{}^{22}\partial_{x^1} f + \partial^2_{x^1} H_1{}^{22} - 2\partial^2_{x^2}\partial_{x^2} f.
\end{aligned} \tag{5.12}$$

Without solving any differential equations, we can substitute these expressions into the components of the curvature tensor and its covariant derivative of the deformed metric. Doing so, the components of the curvature tensor relative to this fixed coframe are equal to the curvature tensor of the original metric relative to the coframe (5.3), and the components of the covariant derivative of the curvature tensor vanish identically. This is sufficient to show that the given deformed Einstein para-Kähler metric is equivalent to the original neutral metric for the complex two-dimensional hyperbolic plane as they have identical sets of Cartan invariants.

Using the same metric deformation we will now consider a different choice of coframe to construct a para-Kähler form by assuming that $H_k{}^{ij} = 0$ and $B^{ij} = 0$ if $i > j$. This reproduces the same metric as the initial coframe but picks out a different para-Kähler form for the deformed space

$$\omega = dx^1 \wedge dy^2 + dx^2 \wedge dy^2 + (B^{12} + y^2 H_2{}^{12} + y^1 H_1{}^{12}) dx^1 \wedge dx^2. \tag{5.13}$$

The almost-para-Kähler condition, $d\omega = 0$, implies that

$$H_1{}^{12} = H_2{}^{12} = 0. \tag{5.14}$$

The components of the para-Kähler condtion $\nabla\omega = 0$ are polynomials in y^i whose coefficients must vanish for all values of y^i and yield differential equations for the remaining functions:

$$\begin{aligned} {H_2}^{11}B^{12} - \partial_{x^2}B^{11} + \partial_{x^1}B^{12} &= 0, \\ -\partial_{x^2}{H_2}^{11} + \frac{1}{2}B^{11}a &= 0, \\ -{H_1}^{22}B^{\,11} + \partial_{x^1}B^{22} &= 0, \\ \partial_{x^1}{H_1}^{22} - \frac{1}{2}B^{22}a &= 0, \end{aligned} \tag{5.15}$$

and two algebraic constraints:

$$B^{12} = 0 \text{ and } {H_1}^{22}{H_2}^{11} = 0. \tag{5.16}$$

The two possible subcases arising from the second condition are related by a discrete transformation $(x^1, y^1) \leftrightarrow (x^2, y^2)$ and so, without loss of generality, we will set ${H_2}^{11} = 0$. The remaining equations are now

$$\begin{aligned} \partial_{x^2}B^{11} &= 0, \\ B^{11}a &= 0, \\ -{H_1}^{22}B^{\,11} + \partial_{x^1}B^{22} &= 0, \\ \partial_{x^1}{H_1}^{22} - \frac{1}{2}B^{22}a &= 0. \end{aligned} \tag{5.17}$$

The only non-zero deformation functions are

$${H_1}^{22} = {h^1}_{22}(x^2)x^1 + {h'^1}_{122}(x^2) \text{ and } B^{22} = \frac{2}{a}{h^1}_{22}(x^2). \tag{5.18}$$

The deformed metric is then

$$\begin{aligned} \mathbf{g} &= 2dx^1(-av^2dx^1 + dy^1 - ay^1y^2dx^2) \\ &+ 2dx^2\left(a(y^1)^2 + y^1({h^1}_{22}x^1 + {h'^1}_{122}) + \frac{2}{a}{h^1}_{22}dx^2 + dy^2\right). \end{aligned} \tag{5.19}$$

The components of the curvature tensor relative to the newly deformed coframe are constant and match those of the curvature tensor of the original space. Similarly, the covariant derivative of the curvature tensor of the deformed space vanishes. This implies that the sets of Cartan invariants for both spaces match and the deformed space must be equivalent to the original space. Thus the original metric is rigid under any deformation that preserves both the set $\mathcal{I}$ and the para-Kähler condition for this choice of deformed coframe.

5.4 *Rigidity of Higher Dimensional Para-Kähler Einstein Neutral Metrics*

The six and eight dimensional cases have been investigated as well. Using the deformed coframe (5.4) with the condition that ${H_k}^{ij} = 0$ for $i > j$, they are rigid under $\mathcal{I}$-preserving deformations. In light of this result, we propose the following conjecture.

Conjecture 1 The neutral signature metrics which are equivalent to the standard metric on the quotient $M = SL(n+1, \mathcal{R})/GL(n)$ for $n > 1$ are rigid under all deformations preserving $\mathcal{I}$ and the para-Kähler condition for the choice of coframe defined in Eq. (5.4) with $B^{ij} = 0$ and ${H_k}^{ij} = 0$ for $i > j$.

5.5 *Almost-Para-Kähler Einstein Neutral Metrics*

Using the coframe defined in Eq. (5.4) with $B^{ij} = 0$ and ${H_k}^{ij} = 0$ for $i > j$, if we relax the para-Kähler condition and instead ask that $d\omega = 0$, then we are able to construct examples of new Einstein neutral metrics. We will consider the four-dimensional case.

Choosing coordinates so that ${H_1}^{11} = {H_2}^{22} = 0$, the almost-para-Kähler condition gives Eq. (5.14) and the Einstein conditions are then

$$aB^{11} = \partial_{x^2} {H_2}^{11},\ aB^{12} = 2{H_2}^{11}{H_1}^{22},\ aB^{22} = \partial_{x^1} {H_1}^{22}. \tag{5.20}$$

Since the para-Kähler condition is not preserved, it is no longer guaranteed that the deformed coframe coincides with the invariantly defined coframe. To show inequivalence with the original metric, we note that for generic choices of the remaining free metric functions, the covariant derivative of the curvature tensor is non-zero. This is sufficient to show the new metrics are not equivalent to the original symmetric para-Kähler metric as their respective sets of Cartan invariants are different.

6 Kähler Metrics

The Riemannian metrics of complex hyperbolic spaces can be Wick-rotated to give pseudo-Riemannian metrics of signature $(2p, 2p+2q+2)$. Relative to the null coframe basis

$$\theta^{\hat{i}} = du^i, \quad (i = 1, \dots, 2p) \tag{6.1}$$

$$\theta^i = dv^i - v^i dw, \tag{6.2}$$

$$\theta^a = e^{w/2} dy^a, \quad (a = 1, \dots, q) \tag{6.3}$$

$$\theta^{\hat{a}} = e^{w/2} dz^a, \tag{6.4}$$

$$\theta^w = dw, \tag{6.5}$$

$$\theta^x = e^w \left(dx - \frac{1}{2} \sum_{a=1}^{q} \left(z^a dy^a - y^a dz^a \right) \right) + \sum_{k=1}^{p} \left(v^{2k} du^{2k-1} - v^{2k-1} du^{2k} \right). \tag{6.6}$$

the Wick-rotated metric and symplectic form are, respectively,

$$\mathbf{g} = 2 \sum_{i=1}^{2p} \theta^{\hat{i}} \theta^i + \sum_{a=1}^{q} \left(\theta^a \theta^a + \theta^{\hat{a}} \theta^{\hat{a}} \right) + \theta^w \theta^w + \theta^x \theta^x \tag{6.7}$$

and

$$\boldsymbol{\omega} = \sum_{k=1}^{p} \left(\theta^{2k} \wedge \theta^{\widehat{2k-1}} - \theta^{2k-1} \wedge \theta^{\widehat{2k}} \right) + \sum_{a=1}^{q} \theta^a \wedge \theta^{\hat{a}} + \theta^w \wedge \theta^x. \tag{6.8}$$

These are symmetric metrics with isotropy group $U(p, p+q+1)$.

6.1 $\mathcal{I}$-Preserving Metric Deformations

Let the metric and symplectic form be as in Eqs. (6.7) and (6.8), while we deform the one-forms in (6.2) to be

$$\theta^i = dv^i - v^i dw + H^{ij}(u^k, x^\beta) du^j + B^{i\alpha}(u^k, x^\beta) dx^\alpha, \tag{6.9}$$

where $x^\alpha = (y^a, z^a, w, x)$, and the other basis one-forms are unchanged. The matrix of functions H^{ij} is symmetric in the indices i and j. We will show that the $\mathcal{I}$-preserving metric deformations can produce new Einstein Kähler metrics and illustrate the differences between the six dimensional and eight dimensional examples. We conclude the section by showing that it is possible to generate new Einstein spaces which are almost-Kähler.

6.2 Example: Kähler Einstein Metrics of Signature (2,4)

The simplest non-trivial example is for $p = 1$ and $q = 0$. Relative to the coordinate basis, the symmetric Kähler metric and the symplectic form are, respectively,

$$\begin{aligned}\mathbf{g} &= 2du^1(dv^1 - v^1dw) + 2du^2(dv^2 - v^2dw) \\ &\quad + dw^2 + \left(e^w dx + v^2du^1 - v^1du^2\right)^2 \end{aligned} \tag{6.10}$$

and

$$\omega = (dv^2 \wedge du^1 - dv^1 \wedge du^2) + e^w dw \wedge dx. \tag{6.11}$$

This space is Einstein with $\mathbf{Ric} = -2\mathbf{g}$. Deforming the metric by introducing the functions H^{ij} and $B^{i\alpha}$, the new metric is no longer Einstein, although the Ricci scalar remains $\mathcal{R} = -8$. In this subsection we will take the metric and symplectic form relative to the coframe basis and show that the deformed coframe in (6.9) can produce new Kähler Einstein metrics.

Proposition 1 *The Kähler metric produced by deforming the symmetric Kähler metric and its symplectic form in Eqs.* (6.10) *and* (6.11)*, respectively, is*

$$\begin{aligned}\mathbf{g} &= 2du^1(dv^1 - v^1dw + H^{1j}du^j) + 2du^2(dv^2 - v^2dw + H^{2j}du^jdx^\alpha) \\ &\quad + dw^2 + \left(e^w dx + v^2du^1 - v^1du^2\right)^2, \end{aligned} \tag{6.12}$$

with H^{ij} of the form

$$H^{12} = F(c_1e^{\epsilon x} + c_2e^{-\epsilon x})[d_1e^w \sin(\epsilon e^{-w}) + d_2e^w \cos(\epsilon e^{-w})], \tag{6.13}$$

$$H^{22} = -F(c_1e^{\epsilon x} - c_2e^{-\epsilon x})[d_2e^w \sin(\epsilon e^{-w}) - d_1e^w \cos(\epsilon e^{-w})] \tag{6.14}$$

$$+ e^w h + \frac{1}{2}(\partial_{u^1}\phi^1 + \partial_{u^2}\phi^2 + \psi), \tag{6.15}$$

$$H^{11} = -H^{22} + \partial_{u^1}\phi^1 + \partial_{u^2}\phi^2 + \psi, \tag{6.16}$$

or

$$H^{12} = F(c_1\sin(x) + c_2\cos(x))[d_1 e^w \sinh(e^{-w}) + d_2 e^w \cosh(e^{-w})], \quad (6.17)$$

$$H^{22} = -F(c_1\sin(x) - c_2\cos(x))[d_2 e^w \sinh(e^{-w}) + d_1 e^w \cosh(e^{-w})] \quad (6.18)$$

$$+ e^w h + \frac{1}{2}(\partial_{u^1}\phi^1 + \partial_{u^2}\phi^2 + \psi), \quad (6.19)$$

$$H^{11} = -H^{22} + \partial_{u^1}\phi^1 + \partial_{u^2}\phi^2 + \psi, \quad (6.20)$$

where c_1, c_2, d_1 *and* d_2 *are real-valued constants, the free functions* ϕ^1, ϕ^2, ψ, F *and* h *depend on* u^1 *and* u^2*, and* $\epsilon = 0$ *or* 1. *The symplectic form for this metric is*

$$\omega = -(\partial_{u^1}\phi^1 + \partial_{u^2}\phi^2 + \psi)du^1 \wedge du^2$$
$$+ (dv^2 \wedge du^1 - dv^1 \wedge du^2) + e^w dw \wedge dx. \quad (6.21)$$

If $\mathbf{g}$ *is an Einstein metric then*

$$\psi = -\partial_{u^1}\phi^1 - \partial_{u^2}\phi^2. \quad (6.22)$$

The deformed Einstein metric $\mathbf{g}$ *is distinct from the metric in* (6.10) *so long as*

$$F \neq 0 \text{ and } \partial^2_{u^1}h \neq \partial^2_{u^2}h. \quad (6.23)$$

Proof The almost-Kähler condition $d\omega = 0$ gives the following conditions on the deformation functions:

$$\partial_w B^{ix} - \partial_x B^{iw} = 0, \quad (6.24)$$

$$\partial_x(H^{11} + H^{22}) - (\partial_{u^1}B^{1x} + \partial_{u^2}B^{2x}) = 0, \quad (6.25)$$

$$\partial_w(H^{11} + H^{22}) - (\partial_{u^1}B^{1w} + \partial_{u^2}B^{2w}) = 0. \quad (6.26)$$

These equations are solved as follows. Let $\phi^i = \phi^i(u^1, u^2, x, w)$ $(i = 1, 2)$ and $\psi = \psi(u^1, u^2)$ be arbitrary functions. Then

$$B^{ix} = \partial_x\phi^i, \quad B^{iw} = \partial_w\phi^i, \quad H^{11} = -H^{22} + \partial_{u^1}\phi^1 + \partial_{u^2}\phi^2 + \psi, \quad (6.27)$$

while H^{12} and H^{22} are arbitrary functions.

Imposing the constraints from the Kähler condition $\nabla\omega = 0$, the v^1 and v^2 linear terms give the following conditions:

$$\partial_x\phi^i = \partial_w\phi^i = 0, \quad (6.28)$$

$$\partial_x H^{22} + e^w(\partial_w H^{12} - H^{12}) = 0, \quad (6.29)$$

$$\frac{1}{2}(\psi + \partial_{u^1}\phi^1 + \partial_{u^1}\phi^2) + \partial_w H^{22} - H^{22} - e^{-w}\partial_x H^{12} = 0. \quad (6.30)$$

A differential equation for H^{12} can be constructed by differentiating equations (6.29) and (6.30) by w and x, respectively, and adding the result together to get

$$-e^{2w}(\partial_w H^{12} - \partial_w^2 H^{12}) + \partial_x^2 H^{12} = 0. \quad (6.31)$$

This equation can be solved by assuming H^{12} is separable, $H^{12} = F(u_1, u_2)H_1(x)$ $H_2(w)$, giving two differential equations:

$$\partial_x^2 H_1 - cH_1 = 0, \;\; \partial_w^2 H_2 - \partial_w H_2 + cH_2 e^{-2w} = 0. \quad (6.32)$$

Thus, there are three distinct solutions depending on the sign of the constant, c, arising from separation of variables. If $c = 0$ then the solution is

$$H^{12} = F(c_1 x + c_2)(d_1 + d_2 e^w), \quad (6.33)$$

if $c > 0$ then

$$H^{12} = F(c_1 e^{\sqrt{c}x} + c_2 e^{-\sqrt{c}x})[d_1 e^w \sin(e^{-w}\sqrt{c}) + d_2 e^w \cos(e^{-w}\sqrt{c})], \quad (6.34)$$

while if $c < 0$

$$\begin{aligned} H^{12} =& F(c_1 \sin(\sqrt{|c|}x) + c_2 \cos(\sqrt{|c|}x))\cdot \\ & [d_1 e^w \sinh(e^{-w}\sqrt{|c|}) + d_2 e^w \cosh(e^{-w}\sqrt{|c|})], \end{aligned} \quad (6.35)$$

where $F = F(u^1, u^2)$ is an arbitrary function. Then by integrating (6.29) and considering the sign of c, the form of H^{22} is then for $c = 0$,

$$H^{22} = \frac{1}{2} F e^w d_1 (c_1 x^2 + 2c_2 x) + +h^{22}(u^1, u^2, w), \quad (6.36)$$

for $c > 0$,

$$\begin{aligned} H^{22} =& -F(c_1 e^{\sqrt{c}x} - c_2 e^{-\sqrt{c}x})[d_2 e^w \sin(e^{-w}\sqrt{c}) - d_1 e^w \cos(e^{-w}\sqrt{c})] \\ & + h^{22}(u^1, u^2, w), \end{aligned} \quad (6.37)$$

while for $c < 0$,

$$\begin{aligned} H^{22} =& - F(c_1 \cos(\sqrt{|c|}x) - c_2 \sin(\sqrt{|c|}x))\cdot \\ & [d_2 e^w \sinh(e^{-w}\sqrt{|c|}) + d_1 e^w \cosh(e^{-w}\sqrt{|c|})] + h^{22}(u^1, u^2, w). \end{aligned} \quad (6.38)$$

where h^{22} is an arbitrary function of integration.

Substituting H^{12} and H^{22} into (6.30) yields an inhomogeneous differential equation for h^{22},

$$2\partial_w h^{22} - h^{22} = -(\partial_{u^1}\phi^1 + \partial_{u^2}\phi^2 + \psi), \tag{6.39}$$

which has the solution

$$h^{22} = e^w h(u^1, u^2) + \frac{1}{2}(\partial_{u^1}\phi^1 + \partial_{u^2}\phi^2 + \psi). \tag{6.40}$$

If $c \neq 0$, the magnitude of c can be scaled to one using the coordinate transformation

$$x' = \frac{x}{\sqrt{|c|}},\ w' = w - \ln(\sqrt{|c|})$$

and a redefinition of the arbitrary functions F and h. In the case that $c = 0$, the Kähler condition further imposes that $H^{12}_{,x} = 0$ and the solutions of H^{12} and H^{22} may be recovered from the form of the functions H^{12} and H^{22} in the $c > 0$ case by setting $c = 0$ and translating the w coordinate appropriately. To summarize the two distinct solutions, we introduce $\epsilon = 0, 1$ in Eqs. (6.13–6.16).

To preserve the Einstein condition, there is only one condition that must be satisfied, namely

$$\partial_{u^1}\phi^1 + \partial_{u^2}\phi^2 + \psi = 0. \tag{6.41}$$

As the Kähler condition ensures the deformed coframe corresponds to the invariantly defined coframe, we can determine conditions to show that the deformed space is not equivalent to the original space. If the metric functions $F(u^1, u^2)$ and $h(u^1, u^2)$ satisfy

$$F = 0 \text{ and } \partial^2_{u^1} h = \partial^2_{u^2} h, \tag{6.42}$$

then the covariant derivative of the curvature tensor vanishes and the components of the deformed space's curvature tensor match those of the original space, implying that the set of Cartan invariants of the two spaces are identical and the two spaces are equivalent. For any choice of metric functions F and h not satisfying (6.42), the covariant derivative of the curvature tensor is non-zero, and so the deformed Kähler Einstein space must be distinct from the original space as the set of Cartan invariants of the two spaces are distinct. Furthermore, as the metric functions in (6.13–6.16) and (6.17–6.20) are solutions to distinct ordinary differential equations, their metrics will be inequivalent. □

6.3 Example: Kähler Einstein Metrics of Signature (2,6)

Keeping $p = 1$ and setting $q = 1$ we can produce a non-trivial example with different constraints on the deformation functions due to the change in the dimension. Explicitly, the symmetric Kähler metric and the symplectic form are, respectively,

$$\begin{aligned}\mathbf{g} &= 2du^1(dv^1 - v^1dw) + 2du^2(dv^2 - v^2dw) + dw^2 \\ &+ \left(e^w dx - \frac{e^w}{2}(ydz - zdy) + v^2du^1 - v^1du^2\right)^2 + e^w(dy^2 + dz^2) \end{aligned} \quad (6.43)$$

and

$$\begin{aligned}\boldsymbol{\omega} &= (dv^2 \wedge du^1 - dv^1 \wedge du^2) \\ &+ e^w\left[dw \wedge \left(dx - \frac{1}{2}(ydz - zdy)\right) + dy \wedge dz\right]. \end{aligned} \quad (6.44)$$

This space is Einstein with $\mathbf{Ric} = -(5/2)\mathbf{g}$. We will take the deformed coframe in (6.9) to construct the metric and symplectic form in this section and construct new Kähler Einstein metrics.

Proposition 2 *The Kähler metric produced by deforming the symmetric Kähler metric and its symplectic form in Eqs. (6.43) and (6.44), respectively, is*

$$\begin{aligned}\mathbf{g} &= 2du^1(dv^1 - v^1dw + H^{1j}du^j) + 2du^2(dv^2 - v^2dw + H^{2j}du^j) \\ &+ e^w dy^2 + e^w dz^2 + dw^2 + \left(e^w dx + v^2du^1 - v^1du^2\right)^2, \end{aligned} \quad (6.45)$$

with H^{ij} of the form

$$H^{12} = Fe^w, \quad (6.46)$$

$$H^{22} = c_0H^{12} + e^w h + \frac{1}{2}(\partial_{u^1}\phi^1 + \partial_{u^2}\phi^2 + \psi), \quad (6.47)$$

$$H^{11} = -H^{22} + \partial_{u^1}\phi^1 + \partial_{u^2}\phi^2 + \psi, \quad (6.48)$$

where $x^\alpha = (y, z, w, x)$, the free functions ϕ^1, ϕ^2, ψ, F and h depend on u^1 and u^2 and c_0 is a real-valued constant. The symplectic form for this metric is then

$$\begin{aligned}\boldsymbol{\omega} &= -(\partial_{u^1}\phi^1 + \partial_{u^2}\phi^2 + \psi)du^1 \wedge du^2 \\ &+ (dv^2 \wedge du^1 - dv^1 \wedge du^2) \\ &+ e^w\left[dw \wedge \left(dx - \frac{1}{2}(ydz - zdy)\right) + dy \wedge dz\right]. \end{aligned} \quad (6.49)$$

If $\mathbf{g}$ *is an Einstein metric then*

$$\psi = -\partial_{u^1}\phi^1 - \partial_{u^2}\phi^2. \tag{6.50}$$

Then, the deformed Einstein metric $\mathbf{g}$ *is distinct from the metric in* (6.43) *so long as*

$$F \neq 0 \text{ and } \partial^2_{u^1} h \neq \partial^2_{u^2} h. \tag{6.51}$$

Proof Deforming the metric with functions $H^{ij}(u^i, x^\alpha)$ and $B^{i\alpha}(u^i, x^\alpha)$ with $x^\alpha = (y, z, w, x)$, the almost-Kähler conditions are now

$$\partial_\alpha B^{i\beta} - \partial_\beta B^{i\alpha} = 0, \tag{6.52}$$

$$\partial_\alpha (H^{11} + H^{22}) - (\partial_{u^1} B^{1\alpha} + \partial_{u^2} B^{2\alpha}) = 0. \tag{6.53}$$

Setting $\phi^i = \phi^i(u^1, u^2, x^\alpha)$ with $i = 1, 2$ and $\psi = \psi(u^1, u^2)$ to be arbitrary functions, the above equations have the following solution:

$$B^{i\alpha} = \partial_\alpha \phi^i, \; H^{11} = -H^{22} + \partial_{u^1}\phi^1 + \partial_{u^2}\phi^2 + \psi, \tag{6.54}$$

where H^{12} and H^{22} are arbitrary functions.

For the sake of presentation we will first display and solve the smaller equations arising from the Kähler condition $\nabla\boldsymbol{\omega} = 0$ before moving onto the larger equations. Considering the v_i-linear coefficients first, the functions ϕ^i must satisfy

$$\partial_x \phi^i = \partial_w \phi^i = 0, \; \partial_y \phi^1 = -\partial_z \phi^2, \; \partial_z \phi^1 = \partial_y \phi^2. \tag{6.55}$$

In addition, there are the two differential equations for H^{12} and H^{22}:

$$\partial_x H^{22} + e^w(\partial_w H^{12} - H^{12}) = 0, \tag{6.56}$$

$$\frac{1}{2}(\psi + \partial_{u^1}\phi^1 + \partial_{u^2}\phi^2) + \partial_w H^{22} - H^{22} - e^{-w}\partial_x H^{12} = 0. \tag{6.57}$$

As in the previous case, a differential equation for H^{12} can be constructed by differentiating equations (6.56) and (6.57) by w and x, respectively, and adding the result together to give

$$-e^{2w}(\partial_w H^{12} - \partial^2_w H^{12}) + \partial^2_x H^{12} = 0. \tag{6.58}$$

Assuming H^{12} is separable, so that $H^{12} = F(u^1, u^2, y, z)H_1(x)H_2(w)$, this equation gives rise to the same ordinary differential equations in (6.31) and can be solved depending on the sign of the constant, c, arising from separation of variables. As the analysis is similar to the (2, 4) Kähler examples, we will combine the results of the $c = 0$ and $c > 0$ cases. If $c \geq 0$ then

$$H^{12} = F(c_1 e^{\sqrt{c}x} + c_2 e^{-\sqrt{c}x})[d_1 e^w \sin(e^{-w}\sqrt{c}) + d_2 e^w \cos(e^{-w}\sqrt{c})], \tag{6.59}$$

while if $c < 0$ then

$$\begin{aligned} H^{12} =& F(c_1 \sin(\sqrt{|c|}x) + c_2 \cos(\sqrt{|c|}x))\cdot \\ & [d_1 e^w \sinh(e^{-w}\sqrt{|c|}) + d_2 e^w \cosh(e^{-w}\sqrt{|c|})], \end{aligned} \tag{6.60}$$

where $F = F(u^1, u^2, y, z)$ is an arbitrary function. The form of H^{22} is determined by integrating equation (6.56), yielding two solutions depending on the sign of c; if $c \geq 0$ then

$$\begin{aligned} H^{22} =& - F(c_1 e^{\sqrt{c}x} - c_2 e^{-\sqrt{c}x})\cdot \\ & [d_2 e^w \sin(e^{-w}\sqrt{c}) - d_1 e^w \cos(e^{-w}\sqrt{c})] + h^{22}(u^1, u^2, w), \end{aligned} \tag{6.61}$$

or if $c < 0$ then

$$\begin{aligned} H^{22} =& - F(c_1 \cos(\sqrt{|c|}x) - c_2 \sin(\sqrt{|c|}x))\cdot \\ & [d_2 e^w \sinh(e^{-w}\sqrt{|c|}) + d_1 e^w \cosh(e^{-w}\sqrt{|c|})] + h^{22}(u^1, u^2, w), \end{aligned} \tag{6.62}$$

The function h^{22} satisfies Eq. (6.57) which has the solution,

$$h^{22} = e^w h(u^1, u^2) + \frac{1}{2}(\partial_{u^1}\phi^1 + \partial_{u^2}\phi^2 + \psi). \tag{6.63}$$

Denoting the components of $\nabla\omega$ as $\omega_{ab;c}$, the remaining algebraically independent equations are $\omega_{51;1} = 0$ and $\omega_{51;3} = 0$. Assuming $c \geq 0$ and substituting the form of H^{12} and H^{22}, the resulting equations are large polynomial equations in terms of the functions dependent on the coordinates x and w. By combining the second derivatives of these equations with respect to x it is possible to show that

$$\sqrt{c}F(d_1^2 + d_2^2) = 0. \tag{6.64}$$

The most general solution occurs when $\sqrt{c} = 0$. The e^x-linear terms in $\omega_{51;1} = 0$ and $\omega_{51;3} = 0$ imply

$$\partial_y F = \partial_z F = 0, \tag{6.65}$$

while the remaining terms give

$$\partial_y \phi^i = \partial_z \phi^i = 0. \tag{6.66}$$

A similar approach can be employed in the case that $c < 0$ leading to large polynomial equations in terms of the functions dependent on the coordinates x and w which lead to the conditions $F = 0$ and $\partial_y\phi^i = \partial_z\phi^i = 0$ implying that H^{12} and H^{22} both vanish and this case may be ignored as it is contained in the $c = 0$ case.

Therefore the metric functions H^{12} and H^{22} are

$$H^{12} = Fc_1e^w, \quad H^{22} = c_0H^{12} + e^w h + \frac{1}{2}(\partial_{u^1}\phi^1 + \partial_{u^2}\phi^2 + \psi), \tag{6.67}$$

where $c_0 = (c_1 - c_2)d_1$ and the remaining arbitrary functions $\phi^1, \phi^2, \psi,\ F$ and h are dependent on u^1 and u^2 only. As in the $(2, 4)$ case, imposing the Einstein condition removes one free function:

$$\partial_{u^1}\phi^1 + \partial_{u^2}\phi^2 + \psi = 0. \tag{6.68}$$

Using the deformed coframe, we may compute the covariant derivative of the curvature tensor to determine when the deformed space is inequivalent to the original space. Requiring that that this tensor vanishes gives the conditions on F and h

$$F = 0 \text{ and } \partial^2_{u^1}h = \partial^2_{u^2}h. \tag{6.69}$$

Imposing these conditions, the deformed space and the original space will have identical sets of Cartan invariants, and so they must be equivalent. For any choice of metric functions F and h not satisfying (6.69), the deformed Kähler Einstein space will be inequivalent to the original space as the sets of Cartan invariants will differ.

□

6.4 Almost-Kähler Einstein Spaces

Taking the coframe defined in Eqs. (6.1- 6.6) with θ^i replaced with the one-form defined in Eq. (6.9) we will consider the signature $(2, 6)$ case and relax the Kähler condition. The almost-Kähler condition will still give Eqs. (6.52) and (6.53) requiring that $B^{i\alpha}$ and H^{11} takes the form given in Eq. (6.54).

The Einstein condition gives three differential equations for the remaining deformation functions:

$$\begin{aligned} 0 =& \partial_{u^1}\phi^1 + \partial_{u^2}\phi^2 + \psi, \\ 0 =& (\partial_x^2 H^{22}(y^2+z^2) + 4\partial_y^2 H^{22} + \partial_z^2 H^{22} - 4y\partial_x\partial_z H^{22} + 4z\partial_x\partial_y H^{22} - 8\partial_x H^{12})e^{-w} \\ &+ 4\partial_w^2 H^{22} + 8\partial_w H^{22} - 12H^{12} + 4\partial_x^2 H^{22}e^{-2w}, \\ 0 =& (\partial_x^2 H^{12}(y^2+z^2) + 4\partial_y^2 H^{12} + \partial_z^2 H^{12} - 4y\partial_x\partial_z H^{12} + 4z\partial_x\partial_y H^{12} + 8\partial_x H^{22})e^{-w} \\ &+ 4\partial_w^2 H^{12} + 8\partial_w H^{12} - 12H^{12} + 4\partial_x^2 H^{12}e^{-2w}. \end{aligned} \tag{6.70}$$

The first condition implies that H^{11} in Eq. (6.54) must be

$$H^{11} = -H^{22}. \tag{6.71}$$

Setting $H^{11} = H^{22} = 0$, the only independent function is H^{12}. The Einstein requirement implies $\partial_x H^{12} = 0$ and gives another differential equation

$$\partial_w^2 H^{12} + 2\partial_w H^{12} - 3H^{12} - e^{-w}\left(\partial_y^2 + \partial_z^2\right) H^{12} = 0. \tag{6.72}$$

Alternatively, setting $H^{22} = H^{12}$, the two equations can be combined to imply that $\partial_x H^{12} = 0$ and the same differential equation as in the $H^{11} = H^{22} = 0$ case:

$$\partial_w^2 H^{12} + 2\partial_w H^{12} - 3H^{12} - e^{-w}\left(\partial_y^2 + \partial_z^2\right) H^{12} = 0. \tag{6.73}$$

Since this equation is linear, a linear combination of solutions is also a solution.

As an illustration, we will construct a simple class of solutions. Let $f(y, z, u^1, u^2)$ be a harmonic function for the y and z coordinates, so that $\left(\partial_y^2 + \partial_z^2\right) f = \lambda f$ (λ may depend on u^i). Then

$$H^{12} = e^w C_1(u^1, u^2) + e^{-3w} C_2(u^1, u^2) - \frac{\lambda}{4} e^{-w} f(y, z, u^1, u^2).$$

Other solutions can be found by standard methods. Since the covariant derivative of the curvature tensor is non-zero for non-trivial H^{12}, we conclude these metrics are distinct from the original space.

7 Conclusions

In this paper, we have considered metric deformations of pseudo-Riemannian symmetric Einstein spaces that leave the set of scalar polynomial curvature invariants $\mathcal{I}$ unchanged. Such deformations are only applicable to the class of $\mathcal{I}$-degenerate pseudo-Riemannian spaces which cannot be characterized by $\mathcal{I}$ [2]. For any $\mathcal{I}$-degenerate space with desirable geometric properties, new pseudo-Riemannian spaces can be generated using such metric deformations which will inherit the relevant geometric properties, such as the Einstein condition. To illustrate this approach we have considered deformations of the metrics of pseudo-Riemannian symmetric Einstein spaces resulting from a Wick-rotation of the Riemannian metric for complex hyperbolic spaces.

The complex hyperbolic spaces are Kähler and so there exists, at each point, a complex structure in terms of a parallel skew-symmetric endomorphism $K : T_pM \to T_pM$ with $K^2 = -Id$ such that

$$\mathbf{g}(KX, KY) = -\mathbf{g}(X, Y), \tag{7.1}$$

where X and Y are arbitrary vector fields at a point. Depending on the signature of the metric, the resulting Wick-rotated signature pseudo-Riemannian spaces can be Kähler if the signature is $(2p, 2p + 2q + 2)$ while for the neutral signature (n, n) the resulting spaces are para-Kähler manifolds. That is, the Wick-rotated parallel complex structure K' is para-complex, $K^2 = Id$, and now satisfies

$$\mathbf{g}'(K'X, K'Y) = \mathbf{g}'(X, Y). \tag{7.2}$$

Using a coframe approach we have considered deformations that preserve the Einstein condition and maintain the (para-)Kähler condition or, at least, the almost (para-)Kähler condition. In order to construct the associated (para-)Kähler form $\boldsymbol{\omega}$ for the deformed manifold, a choice of the deformation for the coframe is required. This approach has the advantage that the deformed coframe must coincide with the invariant coframe needed to construct the set of Cartan invariants, and hence allows for a straightforward procedure to determine if the deformed space is equivalent to the original space.

Explicit examples have been provided in dimensions 4, 6 and 8. We have shown that the four-dimensional para-Kähler Einstein neutral metrics are rigid under metric deformations for two choices of the deformed coframe and argued that the higher dimensional para-Kähler Einstein neutral metrics will be rigid for a particular choice of deformed coframe. We have provided a straightforward approach to construct new almost-para-Kähler Einstein neutral metrics. In a similar manner, for a particular choice of deformed coframe, we have presented new Kähler Einstein metrics and almost-Kähler Einstein metrics with signature $(2p, 2p + 2q + 2)$ in dimension 6 an 8.

It is of interest to apply metric deformations to other symmetric $\mathcal{I}$-degenerate pseudo-Riemannian spaces to construct new spaces which inherit some of the properties of the undeformed spaces. For example, by deforming the class of pseudo-Riemannian Einstein spaces arising from Wick rotations of the metrics associated with the Damek-Ricci spaces [24] we can construct new Einstein spaces. With regards to (para-)Kähler pseudo-Riemannian spaces, the relationship between the choice of deformed coframe and the (para-)Kähler condition must be investigated in order to determine when a given space is rigid under metric deformations; this will be explored in future work.

Acknowledgments We would like to thank Paul de Medeiros for helpful comments at the start of this project. This work was supported through the Research Council of Norway, Toppforsk grant no. 250367: Pseudo- Riemannian Geometry and Polynomial Curvature Invariants: Classification, Characterisation and Applications.

References

1. A. Coley, S. Hervik, N. Pelavas, Lorentzian manifolds and scalar curvature invariants. Classical Quantum Gravity **27**(10), 102001 (2010). arXiv:1003.2373 [gr-qc]
2. S. Hervik, A. Haarr, K. Yamamoto, I-degenerate pseudo-Riemannian metrics. J. Geom. Phys. **98**, 384–399 (2015). arXiv:1410.4347 [gr-qc]
3. F. Prüfer, F. Tricerri, L. Vanhecke, Curvature invariants, differential operators and local homogeneity. Trans. Am. Math. Soc. **348**(11), 4643–4652 (1996)
4. H. Stephani, D Kramer, M. MacCallum, C. Hoenselaers, E. Herlt, *Exact Solutions of Einstein's Field Equations* (Cambridge University Press, Cambridge, 2009)
5. A. Coley, S. Hervik, G. Papadopoulos, N. Pelavas, Kundt spacetimes. Classical and Quantum Gravity **26**(10), 105016 (2009). arXiv:0901.0394 [gr-qc].
6. A.G. Walker, On parallel fields of partially null vector spaces. Q. J. Math. **1**, 135–145 (1949)
7. A.G. Walker, Canonical form for a Riemannian space with a parallel field of null planes. Q. J. Math. **1**(1), 69–79 (1950)
8. C. Helleland, S. Hervik, Wick rotations and real GIT. J. Geom. Phys. **123**, 343–361 (2018). arXiv:1703.04576 [math.DG]
9. C. Helleland, S. Hervik, Real GIT with applications to compatible representations and Wick-rotations. J. Geom. Phys. **142**, 92–110 (2019). arXiv:1807.05879 [math-ph]
10. W.M. Goldman, *Complex Hyperbolic Geometry* (Oxford University Press, Oxford, 1999)
11. G. Fubini, *Sulle metriche definite da una forma hermitiana: nota* (Office graf. C. Ferrari, 1904)
12. E. Study, Kürzeste wege im komplexen gebiet. Math. Ann. **60**, 321–378 (1905)
13. M. Ortaggio, V. Pravda, A. Pravdová, Algebraic classification of higher dimensional spacetimes based on null alignment. Classical Quantum Gravity **30**(1), 013001 (2012). arXiv:1211.7289 [gr-qc]
14. A. Coley, S. Hervik, N. Pelavas, On spacetimes with constant scalar invariants. Classical Quantum Gravity **23**(9), 3053 (2006). arXiv:0509113 [gr-qc]
15. A.A. Coley, S. Hervik, Universality and constant scalar curvature invariants. ISRN Geom. (2011). arXiv:1105.2356 [gr-qc]
16. S. Hervik, T. Málek, V. Pravda, A. Pravdová, Type II universal spacetimes. Classical Quantum Gravity **32**(24), 245012 (2015). arXiv:1503.08448 [gr-qc]
17. A. Coley, S. Hervik, N. Pelavas, Lorentzian spacetimes with constant curvature invariants in three dimensions. Classical Quantum Gravity **25**(2), 025008 (2008). arXiv:0710.3903 [gr-qc]
18. A. Coley, S. Hervik, N. Pelavas, Lorentzian spacetimes with constant curvature invariants in four dimensions. Classical Quantum Gravity **26**(12), 125011 (2009). arXiv:0904.4877 [gr-qc]
19. A. Coley, R. Milson, V. Pravda, A. Pravdová, Vanishing scalar invariant spacetimes in higher dimensions. Classical Quantum Gravity **21**, 5519–5542 (2004). arXiv:0410070 [gr-qc]
20. A. Coley, A. Fuster, S. Hervik, Supergravity solutions with constant scalar invariants. Int. J. Mod. Phys. A **24**, 1119–1133 (2009). arXiv:0707.0957 [gr-qc]
21. S. Hervik, D. McNutt, Locally homogeneous Kundt triples and CSI metrics. Classical Quantum Gravity **36**(18), 185013 (2019). arXiv:1812.11148 [gr-qc]
22. S. Hervik, A new class of infinitesimal group actions on pseudo-Riemannian manifolds (2018). preprint. arXiv:1805.09402 [gr-qc]
23. D.D. McNutt, M.T. Aadne, $\mathcal{I}$-preserving diffeomorphisms of Lorentzian manifolds. J. Math. Phys. **60**, 032501 (2019). arXiv:1901.04728 [gr-qc]
24. J. Berndt, F. Tricerri, L. Vanhecke, *Generalized Heisenberg Groups and Damek-Ricci Harmonic Spaces* (Springer, Berlin, 2006)
25. D.D. McNutt, A.A. Coley, A. Forget, The Cartan algorithm in five dimensions. J. Math. Phys. **58**(3), 032502 (2017). arXiv:1704.03061 [gr-qc]
26. P.J. Olver, *Equivalence, Invariants and Symmetry* (Cambridge University Press, Cambridge, 1995)

27. A. Coley, S. Hervik, D. McNutt, N. Musoke, D. Brooks, Neutral signature Walker-VSI metrics. Classical Quantum Gravity **31**(3), 035015 (2014)
28. D. Brooks, N. Musoke, D. McNutt, A. Coley, Mathematical properties of a class of four-dimensional neutral signature metrics. J. Geom. Phys. **97**, 1–13 (2015). arXiv:1401.2115 [math.DG]

Remarks on Highly Supersymmetric Backgrounds of 11-Dimensional Supergravity

Andrea Santi

Abstract This note focuses on some properties and uses of filtered deformations in the context of $D = 11$ supergravity. We define the concept of abstract symbol and give a strong version of the Reconstruction Theorem, namely a bijective correspondence from the space of highly supersymmetric supergravity backgrounds to the space of abstract symbols. We propose a general strategy to construct highly supersymmetric supergravity backgrounds and present an example in detail, which includes the computation of the ideal generated by the Killing spinors of two known pp-wave backgrounds with $N = 24$ supersymmetry. Finally, we give an alternative proof, based on the isotropy algebra of a supergravity background, of a classical supersymmetry gap result of Gran, Gutowski, Papadopoulos and Roest.

1 Introduction

An important line of research in string and M-theory is the construction of backgrounds of their low-energy effective counterparts, i.e., supergravity, since this is where strings and higher-dimensional branes can propagate. Backgrounds preserving maximal or near to maximal supersymmetry are important classes of such solutions.

In particular, the classical limit of M-theory is supergravity in $D = 11$ dimensions. From a geometric perspective, a background of $D = 11$ supergravity consists of an 11-dimensional Lorentzian spin manifold (M, g) with a closed 4-form $F \in \Omega^4(M)$, satisfying the following coupled system of PDEs:

$$\begin{aligned} d \star F &= \tfrac{1}{2} F \wedge F\ , \\ \mathrm{Ric}(X, Y) &= \tfrac{1}{2} g(\iota_X F, \iota_Y F) - \tfrac{1}{6} \|F\|^2 g(X, Y)\ , \end{aligned} \tag{1}$$

A. Santi (✉)
University of Padova, Mathematics Department "Tullio Levi-Civita", Padova, Italy
e-mail: asanti.math@gmail.com

S. Hervik et al. (eds.), *Geometry, Lie Theory and Applications*, Abel Symposia 16,
https://doi.org/10.1007/978-3-030-81296-6_12

for all $X, Y \in \mathfrak{X}(M)$. These are the bosonic field equations of $D = 11$ supergravity.

Let $S(M) \to M$ be the spinor bundle of (M, g), with typical fiber the irreducible module $S \cong \mathbb{R}^{32}$ of the Clifford algebra $C\ell(V) \cong 2\mathbb{R}(32)$. A spinor field $\varepsilon \in \Gamma(S(M))$ is called a Killing spinor if we have

$$\nabla_X \varepsilon - \tfrac{1}{24}(X \cdot F - 3F \cdot X) \cdot \varepsilon = 0 \tag{2}$$

for all $X \in \mathfrak{X}(M)$, where ∇ is the Levi-Civita connection and $\cdot$ Clifford multiplication. The connection on $S(M)$ defined by the L.H.S. of (2) is called the "superconnection" and the amount of spinors it preserves, i.e., Killing spinors, is an important invariant of a supergravity background.

Backgrounds with $16 < N \leq 32$ Killing spinors are called *highly supersymmetric* and their classification is widely open. Although this might be a distant goal, a number of different solutions have been proposed in the literature:

- Maximally supersymmetric backgrounds, i.e., with $N = 32$ Killing spinors,
- pp-waves backgrounds, i.e., Brinkmann spaces with flat transverse geometry,
- Gödel backgrounds, i.e., solutions admitting closed timelike curves.

Maximally supersymmetric backgrounds were classified in [7] up to local isometry (see [8] for a purely Lie theoretic proof in the spirit of this note): one has Minkowski spacetime, the Freund-Rubin backgrounds $Ad_4 \times S^7$ and $Ad_7 \times S^4$, and the Kowalski-Glikman symmetric pp-wave. A supersymmetry gap result is also known:

Theorem 1 ([13]) *If a background of $D = 11$ supergravity has at least 31 Killing spinors, then it is locally isometric to a maximally supersymmetric background.*

The proof relies on a careful analysis of the integrability conditions on the curvature of the superconnection that arise from Killing spinor equations, bosonic field equations, Bianchi identities of the Riemann curvature and $dF = 0$. The result was further extended to $N = 30$ in [14]. At present, the highest number of Killing spinors known for a non-maximally supersymmetric background is $N = 26$, for the pp-wave of [17]. Highly supersymmetric pp-waves are known with any even number $18 \leq N \leq 24$ of Killing spinors [10], whereas typically $N = 20$ for Gödel backgrounds [15].

Few general structural results were known until recently:

Theorem 2 ([5, 6]) *Let (M, g, F) be a background of $D = 11$ supergravity. Then:*

1. *There exists an associated Lie superalgebra $\mathfrak{k} = \mathfrak{k}_{\bar{0}} \oplus \mathfrak{k}_{\bar{1}}$, where $\mathfrak{k}_{\bar{0}}$ is the space of Killing vectors preserving F and $\mathfrak{k}_{\bar{1}}$ the space of Killing spinors;*
2. *If (M, g, F) is highly supersymmetric then it is locally homogeneous.*

The Lie superalgebra $\mathfrak{k} = \mathfrak{k}_{\bar{0}} \oplus \mathfrak{k}_{\bar{1}}$ is called the Killing superalgebra of the background. Its odd part generates an ideal $[\mathfrak{k}_{\bar{1}}, \mathfrak{k}_{\bar{1}}] \oplus \mathfrak{k}_{\bar{1}}$, which in this paper call the *transvection superalgebra*, following a similar terminology used in the context of symmetric spaces. Clearly $N = \dim \mathfrak{k}_{\bar{1}}$ and $\dim \mathfrak{k}_{\bar{1}} > 16$ precisely in the highly supersymmetric regime. These results paved the way to use Lie algebraic techniques, but unfortunately many homogeneous supergravity backgrounds are not highly supersymmetric.

Inspired by the superspace presentation of supergravity theories in terms of Tanaka structures [19, 20], the precise algebraic structure of Killing superalgebras has been derived and further structural results obtained:

Theorem 3 ([8, 9]) *Let (M, g, F) be an 11-dimensional Lorentzian spin manifold with a closed 4-form $F \in \Omega^4(M)$. If* $\dim \mathfrak{k}_{\bar{1}} > 16$, *then:*

1. *The bosonic field equations (1) of $D = 11$ supergravity are automatically satisfied;*
2. *The associated Killing superalgebra is (isomorphic to) a filtered subdeformation of the Poincaré superalgebra, the same holds for the transvection superalgebra;*
3. *The background is fully determined, up to local isometry, by the pair (φ, S'), where*

$$\varphi = F^\sharp|_o \in \Lambda^4 V \quad \text{and} \quad S' = \mathfrak{k}_{\bar{1}}|_o \subset S\,. \tag{3}$$

In particular if $\varphi = 0$ then (M, g) is locally isometric to Minkowski spacetime.

We call the pair *symb*$(M, g, F) = (\varphi, S')$ given in (3) the *geometric symbol* of (M, g, F).

This note aims to highlight some aspects and uses of the correspondence between highly supersymmetric supergravity backgrounds and transvection superalgebras.

We will present an alternative and relatively short proof of Theorem 1, which does not use the restrictions on the holonomy of the superconnection but it is based on the geometric symbol. More precisely, we do not require the Bianchi identities of the Riemann curvature or $dF = 0$ (equivalently, the bosonic field equations) but resort to one of the first properties satisfied by the isotropy of the transvection superalgebra of the background, and a partial knowledge of the space of bispinors.

The usage of filtered subdeformations to construct backgrounds has so far been hindered by the amount of algebraic data required to reconstruct the deformation from the symbol. We significantly simplify this data in Theorem 5, paving the way to the generation of new highly supersymmetric backgrounds with prescribed symmetries. The proposed method takes $\mathrm{SO}(V)$-orbits of four-vectors $\varphi \in \Lambda^4 V$ as starting point and their rank as a useful organizing principle. To date, up to my knowledge, there is no known highly supersymmetric background for which the rank is maximal, i.e., $\mathrm{rk}(\varphi) = 11$, and all known backgrounds with $\mathrm{rk}(\varphi) = 10$ have $N \leq 22$. On the other hand, those with $N \geq 24$ have rank relatively small: $\mathrm{rk}(\varphi) \leq 8$. Roughly speaking, we might say that the more Killing spinors there are, the more the underlying geometry ought to be rigid and the four-vector less intricate.

The paper has been organized as follows. We first review in Sects. 2 and 3 the basics of filtered subdeformations of the Poincaré superalgebra, including the Reconstruction Theorem [9] and the concepts of Dirac kernel and Lie pair also introduced in [9]. A Lie pair is a pair (φ, S') satisfying a system of coupled algebraic equations, quadratic on φ and cubic on S', which allows to recover the isotropy algebra of the transvection superalgebra. We then prove Theorem 1 using Dirac kernels and Lie pairs in Sect. 4.

In Sect. 5, we turn to study the reconstruction problem for supergravity backgrounds: we prove Theorem 5, introduce the concept of abstract symbol in Definition 3 and state the strong version of the Reconstruction Theorem in Theorem 6.

The last Sects. 6 and 7 give details of the strategy to construct supergravity backgrounds in an example, in particular we describe the transvection superalgebras of two symmetric pp-waves backgrounds with $N = 24$ Killing spinors discovered in [10].

2 Filtered Deformations and the Reconstruction Theorem

Let (V, η) be the Lorentzian vector space with "mostly minus" signature $(1, 10)$ and S the spinor representation of $\mathfrak{so}(V)$. We recall that S has an $\mathfrak{so}(V)$-invariant symplectic structure $\langle -, - \rangle$ such that $\langle v \cdot s_1, s_2 \rangle = - \langle s_1, v \cdot s_2 \rangle$ for all $s_1, s_2 \in S$ and $v \in V$.

The *Poincaré superalgebra* $\mathfrak{p}$ has underlying vector space $\mathfrak{so}(V) \oplus S \oplus V$ and nonzero Lie brackets

$$[A, B] = AB - BA\ , \qquad [A, s] = \sigma(A)s\ , \qquad [A, v] = Av\ , \qquad [s, s] = \kappa(s, s)\ ,$$

where $A, B \in \mathfrak{so}(V)$, $v \in V$ and $s \in S$. Here σ is the spinor representation of $\mathfrak{so}(V)$ and $\kappa : \odot^2 S \to V$ is the so-called *Dirac current*, defined by

$$\eta(\kappa(s, s), v) = \langle s, v \cdot s \rangle\ ,$$

for all $s \in S$ and $v \in V$. One important property of κ is that its restriction to $\odot^2 S'$ is surjective on V, for any subspace $S' \subset S$ with $\dim S' > 16$ [5]. This algebraic fact is usually referred to as the "local Homogeneity Theorem", due to the role it plays in the proof of Theorem 2. If we grade $\mathfrak{p}$ so that $\mathfrak{so}(V)$, S and V have degrees $0, -1$ and -2, respectively, then the above Lie brackets turn $\mathfrak{p}$ into a $\mathbb{Z}$-graded Lie superalgebra

$$\mathfrak{p} = \mathfrak{p}_0 \oplus \mathfrak{p}_{-1} \oplus \mathfrak{p}_{-2}, \qquad \mathfrak{p}_0 = \mathfrak{so}(V), \qquad \mathfrak{p}_{-1} = S, \qquad \mathfrak{p}_{-2} = V.$$

Note that the parity is consistent with the $\mathbb{Z}$-grading, since $\mathfrak{p}_{\bar{0}} = \mathfrak{p}_0 \oplus \mathfrak{p}_{-2}$ and $\mathfrak{p}_{\bar{1}} = \mathfrak{p}_{-1}$.

Let $\mathfrak{a} = \mathfrak{h} \oplus S' \oplus V$ be a graded subalgebra of $\mathfrak{p}$ with $\dim S' > 16$ and $\mathfrak{g}$ a *filtered deformation* of $\mathfrak{a}$, i.e., the Lie brackets of $\mathfrak{g}$ take the following general form [2]:

$$\begin{aligned} &[A, B] = AB - BA && [s, s] = \kappa(s, s) + \gamma(s, s) \\ &[A, s] = \sigma(A)s && [v, s] = \beta(v, s) \\ &[A, v] = Av + \delta(A, v) && [v, w] = \alpha(v, w) + \rho(v, w) \end{aligned} \tag{4}$$

where $A, B \in \mathfrak{h}$, $v, w \in V$ and $s \in S'$. Here $\alpha \in \mathrm{Hom}(\Lambda^2 V, V)$, $\beta \in \mathrm{Hom}(V \otimes S', S')$, $\gamma \in \mathrm{Hom}(\odot^2 S', \mathfrak{h})$ and $\delta \in \mathrm{Hom}(\mathfrak{h} \otimes V, \mathfrak{h})$ are maps of degree 2, and $\rho \in \mathrm{Hom}(\wedge^2 V, \mathfrak{h})$ of degree 4. If we do not mention the subalgebra $\mathfrak{a}$ explicitly, we simply say that $\mathfrak{g}$ is a *highly supersymmetric filtered subdeformation* of $\mathfrak{p}$. Finally, we say that $\mathfrak{g} = \mathfrak{g}_{\bar{0}} \oplus \mathfrak{g}_{\bar{1}}$ is *odd-generated* if $\mathfrak{g}_{\bar{0}} = [\mathfrak{g}_{\bar{1}}, \mathfrak{g}_{\bar{1}}]$.

The notion of isomorphism $\Phi : \mathfrak{g} \to \widetilde{\mathfrak{g}}$ of highly supersymmetric filtered subdeformations is given in [9, Def. 5]: it suffices to know that

$$\Phi(A) = g \cdot A, \qquad \Phi(s) = g \cdot s, \qquad \text{and} \qquad \Phi(v) = g \cdot v + X_v, \tag{5}$$

for some $g \in \mathrm{Spin}(V)$, $X : V \to \widetilde{\mathfrak{h}}$. The notion of embedding is given in [9, Def. 11].

As we now explain, we are interested in filtered deformations that are *realizable*, that is, corresponding to some highly supersymmetric supergravity background.

Associated to any $\varphi \in \Lambda^4 V$, there are two natural maps $\beta^\varphi : V \otimes S \to S$ and $\gamma^\varphi : \odot^2 S \to \mathfrak{so}(V)$, defined by

$$\beta^\varphi(v, s) = \tfrac{1}{24}(v \cdot \varphi - 3\varphi \cdot v) \cdot s\ , \tag{6}$$

$$\gamma^\varphi(s, s)v = -2\kappa(\beta^\varphi(v, s), s)\ , \tag{7}$$

for all $s \in S$, $v \in V$. We will sometimes use also the notation $\beta^\varphi_v(s) = \beta^\varphi(v, s)$. As proven in [8, Proposition 7], these maps are characterized by the "cocyle condition"

$$\sigma(\gamma^\varphi(s, s))s = -\beta^\varphi(\kappa(s, s), s)\ , \tag{8}$$

for all $s \in S$, and it turns out that the Spencer cohomology group $H^{2,2}(\mathfrak{p}_-, \mathfrak{p}) \cong \Lambda^4 V$ precisely via (6)–(7). (In general this cohomology group encodes the Killing spinor equations for supersymmetric field theories, see [3, 4, 8] for details.)

Now, it is well-known that $\odot^2 S \cong \Lambda^1 V \oplus \Lambda^2 V \oplus \Lambda^5 V$ as an $\mathfrak{so}(V)$-module. This decomposition is unique, since all the three summands are $\mathfrak{so}(V)$-irreducible and inequivalent, so we will consider $\Lambda^q V$ directly as a subspace of $\odot^2 S$, for $q = 1, 2, 5$. We decompose any element $\omega \in \odot^2 S$ into $\omega = -\frac{1}{32}(\omega^{(1)} + \omega^{(2)} + \omega^{(5)})$ accordingly, where $\omega^{(q)} \in \Lambda^q V$ for $q = 1, 2, 5$. The overall factor of $-\frac{1}{32}$ is introduced so that $\omega^{(1)}$ coincides exactly with the Dirac current of ω. We may then re-write (7) as:

$$\begin{aligned}\eta(\gamma^{\varphi}(\omega)v, w) &= \tfrac{1}{6}\langle s, \big(2\iota_v\iota_w\varphi - v\wedge w\wedge\varphi\big)\cdot s\rangle \\ &= \tfrac{1}{3}\eta(\iota_v\iota_w\varphi, \omega^{(2)}) + \tfrac{1}{6}\eta(\iota_v\iota_w \star\varphi, \omega^{(5)})\,,\end{aligned} \tag{9}$$

for all $v, w \in V$, where musical isomorphisms have been tacitly used. (This will also be the case throughout the whole paper, without further mention.) Evidently one has $\operatorname{Ker}\gamma^{\varphi} \supset \Lambda^1 V$.

Definition 1 ([9]) A highly supersymmetric filtered subdeformation $\mathfrak{g}$ of $\mathfrak{p}$ is *realizable* if there exists $\varphi \in \Lambda^4 V$ such that:

1. φ is $\mathfrak{h}$-invariant;
2. φ is closed, i.e.,

$$d\varphi(v_0,\ldots,v_4) = \sum_{i<j}(-1)^{i+j}\varphi(\alpha(v_i,v_j), v_0,\ldots,\hat{v}_i,\ldots,\hat{v}_j,\ldots,v_4) = 0 \tag{10}$$

   for all $v_0,\ldots,v_4 \in V$;
3. The components of the Lie brackets of $\mathfrak{g}$ of degree 2 are of the form

$$\begin{aligned} &\alpha(v,w) = X_v w - X_w v && \beta(v,s) = \beta^{\varphi}(v,s) + \sigma(X_v)s \\ &\gamma(s,s) = \gamma^{\varphi}(s,s) - X_{\kappa(s,s)} && \delta(A,v) = [A, X_v] - X_{Av}\end{aligned} \tag{11}$$

   for some linear map $X : V \to \mathfrak{so}(V)$, where $A, B \in \mathfrak{h}$, $v, w \in V$ and $s \in S'$.

It is a non-trivial fact that a realizable $\mathfrak{g}$ has a *unique* $\varphi \in \Lambda^4 V$ that satisfies (1)-(3) of Definition 1 and that the component ρ of degree 4 is fully determined by those of degree 2. (The idea is that φ is an element of the cohomology group $H^{2,2}(\mathfrak{a}_-, \mathfrak{a})$ and, since $H^{4,2}(\mathfrak{a}_-, \mathfrak{a}) = 0$, this group fully determines the subdeformation.) Changing the map $X : V \to \mathfrak{so}(V)$ by some values in $\mathfrak{h}$ gives isomorphic filtered subdeformations.

For simplicity of exposition, we will call *tamed* any filtered subdeformation of $\mathfrak{p}$ that is highly supersymmetric, odd-generated and realizable. We also let

$$\mathcal{SB} = \frac{\{\text{highly supersymmetric supergravity bkgds } (M, g, F)\}}{\text{local isometry}}$$

$$\mathcal{FD} = \frac{\{\text{maximal tamed filtered subdeformations } \mathfrak{g} \text{ of } \mathfrak{p}\}}{\text{isomorphism}}$$

be the moduli spaces of highly supersymmetric backgrounds and maximal tamed filtered subdeformations. The following result was proved under the assumption that the isotropy group is closed, but it can be easily relaxed (see, e.g., [18]).

Theorem 4 (Reconstruction Theorem [9])

1. *The assignment* $\mathcal{SB} \longrightarrow \mathcal{FD}$ *that sends a highly supersymmetric supergravity background to its transvection superalgebra is a* $1:1$ *correspondence;*
2. *The curvature* $R : \Lambda^2 V \to \mathfrak{so}(V)$ *of the supergravity background associated to a* $\mathfrak{g} \in \mathcal{FD}$ *is given by* $R(v, w) = \rho(v, w) - [X_v, X_w] + X_{\alpha(v,w)}$, *for all* $v, w \in V$.

3 Dirac Kernels and Lie Pairs

Let $\mathfrak{g}$ be a tamed filtered subdeformation of $\mathfrak{p}$, with associated graded Lie algebra $\mathfrak{a} = \mathfrak{h} \oplus S' \oplus V$ and $\varphi \in \Lambda^4 V$. Due to Theorem 4, we may call (φ, S') the symbol of $\mathfrak{g}$. Since S' has dimension $\dim S' > 16$, then $\odot^2 S' \subset \odot^2 S$ projects surjectively on $\Lambda^1 V$ via the Dirac current operation. The embedding $\odot^2 S' \subset \odot^2 S \cong \Lambda^1 V \oplus \Lambda^2 V \oplus \Lambda^5 V$ is in general diagonal and one cannot expect $\odot^2 S'$ to contain $\Lambda^q V$, not even if $q = 1$. Restricting the Dirac current to $\odot^2 S'$ gives rise to a short exact sequence

$$0 \longrightarrow \mathfrak{D} \longrightarrow \odot^2 S' \xrightarrow{\kappa} V \longrightarrow 0\,,$$

where

$$\mathfrak{D} = \odot^2 S' \cap (\Lambda^2 V \oplus \Lambda^5 V) = \left\{ \omega \in \odot^2 S' \,\middle|\, \omega^{(1)} = 0 \right\} \tag{12}$$

is called the *Dirac kernel* of S'. The Dirac kernel plays a crucial role in our arguments.

A splitting of the above short exact sequence—that is, a linear map $\Sigma : V \to \odot^2 S'$ such that $\Sigma(v)^{(1)} = v$ for all $v \in V$—is called a *section* associated to S'. A section associated to S' always exists and it is unique up to elements in the Dirac kernel.

Definition 2 Let $\mathfrak{h}_{(\varphi,S')}$ be the subspace of $\mathfrak{so}(V)$ given by

$$\mathfrak{h}_{(\varphi,S')} = \gamma^{\varphi}(\mathfrak{D}) = \left\{ \gamma^{\varphi}(\omega) \,\middle|\, \omega \in \odot^2 S' \text{ with } \omega^{(1)} = 0 \right\}. \tag{13}$$

The pair (φ, S') is called a *Lie pair* if

$$\begin{aligned} A \cdot \varphi &= 0 \\ \sigma(A) S' &\subset S' \end{aligned} \tag{14}$$

for every $A \in \mathfrak{h}_{(\varphi,S')}$.

The name "Lie pair" is motivated by the fact that the associated (13) is, in that case, a Lie subalgebra of $\mathfrak{so}(V)$. The following result gives necessary conditions satisfied by any tamed subdeformation.

Proposition 1 ([9]) *Let $\mathfrak{g}$ be a tamed filtered subdeformation, with underlying graded Lie algebra $\mathfrak{a} = \mathfrak{h} \oplus S' \oplus V$. Then the symbol (φ, S') is a Lie pair and:*

1. *the isotropy algebra $\mathfrak{h} = \mathfrak{h}_{(\varphi,S')}$;*
2. *the map $X : V \to \mathfrak{so}(V)$ is determined, up to elements in $\mathfrak{h}$, by the identity*

$$X = \gamma^{\varphi} \circ \Sigma\,, \tag{15}$$

where Σ is any section associated to S'.

In particular $\mathfrak{g}$ is fully determined, up to isomorphism, by the associated symbol.

Corollary 1 *Any highly supersymmetric supergravity background (M, g, F) is fully determined by its geometric symbol symb(M, g, F), up to local isometry.*

4 An Alternative Proof of Theorem 1

Let (M, g, F) be a background with *exactly* 31 Killing spinors (locally), and (φ, S') the associated geometric symbol. Since S' has dimension 31, we may write $S' = (\mathbb{R}s)^{\perp}$ as the symplectic orthogonal of the line spanned by some spinor $s \in S$ and, as in the original proof of Theorem 1 in [13], there are two cases to examine.

Indeed, as shown, e.g., in [1], there are two orbits of Spin(V) on $\mathbb{P}(S)$, distinguished by the causal character of the associated Dirac current $\xi = \kappa(s, s)$: either timelike or null. We give our proof of Theorem 1 using Dirac kernels only in the timelike case. Although the overall strategy in the null case is very close, the actual details are different and deserve a separate treatment. We leave them to the interested reader.

We first recall some known facts on realizations of the spinor module S [11, 12] (see also e.g. [9, Proposition 20], which follows the conventions used in this paper). We normalise $s \in S$ so that $\eta(\xi, \xi) = 1$ and consider the orthogonal decomposition $V = \mathbb{R}\xi \oplus W$. We set $\Omega^{(q)} = \omega^{(q)}(s, s) \in \Lambda^q V$ for $q = 1, 2, 5$; by definition $\Omega^{(1)} = \xi$. The endomorphism $J = \Omega^{(2)} \in \Lambda^2 V \cong \mathfrak{so}(V)$ acts trivially on ξ and as a complex structure on W. Clearly the stabilizer

$$\begin{aligned}\mathfrak{stab}_{\mathfrak{so}(V)}(s) &= \mathfrak{stab}_{\mathfrak{so}(V)}(\Omega^{(1)}) \cap \mathfrak{stab}_{\mathfrak{so}(V)}(\Omega^{(2)}) \cap \mathfrak{stab}_{\mathfrak{so}(V)}(\Omega^{(5)})\\ &= \mathfrak{u}(W, J) \cap \mathfrak{stab}_{\mathfrak{so}(V)}(\Omega^{(5)})\,,\end{aligned}$$

and it turns out that $\Omega = \Omega^{(5)} + \frac{1}{2}\Omega^{(1)} \wedge \Omega^{(2)} \wedge \Omega^{(2)}$ is the real component of a complex volume form $\Omega - i(\iota_\xi \star \Omega)$ on W. Ultimately $\mathfrak{stab}_{\mathfrak{so}(V)}(s) = \mathfrak{su}(W, J) \cong \mathfrak{su}(5)$.

For our purposes, it is sufficient to work at the complexified level. In particular $\mathbb{W} = W \otimes \mathbb{C}$ decomposes as the direct sum of isotropic subspaces $\mathbb{W} = W^{10} \oplus W^{01}$, where $W^{10} = \left\langle w^{10} = \frac{1}{2}(w - iJw) \;\middle|\; w \in W \right\rangle$ and $W^{01} = \overline{W^{10}}$ are irreducible complex $\mathfrak{su}(5)$-modules. The Dirac spinor module $\mathbb{S} = S \otimes \mathbb{C}$ is then given by

$$\mathbb{S} = \Lambda^{(\bullet,0)} = \bigoplus_{0 \leq p \leq 5} \Lambda^{(p,0)}_{[5-2p]} , \tag{16}$$

where $\Lambda^{(p,0)} = \Lambda^p (W^{10})^*$ is the irreducible $\mathfrak{su}(5)$-module of the $(p, 0)$-forms on $\mathbb{W}$ and the square brackets denote the charges of $\sigma(2J)$, with imaginary units removed.

As an aside, we recall that the modules $\Lambda^{(p,q)}$ of (p, q)-forms are $\mathfrak{su}(5)$-irreducible only for $p = 0$ or $q = 0$ and that for all $0 \leq p, q \leq 5$ we have natural isomorphisms

$$\Lambda^{(p,0)} \cong \Lambda^{(0,5-p)} \qquad \text{and} \qquad \Lambda^{(p,q)} \cong \Lambda^{(5-q,5-p)} .$$

For example $\Lambda^{(1,1)} \cong \Lambda^{(1,1)}_o \oplus \Lambda^{(0,0)}$ and $\Lambda^{(4,3)} \cong \Lambda^{(2,1)} \cong \Lambda^{(2,1)}_o \oplus \Lambda^{(1,0)}$ into irreducible $\mathfrak{su}(5)$-modules. For any $u \in W^{10}$ and $t \in \mathbb{S}$, the identities

$$\begin{aligned} u \cdot t &:= -\sqrt{2}\,\imath_u t , \\ \overline{u} \cdot t &:= \sqrt{2}\,\overline{u}^\flat \wedge t , \\ \xi \cdot t &:= \mathrm{vol}_W \cdot t , \end{aligned} \tag{17}$$

give an irreducible representation of the complex Clifford algebra $\mathbb{C}l(11) \cong 2\mathbb{C}(32)$, where $\mathrm{vol}_W \in \Lambda^{10} W$ is the volume element of W. The only trivial $\mathfrak{su}(5)$-submodules of $\mathbb{S}$ are $\Lambda^{(0,0)}_{[5]}$ and $\Lambda^{(5,0)}_{[-5]}$, more precisely we have

$$\Lambda^{(5,0)}_{[-5]} = \left\langle s^{(5,0)} = \tfrac{1}{2}(s + i\xi \cdot s) \right\rangle \quad \text{and} \quad \Lambda^{(0,0)}_{[5]} = \left\langle s^{(0,5)} = \tfrac{1}{2}(s - i\xi \cdot s) \right\rangle ,$$

whence the two elements $s = s^{(5,0)} + s^{(0,5)}$ and $\xi \cdot s = -i(s^{(5,0)} - s^{(0,5)})$ do *not* have a definite charge. The proof of the following lemma is straightforward.

Lemma 1 *The space of bispinors* $\odot^2\mathbb{S}$ *decomposes into* $\mathfrak{su}(5)$*-modules as*

$$\begin{aligned} \odot^2\mathbb{S} &\cong \Lambda^1\mathbb{V} \oplus \Lambda^2\mathbb{V} \oplus \Lambda^5\mathbb{V} \\ &\cong (\Lambda^1\mathbb{W} \oplus \Lambda^0\mathbb{W}) \oplus (\Lambda^2\mathbb{W} \oplus \Lambda^1\mathbb{W}) \oplus (\Lambda^5\mathbb{W} \oplus \Lambda^4\mathbb{W}) \end{aligned} \tag{18}$$

with

$$
\begin{aligned}
\Lambda^0\mathbb{W} &= \mathbb{C}\Omega^{(1)} \\
\Lambda^1\mathbb{W} &= \Lambda^{(1,0)}_{[-2]} \oplus \Lambda^{(0,1)}_{[2]} \\
\Lambda^2\mathbb{W} &= \Lambda^{(2,0)}_{[-4]} \oplus \Lambda^{(0,2)}_{[4]} \oplus \Lambda^{(1,1)}_{o[0]} \oplus \mathbb{C}\Omega^{(2)} \\
\Lambda^4\mathbb{W} &= \Lambda^{(4,0)}_{[-8]} \oplus \Lambda^{(0,4)}_{[8]} \oplus \Lambda^{(3,1)}_{o[-4]} \oplus \Lambda^{(1,3)}_{o[4]} \oplus \Lambda^{(2,0)}_{[-4]} \oplus \Lambda^{(0,2)}_{[4]} \\
&\quad \oplus \Lambda^{(2,2)}_{o[0]} \oplus \Lambda^{(1,1)}_{o[0]} \oplus \mathbb{C}(\Omega^{(1)} \wedge \Omega^{(2)} \wedge \Omega^{(2)}) \\
\Lambda^5\mathbb{W} &= \mathbb{C}\Omega \oplus \mathbb{C}(\imath_\xi \star \Omega) \oplus \Lambda^{(4,1)}_{o[-6]} \oplus \Lambda^{(1,4)}_{o[6]} \oplus \Lambda^{(3,0)}_{[-6]} \oplus \Lambda^{(0,3)}_{[6]} \\
&\quad \oplus \Lambda^{(3,2)}_{o[-2]} \oplus \Lambda^{(2,3)}_{o[2]} \oplus \Lambda^{(2,1)}_{o[-2]} \oplus \Lambda^{(1,2)}_{o[2]} \oplus \Lambda^{(1,0)}_{[-2]} \oplus \Lambda^{(0,1)}_{[2]}
\end{aligned}
\tag{19}
$$

into $\mathfrak{su}(5)$*-irreducibles. The modules not isomorphic to* $\Lambda^{(p,0)}$ *for some* $0 \leq p \leq 5$ *are colored red. The trivial* $\mathfrak{su}(5)$*-modules generated by* $\Omega^{(1)}$, $\Omega^{(2)}$ *and* $\Omega^{(1)} \wedge \Omega^{(2)} \wedge \Omega^{(2)}$ *have charge zero, whereas* Ω *together with* $\imath_\xi \star \Omega$ *generate the sum* $\Lambda^{(5,0)}_{[-10]} \oplus \Lambda^{(0,5)}_{[10]}$ *of the trivial* $\mathfrak{su}(5)$*-modules with charge* ± 10.

In particular, the isotypic decomposition of $\odot^2\mathbb{S}$ *is*

$$
\begin{aligned}
\odot^2\mathbb{S} \cong\ & 5\Lambda^{(0,0)} \oplus 4\left(\Lambda^{(1,0)} \oplus \Lambda^{(0,1)}\right) \oplus 3\left(\Lambda^{(2,0)} \oplus \Lambda^{(0,2)}\right) \oplus 2\Lambda^{(1,1)}_o \oplus \Lambda^{(2,2)}_o \\
& \oplus \left(\Lambda^{(3,1)}_o \oplus \Lambda^{(1,3)}_o\right) \oplus \left(\Lambda^{(4,1)}_o \oplus \Lambda^{(1,4)}_o\right) \oplus \left(\Lambda^{(3,2)}_o \oplus \Lambda^{(2,3)}_o\right) \oplus \left(\Lambda^{(2,1)}_o \oplus \Lambda^{(1,2)}_o\right),
\end{aligned}
$$

where we have grouped conjugated modules with brackets.

Note that $\mathbb{S} = \langle s, \xi \cdot s \rangle \oplus \widetilde{\mathbb{S}}$ decomposes as the direct sum of orthogonal $\mathfrak{su}(5)$-modules

$$
\langle s, \xi \cdot s \rangle = \Lambda^{(5,0)}_{[-5]} \oplus \Lambda^{(0,0)}_{[5]} \quad \text{and} \quad \widetilde{\mathbb{S}} = \Lambda^{(4,0)}_{[-3]} \oplus \Lambda^{(3,0)}_{[-1]} \oplus \Lambda^{(2,0)}_{[1]} \oplus \Lambda^{(1,0)}_{[3]} . \tag{20}
$$

We would like to estimate the size of the Dirac kernel of $\mathbb{S}' = (\mathbb{C}s)^\perp = \mathbb{C}s \oplus \widetilde{\mathbb{S}}$:

Proposition 2 *The Dirac kernel* $\mathfrak{D}$ *of* $\mathbb{S}'$ *includes at least the following elements:*

1. *Any* $\omega \in \Lambda^{(2,0)} \oplus \Lambda^{(0,2)} \subset \Lambda^2\mathbb{W}$*, modulo* $\Lambda^4\mathbb{W} \subset \Lambda^5\mathbb{V}$*;*
2. *Any* $\omega \in \Lambda^{(4,1)} \oplus \Lambda^{(1,4)} \subset \Lambda^5\mathbb{W}$*, modulo* $\Lambda^2\mathbb{W} \oplus \Lambda^4\mathbb{W} \subset \Lambda^2\mathbb{V} \oplus \Lambda^5\mathbb{V}$*;*
3. *Any* $\omega \in \Lambda^{(3,2)} \oplus \Lambda^{(2,3)} \subset \Lambda^5\mathbb{W}$*, modulo* $\Lambda^4\mathbb{W} \subset \Lambda^5\mathbb{V}$*;*

Proof First of all, we note that

$$
\begin{aligned}
\odot^2\mathbb{S} &\cong \odot^2\mathbb{S}' \oplus \left(\mathbb{S}' \odot (\xi \cdot s)\right) \oplus \mathbb{C}\left(\xi \cdot s \odot \xi \cdot s\right) \\
\odot^2\mathbb{S}' &\cong \odot^2\widetilde{\mathbb{S}} \oplus \left(\widetilde{\mathbb{S}} \odot s\right) \oplus \mathbb{C}\left(s \odot s\right)
\end{aligned}
$$

into $\mathfrak{su}(5)$-modules, and a routine computation using (20) gives

$$\begin{aligned}\odot^2\widetilde{\mathbb{S}} \cong\ & 2\Lambda^{(0,0)}_{[0]} \oplus 2\left(\Lambda^{(1,0)}_{[-2]} \oplus \Lambda^{(0,1)}_{[2]}\right) \oplus \left(\Lambda^{(2,0)}_{[-4]} \oplus \Lambda^{(0,2)}_{[4]}\right) \oplus 2\Lambda^{(1,1)}_{o[0]} \oplus \Lambda^{(2,2)}_{o[0]} \\ &\oplus \left(\Lambda^{(3,1)}_{o[-4]} \oplus \Lambda^{(1,3)}_{o[4]}\right) \oplus \left(\Lambda^{(4,1)}_{o[-6]} \oplus \Lambda^{(1,4)}_{o[6]}\right) \oplus \left(\Lambda^{(3,2)}_{o[-2]} \oplus \Lambda^{(2,3)}_{o[2]}\right) \\ &\oplus \left(\Lambda^{(2,1)}_{o[-2]} \oplus \Lambda^{(1,2)}_{o[2]}\right)\end{aligned} \tag{21}$$

and $\widetilde{\mathbb{S}} \odot s \cong \widetilde{\mathbb{S}} \cong \left(\Lambda^{(1,0)} \oplus \Lambda^{(0,1)}\right) \oplus \left(\Lambda^{(2,0)} \oplus \Lambda^{(0,2)}\right)$ into $\mathfrak{su}(5)$-irreducible modules. The latter ones do not have a definite charge, since $s = s^{(5,0)} + s^{(0,5)} \in \Lambda^{(5,0)}_{[-5]} \oplus \Lambda^{(0,0)}_{[5]}$.

Comparing (21) with the isotypic decomposition of $\odot^2\mathbb{S}$ says that all the $\mathfrak{su}(5)$-submodules colored in red in (19) are contained in $\odot^2\widetilde{\mathbb{S}}$. Clearly they also lie in $\mathfrak{D}$. It remains to deal with the isotypic components of type $\Lambda^{(p,0)}$, where $0 \leq p \leq 4$.

Claim 1 Note that the two copies of $\left(\Lambda^{(2,0)} \oplus \Lambda^{(0,2)}\right)$ with charge ± 4 are contained in $\Lambda^2\mathbb{W}$ and $\Lambda^4\mathbb{W}$. Taking an appropriate $t = t^{(4,0)} + t^{(3,0)} \in \Lambda^{(4,0)}_{[-3]} \oplus \Lambda^{(3,0)}_{[-1]} \subset \widetilde{\mathbb{S}}$ and $u_1, u_2 \in W^{10}$, we may use the first equation in (17) to get

$$\langle t, (u_1 \wedge u_2) \cdot t\rangle = 2\langle t^{(4,0)}, \underbrace{(u_1 \wedge u_2) \cdot t^{(3,0)}}_{\text{element of } \Lambda^{(1,0)}_{[3]}}\rangle \neq 0\ .$$

It follows that $\odot^2\widetilde{\mathbb{S}}$ projects surjectively to $\Lambda^{(2,0)} \subset \Lambda^2\mathbb{W}$ modulo elements in $\Lambda^4\mathbb{W}$, whence to $\Lambda^{(2,0)} \oplus \Lambda^{(0,2)}$ as well (because $\widetilde{\mathbb{S}}$ is a module of real type).

Claim 2 We already established that $\Lambda^{(4,1)}_{o[-6]} \oplus \Lambda^{(1,4)}_{o[6]} \subset \mathfrak{D}$. Now $\Lambda^{(3,0)}_{[-6]} \oplus \Lambda^{(0,3)}_{[6]}$ is the only module of $\odot^2\mathbb{S}$ isomorphic to $\Lambda^{(2,0)} \oplus \Lambda^{(0,2)}$ and with charge different from ± 4. It follows that $\widetilde{\mathbb{S}} \odot s$ projects surjectively to it, modulo elements in $\Lambda^2\mathbb{W} \oplus \Lambda^4\mathbb{W}$.

Claim 3 We already established that the submodules $\Lambda^{(3,2)}_{o[-2]} \oplus \Lambda^{(2,3)}_{o[2]} \oplus \Lambda^{(2,1)}_{o[-2]} \oplus \Lambda^{(1,2)}_{o[2]}$ of $\Lambda^5\mathbb{W}$ belong to $\mathfrak{D}$, so it remains to consider $\Lambda^{(1,0)}_{[-2]} \oplus \Lambda^{(0,1)}_{[2]} \subset \Lambda^5\mathbb{W}$.

There are four modules isomorphic to $\Lambda^{(1,0)}$ in $\odot^2\mathbb{S}$: two copies of $\Lambda^{(1,0)}_{[-2]} \subset \odot^2\widetilde{\mathbb{S}}$ and two other two copies

$$s \odot \Lambda^{(1,0)}_{[3]} = \left\{ s^{(5,0)} \odot \alpha + s^{(0,5)} \odot \alpha \mid \alpha \in \Lambda^{(1,0)}_{[3]} \right\} \subset s \odot \widetilde{\mathbb{S}}\ ,$$

$$(\xi \cdot s) \odot \Lambda^{(1,0)}_{[3]} = \left\{ s^{(5,0)} \odot \alpha - s^{(0,5)} \odot \alpha \mid \alpha \in \Lambda^{(1,0)}_{[3]} \right\} \subset (\xi \cdot s) \odot \widetilde{\mathbb{S}}$$

that do not have a definite charge. It follows that $2\Lambda^{(1,0)}_{[-2]} \oplus \left(s \odot \Lambda^{(1,0)}_{[3]}\right) \subset \odot^2\mathbb{S}'$ projects surjectively to $3\Lambda^{(1,0)}_{[-2]} \subset \odot^2\mathbb{S}$, modulo elements of charge 8. A similar property holds for $3\Lambda^{(0,1)}_{[2]}$ and this readily implies our last claim. □

Let (φ, S') be the geometric symbol of a background with exactly 31 Killing spinors. By Proposition 1 the isotropy $\mathfrak{h} = \mathfrak{h}_{(\varphi,S')} = \gamma^\varphi(\mathfrak{D})$ and the symbol (φ, S') is a Lie pair, in particular $\mathfrak{h} \subset \mathfrak{stab}_{\mathfrak{so}(V)}(S') = \mathfrak{stab}_{\mathfrak{so}(V)}(\mathbb{R}s)^\perp = \mathfrak{stab}_{\mathfrak{so}(V)}(\mathbb{R}s)$. Since ξ is timelike, we actually have

$$\mathfrak{h} \subset \mathfrak{stab}_{\mathfrak{so}(V)}(s) = \mathfrak{su}(5) \subset \mathfrak{stab}_{\mathfrak{so}(V)}(\xi) \ .$$

The idea is very simple: we write $\varphi = \xi \wedge \phi + \Phi$, for some $\phi \in \Lambda^3 W$ and $\Phi \in \Lambda^4 W$ and use the definition (9) of the map γ^φ with $w = \xi$ and $\omega \in \mathfrak{D}$:

$$\begin{aligned} 0 = \eta(\gamma^\varphi(\omega)v, \xi) &= \tfrac{1}{3}\eta\big(\iota_v\iota_\xi\varphi, \omega^{(2)}\big) + \tfrac{1}{6}\eta\big(\iota_v\iota_\xi \star \varphi, \omega^{(5)}\big) \\ &= \tfrac{1}{3}\eta\big(\underbrace{\iota_v\phi}_{\text{element of } \Lambda^2 W}, \omega^{(2)}\big) + \tfrac{1}{6}\eta\big(\underbrace{\iota_v\bigstar\Phi}_{\text{element of } \Lambda^5 W}, \omega^{(5)}\big)\ , \end{aligned} \tag{22}$$

for all $v \in W$, with $\bigstar$ the Hodge star on W. We now exploit that $\mathfrak{D}$ is sufficiently big.

Taking ω as in (1) of Proposition 2, Eq. (22) becomes $\eta(\iota_v\phi, \omega^{(2)}) = 0$ for all $\omega^{(2)} \in \Lambda^{(2,0)} \oplus \Lambda^{(0,2)} \subset \Lambda^2\mathbb{W}$. Decomposing $\phi = \phi^{(3,0)} + \phi^{(2,1)} + \phi^{(1,2)} + \phi^{(0,3)}$ into types and taking $v \in W^{10}$ gives $\phi^{(3,0)} = \phi^{(1,2)} = 0$. Similarly $\phi^{(2,1)} = \phi^{(0,3)} = 0$, so $\phi = 0$. If ω is now as in (2) of Proposition 2, then $\eta(\iota_v\bigstar\Phi, \omega^{(5)}) = 0$ for all $\omega^{(5)} \in \Lambda^{(4,1)} \oplus \Lambda^{(1,4)}$. Decomposing $\bigstar\Phi \in \Lambda^6 W$ into types yields $(\bigstar\Phi)^{(p,q)} = 0$ except for $(\bigstar\Phi)^{(3,3)}$, but this is zero by (3) of Proposition 2 and a similar argument.

Hence $\varphi = 0$, so (M, g) is locally isometric to Minkowski spacetime, in particular it is maximally supersymmetric, a contradiction. The proof is completed.

5 The Reconstruction Problem for Supergravity Backgrounds

It is not true that every Lie pair has a corresponding filtered subdeformation, in other words, it is the geometric symbol of a background. Indeed, the Lie brackets of a tamed filtered subdeformation $\mathfrak{g}$ are given by

$$\begin{aligned} &[A, B] = AB - BA && [s, s] = \kappa(s, s) + \gamma^\varphi(s, s) - X_{\kappa(s,s)} \\ &[A, s] = \sigma(A)s && [v, s] = \beta^\varphi(v, s) + \sigma(X_v)s \\ &[A, v] = Av + [A, X_v] - X_{Av} && [v, w] = X_v w - X_w v + [X_v, X_w] - X_{X_v w - X_w v} + R(v, w) \end{aligned}$$

for all $A, B \in \mathfrak{h}$, $v, w \in V$, $s \in S'$. Here $\mathfrak{h} = \mathfrak{h}_{(\varphi,S')}$ and $X : V \to \mathfrak{so}(V)$ is as in (15). The rest of the data $R : \Lambda^2 V \to \mathfrak{so}(V)$ also depends on the Lie pair, as we now recall.

First of all, it bears reminding that the right-hand sides of the above Lie brackets take values in $\mathfrak{a} = \mathfrak{h} \oplus S' \oplus V$, but the individual terms may not. Explicitly,

$$\beta^{\varphi}(v, s) + \sigma(X_v)s \in S' \tag{23}$$

$$[A, X_v] - X_{Av} \in \mathfrak{h} \tag{24}$$

$$\gamma^{\varphi}(s, s) - X_{\kappa(s,s)} \in \mathfrak{h} \tag{25}$$

$$[X_v, X_w] - X_{X_v w - X_w v} + R(v, w) \in \mathfrak{h} \tag{26}$$

for all $A \in \mathfrak{h}$, $v, w \in V$, $s \in S'$. Moreover the Lie brackets are subject to the Jacobi identities. There are ten components of them, and five are simply satisfied by equivariance and the fact that $\mathfrak{h} \subset \mathfrak{stab}_{\mathfrak{so}(V)}(S') \cap \mathfrak{stab}_{\mathfrak{so}(V)}(\varphi)$ [9, §4.1]. The Jacobi identity with three odd elements is the cocycle condition (8).

The remaining four components of the Jacobi identities are more involved:

- The $[\mathfrak{h} V V]$ Jacobi identity is satisfied if and only if

$$R : \Lambda^2 V \to \mathfrak{so}(V) \quad \text{is } \mathfrak{h}\text{-equivariant}; \tag{27}$$

- The $[S'S'V]$ Jacobi is equivalent to

$$\begin{aligned} \tfrac{1}{2} R(v, \kappa(s, s))w &= \kappa((X_v\beta^{\varphi})(w, s), s) + \gamma^{\varphi}(\beta^{\varphi}_v(s), s)w \\ &= \kappa((X_v\beta^{\varphi})(w, s), s) - \kappa(\beta^{\varphi}_v(s), \beta^{\varphi}_w(s)) - \kappa(\beta^{\varphi}_w\beta^{\varphi}_v(s), s) \end{aligned} \tag{28}$$

for all $s \in S'$, $v \in V$ and $w \in V$;

- The $[S'VV]$ Jacobi identity expands to the following condition

$$R(v, w)s = (X_v\beta^{\varphi})(w, s) - (X_w\beta^{\varphi})(v, s) + [\beta^{\varphi}_v, \beta^{\varphi}_w](s), \tag{29}$$

for all $s \in S'$ and $v, w \in V$;

- Finally, the $[VVV]$ Jacobi identity expands to the algebraic and differential Bianchi identities

$$R(u, v)w + R(v, w)u + R(w, u)v = 0\,, \tag{30}$$

$$(X_u R)(v, w) + (X_v R)(w, u) + (X_w R)(u, v) = 0\,, \tag{31}$$

for all $u, v, w \in V$.

Using the local Homogeneity Theorem, it is not difficult to see that each of Eqs. (28)–(29) determines uniquely the curvature tensor in terms of the Lie pair.

In summary, if we aim to construct backgrounds via filtered subdeformations and Theorem 4, a long task awaits us: find a Lie pair (φ, S'), compute the isotropy $\mathfrak{h}$ and the map X via the Dirac kernel, check that Eqs. (23)–(31) are satisfied, for a putative curvature tensor to be determined, and don't forget also $d\varphi = 0$.

The following result drastically simplifies the situation:

Theorem 5 *Assume* (φ, S') *is a Lie pair satisfying* (23) *and there is* $R : \Lambda^2 V \to \mathfrak{so}(V)$ *so that* (28) *and* (29) *hold. Then the identities* (24)–(27) *and* (30)–(31) *automatically hold.*

Proof We split the proof in six steps, one for each identity.

Step I For all $A \in \mathfrak{h}$, $v \in V$, we have

$$[A, X_v] - X_{Av} = [A, \gamma^{\varphi}(\Sigma v)] - \gamma^{\varphi}(\Sigma A v) = \gamma^{\varphi}(A\Sigma v - \Sigma A v)$$

and $A\Sigma v - \Sigma A v \in \mathfrak{D}$, since $(A\Sigma v)^{(1)} = A(\Sigma v)^{(1)} = Av = (\Sigma A v)^{(1)}$ and $\Sigma : V \to \odot^2 S'$. Identity (24) follows then directly from $\mathfrak{h} = \mathfrak{h}_{(\varphi,S')} = \gamma^{\varphi}(\mathfrak{D})$.

Step II In a similar way $\gamma^{\varphi}(\omega) - X_{\omega^{(1)}} = \gamma^{\varphi}(\omega - \Sigma\omega^{(1)})$ is an element of $\mathfrak{h}$ for all $\omega \in \odot^2 S'$, proving (25).

Step III We now establish (26). Using (28) and the definition of γ^{φ} we compute

$$\begin{aligned}
R(v, \kappa(s, s))w &= 2\kappa\big((X_v\beta^{\varphi})(w, s), s\big) + 2\gamma^{\varphi}\big(\beta^{\varphi}_v(s), s\big)w \\
&= 2\kappa\big(\sigma(X_v)(\beta^{\varphi}(w, s)), s\big) + \gamma^{\varphi}(s, s)X_v w - 2\kappa\big(\beta^{\varphi}(w, \sigma(X_v)s), s\big) \\
&\quad + 2\gamma^{\varphi}\big(\beta^{\varphi}_v(s), s\big)w \\
&= -X_v\big(\gamma^{\varphi}(s, s)w\big) - 2\kappa\big(\beta^{\varphi}(w, s), \sigma(X_v)s\big) - 2\kappa\big(\beta^{\varphi}(w, \sigma(X_v)s), s\big) \\
&\quad + \gamma^{\varphi}(s, s)X_v w + 2\gamma^{\varphi}\big(\beta^{\varphi}_v(s), s\big)w \\
&= -X_v\big(\gamma^{\varphi}(s, s)w\big) + 2\gamma^{\varphi}\big(s, \sigma(X_v)s\big)w + \gamma^{\varphi}(s, s)X_v w \\
&\quad + 2\gamma^{\varphi}\big(\beta^{\varphi}_v(s), s\big)w \\
&= [\gamma^{\varphi}(s, s), X_v]w + 2\gamma^{\varphi}\big(s, \underbrace{(\beta^{\varphi}_v(s) + \sigma(X_v)s)}_{\text{element of } S' \text{ (by (23))}}\big)w
\end{aligned}$$

for all $s \in S'$, $v, w \in V$. On the other hand

$$[X_v, X_{\kappa(s,s)}] = [X_v, \gamma^{\varphi}(\Sigma\kappa(s, s))]$$

and

$$X_{X_{\kappa(s,s)}v} - X_{X_v\kappa(s,s)} = X_{\gamma^{\varphi}(\Sigma\kappa(s,s))v} - X_{X_v\kappa(s,s)} - X_{\gamma^{\varphi}(s,s)v} + X_{\gamma^{\varphi}(s,s)v}\,,$$

for all $s \in S'$ and $v \in V$.

We now sum up the three contributions to the identity (26) and regroup the various terms into the sum of

$$[\gamma^{\varphi}(s,s) - \gamma^{\varphi}(\Sigma\kappa(s,s)), X_v] + X_{\gamma^{\varphi}(\Sigma\kappa(s,s))v} - X_{\gamma^{\varphi}(s,s)v} \tag{32}$$

and

$$2\gamma^{\varphi}\big(s, \beta_v^{\varphi}(s) + \sigma(X_v)s\big) + X_{\gamma^{\varphi}(s,s)v} - X_{X_v\kappa(s,s)} \,, \tag{33}$$

where $\gamma^{\varphi}(s,s) - \gamma^{\varphi}(\Sigma\kappa(s,s)) \in \mathfrak{h} = \gamma^{\varphi}(\mathfrak{D})$. The term (32) belongs to $\mathfrak{h}$ due to (24), which we established in step I. Finally, the Dirac current

$$\kappa(s, \beta_v^{\varphi}(s) + \sigma(X_v)s) = -\tfrac{1}{2}\gamma^{\varphi}(s,s)v + \tfrac{1}{2}X_v(\kappa(s,s)) \,,$$

so the term (33) is also in $\mathfrak{h}$, thanks to the identity (25) established in step II.

Step IV We prove that R is $\mathfrak{h}$-equivariant. A direct computation using (28) yields

$$\tfrac{1}{2}(A\cdot R)(v,\kappa(s,s))w = \kappa\big((A\cdot X)_v\beta^{\varphi})(w,s),s\big) = \kappa\big(\beta^{(A\cdot X)_v\cdot\varphi}(w,s),s\big) = 0 \,,$$

for all $A \in \mathfrak{h}$, $s \in S'$ and $v, w \in V$. Here we used that $(A\cdot X)_v = [A, X_v] - X_{Av}$ is an element of $\mathfrak{h}$ by step I, hence it annihilates φ.

Step V It is sufficient to establish (30) with $w = \kappa(s,s)$ for all $s \in S'$. Now

$$\begin{aligned}\tfrac{1}{2}\big(R(v,\kappa(s,s))u + R(\kappa(s,s),u)v\big) &= \kappa((X_v\beta^{\varphi})(u,s),s) - \kappa((X_u\beta^{\varphi})(v,s),s)\\ &\quad - \kappa([\beta_u^{\varphi},\beta_v^{\varphi}](s),s)\end{aligned}$$

by Eq. (28) and this term is also equal to $\frac{1}{2}R(v,u)\kappa(s,s) = \kappa(R(v,u)s,s)$ by (29).

Step VI The last step is the most involved and needs a preliminary crucial observation: it is enough to establish (31) when one of the elements of V is of the form $\omega^{(1)}$ for some $\omega \in \odot^2 S'$ in the image of the section $\Sigma : V \to \odot^2 S'$. For simplicity of exposition, we denote $\omega = s \odot s$ with $s \in S'$, although it is really a sum of decomposable bispinors. Our assumption on ω reads then $s \odot s = \omega = \Sigma\omega^{(1)} = \Sigma(\kappa(s,s))$, whence

$$X_{\kappa(s,s)} = \gamma^{\varphi}\big(\Sigma(\kappa(s,s))\big) = \gamma^{\varphi}(s,s) \,. \tag{34}$$

We will crucially use this property in the proof.

Now $A\cdot\beta^{\varphi} = \beta^{A\cdot\varphi}$ for all $A \in \mathfrak{so}(V)$ by $\mathfrak{so}(V)$-equivariance. A direct computation using this fact and (28) yields

$$\begin{aligned}\tfrac{1}{2}(A\cdot R)(v,\kappa(s,s))w &= \kappa(\beta^{(A\cdot X_v\cdot\varphi)}(w,s),s) - \kappa(\beta^{(X_{Av}\cdot\varphi)}(w,s),s)\\ &\quad + \gamma^{A\cdot\varphi}(\beta_v^{\varphi}(s),s)w + \gamma^{\varphi}(\beta_v^{(A\cdot\varphi)}(s),s)w \,,\end{aligned} \tag{35}$$

for all $A \in \mathfrak{so}(V)$. Applying (35) with $A = X_u$ and skew-symmetrizing in u and v says that the contribution

$$\tfrac{1}{2}(X_u \cdot R)(v, \kappa(s,s))w - \tfrac{1}{2}(X_v \cdot R)(u, \kappa(s,s))w \tag{36}$$

to the differential Bianchi identity is given by

$$\begin{aligned}\kappa\big(\beta^{([X_u,X_v]\cdot\varphi)}(w,s),s\big) - \kappa\big(\beta^{(X_{X_u v - X_v u}\cdot\varphi)}(w,s),s\big) + \gamma^{\varphi}\big(\beta_v^{(X_u\cdot\varphi)}(s),s\big)w \\ - \gamma^{\varphi}\big(\beta_u^{(X_v\cdot\varphi)}(s),s\big)w + \gamma^{X_u\cdot\varphi}\big(\beta_v^{\varphi}(s),s\big)w - \gamma^{X_v\cdot\varphi}\big(\beta_u^{\varphi}(s),s\big)w\,.\end{aligned} \tag{37}$$

Using identity (26), established in step III, and the definition of a Lie pair, we directly see that the first two terms in (37) are equal to

$$-\kappa\big(\beta^{(R(u,v)\cdot\varphi)}(w,s),s\big) = \tfrac{1}{2}\gamma^{(R(u,v)\cdot\varphi)}(s,s)w\,. \tag{38}$$

We recall our crucial assumption (34) and turn to compute the last contribution to the differential Bianchi identity:

$$\begin{aligned}\tfrac{1}{2}(X_{\kappa(s,s)}\cdot R)(u,v) &= \tfrac{1}{2}(\gamma^{\varphi}(s,s)\cdot R)(u,v) \\ &= -\tfrac{1}{2}[R(u,v),\gamma^{\varphi}(s,s)] - \tfrac{1}{2}R\big(\gamma^{\varphi}(s,s)u,v\big) - \tfrac{1}{2}R\big(u,\gamma^{\varphi}(s,s)v\big) \\ &= -\tfrac{1}{2}\gamma^{(R(u,v)\cdot\varphi)}(s,s) - \gamma^{\varphi}\big(R(u,v)s,s\big) - \tfrac{1}{2}R\big(\gamma^{\varphi}(s,s)u,v\big) \\ &\quad - \tfrac{1}{2}R\big(u,\gamma^{\varphi}(s,s)v\big)\,,\end{aligned} \tag{39}$$

where we used that $A\cdot\gamma^{\varphi} = \gamma^{A\cdot\varphi}$ for all $A \in \mathfrak{so}(V)$. Thanks to (29) we may expand one of the terms in the next-to-last line:

$$\begin{aligned}-\gamma^{\varphi}\big(R(u,v)s,s\big) &= -\gamma^{\varphi}\big(\beta^{(X_u\cdot\varphi)}(v,s),s\big) + \gamma^{\varphi}\big(\beta^{(X_v\cdot\varphi)}(u,s),s\big) \\ &\quad - \gamma^{\varphi}\big([\beta_u^{\varphi},\beta_v^{\varphi}](s),s\big)\,.\end{aligned} \tag{40}$$

Let us collect what we obtained so far: summing up (37) and (39), and using (38) and (40), we are left with

$$\begin{aligned}\gamma^{X_u\cdot\varphi}\big(\beta_v^{\varphi}(s),s\big)w - \gamma^{X_v\cdot\varphi}\big(\beta_u^{\varphi}(s),s\big)w - \gamma^{\varphi}\big([\beta_u^{\varphi},\beta_v^{\varphi}](s),s\big) \\ - \tfrac{1}{2}R\big(\gamma^{\varphi}(s,s)u,v\big)w - \tfrac{1}{2}R\big(u,\gamma^{\varphi}(s,s)v\big)w\,.\end{aligned} \tag{41}$$

Using the definition of γ^{φ}, Eq. (28) and a direct computation, we see that the second line of (41) is equal to

$$R\big(\kappa(\beta_u^{\varphi}(s), s), v\big)w - R\big(\kappa(\beta_v^{\varphi}(s), s), u\big)w = -\gamma^{X_u\cdot\varphi}\big(\beta_v^{\varphi}(s), s\big)w + \gamma^{X_v\cdot\varphi}\big(\beta_u^{\varphi}(s), s\big)w$$
$$+ \gamma^{\varphi}\big([\beta_u^{\varphi}, \beta_v^{\varphi}](s), s\big) ,$$

so (41) vanishes and the proof is completed. □

Motivated by this result, we give the following definition.

Definition 3 Let (φ, S') be a Lie pair, that is, S' is a subspace of S with $\dim S' > 16$, $\varphi \in \Lambda^4 V$ and $\mathfrak{h} = \mathfrak{h}_{(\varphi,S')} \subset \mathfrak{stab}_{\mathfrak{so}(V)}(S') \cap \mathfrak{stab}_{\mathfrak{so}(V)}(\varphi)$ (see Definition 2 for details). Let $\Sigma : V \to \odot^2 S'$ be any section associated with S' and set $X = \gamma^{\varphi} \circ \Sigma : V \to \mathfrak{so}(V)$ and $\alpha(v, w) = X_v w - W_w v$ for all $v, w \in V$. Then (φ, S') is called an *abstract symbol* if

$$\beta^{\varphi}(v, s) + \sigma(X_v)s \in S' \tag{42}$$

$$d\varphi(v_0, \ldots, v_4) = \sum_{i<j}(-1)^{i+j}\varphi(\alpha(v_i, v_j), v_0, \ldots, \hat{v}_i, \ldots, \hat{v}_j, \ldots, v_4) = 0 \tag{43}$$

$$\tfrac{1}{2}R\big(v, \kappa(s, s)\big)w = \kappa\big((X_v\beta^{\varphi})(w, s), s\big) + \gamma^{\varphi}\big(\beta_v^{\varphi}(s), s\big)w \tag{44}$$

$$R(v, w)s = (X_v\beta^{\varphi})(w, s) - (X_w\beta^{\varphi})(v, s) + [\beta_v^{\varphi}, \beta_w^{\varphi}](s) \tag{45}$$

for some $R : \Lambda^2 V \to \mathfrak{so}(V)$ and all $v, w, v_0, \ldots, v_4 \in V$, $s \in S'$.

Combining Theorem 4, Corollary 1 and the discussion carried out in this section, we arrive at the following version of the Reconstruction Theorem. Therein

$$\mathcal{AS} = \frac{\{\text{abstract symbols } (\varphi, S')\}}{\mathrm{Spin}(V)}$$

is the moduli space of abstract symbols.

Theorem 6 (Reconstruction Theorem: Strong Version) *The map* $\mathcal{SB} \longrightarrow \mathcal{AS}$ *that sends a highly supersymmetric supergravity background to its geometric symbol is a* $1:1$ *correspondence, with image the space of abstract symbols.*

The construction of supergravity backgrounds breaks then into three steps:

1. Describe $\mathrm{Spin}(V)$-orbits of Lie pairs (φ, S') (and therefore the associated graded subalgebras $\mathfrak{a} = \mathfrak{h}_{(\varphi,S')} \oplus S' \oplus V$ of the Poincaré superalgebra);
2. Check (42) and (43) (roughly speaking, this gives the transvection superalgebra of the background at the level of infinitesimal deformation);
3. See if there exists $R : \Lambda^2 V \to \mathfrak{so}(V)$ satisfying (45) and then check (44) (at this stage we fully reconstructed the transvection superalgebra as a deformation).

Remark 1 I have some evidence that (44) is a consequence of the other identities, but I don't have a complete proof so far.

Remark 2 If $\varphi \in \Lambda^4 V$ appears in a Lie pair (φ, S') with $S' = S$ then φ is decomposable, cf. [8]. It would be desirable to have an a priori understanding of which $\varphi \in \Lambda^4 V$ appear in Lie pairs at all. The variety of such four-vectors is $\mathrm{SO}(V)$-stable and in most likelihood properly contained in $\Lambda^4 V$.

It is clear that constructing Lie pairs remains the most difficult step, since they are defined by a rather complicated system of coupled algebraic equations, quadratic on φ and cubic on S'. To a certain extent, this can be regarded as the algebraic counterpart of the bosonic field equations (1) for highly supersymmetric backgrounds. We here propose a "separation of variables" technique to settle it:

- For a given $\varphi \in \Lambda^4 V$, we consider the operator $\gamma^\varphi : \odot^2 S \to \mathfrak{so}(V)$ and determine the intersection

$$\widetilde{\mathfrak{h}} = \operatorname{Im} \gamma^\varphi \cap \mathfrak{stab}_{\mathfrak{so}(V)}(\varphi)$$

  of its *full* image with the stabilizer of φ. It is easy to see that this is a Lie subalgebra of $\mathfrak{so}(V)$. We also determine the $\widetilde{\mathfrak{h}}$-submodule $K = (\gamma^\varphi)^{-1}\big(\mathfrak{stab}_{\mathfrak{so}(V)}(\varphi)\big)$ of $\odot^2 S$;
- We choose a presentation of S adapted to $\widetilde{\mathfrak{h}}$ and look for $\widetilde{\mathfrak{h}}$-submodules S' of S. We identify those which satisfy the inclusion $\mathfrak{D} \subset K$ using $\widetilde{\mathfrak{h}}$-equivariance.

Note that the subspace S' only enters at the last stage and that the inclusion there is a quadratic relation on spinors, and not cubic. A pair (φ, S') so obtained is a Lie pair. Indeed

$$\mathfrak{h} = \gamma^\varphi(\mathfrak{D}) \subset \gamma^\varphi(K) \subset \mathfrak{stab}_{\mathfrak{so}(V)}(\varphi) \tag{46}$$

and $\mathfrak{h} \subset \mathfrak{stab}_{\mathfrak{so}(V)}(S')$, since S' is an $\widetilde{\mathfrak{h}}$-module by construction and $\mathfrak{h} \subset \widetilde{\mathfrak{h}}$ by (46).

In Sect. 7, we work out an example where $V = \mathbb{R}^{1,1} \oplus \mathbb{R}^9$ splits into the ortohognal direct sum of $\mathbb{R}^{1,1} = \{\boldsymbol{e}_+, \boldsymbol{e}_-\}$ and $\mathbb{R}^9 = \{\boldsymbol{e}_1, \ldots, \boldsymbol{e}_9\}$, and $\varphi = \boldsymbol{e}_+ \wedge \phi$ for some $\phi \in \Lambda^3\mathbb{R}^9$ of small rank. The relevant orbits are described in Sect. 6.

6 The $\mathrm{SO}_9(\mathbb{R})$-Orbits in $\Lambda^3\mathbb{R}^9$ of Subminimal Rank

In this section, we set $G = \mathrm{SL}_9(\mathbb{R})$ and let $G^\theta = \mathrm{SO}_9(\mathbb{R})$ be the special orthogonal subgroup, i.e., the fixed point set of the Cartan involution $\theta : G \to G$ of G. We are interested in the stratification under the action of G^θ of small orbits of G on $\Lambda^3\mathbb{R}^9$.

The *support* of a trivector $\phi \in \Lambda^3\mathbb{R}^9$ is the unique minimal subspace $\mathbb{E} \subset \mathbb{R}^9$ such that $\phi \in \Lambda^3\mathbb{E}$. Its dimension is called the *rank* of ϕ and it is a G-invariant, in particular it is one of the simplest G^θ-invariants, together with the trivector's norm.

The trivectors of minimal (non-zero) rank are decomposable and form a G-orbit, which is stratified by the level sets of the norm in a 1-parameter family of G^θ-orbits. The associated filtered subdeformations have been studied in [8]:

Proposition 3 *Let $\phi = \lambda \boldsymbol{e}_{123}$ for some $\lambda > 0$ and (M, g, F) a highly supersymmetric supergravity background with $symb(M, g, F) = (\varphi = \boldsymbol{e}_+ \wedge \phi, S')$ for some $S' \subset S$. Then $S' = S$ and (M, g, F) is locally isometric to the Kowalski-Glikman background.*

Proof The transvection superalgebra of the Kowalski-Glikman background is the filtered subdeformation with symbol (φ, S) [8]. By maximality, the symbol (φ, S') of the transvection superalgebra of our background has to coincide with (φ, S). □

The next step in the analysis of filtered deformations and supergravity backgrounds is the rank 5 orbit, due to the following well-known result:

Lemma 2 *An indecomposable $\phi \in \Lambda^3\mathbb{R}^9$ has rank at least 5. The rank 5 trivectors constitute a G-orbit, with the representative, e.g., $\phi = \boldsymbol{e}_{123} + \boldsymbol{e}_{145}$.*

This is also a subminimal orbit, in the sense that its Zariski-closure consists of the orbit itself, the minimal orbit of the non-zero decomposable trivectors and the zero trivector, see for instance [16, page 104]. We will therefore denote it by $\mathcal{O}^G_{submin}$.

Proposition 4 (G^θ-Orbits of Rank 5 Trivectors) *The subminimal orbit $\mathcal{O}^G_{submin}$ admits a stratification*

$$\mathcal{O}^G_{submin} = \bigcup_{0<\lambda\leq\mu} \mathcal{O}^{G^\theta}_{\phi(\lambda,\mu)} ,$$

where $\mathcal{O}^{G^\theta}_{\phi(\lambda,\mu)}$ is the G^θ-orbit of the trivector $\phi_{(\lambda,\mu)} = \lambda \boldsymbol{e}_{123} + \mu \boldsymbol{e}_{145}$. The Lie algebra of the stabiliser H^θ of $\phi_{(\lambda,\mu)}$ in G^θ is

$$\mathfrak{h}^\theta = \begin{cases} \langle \boldsymbol{e}_{23}, \boldsymbol{e}_{45}\rangle \oplus \mathfrak{so}(\mathbb{E}^\perp) \cong \mathfrak{u}(1) \oplus \mathfrak{u}(1) \oplus \mathfrak{so}(\mathbb{E}^\perp) & \text{if } \lambda < \mu, \\ \langle \boldsymbol{e}_{23}, \boldsymbol{e}_{45}, \boldsymbol{e}_{24} + \boldsymbol{e}_{35}, \boldsymbol{e}_{25} - \boldsymbol{e}_{34}\rangle \oplus \mathfrak{so}(\mathbb{E}^\perp) \cong \mathfrak{u}(2) \oplus \mathfrak{so}(\mathbb{E}^\perp) & \text{if } \lambda = \mu, \end{cases}$$

where $\mathbb{E} = \langle \boldsymbol{e}_1, \dots, \boldsymbol{e}_5\rangle$ is the support of $\phi_{(\lambda,\mu)}$ and $\mathbb{E}^\perp = \langle \boldsymbol{e}_6, \dots, \boldsymbol{e}_9\rangle$.

Proof Let $\phi \in \Lambda^3\mathbb{R}^9$ be a rank 5 trivector. Then $g \cdot \phi$ has support $\mathbb{E} = \langle e_1, \dots, e_5\rangle$ for some $g \in G^\theta$, and two trivectors with support $\mathbb{E}$ are in the same G^θ-orbit if and only if they are in the same $\mathrm{O}(\mathbb{E})$-orbit. Therefore, it is enough to describe the $\mathrm{O}(\mathbb{E})$-orbits of trivectors $\phi \in \Lambda^3\mathbb{E}$ of rank 5. (We will see that they coincide with $\mathrm{SO}(\mathbb{E})$-orbits.)

We fix a volume element $\mathrm{vol} \in \Lambda^5\mathbb{E}^*$ and consider the bijection

$$\Lambda^3\mathbb{E} \to \Lambda^2\mathbb{E}^* , \qquad \phi \mapsto \iota_\phi \mathrm{vol} ,$$

which is an isomorphism of SO($\mathbb{E}$)-modules. Now $\Lambda^2\mathbb{E}^* \cong \mathfrak{so}(\mathbb{E})$, so the SO($\mathbb{E}$)-orbits on $\Lambda^3\mathbb{E}$ are in bijective correspondence with the adjoint orbits. The group SO($\mathbb{E}$) is compact, hence any adjoint orbit has a representative in the Cartan subalgebra

$$\mathfrak{t} = \left\{ \left(\begin{array}{c|cc|cc} 0 & & & & \\ \hline & 0 & \mu & & \\ & -\mu & 0 & & \\ \hline & & & 0 & \lambda \\ & & & -\lambda & 0 \end{array} \right) \mid \lambda, \mu \in \mathbb{R} \right\}$$

of $\mathfrak{so}(\mathbb{E})$. Using the Weyl group of SO($\mathbb{E}$), we may uniquely arrange for $0 \leq \lambda \leq \mu$. The corresponding trivector is $\phi = \lambda \boldsymbol{e}_{123} + \mu \boldsymbol{e}_{145}$ and has rank 5 if and only if $\lambda \neq 0$.

The last claim follows from a direct computation, which we omit. □

Remark 3 A similar analysis for higher rank orbits is possible but it is more involved, as the connection with the adjoint orbits of a compact Lie group is not available in general. A possible strategy is outlined here:

1. Let $\mathcal{O}^G_\phi \cong G/H$ be the G-orbit with representative $\phi \in \Lambda^3\mathbb{R}^9$ of rank $6 \leq k \leq 9$, where H is the stabiliser of ϕ in G. We then have a stratification

$$\mathcal{O}^G_\phi = \bigcup_i \mathcal{O}^{G^\theta}_{\phi_i} \tag{47}$$

   under the action of G^θ, which is parametrised by the double cosets in $G^\theta\backslash G/H$;
2. The H-equivariant map $G^\theta g \mapsto g^t \cdot g$ identifies $G^\theta\backslash G$ with the space Sym_9^+ of positive-definite symmetric matrices with unit determinant. (Surjectivity follows from Sylvester's law of inertia, the right action of H on Sym_9^+ is by congruence.) Hence

$$G^\theta\backslash G/H \cong Sym_9^+/H \tag{48}$$

   parametrises the stratification (47);
3. Tipically we have a non-trivial Levi decomposition

$$H = L \ltimes U \tag{49}$$

   of H with reductive subgroup L and unipotent radical U. To compute Sym_9^+/H, we may first consider Sym_9^+/U and then the residual action of $L \cong H/U$ on it.

The orbits $\mathcal{O}^G_\phi$ of trivectors of rank at most 9 have been determined in [21] (upon complexification). The description of those of rank 9 is extremely involved but those of rank ≤ 8 are automatically nilpotent (in the sense of Vinberg's theory of θ-

groups) and relatively few: upon complexification, there are 13 orbits $\mathcal{O}_\phi^G$ of rank 8 and the orbits of rank ≤ 7 are given by:

Class	Representative ϕ	Rank	Reductive part $\mathfrak{l}$ of $\mathfrak{h}$	dim $\mathfrak{l}$	dim $\mathfrak{h}$
90	$e_{123} + e_{147} + e_{257} + e_{367} + e_{456}$	7	$G_2 \oplus A_1$	17	31
93	$e_{125} + e_{137} + e_{247} + e_{346}$	7	$3A_1 \oplus \mathbb{C}$	10	32
94	$e_{127} + e_{134} + e_{256}$	7	$3A_1 \oplus 2\mathbb{C}$	11	35
95	$e_{125} + e_{136} + e_{147} + e_{234}$	7	$A_2 \oplus A_1 \oplus \mathbb{C}$	12	38
96	$e_{123} + e_{456}$	6	$3A_2$	24	42
97	$e_{124} + e_{135} + e_{236}$	6	$2A_2 \oplus \mathbb{C}$	17	43
99	$e_{123} + e_{145} + e_{167}$	7	$C_3 \oplus A_1 \oplus \mathbb{C}$	25	45
100	$e_{123} + e_{145}$	5	$A_3 \oplus C_2 \oplus \mathbb{C}$	26	50
101	e_{123}	3	$A_5 \oplus A_2$	43	61

where a couple of misprints in [21] have been corrected. Some orbits (but not all) have been investigated in [10] in the context of pp-waves.

By previous steps, the stratification (47) of an orbit $\mathcal{O}_\phi^G$ is parametrised by unit volume scalar products up to H-equivalence. The study of the associated supergravity backgrounds will be considered in future work, we now work out our example. We will omit most of the actual details of this calculation.

7 Example

The proof of the following lemma is straightforward.

Lemma 3 *Let $\varphi \in \Lambda^4 V$ be of the form $\varphi = \boldsymbol{e}_+ \wedge \phi$ for some non-zero $\phi \in \Lambda^3\mathbb{R}^9$. Then*

$$\mathfrak{stab}_{\mathfrak{so}(V)}(\varphi) = \mathfrak{stab}_{\mathfrak{so}(\mathbb{R}^9)}(\phi) \ltimes (\boldsymbol{e}_+ \wedge \mathbb{R}^9) ,$$

in particular $\mathfrak{stab}_{\mathfrak{so}(V)}(\varphi) \subset \mathfrak{stab}_{\mathfrak{so}(V)}(\boldsymbol{e}_+)$.

We consider the case where ϕ has rank 5 with maximum stabilizer, i.e., $\phi = \boldsymbol{e}_{123} + \boldsymbol{e}_{145}$, up to positive multiples. It is well known that the field equations (1) are invariant under a homothety that rescales both the metric and the 4-form and that the associated transvection superalgebras are not isomorphic as filtered subdeformations. However, they are isomorphic if we simply allow for $g \in \mathrm{CSpin}(V)$ in Eq. (5), so we may indeed restrict to $\phi = \boldsymbol{e}_{123}+\boldsymbol{e}_{145}$. We recall that $\mathbb{R}^9 = \mathbb{E} \oplus \mathbb{E}^\perp$, where $\mathbb{E} = \langle \boldsymbol{e}_1, \ldots, \boldsymbol{e}_5\rangle$ is the support of ϕ and $\mathbb{E}^\perp = \langle \boldsymbol{e}_6, \ldots, \boldsymbol{e}_9\rangle$.

Now $\mathfrak{stab}_{\mathfrak{so}(V)}(\varphi) = \big(\mathfrak{u}(2) \oplus \mathfrak{so}(\mathbb{E}^\perp)\big) \ltimes (\boldsymbol{e}_+ \wedge \mathbb{R}^9)$ according to Proposition 4 and the image $\operatorname{Im}\gamma^\varphi$ is a 36-dimensional subspace of $\mathfrak{so}(V)$ satisfying

$$\widetilde{\mathfrak{h}} = \operatorname{Im}\gamma^\varphi \cap \mathfrak{stab}_{\mathfrak{so}(V)}(\varphi) = \big(\mathfrak{u}(1) \oplus \mathfrak{so}(\mathbb{E}^\perp)\big) \ltimes (\boldsymbol{e}_+ \wedge \mathbb{R}^9) ,$$

with $\mathfrak{u}(1) = \mathbb{R}(\boldsymbol{e}_{23} + \boldsymbol{e}_{45})$. The kernel $\operatorname{Ker} \gamma^{\varphi}$ is 492-dimensional.

We use the isomorphism of algebras $C\ell(V) \cong C\ell(\mathbb{R}^9) \otimes C\ell(\mathbb{R}^{1,1}) \cong 2\mathbb{R}(16) \otimes \mathbb{R}(2)$ (we recall that $\mathbb{R}^9$ is negative definite for us) to write

$$S \cong \mathbb{R}^{16} \otimes \mathbb{R}^2 \cong S_+ \oplus S_-$$

with $S_\pm \cong \mathbb{R}^{16}$ and define the Gamma matrices as block matrices with square blocks of order 16:

$$\Gamma_+ = \sqrt{2}\begin{pmatrix} 0 & \mathrm{Id} \\ 0 & 0 \end{pmatrix}, \quad \Gamma_- = -\sqrt{2}\begin{pmatrix} 0 & 0 \\ \mathrm{Id} & 0 \end{pmatrix}, \quad \Gamma_i = \begin{pmatrix} \gamma_i & 0 \\ 0 & -\gamma_i \end{pmatrix} \quad (i = 1, \ldots, 9).$$

The Gamma matrices $\gamma_i \in C\ell(\mathbb{R}^9)$ will be described via a quaternionic formalism.

First of all $C\ell(\mathbb{R}^5) \cong 2\mathbb{H}(2)$ and an isomorphism is given by the quaternionic matrices

$$A_1 = \begin{pmatrix} 1 & 0 \\ 0 & -1 \end{pmatrix}, \; A_2 = \begin{pmatrix} 0 & 1 \\ 1 & 0 \end{pmatrix}, \; A_3 = \begin{pmatrix} 0 & L_i \\ -L_i & 0 \end{pmatrix}, \; A_4 = \begin{pmatrix} 0 & L_j \\ -L_j & 0 \end{pmatrix}, \; A_5 = \begin{pmatrix} 0 & L_k \\ -L_k & 0 \end{pmatrix},$$

where $L_q : \mathbb{H} \to \mathbb{H}$ is left multiplication by $q \in \mathbb{H}$. We set $\mathbb{R}^{16} \cong \mathbb{H}^2 \otimes \mathbb{R}^2$ and define block matrices with square blocks of order 8 by

$$\gamma_i = \begin{pmatrix} A_i & 0 \\ 0 & -A_i \end{pmatrix}, \; \gamma_6 = \begin{pmatrix} 0 & \mathrm{Id} \\ \mathrm{Id} & 0 \end{pmatrix}, \; \gamma_7 = \begin{pmatrix} 0 & R_i \\ -R_i & 0 \end{pmatrix}, \; \gamma_8 = \begin{pmatrix} 0 & R_j \\ -R_j & 0 \end{pmatrix}, \; \gamma_9 = \begin{pmatrix} 0 & R_k \\ -R_k & 0 \end{pmatrix}$$

where $i = 1, \ldots, 5$. Here $R_q : \mathbb{H} \to \mathbb{H}$ is right multiplication and we used the same symbol for its natural action on $\mathbb{H}^2$.

The subspaces $S_\pm \cong \mathbb{H}^4$ are isotropic w.r.t. the canonical symplectic form, which dually pairs them via the standard inner product on $\mathbb{H}^4$. Thanks to this and the above Gamma matrices, one easily recovers the Dirac current; here we will just mention that $\kappa(S_\pm, S_\pm) = \mathbb{R}\boldsymbol{e}_\pm$ and $\kappa(S_+, S_-) = \mathbb{R}^9$.

Lemma 4 *The spinorial action of $\widetilde{\mathfrak{h}}$ is given by*

$$\boldsymbol{e}_+ \wedge \mathbb{R}^9 = \left\langle \begin{pmatrix} 0 & \gamma_i \\ 0 & 0 \end{pmatrix} \mid i = 1, \ldots, 9 \right\rangle, \tag{50}$$

$$\mathfrak{so}(\mathbb{E}^\perp) = \left\langle \begin{pmatrix} \gamma_i\gamma_j & 0 \\ 0 & \gamma_i\gamma_j \end{pmatrix} \mid 6 \le i < j \le 9 \right\rangle, \tag{51}$$

$$\mathfrak{u}(1) = \left\langle \begin{pmatrix} \gamma_2\gamma_3 + \gamma_4\gamma_5 & 0 \\ 0 & \gamma_2\gamma_3 + \gamma_4\gamma_5 \end{pmatrix} \right\rangle, \tag{52}$$

where $\gamma_2\gamma_3+\gamma_4\gamma_5 = -2\left(\begin{array}{c|c} \begin{array}{c|c} L_i & 0 \\ \hline 0 & 0 \end{array} & 0 \\ \hline 0 & \begin{array}{c|c} L_i & 0 \\ \hline 0 & 0 \end{array} \end{array}\right)$ *as a quaternionic matrix of* $\mathbb{H}^4$. *In particular the subspaces of* S *given by*

$$S_1' = S_+ \oplus \mathbb{H} \oplus (0) \oplus \mathbb{H} \oplus (0) \tag{53}$$

$$S_2' = S_+ \oplus (0) \oplus \mathbb{H} \oplus (0) \oplus \mathbb{H} \tag{54}$$

are $\widetilde{\mathfrak{h}}$*-stable.*

Now $K = (\gamma^\varphi)^{-1}\big(\mathfrak{stab}_{\mathfrak{so}(V)}(\varphi)\big)$ clearly contains $\operatorname{Ker}\gamma^\varphi$ and since $\Lambda^1 V \subset \operatorname{Ker}\gamma^\varphi$ we may write $K = \operatorname{Ker}\gamma^\varphi \oplus C$ for some $C \subset \Lambda^2 V \oplus \Lambda^5 V$. Clearly $\dim C = \dim\widetilde{\mathfrak{h}} = 16$ and it turns out that

$$e_{23}+e_{45}\,,\quad 2e_{13}+e_{36789}\,,\quad -2e_{12}-e_{26789}\,,\quad 2e_{15}+e_{56789}\,,\quad -2e_{14}-e_{46789}\,,$$

$$e_{45789}+e_{23789}\,,\quad e_{45689}+e_{23689}\,,\quad e_{45679}+e_{23679}\,,\quad e_{45678}+e_{23678}\quad \text{and}$$

$$e_-\wedge\big\{-2e_1+e_{6789},\ e_{4589}+e_{2389},\ e_{4579}+e_{2379},\ e_{4578}+e_{2378},\ e_{4569}+e_{2369}$$

$$e_{4568}+e_{2368},\ e_{4567}+e_{2367}\big\}$$

are generators of C. Using this decomposition one may check that $\odot^2 S' \subset K$ in both cases—in particular $\mathfrak{D} \subset K$—and that S' cannot be enlarged (otherwise one constructs $s\odot t \in \odot^2 S'$ with vanishing Dirac current and such that $\gamma^\varphi(s,t) \notin \widetilde{\mathfrak{h}}$).

Corollary 2 *The pairs* (φ, S_1') *and* (φ, S_2') *are Lie pairs, and they are maximal.*

The Dirac kernel $\mathfrak{D}$ has dimension $\dim\mathfrak{D} = \dim\odot^2 S' - \dim V = 289$ in both cases but the two spaces are different; indeed we have

$$\mathfrak{h} = \gamma^\varphi(\mathfrak{D}) = \begin{cases} \widetilde{\mathfrak{h}} = \big(\mathfrak{u}(1)\oplus\mathfrak{so}(\mathbb{E}^\perp)\big) \ltimes (e_+\wedge\mathbb{R}^9) & \text{if } S' = S_1', \\ \mathfrak{u}(1) \ltimes (e_+\wedge\mathbb{E}^\top) & \text{if } S' = S_2', \end{cases} \tag{55}$$

where $\mathbb{E}^\top = \langle e_2, e_3, e_4, e_5\rangle \subset \mathbb{E}$. In the first case

$$X(v) = \gamma^\varphi(\Sigma v) \subset \gamma^\varphi(\odot^2 S') \subset \widetilde{\mathfrak{h}} = \mathfrak{h}$$

for any section $\Sigma : V \to \odot^2 S'$, so we may arrange $X = 0$ without loss of generality. It turns out that $\gamma^\varphi(\odot^2 S') \subset \mathfrak{h}$ in the second case too, so $X = 0$. In particular $d\varphi = 0$.

Using Gamma matrices, it is straightforward to see

$$\beta_{e_+} = 0\,,\qquad \beta_{e_i} = \begin{pmatrix} 0 & * \\ 0 & 0 \end{pmatrix}\quad (i = 1,\ldots,9)\,,$$

as endomorphisms of $S = S_+ \oplus S_-$, and that β_{e_-} preserves the decomposition of S into eight copies of $\mathbb{H}$. It follows that $\beta_v(S') \subset S'$ for all $v \in V$ in both cases.

We are left with the identities involving the curvature tensor $R : \Lambda^2 V \to \mathfrak{so}(V)$. To solve them, we decompose

$$V = \mathbb{R}\boldsymbol{e}_+ \oplus \mathbb{R}\boldsymbol{e}_- \oplus \mathbb{R}\boldsymbol{e}_1 \oplus \mathbb{E}^\top \oplus \mathbb{E}^\perp$$

and accordingly write $v = v_+ + v_- + v_1 + v_\top + v_\perp$ for any $v \in V$. The general strategy to solve (45) is analogous to [8, §4.3], although a tad more involved: we give the final results and briefly comment on how we derived them. In the first and the second case, respectively, we get:

$$\begin{aligned} R(v, w) &= -\eta(\boldsymbol{e}_+, v_-)\left(\tfrac{4}{9} w_1 \wedge \boldsymbol{e}_+ + \tfrac{1}{36} w_\top \wedge \boldsymbol{e}_+ + \tfrac{1}{9} w_\perp \wedge \boldsymbol{e}_+\right) \\ &\quad + \eta(\boldsymbol{e}_+, w_-)\left(\tfrac{4}{9} v_1 \wedge \boldsymbol{e}_+ + \tfrac{1}{36} v_\top \wedge \boldsymbol{e}_+ + \tfrac{1}{9} v_\perp \wedge \boldsymbol{e}_+\right), \\ R(v, w) &= -\tfrac{1}{4}\eta(\boldsymbol{e}_+, v_-) w_\top \wedge \boldsymbol{e}_+ + \tfrac{1}{4}\eta(\boldsymbol{e}_+, w_-) v_\top \wedge \boldsymbol{e}_+ , \end{aligned} \tag{56}$$

for all $v, w \in V$. A subtle point to obtain Eqs. (56) is that the R.H.S. of (45) makes sense for all $s \in S$ but the action in there is *not* that of an element of $\mathfrak{so}(V)$. Upon restriction to S', one may re-absorb to $\Lambda^2 V$ the contributions coming from the elements in $\mathrm{End}(S) \cong \bigoplus_{0 \le p \le 5} \Lambda^p V$ with $p \neq 2$, and this is how we arrived at (56). Checking Eq. (44) is then a direct matter, if tedious, of Clifford identities, as in e.g. [8, Theorem 16].

We recovered two symmetric pp-waves backgrounds that were discovered in [10], one indecomposable and the other decomposable. The proof of the existence comes at the same time with the construction of their transvection superalgebras—they are the filtered deformations of $\mathfrak{a} = \mathfrak{h} \oplus S' \oplus V$ determined by φ—in particular it comes with the knowledge that $N = 24$ is the amount of Killing spinors preserved.

Acknowledgments A.S. is grateful to G. Carnovale, F. Esposito and J. Figueroa-O'Farrill for useful discussions. The research of A. S. is supported by the project "Supergravity backgrounds and Invariant theory" at the University of Padova and partly supported by project BIRD179758/17 "Stratifications in algebraic groups, spherical varieties, Kac Moody algebras and Kac Moody groups" and project DOR1717189/17 "Algebraic, geometric and combinatorial properties of conjugacy classes".

References

1. R. Bryant, Pseudo-Riemannian metrics with parallel spinor fields and vanishing Ricci tensor, in *Global Analysis and Harmonic Analysis (Marseille-Luminy 1999)*, ed. by J.P. Bourgignon, T. Branson, O. Hijazi, vol. 4. Sémin. Congr. (Soc. Math. France, Paris, 2000), pp 53–94
2. S.-J. Cheng, V.G. Kac, Generalized Spencer cohomology and filtered deformations of $\mathbb{Z}$-graded Lie superalgebras. Adv. Theor. Math. Phys. **2**, 1141–1182 (1998)

3. P. de Medeiros, J. Figueroa-O'Farrill, A. Santi, Killing superalgebras for Lorentzian four-manifolds. J. High Energy Phys. **6**, 49 pp (2016)
4. P. de Medeiros, J. Figueroa-O'Farrill, A. Santi, Killing superalgebras for lorentzian six-manifolds J. Geom. Phys. **132**, 13–44 (2018)
5. J. Figueroa-O'Farrill, N. Hustler, The homogeneity theorem for supergravity backgrounds. J. High Energy Phys. **10**, 8 pp (2012)
6. J. Figueroa-O'Farrill, P. Meessen, S. Philip, Supersymmetry and homogeneity of M-theory backgrounds. Classical Quantum Gravity **22**, 207–226 (2005)
7. J. Figueroa-O'Farrill, G. Papadopoulos, Maximally supersymmetric solutions of ten and eleven-dimensional supergravity. J. High Energy Phys. **3**, 25 pp (2003)
8. J. Figueroa-O'Farrill, A. Santi, Spencer cohomology and 11-dimensional supergravity. Commun. Math. Phys. **349**, 627–660 (2017)
9. J. Figueroa-O'Farrill, A. Santi, On the algebraic structures of Killing superalgebras. Adv. Theor. Math. Phys. **21**, 1115–1160 (2017)
10. J. Gauntlett, C. Hull, pp-waves in 11-dimensions with extra supersymmetry. J. High Energy Phys. **6**, 13 pp (2002)
11. J. Gauntlett, S. Pakis, The geometry of D =11 Killing spinors. J. High Energy Phys. **4**, 33 pp (2003)
12. J. Gauntlett, J. Gutowski, S. Pakis, The geometry of D=11 null Killing spinors. J. High Energy Phys. **12**, 29 pp (2003)
13. U. Gran, J. Gutowski, G. Papadopoulos, D. Roest, $N = 31$, $D = 11$. J. High Energy Phys. **2**, 17 pp (2007)
14. U. Gran, J. Gutowski, G. Papadopoulos, M-theory backgrounds with 30 Killing spinors are maximally supersymmetric. J. High Energy Phys. **3**, 38 pp (2010)
15. T. Harmark, T. Takayanagi, Supersymmetric Gödel universes in string theory. Nucl. Phys. B **662**, 3–39 (2003)
16. J. Martinet, Sur les singularités des formes différentielles. Ann. Inst. Fourier (Grenoble) **20**, 95–178 (1970)
17. J. Michelson, A pp-wave with 26 supercharges. Classical Quantum Gravity **19**, 5935–5949 (2002)
18. V. Patrangenaru, Locally homogeneous pseudo-Riemannian manifolds. J. Geom. Phys. **17**, 59–72 (1995)
19. A. Santi, A. Spiro, Super-Poincaré algebras, space-times and supergravities (I). Adv. Theor. Math. Phys. **16**, 1411–1441 (2012)
20. A. Santi, A. Spiro, Super-Poincaré algebras, space-times, and supergravities. II. J. Math. Phys. **53**, 24 pp (2012)
21. E. Vinberg, A. Elashvili, A classification of the three-vectors of nine-dimensional space. Trudy Sem. Vektor. Tenzor. Anal. **18**, 197–233 (1978)

Krichever–Novikov Type Algebras. A General Review and the Genus Zero Case

Martin Schlichenmaier

Abstract In the first part of this survey we recall the definition and some of the constructions related to Krichever–Novikov type algebras. Krichever and Novikov introduced them for higher genus Riemann surfaces with two marked points in generalization of the classical algebras of Conformal Field Theory. Schlichenmaier extended the theory to the multi-point situation and even to a larger class of algebras. The almost-gradedness of the algebras and the classification of almost-graded central extensions play an important role in the theory and in applications. In the second part we specialize the construction to the genus zero multi-point case. This yields beside instructive examples also additional results. In particular, we construct universal central extensions for the involved algebras, which are vector field algebras, differential operator algebras, current algebras and Lie superalgebras. We point out that the recently (re-)discussed N-Virasoro algebras are nothing else as multi-point genus zero Krichever-Novikov type algebras. The survey closes with structure equations and central extensions for the three-point case.

1 Introduction

Krichever-Novikov (KN) type algebras constitutes an important class of infinite dimensional algebras. Roughly speaking, they are defined as algebras of meromorphic objects on compact Riemann surfaces, or equivalently on projective curves. The non-holomorphicity is controlled by a fixed finite set of points where algebraic poles are allowed. A splitting of this set of possible points of poles into two disjoint subsets will induce an "almost-grading" (see Definition 5.1 below). It is a weaker concept as a grading, but still powerful enough to act as a basic tool in representation theory. For example, highest weight representations still can be defined. Of course, central

M. Schlichenmaier (✉)
University of Luxembourg, Department of Mathematics, FSTC, Esch-sur-Alzette, Luxembourg
e-mail: martin.schlichenmaier@uni.lu

S. Hervik et al. (eds.), *Geometry, Lie Theory and Applications*, Abel Symposia 16,
https://doi.org/10.1007/978-3-030-81296-6_13

extensions of these algebras are also needed. They are forced, e.g. by representation theory and by quantization.

Examples of KN type algebras are the well-known algebras of Conformal Field Theory (CFT) [4, 22] the Witt algebra, the Virasoro algebra, the affine Lie algebras (affine Kac-Moody algebras), etc. They appear when the geometric setting consists of the Riemann Sphere, i.e. the genus zero Riemann surface, and the points of possible poles are $\{0\}$ and $\{\infty\}$. The almost-grading is now a honest grading.

Historically, starting from these well-known genus zero algebras, in 1986 Krichever and Novikov [33, 34] suggested a global operator approach via KN objects. Still they only considered two possible points where poles are allowed and were dealing with the vector field and the function algebra. For work on affine algebras Sheinman [54–57]. should be mentioned.

From the applications in CFT (e.g. string theory) but also from purely mathematical reasons, the need of a multi-point theory is evident. In 1990 Schlichenmaier developed a systematic theory valid for all genera (including zero) and any fixed finite set of points where poles are allowed [40–43]. These extensions were not at all straight-forward. The main point was to introduce a replacement of the graded algebra structure present in the "classical" case. Krichever and Novikov found that the already mentioned almost-grading (Definition 5.1) will be enough to allow for the standard constructions in representation theory. In [42, 43] it was realized that a splitting of the set A of points where poles are allowed, into two disjoint non-empty subsets $A = I \cup O$ is crucial for introducing an almost-grading in the multi-point situation. The corresponding almost-grading was explicitly given. In contrast to the classical situation and original KN situation, where there is only one grading, we will have a finite set of non-equivalent gradings and new interesting phenomena show up. This is already true for the genus zero case (i.e. the Riemann sphere case) with more than two points where poles are allowed. These algebras will be only almost-graded, see e.g. [19, 20, 44, 53].

Also other (Lie) algebras were introduced. In fact most of them come from a *Mother Poisson Algebra* [45], the algebra of meromorphic form of all weight, see Sect. 4.2. This algebra carries a (weak) almost-grading which gives the almost-grading for the other algebras. For the relevant algebras almost-graded central extensions are constructed and classified. In the first part of the survey we present the basic definitions and structural results for the KN type algebras.

In the second part, starting in Sect. 7 we have a closer look at the genus zero multi-point situation. In this case the meromorphic objects can be given via rational functions. We obtain by them illustrative examples of KN type algebras. Furthermore, from the applications, e.g. in conformal field theory, resp. string theory, they correspond to the tree-level. Also we obtain additional results. For example we obtain universal central extensions of those algebras which are perfect.[1] We show that all cocycles are bounded cocycle classes with respect to the standard

[1] Most of them are perfect algebras.

splitting, see Eq. (90) for its definition. The classification result of the author for bounded cocycles gives now the universal central extension.

In Sect. 8 the genus zero 3-point situation is covered in detail. Explicit structure equations (including central extensions) are given for the algebras.

KN type algebras have a lot of interesting applications. They show up in the context of deformations of algebras, moduli spaces of marked curves, Wess-Zumino-Novikov-Witten (WZNW) models, Knizhnik-Zamolodchikov (KZ) equations, integrable systems, quantum field theories, symmetry algebras, and in many more domains of mathematics and theoretical physics. The KN type algebras carry a very rich representation theory. We have Verma modules, highest weight representations, Fermionic and Bosonic Fock representations, semi-infinite wedge forms, $b - c$ systems, Sugawara representations and vertex algebras. Unfortunately, this survey does not allow to touch on these applications and the presentation of the representation theory. Instead I refer to my book from 2014 with the title *Krichever–Novikov type algebras. Theory and applications,* [51] which collects all the results, proofs and some applications of the multi-point KN algebras. There also a quite extensive list of references can be found, including articles published by physicists on applications in the field-theoretical context. For some applications in the context of integrable systems see also Sheinman, *Current algebras on Riemann surfaces* [58].

For the proofs of the statements and more material we have to refer to the original articles and the corresponding parts of [51]. There is a certain overlap of the first part with a previous survey of mine [52].

2 The Witt and the Virasoro Algebra and Their Relatives

These algebras supply important examples of non-trivial infinite dimensional Lie algebras. They are widely used in Conformal Field Theory and String Theory. We recall their conventional algebraic definitions.

The *Witt algebra* $\mathcal{W}$, sometimes also called Virasoro algebra without central term, is the Lie algebra generated as vector space over $\mathbb{C}$ by the basis elements $\{e_n \mid n \in \mathbb{Z}\}$ with Lie structure

$$[e_n, e_m] = (m - n)e_{n+m}, \qquad n, m \in \mathbb{Z}. \tag{1}$$

The algebra $\mathcal{W}$ is more than just a Lie algebra. It is a graded Lie algebra. If we set for the degree $\deg(e_n) := n$ then

$$\mathcal{W} = \bigoplus_{n \in \mathbb{Z}} \mathcal{W}_n, \qquad \mathcal{W}_n = \langle e_n \rangle_{\mathbb{C}}. \tag{2}$$

Obviously, $\deg([e_n, e_m]) = \deg(e_n) + \deg(e_m)$.

Remark 2.1 Algebraically $\mathcal{W}$ can also be given as Lie algebra of derivations of the algebra of Laurent polynomials $\mathbb{C}[z, z^{-1}]$. Moreover, in the purely algebraic context our field of definition $\mathbb{C}$ can be replaced by an arbitrary field $\mathbb{K}$ of characteristics 0.

For the Witt algebra the universal one-dimensional central extension is the *Virasoro algebra* $\mathcal{V}$. As vector space it is the direct sum $\mathcal{V} = \mathbb{C} \oplus \mathcal{W}$. If we set for $x \in \mathcal{W}$, $\hat{x} := (0, x)$, and $t := (1, 0)$ then its basis elements are $\hat{e}_n,\ n \in \mathbb{Z}$ and t with the Lie product:[2]

$$[\hat{e}_n, \hat{e}_m] = (m-n)\hat{e}_{n+m} + \frac{1}{12}(n^3 - n)\delta_n^{-m}\, t, \qquad [\hat{e}_n, t] = [t, t] = 0, \tag{3}$$

for all $n, m \in \mathbb{Z}$. By setting $\deg(\hat{e}_n) := \deg(e_n) = n$ and $\deg(t) := 0$ the Lie algebra $\mathcal{V}$ becomes a graded algebra. The algebra $\mathcal{W}$ will only be a subspace, not a subalgebra of $\mathcal{V}$. But it will be a quotient. Up to equivalence of central extensions and rescaling the central element t, this is beside the trivial (splitting) central extension, the only central extension of $\mathcal{W}$.

Given $\mathfrak{g}$ a finite-dimensional Lie algebra (e.g. a finite-dimensional simple Lie algebra) then the tensor product of $\mathfrak{g}$ with the associative algebra of Laurent polynomials $\mathbb{C}[z, z^{-1}]$ carries a Lie algebra structure via

$$[x \otimes z^n, y \otimes z^m] := [x, y] \otimes z^{n+m}. \tag{4}$$

This algebra is called *current algebra* or *loop algebra* and denoted by $\overline{\mathfrak{g}}$. Again we consider central extensions. For this let β be a symmetric, bilinear form for $\mathfrak{g}$ which is invariant (i.e. $\beta([x, y], z) = \beta(x, [y, z])$ for all $x, y, z \in \mathfrak{g}$). Then a central extension is given by

$$[\widehat{x \otimes z^n}, \widehat{y \otimes z^m}] := \widehat{[x, y] \otimes z^{n+m}} - \beta(x, y) \cdot m\, \delta_n^{-m} \cdot t. \tag{5}$$

This algebra is denoted by $\widehat{\mathfrak{g}}$ and called *affine Lie algebra*. With respect to the classification of Kac-Moody Lie algebras, in the case of a simple $\mathfrak{g}$ they are exactly the Kac-Moody algebras of untwisted affine type, [28, 29, 36].

To complete the description let me introduce the Lie superalgebra of Neveu-Schwarz type. The centrally extended superalgebra has as basis (we drop the ^)

$$e_n,\ n \in \mathbb{Z}, \quad \varphi_m,\ m \in \mathbb{Z} + \frac{1}{2}, \quad t \tag{6}$$

[2]Here δ_k^l is the Kronecker delta which is equal to 1 if $k = l$, otherwise zero.

with structure equations

$$
\begin{aligned}
[e_n, e_m] &= (m-n)e_{m+n} + \frac{1}{12}(n^3 - n)\,\delta_n^{-m}\, t, \\
[e_n, \varphi_m] &= (m - \frac{n}{2})\,\varphi_{m+n}, \\
[\varphi_n, \varphi_m] &= e_{n+m} - \frac{1}{6}\,(n^2 - \frac{1}{4})\,\delta_n^{-m}\, t.
\end{aligned}
\tag{7}
$$

By "setting $t = 0$" we obtain the non-extended superalgebra. The elements e_n (and t) are a basis of the subspace of even elements, the elements φ_m are a basis of the subspace of odd elements.

These algebras are Lie superalgebras, defined as follows.

Definition 2.2 Let $\mathcal{S}$ be a vector space which is decomposed into even and odd elements $\mathcal{S} = \mathcal{S}_{\bar{0}} \oplus \mathcal{S}_{\bar{1}}$, i.e. $\mathcal{S}$ is a $\mathbb{Z}/2\mathbb{Z}$-graded vector space. Furthermore, let $[.,.]$ be a $\mathbb{Z}/2\mathbb{Z}$-graded bilinear map $\mathcal{S} \times \mathcal{S} \to \mathcal{S}$ such that for elements x, y of pure parity

$$
[x, y] = -(-1)^{\bar{x}\bar{y}}[y, x]. \tag{8}
$$

Here $\bar{x}$ is the parity of x, etc. These conditions say that

$$
[\mathcal{S}_{\bar{0}}, \mathcal{S}_{\bar{0}}] \subseteq \mathcal{S}_{\bar{0}}, \qquad [\mathcal{S}_{\bar{0}}, \mathcal{S}_{\bar{1}}] \subseteq \mathcal{S}_{\bar{1}}, \qquad [\mathcal{S}_{\bar{1}}, \mathcal{S}_{\bar{1}}] \subseteq \mathcal{S}_{\bar{0}}, \tag{9}
$$

and that $[x, y]$ is symmetric for x and y odd, otherwise anti-symmetric. Now $\mathcal{S}$ is a *Lie superalgebra* if in addition the *super-Jacobi identity* (for x, y, z of pure parity)

$$
(-1)^{\bar{x}\bar{z}}[x, [y, z]] + (-1)^{\bar{y}\bar{x}}[y, [z, x]] + (-1)^{\bar{z}\bar{y}}[z, [x, y]] = 0 \tag{10}
$$

is valid. As long as the type of the arguments is different from *(even, odd, odd)* all signs can be put to $+1$ and we obtain the form of the usual Jacobi identity. In the remaining case we get

$$
[x, [y, z]] + [y, [z, x]] - [z, [x, y]] = 0. \tag{11}
$$

By the definitions $\mathcal{S}_0$ is a Lie algebra.

3 The Geometric Picture

A geometric description of the Witt algebra over $\mathbb{C}$ can be given as follows. Let W be the algebra of those meromorphic vector fields on the Riemann sphere $S^2 = \mathbb{P}^1(\mathbb{C})$ which are holomorphic outside $\{0\}$ and $\{\infty\}$. Its elements can be given as

$$
v(z) = \tilde{v}(z)\frac{d}{dz} \tag{12}
$$

where $\tilde{v}$ is a meromorphic function on $\mathbb{P}^1(\mathbb{C})$, which is holomorphic outside $\{0, \infty\}$. Those are exactly the Laurent polynomials $\mathbb{C}[z, z^{-1}]$. Consequently, this subalgebra has the set $\{e_n, n \in \mathbb{Z}\}$ with $e_n = z^{n+1}\frac{d}{dz}$ as vector space basis. The Lie bracket of vector fields calculates as

$$[v, u] = \left(\tilde{v}\frac{d}{dz}\tilde{u} - \tilde{u}\frac{d}{dz}\tilde{v}\right)\frac{d}{dz}. \tag{13}$$

Evaluated for the basis elements e_n this gives (1) and the algebra can be identified with the Witt algebra defined purely algebraically.

The subalgebra of global holomorphic vector fields is the 3-dimensional subspace $\langle e_{-1}, e_0, e_1\rangle_{\mathbb{C}}$. It is isomorphic to the Lie algebra $\mathfrak{sl}(2, \mathbb{C})$.

Similarly, the algebra $\mathbb{C}[z, z^{-1}]$ can be given as the algebra of meromorphic functions on $S^2 = \mathbb{P}^1(\mathbb{C})$ holomorphic outside of $\{0, \infty\}$.

Recall that the Riemann sphere is the (compact) Riemann surface of genus zero. In the geometric setup for the Virasoro algebra the objects are defined on the Riemann sphere and might have poles at most at two fixed points. For a global operator approach to conformal field theory and its quantization this is not sufficient. One needs Riemann surfaces of arbitrary genus. Moreover, one needs more than two points were singularities are allowed.[3] This higher genus multi-point case was systematically examined by the Schlichenmaier [40–47] and is presented in a current book [51]. For some related approach and partial results, see also Sadov [39].

For the whole contribution let $\Sigma = \Sigma_g$ be a compact Riemann surface without any restriction for the genus $g = g(\Sigma)$. Furthermore, let A be a finite subset of Σ. Later we will need a splitting of A into two non-empty disjoint subsets I and O, i.e. $A = I \cup O$. Set $N := \#A$, $K := \#I$, $M := \#O$, with $N = K + M$. More precisely, let

$$I = (P_1, \ldots, P_K), \quad \text{and} \quad O = (Q_1, \ldots, Q_M) \tag{14}$$

be disjoint ordered tuples of distinct points ("marked points", "punctures") on the Riemann surface. In particular, we assume $P_i \neq Q_j$ for every pair (i, j). The points in I are called the *in-points*, the points in O the *out-points*. .

Sometimes we refer to the *classical situation*. By this we understand

$$\Sigma_0 = \mathbb{P}^1(\mathbb{C}) = S^2, \quad I = \{z = 0\}, \quad O = \{z = \infty\}. \tag{15}$$

The following figures should indicate the geometric picture. Fig. 1 shows the classical situation.

[3]The singularities correspond to points where free fields are entering the region of interaction or leaving it. In particular, from the very beginning there is a natural decomposition of the set of points into two disjoint subsets.

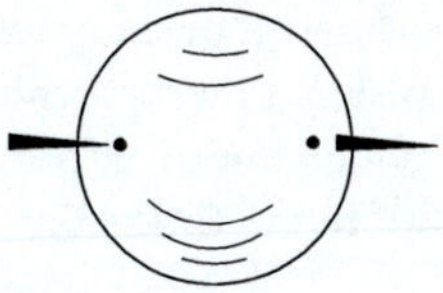

Fig. 1 Riemann surface of genus zero with one incoming and one outgoing point

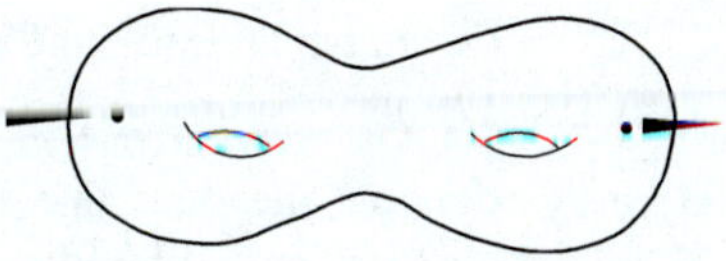

Fig. 2 Riemann surface of genus two with one incoming and one outgoing point

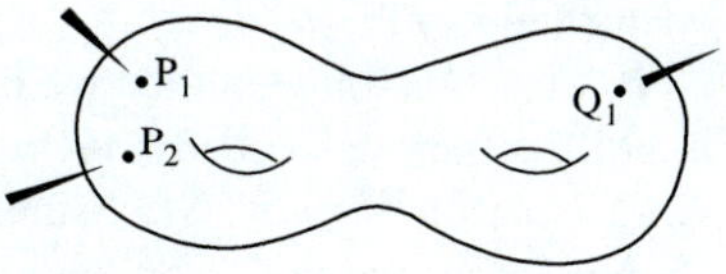

Fig. 3 Riemann surface of genus two with two incoming points and one outgoing point

Figure 2 is genus 2, but still the two-point situation.

Finally, in Fig. 3 the case of a Riemann surface of genus 2 with two incoming points and one outgoing point is visualized.

Remark 3.1 We stress the fact, that these multi-point generalizations are needed also in the case of genus zero. Even in the case of genus zero and three points interesting algebras show up. See Sect. 8, [53].

3.1 Meromorphic Forms

To introduce the elements of the generalized algebras (later called Krichever-Novikov type algebras) we first have to discuss forms of certain (conformal) weights.

Let $\mathcal{K} = \mathcal{K}_\Sigma$ be the canonical line bundle of Σ. Its local sections are the local holomorphic differentials.

If $P \in \Sigma$ is a point and z a local holomorphic coordinate at P then a local holomorphic differential can be written as $f(z)dz$ with a local holomorphic function f defined in a neighborhood of P. A global holomorphic differential can be described locally in coordinates $(U_i, z_i)_{i \in J}$ by a system of local holomorphic functions $(f_i)_{i \in J}$, which are related by the transformation rule induced by the coordinate change map $z_j = z_j(z_i)$ and the condition $f_i dz_i = f_j dz_j$. This yields

$$f_j = f_i \cdot \left(\frac{dz_j}{dz_i}\right)^{-1}. \tag{16}$$

A meromorphic section of $\mathcal{K}$, i.e. a *meromorphic differential* is given as a collection of local meromorphic functions $(h_i)_{i\in J}$ (with respect to a coordinate covering) for which the transformation law (16) remains true.

In the following λ is either an integer or a half-integer. If λ is an integer then

1. $\mathcal{K}^\lambda := \mathcal{K}^{\otimes\lambda}$ for $\lambda > 0$,
2. $\mathcal{K}^0 := \mathcal{O}$, the trivial line bundle, and
3. $\mathcal{K}^\lambda := (\mathcal{K}^*)^{\otimes(-\lambda)}$ for $\lambda < 0$.

Here $\mathcal{K}^*$ denotes the dual line bundle of the canonical line bundle. This is the holomorphic tangent line bundle, whose local sections are the holomorphic tangent vector fields $f(z)(d/dz)$.

If λ is a half-integer, then we first have to fix a "square root" of the canonical line bundle, sometimes called a *theta characteristics*. . This means we fix a line bundle L for which $L^{\otimes 2} = \mathcal{K}$. After such a choice of L is done we set $\mathcal{K}^\lambda := \mathcal{K}^\lambda_L := L^{\otimes 2\lambda}$. In most cases we will drop the mentioning of L, but we have to keep the choice in mind. The fine-structure of the algebras we are about to define will depend on the choice. But the main properties will remain the same.

Remark 3.2 A Riemann surface of genus g has exactly 2^{2g} non-isomorphic square roots of $\mathcal{K}$. The choice of a theta characteristic is also called a spin structure on Σ [3]. Only for $g = 0$ we have a unique unique square root.

We set

$$\mathcal{F}^\lambda(A) := \{f \text{ is a global meromorphic section of } \mathcal{K}^\lambda \mid \\ f \text{ is holomorphic on } \Sigma \setminus A\}. \tag{17}$$

Obviously, this is a $\mathbb{C}$-vector space. To avoid cumbersome notation, we will often drop the set A in the notation if A is fixed and clear from the context. Recall that in the case of half-integer λ everything depends on the theta characteristic L.

Definition 3.3 The elements of the space $\mathcal{F}^\lambda(A)$ are called *meromorphic forms of weight* λ (with respect to the theta characteristic L).

If f is a meromorphic λ-form it can be represented locally by meromorphic functions f_i via $f = f_i(dz_i)^{\otimes\lambda}$. If $f \not\equiv 0$ the local representing functions have only finitely many zeros and poles. Whether a point P is a zero or a pole of f does not depend on the coordinate z_i chosen. We can define for $P \in \Sigma$ the *order*

$$\mathrm{ord}_P(f) := \mathrm{ord}_P(f_i), \tag{18}$$

where $\mathrm{ord}_P(f_i)$ is the lowest nonvanishing order in the Laurent series expansion of f_i in the variable z_i around P. It will not depend on the coordinate z_i chosen.

The order $\mathrm{ord}_P(f)$ is (strictly) positive if and only if P is a zero of f. It is negative if and only if P is a pole of f. Moreover, its value gives the order of the zero and pole respectively.

By compactness of our Riemann surface Σ our $f \neq 0$ can only have finitely many zeros and poles. We define the *(sectional) degree* of f to be

$$\mathrm{sdeg}(f) := \sum_{P \in \Sigma} \mathrm{ord}_P(f). \tag{19}$$

Proposition 3.4 *Let $f \in \mathcal{F}^\lambda$, $f \not\equiv 0$ then*

$$\mathrm{sdeg}(f) = 2\lambda(g-1). \tag{20}$$

For this and related results see e.g. [48].

Later we will need the additional geometric data of a coordinate z_i at every point $P_i \in A$. In fact, only the first order infinitesimal neighbourhood will play a role.

4 Algebraic Structures

Next we introduce algebraic operations on the vector space of meromorphic forms of arbitrary weights. This space is obtained by summing over all weights

$$\mathcal{F} := \bigoplus_{\lambda \in \frac{1}{2}\mathbb{Z}} \mathcal{F}^\lambda. \tag{21}$$

The basic operations will allow us to introduce finally our intended algebras. We will drop the subset A in the notation.

4.1 Associative Structure

The natural map of the locally free sheaves of rang one

$$\mathcal{K}^\lambda \times \mathcal{K}^\nu \to \mathcal{K}^\lambda \otimes \mathcal{K}^\nu \cong \mathcal{K}^{\lambda+\nu}, \quad (s,t) \mapsto s \otimes t, \tag{22}$$

defines a bilinear map

$$\cdot : \mathcal{F}^\lambda \times \mathcal{F}^\nu \to \mathcal{F}^{\lambda+\nu}. \tag{23}$$

With respect to local trivialisations this corresponds to the multiplication of the local representing meromorphic functions

$$(s\,dz^\lambda, t\,dz^\nu) \mapsto s\,dz^\lambda \cdot t\,dz^\nu = s \cdot t\,dz^{\lambda+\nu}. \tag{24}$$

If there is no danger of confusion then we will mostly use the same symbol for the section and for the local representing function.

The following is obvious

Proposition 4.1 *The space $\mathcal{F}$ is an associative and commutative graded (over $\frac{1}{2}\mathbb{Z}$) algebra. Moreover, $\mathcal{A} = \mathcal{F}^0$ is a subalgebra and the $\mathcal{F}^\lambda$ are modules over $\mathcal{A}$.*

Of course, $\mathcal{A}$ is the algebra of those meromorphic functions on Σ which are holomorphic outside of A.

4.2 Lie and Poisson Algebra Structure

Next we define a Lie algebra structure on the space $\mathcal{F}$. The structure is induced by the map

$$\mathcal{F}^\lambda \times \mathcal{F}^\nu \to \mathcal{F}^{\lambda+\nu+1}, \qquad (e, f) \mapsto [e, f], \tag{25}$$

which is defined in local representatives of the sections by

$$(e\,dz^\lambda, f\,dz^\nu) \mapsto [e\,dz^\lambda, f\,dz^\nu] := \left((-\lambda)e\frac{df}{dz} + \nu f\frac{de}{dz}\right) dz^{\lambda+\nu+1}, \tag{26}$$

and bilinearly extended to $\mathcal{F}$.

Proposition 4.2 ([51, Prop. 2.6 and 2.7]) *The prescription $[.,.]$ given by (26) is well-defined and defines a Lie algebra structure on the vector space $\mathcal{F}$.*

Proposition 4.3 ([51, Prop. 2.8]) *The subspace $\mathcal{L} = \mathcal{F}^{-1}$ is a Lie subalgebra, and the $\mathcal{F}^\lambda$'s are Lie modules over $\mathcal{L}$.*

The $\mathcal{L}$ consists of meromorphic vector fields on Σ and the Lie module structure is the Lie derivative. of forms.

We have the Leibniz rule

$$\forall e, f, g \in \mathcal{F} : [e, f \cdot g] = [e, f] \cdot g + f \cdot [e, g]. \tag{27}$$

relating the associative and the Lie structure. Hence, by definition

Theorem 4.4 *[51, Thm. 2.10] The space $\mathcal{F}$ with respect to $\cdot$ and $[.,.]$ is a Poisson algebra.*

Next we consider important substructures. We already encountered the subalgebras $\mathcal{A}$ and $\mathcal{L}$. But there are more structures around.

4.3 The Algebra of Differential Operators

If we look at $\mathcal{F}$, considered as Lie algebra, more closely, we see that $\mathcal{F}^0$ is an abelian Lie subalgebra and the vector space sum $\mathcal{F}^0 \oplus \mathcal{F}^{-1} = \mathcal{A} \oplus \mathcal{L}$ is also a Lie subalgebra. In an equivalent way it can also be constructed as semidirect sum of $\mathcal{A}$ considered as abelian Lie algebra and $\mathcal{L}$ operating on $\mathcal{A}$ by taking the derivative.

Definition 4.5 The *Lie algebra of differential operators of degree* ≤ 1 is defined as the semidirect sum of $\mathcal{A}$ with $\mathcal{L}$ and is denoted by $\mathcal{D}^1$.

In terms of elements the Lie product is

$$[(g, e), (h, f)] = (e\,.\,h - f\,.\,g\,,\ [e, f]). \tag{28}$$

The projection on the second factor $(g, e) \mapsto e$ is a Lie homomorphism and we obtain a short exact sequences of Lie algebras

$$0 \longrightarrow \mathcal{A} \longrightarrow \mathcal{D}^1 \longrightarrow \mathcal{L} \longrightarrow 0\,. \tag{29}$$

Hence, $\mathcal{A}$ is an (abelian) Lie ideal of $\mathcal{D}^1$ and $\mathcal{L}$ a quotient Lie algebra. Obviously, $\mathcal{L}$ is also a subalgebra of $\mathcal{D}^1$.

The vector spaces $\mathcal{F}^\lambda$ become Lie modules over $\mathcal{D}^1$ by the operation

$$(g, e).f := g \cdot f + e.f, \quad (g, e) \in \mathcal{D}^1(A),\ f \in \mathcal{F}^\lambda(A). \tag{30}$$

Differential operators of arbitrary degree can be constructed via universal constructions, see e.g. [51].

4.4 Lie Superalgebras of Half Forms

Recall from Definition 2.2 the notion of a Lie superalgebra.

With the help of our associative product (22) we will obtain examples of Lie superalgebras. First we consider

$$\cdot\, \mathcal{F}^{-1/2} \times \mathcal{F}^{-1/2} \to \mathcal{F}^{-1} = \mathcal{L}\,, \tag{31}$$

and introduce the vector space $\mathcal{S}$ with the product

$$\mathcal{S} := \mathcal{L} \oplus \mathcal{F}^{-1/2}, \quad [(e, \varphi), (f, \psi)] := ([e, f] + \varphi \cdot \psi, e\,.\,\varphi - f\,.\,\psi). \tag{32}$$

The elements of $\mathcal{L}$ are denoted by $e, f, \ldots$, and the elements of $\mathcal{F}^{-1/2}$ by $\varphi, \psi, \ldots$.

The definition (32) can be reformulated as an extension of [., .] on $\mathcal{L}$ to a super-bracket (denoted by the same symbol) on $\mathcal{S}$ by setting

$$[e, \varphi] := -[\varphi, e] := e \,.\, \varphi = (e\frac{d\varphi}{dz} - \frac{1}{2}\varphi\frac{de}{dz})(dz)^{-1/2} \tag{33}$$

and

$$[\varphi, \psi] := \varphi \cdot \psi \,. \tag{34}$$

The elements of $\mathcal{L}$ are elements of even parity, and the elements of $\mathcal{F}^{-1/2}$ are elements of odd parity. For such elements x we denote by $\bar{x} \in \{\bar{0}, \bar{1}\}$ their parity.

The sum (32) can also be described as $\mathcal{S} = \mathcal{S}_{\bar{0}} \oplus \mathcal{S}_{\bar{1}}$, where $\mathcal{S}_{\bar{i}}$ is the subspace of elements of parity $\bar{i}$.

Proposition 4.6 *[50, 51, Prop. 2.15] The space $\mathcal{S}$ with the above introduced parity and product is a Lie superalgebra.*

Leidwanger and Morier-Genoux introduced in [35] a *Jordan superalgebra* in our geometric setting. They put

$$\mathcal{J} := \mathcal{F}^0 \oplus \mathcal{F}^{-1/2} = \mathcal{J}_{\bar{0}} \oplus \mathcal{J}_{\bar{1}}. \tag{35}$$

Recall that $\mathcal{A} = \mathcal{F}^0$ is the associative algebra of meromorphic functions. They define the (Jordan) product $\circ$ via the algebra structures for the spaces $\mathcal{F}^\lambda$ by

$$\begin{aligned} f \circ g &:= f \cdot g \quad && \in \mathcal{F}^0, \\ f \circ \varphi &:= f \cdot \varphi \quad && \in \mathcal{F}^{-1/2}, \\ \varphi \circ \psi &:= [\varphi, \psi] \quad && \in \mathcal{F}^0. \end{aligned} \tag{36}$$

By rescaling the second definition with the factor 1/2 one obtains a *Lie anti-algebra* as introduced by Ovsienko [37]. See [35] for more details and additional results on representations.

4.5 Higher Genus Current Algebras

We fix an arbitrary finite-dimensional complex Lie algebra $\mathfrak{g}$. Our goal is to generalize the classical current algebra to higher genus. For this let (Σ, A) be the geometric data consisting of the Riemann surface Σ and the subset of points A used to define $\mathcal{A}$, the algebra of meromorphic functions which are holomorphic outside of the set $A \subseteq \Sigma$.

Definition 4.7 The *higher genus current algebra* associated to the Lie algebra $\mathfrak{g}$ and the geometric data (Σ, A) is the Lie algebra $\overline{\mathfrak{g}} = \overline{\mathfrak{g}}(A) = \overline{\mathfrak{g}}(\Sigma, A)$ given as vector space by $\overline{\mathfrak{g}} = \mathfrak{g} \otimes_{\mathbb{C}} \mathcal{A}$ with the Lie product

$$[x \otimes f, y \otimes g] = [x, y] \otimes f \cdot g, \qquad x, y \in \mathfrak{g}, \quad f, g \in \mathcal{A}. \tag{37}$$

Sometimes this algebra is also called *loop algebra*.

Proposition 4.8 *$\overline{\mathfrak{g}}$ is a Lie algebra.*

As usual we will suppress the mentioning of (Σ, A) if not needed. The elements of $\overline{\mathfrak{g}}$ can be interpreted as meromorphic functions $\Sigma \to \mathfrak{g}$ from the Riemann surface Σ to the Lie algebra $\mathfrak{g}$, which are holomorphic outside of A.

Later we will introduce central extensions for these current algebras. They will generalize affine Lie algebras, respectively affine Kac-Moody algebras of untwisted type.

For some applications it is useful to extend the definition by considering differential operators (of degree ≤ 1) associated to $\overline{\mathfrak{g}}$. We define $\mathcal{D}^1_{\mathfrak{g}} := \overline{\mathfrak{g}} \oplus \mathcal{L}$ and take in the summands the Lie product defined there and put additionally

$$[e, x \otimes g] := -[x \otimes g, e] := x \otimes (e.g). \tag{38}$$

This operation can be described as semidirect sum of $\overline{\mathfrak{g}}$ with $\mathcal{L}$ and we get

Proposition 4.9 ([51, Prop. 2.15]) *$\mathcal{D}^1_{\mathfrak{g}}$ is a Lie algebra.*

4.6 *Krichever–Novikov Type Algebras*

Above we could even allow that the set A of points where poles are allowed is arbitrary, even non-finite. In case that A is finite and moreover $\#A \geq 2$ the constructed algebras we a called Krichever–Novikov (KN) type algebras. In this way we get the KN vector field algebra, the function algebra, the current algebra, the differential operator algebra, the Lie superalgebra, etc. The reader might ask what is so special about this situation so that these algebras deserve special names. In fact in this case we can endow the algebra with a (strong) almost-graded structure. This will be discussed in the next section. The almost-grading is a crucial tool for representation theory and for extending the classical result to higher genus. Recall that in the classical case we have genus zero and $\#A = 2$.

Strictly speaking, a KN type algebra should be considered to be one of the above algebras for $2 \leq \#A < \infty$ together with a fixed chosen almost-grading, induced by the splitting $A = I \cup O$ into two disjoint non-empty subset, see Definition 5.1.

5 Almost-Graded Structure

5.1 Definition of Almost-Gradedness

In the classical situation discussed in Sect. 2 the algebras introduced in the last section are graded algebras. In the higher genus case and even in the genus zero case with more than two points where poles are allowed there is no non-trivial grading anymore. As realized by Krichever and Novikov [33] there is a weaker concept, an almost-grading, which to a large extend is a valuable replacement of a honest grading. Such an almost-grading is induced by a splitting of the set A into two non-empty and disjoint sets I and O. The (almost-)grading is fixed by exhibiting certain basis elements in the spaces $\mathcal{F}^\lambda$ as homogeneous.

Definition 5.1 Let $\mathcal{L}$ be a Lie or an associative algebra such that $\mathcal{L} = \oplus_{n\in\mathbb{Z}}\mathcal{L}_n$ is a vector space direct sum, then $\mathcal{L}$ is called an *almost-graded* (Lie-) algebra if

(i) $\dim \mathcal{L}_n < \infty$,
(ii) There exists constants $L_1, L_2 \in \mathbb{Z}$ such that

$$\mathcal{L}_n \cdot \mathcal{L}_m \subseteq \bigoplus_{h=n+m-L_1}^{n+m+L_2} \mathcal{L}_h, \qquad \forall n, m \in \mathbb{Z}.$$

The elements in $\mathcal{L}_n$ are called *homogeneous* elements of degree n, and $\mathcal{L}_n$ is called *homogeneous subspace* of degree n.

If $\dim \mathcal{L}_n$ is bounded with a bound independent of n we call $\mathcal{L}$ *strongly almost-graded*. If we drop the condition that $\dim \mathcal{L}_n$ is finite dimensional we call $\mathcal{L}$ *weakly almost-graded*.

In a similar manner almost-graded modules over almost-graded algebras are defined. We can extend in an obvious way the definition to superalgebras, respectively even to more general algebraic structures. Note that this definition makes complete sense also for more general index sets $\mathbb{J}$. In fact we will consider the index set $\mathbb{J} = (1/2)\mathbb{Z}$ in the case of superalgebras. The even elements (with respect to the super-grading) will have integer degree, the odd elements half-integer degree.

5.2 Separating Cycle and Krichever-Novikov Pairing

Before we give the almost-grading we introduce an important structure.

First we recall the splitting of A (14) into two non-empty disjoint subsets I and O. Let C_i be positively oriented (deformed) circles on Σ around the points P_i in I, $i = 1, \ldots, K$ and C_j^* positively oriented circles on Σ around the points Q_j in O, $j = 1, \ldots, M$.

A cycle C_S on Σ is called a *separating cycle* if it is smooth, positively oriented of multiplicity one and if it separates the in-points from the out-points. It might have more than one component. In the following we will integrate meromorphic differentials on Σ without poles in $\Sigma \setminus A$ over closed curves C. Hence, we might consider C and C' as equivalent if $[C] = [C']$ in $\mathrm{H}_1(\Sigma \setminus A, \mathbb{Z})$. In this sense we write for every separating cycle

$$[C_S] = \sum_{i=1}^{K} [C_i] = -\sum_{j=1}^{M} [C_j^*]. \tag{39}$$

The minus sign appears due to the opposite orientation. Another way for giving such a C_S is via level lines of a "proper time evolution", for which I refer to [51, Section 3.9].

Given such a separating cycle C_S (respectively cycle class) we define a linear map

$$\mathcal{F}^1 \to \mathbb{C}, \qquad \omega \mapsto \frac{1}{2\pi \mathrm{i}} \int_{C_S} \omega. \tag{40}$$

The map will not depend on the separating line C_S chosen, as two of such will be homologous and the poles of ω are only located in I and O.

Consequently, the integration of ω over C_S can also be described over the special cycles C_i or equivalently over C_j^*. This integration corresponds to calculating residues

$$\omega \quad \mapsto \quad \frac{1}{2\pi \mathrm{i}} \int_{C_S} \omega \; = \; \sum_{i=1}^{K} \mathrm{res}_{P_i}(\omega) \; = \; -\sum_{l=1}^{M} \mathrm{res}_{Q_l}(\omega). \tag{41}$$

Definition 5.2 The pairing

$$\mathcal{F}^{\lambda} \times \mathcal{F}^{1-\lambda} \to \mathbb{C}, \quad (f, g) \mapsto \langle f, g \rangle := \frac{1}{2\pi \mathrm{i}} \int_{C_S} f \cdot g, \tag{42}$$

between λ and $1 - \lambda$ forms is called *Krichever-Novikov (KN) pairing* .

Note that the pairing depends not only on A (as the $\mathcal{F}^{\lambda}$ depend on it) but also critically on the splitting of A into I and O as the integration path will depend on it. Once the splitting is fixed the pairing will be fixed too.

Below we will see that the pairing is non-degenerate.

5.3 The Homogeneous Subspaces

Given the vector spaces $\mathcal{F}^\lambda$ of forms of weight λ we will now single out subspaces $\mathcal{F}^\lambda_m$ of degree m by giving a basis of these subspaces. This has been done in the 2-point case by Krichever and Novikov [33] and in the multi-point case by the author [40–43], see also Sadov [39]. See in particular [51, Chapters 3, 4, 5] for a complete treatment. All proofs of the statements to come can be found there.

Depending on whether the weight λ is integer or half-integer we set $\mathbb{J}_\lambda = \mathbb{Z}$ or $\mathbb{J}_\lambda = \mathbb{Z} + 1/2$. For $\mathcal{F}^\lambda$ we introduce for $m \in \mathbb{J}_\lambda$ subspaces $\mathcal{F}^\lambda_m$ of dimension K, where $K = \#I$, by exhibiting certain elements $f^\lambda_{m,p} \in \mathcal{F}^\lambda$, $p = 1, \ldots, K$ which constitute a basis of $\mathcal{F}^\lambda_m$. The elements are the elements of degree m. As explained in the following, the degree is in an essential way related to the zero orders of the elements at the points in I.

Let $I = (P_1, P_2, \ldots, P_K)$ then we require for the zero-order at the point $P_i \in I$ of the element $f^\lambda_{n,p}$

$$\operatorname{ord}_{P_i}(f^\lambda_{n,p}) = (n + 1 - \lambda) - \delta^p_i, \quad i = 1, \ldots, K\,. \tag{43}$$

The prescription at the points in O is made in such a way that the element $f^\lambda_{m,p}$ is essentially uniquely given. Essentially unique means up to multiplication with a constant.[4] After fixing as additional geometric data a system of coordinates z_l centered at P_l for $l = 1, \ldots, K$ and requiring that

$$f^\lambda_{n,p}(z_p) = z_p^{n-\lambda}(1 + O(z_p))(dz_p)^\lambda \tag{44}$$

the element $f^\lambda_{n,p}$ is uniquely fixed. In fact, the element $f^\lambda_{n,p}$ only depends on the first order jet of the coordinate z_p.

The element $f^\lambda_{n,p}$ has outside of A exactly g zeros (counted with multiplicities).

Example 5.3 Here we will not give the general recipe for the prescription at the points in O. Just to give an example which is also an important special case, assume $O = \{Q\}$ is a one-element set. If either the genus $g = 0$, or $g \geq 2$, $\lambda \neq 0, 1/2, 1$ and the points in A are in generic position then we require

$$\operatorname{ord}_Q(f^\lambda_{n,p}) = -K \cdot (n + 1 - \lambda) + (2\lambda - 1)(g - 1). \tag{45}$$

In the other cases (e.g. for $g = 1$) there are some modifications at the point in O necessary for finitely many n.

[4] Strictly speaking, there are some special cases where some constants have to be added such that the Krichever-Novikov duality (47) is valid.

Theorem 5.4 ([51, Thm. 3.6]) *Set*

$$\mathcal{B}^\lambda := \{ f^\lambda_{n,p} \mid n \in \mathbb{J}_\lambda,\ p = 1, \ldots, K \}. \tag{46}$$

Then (a) $\mathcal{B}^\lambda$ is a basis of the vector space $\mathcal{F}^\lambda$.

(b) The introduced basis $\mathcal{B}^\lambda$ of $\mathcal{F}^\lambda$ and $\mathcal{B}^{1-\lambda}$ of $\mathcal{F}^{1-\lambda}$ are dual to each other with respect to the Krichever-Novikov pairing (42), i.e.

$$\langle f^\lambda_{n,p}, f^{1-\lambda}_{-m,r} \rangle = \delta^r_p\, \delta^m_n, \quad \forall n, m \in \mathbb{J}_\lambda, \quad r, p = 1, \ldots, K. \tag{47}$$

In particular, from part (b) of the theorem it follows that the Krichever-Novikov pairing is non-degenerate.

An important consequence from the KN duality decomposition with respect to the basis given above is that we can write

$$v \in \mathcal{F}^\lambda, \quad v = \sum a_{n,p} f^\lambda_{n,p} \tag{48}$$

with

$$a_{n,p} = \langle v, f^{1-\lambda}_{-n,p} \rangle = \sum_{i=1}^{K} \operatorname{res}_{P_i}(v \cdot f^{1-\lambda}_{-n,p}) = -\sum_{j=1}^{M} \operatorname{res}_{Q_j}(v \cdot f^{1-\lambda}_{-n,p}). \tag{49}$$

It is quite convenient to use special notations for elements of some important weights:

$$\begin{aligned} e_{n,p} &:= f^{-1}_{n,p}, \quad \varphi_{n,p} := f^{-1/2}_{n,p}, \quad A_{n,p} := f^0_{n,p}, \\ \omega^{n,p} &:= f^1_{-n,p}, \quad \Omega^{n,p} := f^2_{-n,p}. \end{aligned} \tag{50}$$

In view of (47) for the forms of weight 1 and 2 we invert the index n and write it as a superscript.

Remark 5.5 It is also possible (and for certain applications necessary) to write explicitely down the basis elements $f^\lambda_{n,p}$ in terms of "usual" objects defined on the Riemann surface Σ. For genus zero they can be given with the help of rational functions in the quasi-global variable z. Indeed the second half of this survey will deal with it. For genus one (i.e. the torus case) representations with the help of Weierstraß σ and Weierstraß $\wp$ functions exists. For genus ≥ 1 there exists expressions in terms of theta functions (with characteristics) and prime forms. Here the Riemann surface has first to be embedded into its Jacobian via the Jacobi map. See [41, 44, 51, Chapter 5] for more details.

5.4 The Algebras

Theorem 5.6 ([51, Thm. 3.8]) *There exists constants R_1 and R_2 (depending on the number and splitting of the points in A and on the genus g) independent of λ and ν and independent of $n, m \in \mathbb{J}$ such that for the basis elements*

$$
\begin{aligned}
f^{\lambda}_{n,p} \cdot f^{\nu}_{m,r} = \;& f^{\lambda+\nu}_{n+m,r}\,\delta^r_p \\
&+ \sum_{h=n+m+1}^{n+m+R_1} \sum_{s=1}^{K} a^{(h,s)}_{(n,p)(m,r)} f^{\lambda+\nu}_{h,s}, \qquad a^{(h,s)}_{(n,p)(m,r)} \in \mathbb{C}, \\
[f^{\lambda}_{n,p}, f^{\nu}_{m,r}] = \;& (-\lambda m + \nu n)\, f^{\lambda+\nu+1}_{n+m,r}\,\delta^r_p \\
&+ \sum_{h=n+m+1}^{n+m+R_2} \sum_{s=1}^{K} b^{(h,s)}_{(n,p)(m,r)} f^{\lambda+\nu+1}_{h,s}, \quad b^{(h,s)}_{(n,p)(m,r)} \in \mathbb{C}.
\end{aligned}
\tag{51}
$$

For $N = 2$ points and in generic situations one obtains $R_1 = g$ and $R_2 = 3g$.

The theorem says in particular that with respect to both the associative and Lie structure the algebra $\mathcal{F}$ is weakly almost-graded. The reason why we only have weakly almost-gradedness is that

$$
\mathcal{F}^{\lambda} = \bigoplus_{m \in \mathbb{J}_\lambda} \mathcal{F}^{\lambda}_m, \qquad \text{with} \quad \dim \mathcal{F}^{\lambda}_m = K, \tag{52}
$$

and if we add up for a fixed m all λ we get that our homogeneous spaces are infinite dimensional.

In the proof of the theorem the KN duality property of the basis plays a crucial role. By (48) we obtain e.g.

$$
a^{(h,s)}_{(n,p),(m,r)} = \langle f^{\lambda}_{n,p} \cdot f^{\nu}_{m,r}, f^{1-(\lambda+\nu)}_{-h,s} \rangle\,. \tag{53}
$$

By considering the orders of the 3 basis elements at the points in I and O separately we can give finite ranges for the index (h, s) for which there are poles both at the points in I and at the points in O. Only in those ranges there will be a residue. Recall that for a compact Riemann surface the total residue has to be zero. Hence, there can only be a non-vanishing residue if summed over I if it compensated by a non-vanishing residue summed over O and vice versa.

In the definition of our KN type algebra only finitely many λs are involved, hence the following is immediate

Theorem 5.7 *The Krichever-Novikov type vector field algebras $\mathcal{L}$, function algebras $\mathcal{A}$, differential operator algebras $\mathcal{D}^1$, Lie superalgebras $\mathcal{S}$, and Jordan*

superalgebras $\mathcal{J}$ are (strongly) almost-graded algebras and the corresponding modules $\mathcal{F}^\lambda$ are almost-graded modules.

We obtain with $n \in \mathbb{J}_\lambda$

$$\begin{aligned} \dim \mathcal{L}_n = \dim \mathcal{A}_n = \dim \mathcal{F}^\lambda_n = K, \\ \dim \mathcal{S}_n = \dim \mathcal{J}_n = 2K, \quad \dim \mathcal{D}^1_n = 3K\,. \end{aligned} \tag{54}$$

If $\mathcal{U}$ is any of these algebras, with product denoted by $[\ ,\ ]$ then

$$[\mathcal{U}_n, \mathcal{U}_m] \subseteq \bigoplus_{h=n+m}^{n+m+R_i} \mathcal{U}_h, \tag{55}$$

with $R_i = R_1$ for $\mathcal{U} = \mathcal{A}$ and $R_i = R_2$ otherwise.

The lowest degree term component in (51) calculates for certain special cases.

$$\begin{aligned} A_{n,p} \cdot A_{m,r} &= A_{n+m,r}\,\delta_r^p \ + \text{ h.d.t.} \\ A_{n,p} \cdot f^\lambda_{m,r} &= f^\lambda_{n+m,r}\,\delta_r^p \ + \text{ h.d.t.} \\ [e_{n,p}, e_{m,r}] &= (m-n)\cdot e_{n+m,r}\,\delta_r^p \ + \text{ h.d.t.} \\ e_{n,p} \cdot f^\lambda_{m,r} &= (m+\lambda n)\ \cdot f^\lambda_{n+m,r}\,\delta_r^p \ + \text{ h.d.t.} \end{aligned} \tag{56}$$

Here h.d.t. denote linear combinations of basis elements of degree between $n+m+1$ and $n+m+R_i$,

Finally, the almost-grading of $\mathcal{A}$ induces an almost-grading of the current algebra $\overline{\mathfrak{g}}$ by setting $\overline{\mathfrak{g}}_n = \mathfrak{g} \otimes \mathcal{A}_n$. We obtain

$$\overline{\mathfrak{g}} = \bigoplus_{n\in\mathbb{Z}} \overline{\mathfrak{g}}_n, \quad \dim \overline{\mathfrak{g}}_n = K \cdot \dim \mathfrak{g}. \tag{57}$$

5.5 *Triangular Decomposition and Filtrations*

Let $\mathcal{U}$ be one of the above introduced algebras (including the current algebra). On the basis of the almost-grading we obtain a triangular decomposition of the algebras

$$\mathcal{U} = \mathcal{U}_{[+]} \oplus \mathcal{U}_{[0]} \oplus \mathcal{U}_{[-]}, \tag{58}$$

where

$$\mathcal{U}_{[+]} := \bigoplus_{m>0} \mathcal{U}_m, \quad \mathcal{U}_{[0]} = \bigoplus_{m=-R_i}^{m=0} \mathcal{U}_m, \quad \mathcal{U}_{[-]} := \bigoplus_{m<-R_i} \mathcal{U}_m. \tag{59}$$

By the almost-gradedness the [+] and [−] subspaces are (infinite dimensional) subalgebras. The [0] spaces in general not. Sometimes we call them *critical strips*.

With respect to the almost-grading of $\mathcal{F}^\lambda$ we introduce a filtration

$$\begin{gathered} \mathcal{F}^\lambda_{(n)} := \bigoplus_{m\ge n} \mathcal{F}^\lambda_m, \\ \dots \supseteq \mathcal{F}^\lambda_{(n-1)} \supseteq \mathcal{F}^\lambda_{(n)} \supseteq \mathcal{F}^\lambda_{(n+1)} \dots \end{gathered} \tag{60}$$

Proposition 5.8 *[51, Prop. 3.15]*

$$\mathcal{F}^\lambda_{(n)} = \{ f \in \mathcal{F}^\lambda \mid \mathrm{ord}_{P_i}(f) \ge n - \lambda, \forall i = 1, \dots, K \}. \tag{61}$$

6 Central Extensions

Central extensions of our algebras appear naturally in the context of quantization and regularization of actions. Of course they are also of independent mathematical interest. Recall that a projective action of a Lie algebra L defines a honest Lie action of a centrally extended algebra $\hat{L}$.

6.1 Central Extensions and Cocycles

For the convenience of the reader let us repeat the relation between central extensions and the second Lie algebra cohomology with values in the trivial module. A central extension of a Lie algebra W is a special Lie algebra structure on the vector space direct sum $\widehat{W} = \mathbb{C} \oplus W$. If we denote $\hat{x} := (0, x)$ and $t := (1, 0)$ then the Lie structure is given by

$$[\hat{x}, \hat{y}] = \widehat{[x, y]} + \psi(x, y) \cdot t, \quad [t, \widehat{W}] = 0, \quad x, y \in W, \tag{62}$$

with a bilinear form ψ. The map $x \mapsto \hat{x} = (0, x)$ is a linear splitting map. $\widehat{W}$ will be a Lie algebra, e.g. will fulfill the Jacobi identity, if and only if ψ is an antisymmetric bilinear form and fulfills the Lie algebra 2-cocycle condition

$$0 = d_2\psi(x, y, z) := \psi([x, y], z) + \psi([y, z], x) + \psi([z, x], y). \tag{63}$$

Two central extensions are equivalent if they essentially correspond only to the choice of different splitting maps. A 2-cochain ψ is a coboundary if there exists a linear form $\varphi : W \to \mathbb{C}$ such that

$$\psi(x, y) = \varphi([x, y]). \tag{64}$$

Every coboundary is a cocycle. The second Lie algebra cohomology $\mathrm{H}^2(W, \mathbb{C})$ of W with values in the trivial module $\mathbb{C}$ is defined as the quotient of the space of 2-cocycles modulo coboundaries. Moreover, two central extensions are equivalent if and only if the difference of their defining 2-cocycles ψ and ψ' is a coboundary. In this way the second Lie algebra cohomology $\mathrm{H}^2(W, \mathbb{C})$ classifies equivalence classes of central extensions. The class [0] corresponds to the trivial central extension. In this case the splitting map is a Lie homomorphism. We construct central extensions of our algebras by exhibiting such Lie algebra 2-cocycles.

Clearly, equivalent central extensions are isomorphic. The opposite is not true. In our case we can always rescale the central element by multiplying it with a nonzero scalar. This is an isomorphism but not an equivalence of central extensions. Nevertheless, it is an irrelevant modification. Hence we will be mainly interested in central extensions modulo equivalence and rescaling. They are classified by [0] and the elements of the projectivized cohomology space $\mathbb{P}(\mathrm{H}^2(W, \mathbb{C}))$.

In the classical case we have $\dim \mathrm{H}^2(\mathcal{W}, \mathbb{C}) = 1$, hence there are only two essentially different central extensions, the splitting one given by the direct sum $\mathbb{C} \oplus \mathcal{W}$ of Lie algebras and the up to equivalence and rescaling unique non-trivial one, the Virasoro algebra $\mathcal{V}$.

6.2 Geometric Cocycles

The cocycle of the Witt algebra $1/12\ (n^3 - n)\delta_n^{-m}$ used to define the Virasoro algebra is very special. Obviously, it does not make any sense in the higher genus and/or multi-point case. We need to find a geometric description. For this we have first to introduce connections.

6.2.1 Projective and Affine Connections

Let $(U_\alpha, z_\alpha)_{\alpha \in J}$ be a covering of the Riemann surface by holomorphic coordinates with transition functions $z_\beta = f_{\beta\alpha}(z_\alpha)$.

Definition 6.1

(a) A system of local (holomorphic, meromorphic) functions $R = (R_\alpha(z_\alpha))$ is called a (holomorphic, meromorphic) *projective connection* if it transforms as

$$R_\beta(z_\beta)\cdot(f'_{\beta,\alpha})^2 = R_\alpha(z_\alpha) + S(f_{\beta,\alpha}), \qquad \text{with} \quad S(h) = \frac{h'''}{h'} - \frac{3}{2}\left(\frac{h''}{h'}\right)^2, \tag{65}$$

the Schwartzian derivative. Here $'$ denotes differentiation with respect to the coordinate z_α.

(b) A system of local (holomorphic, meromorphic) functions $T = (T_\alpha(z_\alpha))$ is called a (holomorphic, meromorphic) *affine connection* if it transforms as

$$T_\beta(z_\beta)\cdot(f'_{\beta,\alpha}) = T_\alpha(z_\alpha) + \frac{f''_{\beta,\alpha}}{f'_{\beta,\alpha}}. \tag{66}$$

Every Riemann surface admits a holomorphic projective connection [23, 25]. Given a point P then there exists always a meromorphic affine connection holomorphic outside of P and having maximally a pole of order one there [43].

From their very definition it follows that the difference of two affine (projective) connections will be a (quadratic) differential. Hence, after fixing one affine (projective) connection all others are obtained by adding (quadratic) differentials.

Next we introduce in a geometric way cocycles by integration of certain differentials, associated to pairs of Lie algebra elements, over arbitrary smooth curves. Such cocycles we call geometric cocycles. For the proofs that the following expressions are indeed 2-cocycles we refer to [43, 47] (and [51]).

6.2.2 The Function Algebra $\mathcal{A}$

We consider it as abelian Lie algebra. Let C be an arbitrary smooth but not necessarily connected curve. We set

$$\psi^1_C(g,h) := \frac{1}{2\pi \mathrm{i}}\int_C g\,dh, \qquad g,h\in\mathcal{A}. \tag{67}$$

6.2.3 The Current Algebra $\overline{\mathfrak{g}}$

For $\overline{\mathfrak{g}} = \mathfrak{g}\otimes\mathcal{A}$ we fix a symmetric, invariant, bilinear form β (not necessarily non-degenerate) on the finite-dimensional Lie algebra $\mathfrak{g}$. Recall, that invariance means that we have $\beta([x,y],z) = \beta(x,[y,z])$ for all $x, y, z \in \mathfrak{g}$. Then the following expressions define a cocycle

$$\psi^2_{C,\beta}(x \otimes g, y \otimes h) := \beta(x,y) \cdot \frac{1}{2\pi \mathrm{i}} \int_C g dh, \quad x, y \in \mathfrak{g},\ g, h \in \mathcal{A}. \tag{68}$$

6.2.4 The Vector Field Algebra $\mathcal{L}$

Here it is a little bit more delicate. First we have to choose a (holomorphic) projective connection R. We define

$$\psi^3_{C,R}(e,f) := \frac{1}{24\pi \mathrm{i}} \int_C \left(\frac{1}{2}(e'''f - ef''') - R \cdot (e'f - ef') \right) dz\,. \tag{69}$$

Only by the term coming with the projective connection it will be a well-defined differential, i.e. independent of the coordinate chosen. Another choice of a projective connection will result in a cohomologous one. Hence, the equivalence class of the central extension will be the same.

6.2.5 The Differential Operator Algebra $\mathcal{D}^1$

For the differential operator algebra the cocycles of type (67) for $\mathcal{A}$ can be extended by zero on the subspace $\mathcal{L}$. The cocycles for $\mathcal{L}$ can be pulled back. In addition there is a third type of cocycles mixing $\mathcal{A}$ and $\mathcal{L}$:

$$\psi^4_{C,T}(e,g) := \frac{1}{24\pi \mathrm{i}} \int_C (eg'' + Teg')dz, \qquad e \in \mathcal{L},\ g \in \mathcal{A}, \tag{70}$$

with an affine connection T, with at most a pole of order one at a fixed point in O. Again, a different choice of the connection will not change the cohomology class.

6.2.6 The Lie Superalgebra $\mathcal{S}$

Here we have to take into account that it is not a Lie algebra. Hence, the Jacobi identity has to be replaced by the super-Jacobi identity. The conditions for being a cocycle for the superalgebra cohomology will change too. Recall the definition of the algebra from Sect. 4.4, in particular that the even elements (parity 0) are the vector fields and the odd elements (parity 1) are the half-forms. A bilinear form c is a cocycle if the following is true. The bilinear map c will be symmetric if both x and y are odd, otherwise it will be antisymmetric:

$$c(x,y) = -(-1)^{\bar{x}\bar{y}} c(x,y). \tag{71}$$

The super-cocycle condition reads as

$$(-1)^{\bar{x}\bar{z}}c(x,[y,z]) + (-1)^{\bar{y}\bar{x}}c(y,[z,x]) + (-1)^{\bar{z}\bar{y}}c(z,[x,y]) = 0. \tag{72}$$

With the help of c we can define central extensions in the Lie superalgebra sense. If we put the condition that the central element is even then the cocycle c has to be an even map and c vanishes for pairs of elements of different parity.

By convention we denote vector fields by $e, f, g, \ldots$ and $-1/2$-forms by $\varphi, \psi, \chi, \ldots$ and get

$$c(e,\varphi) = 0, \quad e \in \mathcal{L},\ \varphi \in \mathcal{F}^{-1/2}. \tag{73}$$

The super-cocycle conditions for the even elements is just the cocycle condition for the Lie subalgebra $\mathcal{L}$. The only other nonvanishing super-cocycle condition is for the *(even,odd,odd)* elements and reads as

$$c(e,[\varphi,\psi]) - c(\varphi, e\,.\,\psi) - c(\psi, e\,.\,\varphi) = 0. \tag{74}$$

Here the definition of the product $[e,\psi] := e\,.\,\psi$ was used.

If we have a cocycle c for the algebra $\mathcal{S}$ we obtain by restriction a cocycle for the algebra $\mathcal{L}$. For the mixing term we know that $c(e,\psi) = 0$. A naive try to put just anything for $c(\varphi,\psi)$ (for example 0) will not work as (74) relates the restriction of the cocycle on $\mathcal{L}$ with its values on $\mathcal{F}^{-1/2}$.

Proposition 6.2 ([50]) *Let C be any closed (differentiable) curve on Σ not meeting the points in A, and let R be any (holomorphic) projective connection then the bilinear extension of*

$$\begin{aligned}
\Phi_{C,R}(e,f) &:= \frac{1}{24\pi \mathrm{i}}\int_C \left(\frac{1}{2}(e'''f - ef''') - R\cdot(e'f - ef')\right)dz \\
\Phi_{C,R}(\varphi,\psi) &:= -\frac{1}{24\pi \mathrm{i}}\int_C \left(\varphi''\cdot\psi + \varphi\cdot\psi'' - R\cdot\varphi\cdot\psi\right)dz \\
\Phi_{C,R}(e,\varphi) &:= 0
\end{aligned} \tag{75}$$

gives a Lie superalgebra cocycle for $\mathcal{S}$, hence defines a central extension of $\mathcal{S}$. A different projective connection will yield a cohomologous cocycle.

Note that the $\Phi_{C,R}$ restricted to $\mathcal{L}$ gives $\Psi^3_{C,R}$.

A similar formula was given by Bryant in [12]. By adding the projective connection in the second part of (75) he corrected some formula appearing in [6]. He only considered the two-point case and only the integration over a separating cycle. See also [32] for the multi-point case, where still only the integration over a separating cycle is considered.

In contrast to the differential operator algebra case the two parts cannot be prescribed independently. Only with the same integration path (more precisely, homology class) and the given factors in front of the integral it will work. The reason for this is that (74) relates both.

6.3 Uniqueness and Classification of Central Extensions

The above introduced cocycles depend on the choice of the connections R and T. Different choices will not change the cohomology class. Hence, this ambiguity does not disturb us. What really matters is that they depend on the integration curve C chosen.

In contrast to the classical situation, for the higher genus and/or multi-point situation there are many essentially different closed curves and hence many non-equivalent central extensions defined by the integration.

But we should take into account that we want to extend the almost-grading from our algebras to the centrally extended ones. This means we take $\deg \hat{x} := \deg x$ and assign a degree $deg(t)$ to the central element t, and still we want to obtain almost-gradedness.

This is possible if and only if our defining cocycle ψ is *local*, or *almost-graded* in the following sense, There exists $M_1, M_2 \in \mathbb{Z}$ such that

$$\forall n, m : \quad \psi(W_n, W_m) \neq 0 \implies M_1 \leq n + m \leq M_2. \tag{76}$$

Here W stands for any of our algebras (including the supercase). The name "local" was introduced in the two point case by Krichever and Novikov in [33]).

Very important, "local" is defined in terms of the almost-grading, and the almost-grading itself depends on the splitting $A = I \cup O$. Hence what is "local" depends on the splitting too.

We will call a cocycle *bounded* (from above) if there exists $M \in \mathbb{Z}$ such that

$$\forall n, m : \quad \psi(W_n, W_m) \neq 0 \implies n + m \leq M. \tag{77}$$

Similarly bounded from below can be defined. Locality means bounded from above and from below.

Given a cocycle class we call it *bounded* (respectively *local*) if and only if it contains a representing cocycle which is bounded (respectively local). Not all cocycles in a bounded class have to be bounded. If we choose as integration path a separating cocycle C_S, or one of the C_i then the above introduced geometric cocycles are local, respectively bounded (from above). Recall that in this case integration can be done by calculating residues at the in-points or at the out-points. All these cocycles are cohomologically nontrivial. The theorems in the following concern the opposite direction. They were treated in my works [46, 47, 50]. See also [51] for a complete and common treatment.

The following result for the vector field algebra $\mathcal{L}$ gives the principal structure of the classification results.

Theorem 6.3 ([47], [51, Thm. 6.41]) *Let $\mathcal{L}$ be the Krichever–Novikov vector field algebra with a given almost-grading induced by the splitting $A = I \cup O$.*

(a) The space of bounded cohomology classes is K-dimensional ($K = \#I$). A basis is given by setting the integration path in (69) to C_i, $i = 1, \ldots, K$ the little (deformed) circles around the points $P_i \in I$.
(b) The space of local cohomology classes is one-dimensional. A generator is given by integrating (69) over a separating cocycle C_S, i.e.

$$\psi^3_{C_S,R}(e,f) = \frac{1}{24\pi\mathrm{i}} \int_{C_S} \left(\frac{1}{2}(e'''f - ef''') - R \cdot (e'f - ef') \right) dz \,. \tag{78}$$

(c) Up to equivalence and rescaling there is only one non-trivial one-dimensional central extension $\widehat{\mathcal{L}}$ of the vector field algebra $\mathcal{L}$ which allows an extension of the almost-grading.

Remark 6.4 In the classical situation, Part (c) shows also that the Virasoro algebra is the unique non-trivial central extension of the Witt algebra (up to equivalence and rescaling). This result extends to the more general situation under the condition that one fixes the almost-grading, hence the splitting $A = I \cup O$. Here I like to repeat the fact that for $\mathcal{L}$ depending on the set A and its possible splittings into two disjoint subsets there are different almost-gradings. Hence, the "unique" central extension finally obtained will also depend on the splitting. Only in the two point case there is only one splitting possible. In the case that the genus $g \geq 1$ there are even integration paths possible in the definition of (69) which are not homologous to a separating cycle of any splitting. Hence, there are other central extensions possible not corresponding to any almost-grading.

The above theorem is a model for all other classification results. We will always obtain a statement about the bounded (from above) cocycles and then for the local cocycles.

If we consider the function algebra $\mathcal{A}$ as an abelian Lie algebra then every skew-symmetric bilinear form will be a non-trivial cocycle. Hence, there is no hope of uniqueness. But if we add the condition of $\mathcal{L}$-invariance to the cocycle, which is given as

$$\psi(e.g, h) + \psi(g, e.h) = 0, \quad \forall e \in \mathcal{L},\ g, h \in \mathcal{A} \tag{79}$$

things will change. Note that the cocycle (67) is $\mathcal{L}$-invariant.

Let us denote the subspace of local cohomology classes by H^2_{loc}, and the subspace of local and $\mathcal{L}$-invariant cohomology classes by $\mathrm{H}^2_{\mathcal{L},loc}$. Note that the conditions are only required for at least one representative in the cohomology class. We collect a part of the results for the cocycle classes of the other algebras in the following theorem.

Theorem 6.5 ([51, Cor. 6.48])

(a) $\dim \mathrm{H}^2_{\mathcal{L},loc}(\mathcal{A}, \mathbb{C}) = 1$,
(b) $\dim \mathrm{H}^2_{loc}(\mathcal{L}, \mathbb{C}) = 1$,
(c) $\dim \mathrm{H}^2_{loc}(\mathcal{D}^1, \mathbb{C}) = 3$,
(d) $\dim \mathrm{H}^2_{loc}(\overline{\mathfrak{g}}, \mathbb{C}) = 1$ *for* $\mathfrak{g}$ *a simple finite-dimensional Lie algebra,*
(e) $\dim \mathrm{H}^2_{loc}(\mathcal{S}, \mathbb{C}) = 1$,

A basis of the cohomology spaces are given by taking the cohomology classes of the cocycles (67), (69), (70), (68), (75) obtained by integration over a separating cycle C_S.

Consequently, we obtain also for these algebras the corresponding result about uniqueness of almost-graded central extensions. For the differential operator algebra we get three independent cocycles. This generalizes results obtained in [2] for the classical case.

For results on the bounded cocycle classes we have to multiply the dimensions above by $K = \#I$. For the supercase with odd central element the bounded cohomology vanishes.

For $\mathfrak{g}$ a reductive Lie algebra and if the cocycle is $\mathcal{L}$-invariant if restricted to the abelian part, a complete classification of local cocycle classes for both $\overline{\mathfrak{g}}$ and $\mathcal{D}^1_{\mathfrak{g}}$ can be found in [46], [51, Chapter 9].

I like to mention that in all the applications I know of, the cocycles coming from representations, regularizations, etc. are local. Hence, the uniqueness or classification result above can be applied.

Remark 6.6 (On $\mathcal{L}$-Invariance) The $\mathcal{L}$-invariance is a natural condition. If we restrict a cocycle for the differential operator algebra $\mathcal{D}^1$ to the subalgebra $\mathcal{A}$ of functions we obtain automatically an $\mathcal{L}$-invariant cocycle for $\mathcal{A}$. Vice versa, every $\mathcal{L}$-invariant cocycle for A can be extended by zero on the complementary space $\mathcal{L}$ to a cocycle for $\mathcal{D}^1$.

The cocycle (67) has another interesting property. It is *multiplicative*, i.e.

$$\psi(a \cdot b, c) + \psi(b \cdot c, a) + \psi(c \cdot a, b) = 0, \qquad a, b, c \in \mathcal{A}. \tag{80}$$

It is this property which is needed that (68) is a cocycle for $\overline{\mathfrak{g}}$, see also [46]. It is shown in [47] that the classes of multiplicative bounded cocycles equals the classes of bounded $\mathcal{L}$-invariant cocycles.

7 The Genus Zero and Multi-Point Case

In the second half of this survey we consider the situation that our Riemann surface is the Riemann sphere (i.e. of genus zero) and that we have a finite number of marked points where poles are allowed. Again the classical situation is a special case.

There are many reasons to study this situation in detail. First, of course we give an illustration of the definition and construction done in arbitrary genus in an explicit way using explicit forms of generators. Second, we are able to deduce additional properties in the genus zero case. In particular, we will construct the universal central extensions of those algebras which admit an universal central extension.

The third reason is that recently there was a revived interest in the genus zero multi-point situation by a number of people coming from field theory and representation theory. In these communities the algebras are called *N-Virasoro algebras*.

But in their work is no almost-graded structure used. Now, in the case of infinite dimensional Lie algebra the existence of a grading respectively the weaker version of an almost-grading is crucial for understanding and constructions. It was my intention in [53] to show that these N-Virasoro algebras are nothing else as multi-point genus zero KN type algebras and to use the theory of KN type (in particular the almost-gradedness) to deduce faster and more conceptual their results and even more, giving explanations of observed results.

In the following I will review our results. For the proofs I mainly refer to [53].

7.1 The Geometric Set-up

Now let Σ_0 be the Riemann sphere S^2, or equivalently $\mathbb{P}^1(\mathbb{C})$ with the quasi-global coordinate z. We call it quasi-global, as it is not defined at the point ∞. Let us denote the set of points

$$A = \{P_1, P_2, \ldots, P_N\}, \qquad P_i \neq P_j, \text{ for } i \neq j. \tag{81}$$

For notational simplicity we single out the point P_N as reference point. By an automorphism of $\mathbb{P}^1(\mathbb{C})$, i.e. a fractional linear transformation or equivalently an element of PGL$(2, \mathbb{C})$, the point P_N can be brought to ∞. In fact two more points could be normalized to be 0 and 1. In this section we will not do so, but see Sect. 8.

Our points are given by their global coordinates

$$P_i = a_i, \quad a_i \in \mathbb{C}, \ i = 1, \ldots, N-1, \qquad P_N = \infty. \tag{82}$$

At these points we have the local coordinates

$$z - a_i, \ i = 1, \ldots, N-1, \qquad w = 1/z. \tag{83}$$

Recall that the canonical line bundle $\mathcal{K}$ of Σ_0 is the holomorphic line bundle whose local sections are the local holomorphic differentials. For $\mathbb{P}^1(\mathbb{C})$, in the language of algebraic geometry, we have that $\mathcal{K} = \mathcal{O}(-2)$. This bundle has a unique square root $L = \mathcal{O}(-1)$, which is the tautological bundle, respectively the dual of the hyperplane section bundle. We denote this bundle also by $\mathcal{K}^{1/2}$.[5] We introduced meromorphic forms of weight λ as sections of the bundle $\mathcal{K}^{\lambda}$. In our genus zero situation we can describe the form by a meromorphic function on the affine part $\mathbb{C}$ with respect to the coordinate z. By this description its behaviour at the point ∞ is uniquely fixed by the fundamental transformation $dz = -w^{-2}dw$. Moreover, by the fixing $P_N = \infty$ the set of meromorphic forms f of weight λ on $\mathbb{P}^1(\mathbb{C})$ holomorphic outside of A correspond 1:1 to meromorphic functions $a(z)$ holomorphic outside of A via $f(z) = a(z)dz^{\lambda}$. Both a and f will have the same orders at the points in $\mathbb{C}$. For the order at the point ∞ we have

$$\mathrm{ord}_{\infty}(f) = \mathrm{ord}_{\infty}(a) - 2\lambda. \tag{84}$$

Recall that on a compact Riemann surface the sum of the orders (summed over all points) of a meromorphic function $f \not\equiv 0$ equals zero. Hence, more generally

Proposition 7.1 *Let $f \in \mathcal{F}^{\lambda}$, $f \not\equiv 0$ then*

$$\sum_{P \in \Sigma_0} \mathrm{ord}_P(f) = -2\lambda. \tag{85}$$

For this and related results see e.g. [48]. Also recall that the meromorphic functions in our case are nothing else as rational functions with respect to the variable z.

Recall from Sect. 4 that

$$\mathcal{F} := \bigoplus_{\lambda \in \frac{1}{2}\mathbb{Z}} \mathcal{F}^{\lambda}. \tag{86}$$

We specialize from Sect. 5 the almost-grading for genus zero as follows. We numerate the points in the splitting like

$$A = I \cup O, \qquad I = (P_1, P_2, \ldots, P_K), \qquad O = (P_{K+1}, \ldots, P_N = \infty). \tag{87}$$

As $P_N = \infty \in O$ it is enough to construct a basis $\{A_{n,p} \mid n \in \mathbb{Z}, p = 1, \ldots, K\}$ for $\mathcal{A} = \mathcal{F}^0$. The decomposition of $\mathcal{A}$ into homogeneous subspaces induces a decomposition of $\mathcal{F}^{\lambda}$ by

$$\mathcal{F}^{\lambda}_n = \mathcal{A}_{n-\lambda} dz^{\lambda}, \quad \text{respectively} \quad f^{\lambda}_{n,p} = A_{n-\lambda,p} dz^{\lambda}. \tag{88}$$

[5] There is no ambiguity in choosing the square root.

The shift by $-\lambda$ is quite convenient and is beside other things related to the duality property. The recipe for constructing the $A_{n,p}$ is given in [41, 42], see also [51]. As a principal property we have (like in the arbitrary genus case)

$$\mathrm{ord}_{P_i}(A_{n,p}) = (n+1) - \delta_i^p\,,\ i = 1, \ldots, K\,. \tag{89}$$

At the points in O corresponding orders are set to make the element unique up to multiplication by a non-zero scalar.

Example 7.2 We call the splitting

$$I = (P_1, P_2, \ldots, P_K), \quad O = (P_N = \infty), \quad K = N - 1 \tag{90}$$

the *standard splitting*. Recall that P_i corresponds to the point given by the coordinate $z = a_i$. We set

$$\alpha(p) := \Big(\prod_{\substack{i=1\\ i\neq p}}^{K} (a_p - a_i)\Big)^{-1}. \tag{91}$$

and define

$$A_{n,p}(z) := (z - a_p)^n \cdot \prod_{\substack{i=1\\ i\neq p}}^{K} (z - a_i)^{n+1} \cdot \alpha(p)^{n+1}\,. \tag{92}$$

The last factor is a normalization factor yielding

$$A_{n,p}(z) = (z - a_p)^n (1 + O(z - a_p)). \tag{93}$$

With this description the order at ∞ is fixed as

$$-(Kn + K - 1)\,. \tag{94}$$

In particular, we obtain for $\mathcal{L}$ the basis

$$e_{n,p} = f_{n,p}^{-1} = A_{n+1,p}\frac{d}{dz} = (z - a_p)^{n+1} \cdot \prod_{\substack{i=1\\ i\neq p}}^{K} (z - a_i)^{n+2} \cdot \alpha(p)^{n+2}\,\frac{d}{dz}. \tag{95}$$

7.2 The Algebras for the Standard Splitting

Theorem 5.6 is of course also valid in our special genus zero situation. We will have a closer look on the result for the standard splitting. Recall that in case of the standard splitting the set O consists only of the point ∞.

For illustration we give the bounds R_1 and R_2 in Theorem 5.6

Proposition 7.3

$$R_1 = \begin{cases} 0, & K = 1, \\ 1, & K > 1, \end{cases} \qquad R_2 = \begin{cases} 0, & K = 1, \\ 1, & K = 2, \\ 2, & K > 2. \end{cases} \tag{96}$$

Proof For the upper bound we have to check the order at ∞ with respect to the variable w. We use for the individual factors in the expression (53) the value (94) and sum over all factors and do not forget to decrease the order by 2 coming from dz. If we do this for the algebra $\mathcal{A}$ we get as order at ∞ for $\omega := A_{n,p} \cdot A_{m,r} \cdot A_{-(n+m+k)-1,s} dz$ the value $-2K + K \cdot k + 1$. A pole is only possible if this value is ≤ -1. Hence only for

$$k \le -\frac{2}{K} + 2\,. \tag{97}$$

This yields the claimed value for R_1. For the Lie algebra $\mathcal{L}$, respectively for the Lie module we have to consider $\omega := [A_{n+1,p}, A_{m-\lambda,r}] \cdot A_{-(n+m+k)-(1-\lambda)}\, dz$. Here the As should be taken as the local expressions of the vector fields, resp. of the forms. For the order at ∞ we obtain $-3K + K \cdot k + 2$. Which yields that a pole is only possible for

$$k \le -\frac{3}{K} + 3\,, \tag{98}$$

and hence the claimed value for R_2. □

Indeed these bounds are also valid for $K = M$.

For the algebra $\mathcal{A}$ we obtain by using (96) and (56)

$$A_{n,i} \cdot A_{m,j} = \delta_i^j\, A_{n+m,j} + \sum_{s=1}^{K} \alpha_{(n,i)(m,j)}^{(n+m+1,s)} A_{n+m+1,s}\,. \tag{99}$$

The unknown structure coefficients in (99) of the algebras can be directly calculated by calculating the residues of the rational expression for ω.

Example 7.4 ($N = 2$) By a PGL(2, $\mathbb{C}$) action the two points can be transported to 0 and ∞. This is the classical situation and there is only one splitting. Hence, everything is fixed. The above basis gives back the conventional one.

Example 7.5 ($N = 3$) Here by a PGL(2, $\mathbb{C}$) action the three points can be normalized to $\{0, 1, \infty\}$. Hence, up to isomorphy there are only one three-point algebra (for each type of algebras). If we fix such an algebra we obtain three different splittings of the set $\{0, 1, \infty\}$ and consequently also 3 essentially different almost-gradings, triangular decompositions, etc. The three-point case is in a certain sense special as there are still the automorphism of $\mathbb{P}^1(\mathbb{C})$ permuting these three points. They induce automorphisms of the algebras which permute the almost-gradings. We will consider this situation in detail in Sect. 8.

Example 7.6 ($N = 4$) This is the first case where we have a moduli parameter for the geometric situation. We normalize our A to

$$\{0, 1, a, \infty\}, \qquad a \in \mathbb{C},\ a \neq 0, 1\,. \tag{100}$$

We have 2 different types of splittings, i.e. the type $4 = 3+1$ and the type $4 = 2+2$. For example

$$\{0, 1, a\} \cup \{\infty\}, \quad \text{and} \quad \{0, 1\} \cup \{a, \infty\}\,. \tag{101}$$

The first type is the standard splitting for which we gave the basis above (92). For the second splitting a basis of $\mathcal{A}$ and hence of all $\mathcal{F}^\lambda$ is

$$\begin{aligned} A_{n,1}(z) &= z^n (z-1)^{n+1} (z-a)^{-(n+1)} a^{(n+1)}, \\ A_{n,2}(z) &= z^{n+1} (z-1)^n (z-a)^{-(n+1)} (1-a)^{(n+1)}, \end{aligned} \tag{102}$$

where $n \in \mathbb{Z}$. The last factor is again a normalization constant. This basis defines an almost-grading for the four-point algebra which is not equivalent to the standard almost-grading. Again the algebra coefficients can be calculated easily via residues. Using (96) for $K = M = 2$ the general expression is given also by (99).

7.3 Central Extensions

Recall that in Sect. 6 we gave expressions for the central extensions by integrating differential forms $\widehat{\gamma}$ over cycles on the Riemann surface yielding cocycles. Such cocycles we called geometric cocycles. The relevant space is the homology space $\mathrm{H}_1(\Sigma \setminus A, \mathbb{C})$.

For genus zero and $N \geq 1$ we have

$$\dim \mathrm{H}_1(\Sigma_0 \setminus A, \mathbb{C}) = (N-1). \tag{103}$$

A basis of this cohomology space is given by circles C_i around the points P_i where we leave out one of them. For example we can take $[C_i]$, $i = 1, \ldots, N-1$. We have the relation

$$\sum_{i=1}^{N-1} [C_i] = -[C_N]. \tag{104}$$

But there is another choice. As explained above after choosing a splitting with separating cycle $[C_S]$ we take it as one of the basis elements and $N-2$ other $[C_i]$s which are linearly independent. For the standard splitting with $P_N = \{\infty\}$ we have

$$[C_S] = -[C_\infty], \qquad [C_i],\ i = 1, \ldots, N-2. \tag{105}$$

There is a crucial difference to the case of higher genus. In genus zero our cycles are always build from circles around our points where poles are allowed. This is not the case in higher genus. And this makes a big difference. Integration around the C_i can be done via calculations of residues. Hence, we always get for our geometric cocycles (for the standard splitting)

$$\gamma_{[C_S]} = \sum_{i=1}^{N-1} \operatorname{res}_{P_i}(\widehat{\gamma}) = -\operatorname{res}_\infty(\widehat{\gamma}), \qquad \gamma_{[C_i]} = \operatorname{res}_{P_i}(\widehat{\gamma}),\ i = 1, \ldots, N-2. \tag{106}$$

Another difference to higher genus is that that in genus zero and our system of coordinates z_i the coordinate change maps are projective transformation and the Schwartzian derivative vanishes, see (65). Hence, the projective connection R appearing in the definition of the cocycle for the vector field algebra can be set to $R = 0$ on all our coordinate patches. In the mixing cocycle for the differential operator algebra there appears the affine connection. We cannot take $T = 0$ globally, but from (66) it follows that we may take $T_i = 0$ on the affine part and $T(w) = -2/w$ around the point ∞.

If we consider the standard splitting and specialize the cocycles γ introduced in Sect. 6 then the theorems presented there show that $\gamma_{[C_S]}$ will be local, whereas the $\gamma_{[C_i]}$ are bounded. We also know that the $\gamma_{[C_i]}$, $i = 1, \ldots, N-1$ gives a basis of the space of bounded cocycle classes. Here bounded means bounded with respect to the standard splitting where the only point in O is the point ∞. For simplicity we will sometimes use γ_i for $\gamma_{[C_i]}$ and γ_S for $\gamma_{[C_S]}$.

7.4 Universal Central Extensions

In the following we will show that in the genus zero case all cocycle classes will be bounded. Moreover, we will describe with this result the universal central extensions in case that they exist.

Theorem 7.7 *In the case of genus zero all cocycle classes for the algebras in Sect. 4 are bounded classes with respect to the standard splitting. In particular, a basis of all cocycle classes is given by* γ_i, $i = 1, \ldots, N-1$. *This is also true for the function algebra* $\mathcal{A}$ *if we assume that the cocycle is* $\mathcal{L}$*-invariant or multiplicative.*

Note that we do not make any reference on any splitting at the beginning. Also we remark that this theorem is only true in genus zero. Below we will make some comments on the proof. But first we draw some consequences.

First we take a look on the function algebra $\mathcal{A}$.

Theorem 7.8 *In the* N*-point genus zero situation the space of* $\mathcal{L}$*-invariant (or multiplicative) cocycles is* $(N-1)$ *dimensional and is isomorphic to* $\mathrm{H}_1(\Sigma_0 \setminus A, \mathbb{C})$ *via*

$$[C] \mapsto \gamma_C^{\mathcal{A}}; \qquad \gamma_C^{\mathcal{A}}(f, g) = \frac{1}{2\pi \mathrm{i}} \int_C f dg. \tag{107}$$

In particular, every $\mathcal{L}$*-invariant cocycle is multiplicative and vice versa. These cocycles are geometric.*

Proposition 7.9 *Up to rescaling the cocycle*

$$\gamma_\infty^{\mathcal{A}} = -\sum_{i=1}^{N-1} \gamma_{C_i}^{\mathcal{A}}(f, g) = \mathrm{res}_\infty(f dg) \tag{108}$$

is the unique $\mathcal{L}$*-invariant (and equivalently multiplicative) cocycle which is local with respect to the standard splitting.*

Remark 7.10 (Heisenberg Algebra) We consider the central extensions of $\mathcal{A}$ with central terms coming from the $\mathcal{L}$-invariant or equivalently multiplicative cocycles. The corresponding central extension $\widehat{\mathcal{A}}$ of $\mathcal{A}$ will have a $(N-1)$-dimensional center and will be given as

$$[\widehat{f}, \widehat{g}] = \sum_{i=1}^{N-1} \alpha_i \cdot \gamma_{C_i}^{\mathcal{A}}(f, g) \cdot t_i, \quad \alpha_i \in \mathbb{C}, \quad [t_i, \widehat{\mathcal{A}}] = 0\,. \tag{109}$$

The local cocycle $\gamma_{C_S}^{\mathcal{A}}$ will yield a one-dimensional central extension which is almost-graded. One might either call the $(N-1)$-dimensional central extension or the almost-graded one-dimensional central extension *(multi-point) Heisenberg algebra*.

Now we turn to the other algebras. A Lie algebra is called perfect if the commutator ideal of L equals L, i.e. $L = [L, L]$. Recall that a Lie algebra L has a universal central extension if and only if it is perfect. In this case the (higher-dimensional) center can be given by the space of cocycle classes $\mathrm{H}^2(L, \mathbb{C})$.

Proposition 7.11 *The genus zero KN type vector field algebras, differential operator algebras, the super algebras, and current algebras for semi-simple finite Lie algebras are perfect Lie algebras and consequently admit central extensions.*

For the proofs see e.g. [53], resp. the Remark 7.13 below. For the general genus vector field case see also [59] and for the general current algebra case see also [30, 31]. Obviously, the function algebra as abelian Lie algebra is not perfect.

Combining Proposition 7.11 and Theorem 7.7 we obtain

Theorem 7.12

(a) In genus zero with N marked points the vector field algebra $\mathcal{L}$ has a universal central extension with $(N-1)$-dimensional center.
(b) The differential operator algebra $\mathcal{D}^1$ has a universal central extension with $3(N-1)$-dimensional center.
(c) The Lie superalgebra $\mathcal{S}$ has a universal central extension with $(N-1)$-dimensional center.
(d) The current algebra $\overline{\mathfrak{g}}$ associated to a simple Lie algebra $\mathfrak{g}$ has a universal central extension with $(N-1)$ dimensional center.

The corresponding cocycles and hence the universal central extension can be explicitly given by the cocycles introduced in Sect. 6, in Equations (69), (67), (70), (75), (68), where the integration runs over C_i, $i=1,\ldots,N-1$. Equivalently they can be calculated via residues at the point P_i, $i=1,\ldots,N-1$.

***Proof of Theorem* 7.7** We only give some rough ideas of the proof. The complete proof needs several pages and can be found in the original publication [53]. Also we will only deal with the vector field algebra as a typical example.

The following elements

$$e_n^{(i)} = (z-a_i)^{n+1}\frac{d}{dz}, \qquad n\in\mathbb{Z},\ i=1,\ldots,N-1. \tag{110}$$

give a generating set of $\mathcal{L}$. For each i separately, we get the usual Witt algebra as subalgebra of $\mathcal{L}$. Now we fix one of the superscripts, e.g. $i=1$, then

$$e_n^{(1)},\ n\in\mathbb{Z}, \qquad e_m^{(i)}, \quad m\le -2,\ i=2,\ldots,N-1\,. \tag{111}$$

is a basis [40]. This basis does not define any almost-grading. But we can relate it to the filtration (60) defined with respect to the almost-grading given by the standard splitting. With respect to this splitting we denote the basis elements as usual by $e_{n,p}$ We repeat the definition and the property of the filtration for $\lambda=-1$

$$\begin{aligned} \mathcal{L}_{(n)} &= \bigoplus_{m\ge n}\mathcal{L}_m, \qquad \mathcal{L}_m := \langle e_{m,p} \mid p=1,\ldots,K\rangle\,, \\ \mathcal{L}_{(n)} &= \{f\in\mathcal{L}\mid \operatorname{ord}_{P_i}(f)\ge n+1,\ i=1,\,.\,,N-1\}\,. \end{aligned} \tag{112}$$

Now we take the alternate basis into account and get

$$
\begin{aligned}
\mathcal{L}_{(n)} &= \bigcap_{i=1,\dots,N-1} \langle e_k^{(i)} \mid k \geq n \rangle_{\mathbb{C}}, \qquad \text{for } n \geq -1 \\
\mathcal{L}_{(n)} &= \sum_{i=1}^{N-1} \langle e_k^{(i)} \mid k \geq n \rangle_{\mathbb{C}}, \qquad \text{for } n < -1\,.
\end{aligned}
\tag{113}
$$

Attention, the sum in the second line is not a direct sum. Also note that $\mathcal{L}_{-1}$ consists of those vector fields which are holomorphic on the affine part. In particular for Ł$_{-1}$ all sets in the intersection are equal, hence it is enough to consider just one i.

We make a cohomological change via the linear form $\phi : \mathcal{L} \to \mathbb{C}$ with

$$
\begin{aligned}
\phi(e_n^{(1)}) &:= \frac{1}{n}\gamma(e_0^{(1)}, e_n^{(1)}), \quad n \in \mathbb{Z}, n \neq 0 \qquad \phi(e_0^{(1)}) := \frac{1}{2}\gamma(e_{-1}^{(1)}, e_1^{(1)}), \\
\phi(e_n^{(i)}) &:= \frac{1}{n}\gamma(e_0^{(i)}, e_n^{(i)}), \quad n \leq -2\,.
\end{aligned}
\tag{114}
$$

The cohomologous cocycle is now

$$
\gamma'(e, f) = \gamma(e, f) - \phi([e, f]). \tag{115}
$$

With the help of the filtration property and (113) we are able to show that

$$
\gamma'(e, f) = 0 \qquad e \in \text{Ł}_k,\ f \in \mathcal{L}_l \qquad \text{if } k + l > 0. \tag{116}
$$

Hence, $[\gamma]$ is bounded class class for the standard splitting and we can use Theorem 6.3 to conclude the proof. □

Remark 7.13 The alternative generating set introduced above shows also that $\mathcal{L}$ is a perfect algebra. As $[e_o^{(i)}, e_n^{(i)}] = n \cdot e_n^{(i)}$ and $[e_{-1}^{(i)}, e_1^{(i)}] = 2 \cdot e_0^{(i)}$ all generators can be written as commutators, i.e. [Ł, Ł] = Ł.

Remark 7.14 Let me give here some references to other works on multi-point genus zero work: [1, 5, 7–9, 13–18, 24, 26, 27, 44].

Remark 7.15 In the genus one case, i.e. the elliptic curve case, also interesting algebras show up, see e.g. [10, 11, 19–21, 38, 44]. In joint work with Alice Fialowski the current author examined deformations of the elliptic algebras to genus zero algebras, respectively subalgebras of them.

8 The Three-Point and Genus Zero Case

8.1 Symmetries

The case of genus zero with only three points where poles are allowed is to a certain extend special as we have additional symmetries. These symmetries can be used to simplify the calculations of structure constants even further.

Additionally, the three-point cases play a role in quite a number of applications. See e.g. the tetrahedron algebra appearing in statistical mechanics, in particular the work of Terwilliger and collaborators [5, 24].

By a complex automorphism of the Riemann sphere, i.e. by a fractional linear transformation, respectively by an PGL(2) action the three points can be brought to the points 0, 1 and ∞. The corresponding automorphism will yield an isomorphism of the involved algebras. Even after this is done there are still automorphisms of $\mathbb{P}^1(\mathbb{C})$ permuting the 3 points $\{0, 1, \infty\}$. Hence, we still have the action of the symmetric group S_3 of 3 elements. The corresponding algebraic maps are now automorphisms of the algebras.

Recall that in the previous section we discussed for every splitting of the set A, (here $\{0, 1, \infty\}$), into two disjoint non-empty subset I and O a distinguished basis which yields an almost-graded structure for the algebras. Essentially different splittings will yield essentially different basis elements respectively essentially different almost-gradings.

Here the only type of splitting is into a subset consisting of two points and a subset consisting of one point. After having fixed $A = \{0, 1, \infty\}$ we apply an automorphism from the remaining group S_3 such that

$$A = I \cup O, \qquad I := \{0, 1\}, \quad \text{and} \quad O := \{\infty\}. \tag{117}$$

This is exactly the situation which we will consider here.

I like to stress the fact, that this does not say, that there is only one possible choice of an almost-grading. In fact, given the set A and hence a unique algebra, we have 3 essentially different splitting, hence also 3 essentially different set of basis elements and consequently 3 almost-gradings. But in the three-point situation there will be always an automorphism of our algebra mapping the different almost-gradings to each other.

Remark 8.1 In [44] the situation $A = \{\alpha, -\alpha, \infty\}$ was considered for $\alpha \in \mathbb{C}$, $\alpha \neq 0$. This changes nothing. The corresponding algebras are isomorphic to the algebras considered here. Even our basis elements can be identified (up to some rescaling and re-indexing) and the structure equations remain the same (again up to scaling factors). The reason for the choice there was that we wanted to introduce a free parameter α which can be used to study degeneration process, respectively deformation families. For $\alpha \to 0$ we "degenerate" to the two-point situation. In this respect see also our joint work with Fialowski [19–21]. Clearly, everything that will be done in this section could be formulated also for $\{\alpha, -\alpha, \infty\}$, $\alpha \neq 0$.

8.2 The Associative Algebra

The basis elements of degree n of the algebra $\mathcal{A}$ with respect to our standard splitting are (see Sect. 7)

$$A_{n,1}(z) = z^n(z-1)^{n+1}, \qquad A_{n,2}(z) = z^{n+1}(z-1)^n, \quad n \in \mathbb{Z}, \tag{118}$$

where we ignore the scaling factors. We "symmetrize" and "anti-symmetrize" them for each degree separately by taking

$$\begin{aligned} A_n(z) &= A_{n,2}(z) - A_{n,1}(z) &&= z^n(z-1)^n, \\ B_n(z) &= A_{n,2}(z) + A_{n,1}(z) &&= z^n(z-1)^n(2z-1), \end{aligned} \tag{119}$$

Proposition 8.2 *The associative algebra* $\mathcal{A}$ *of meromorphic functions on the Riemann sphere* $\mathbb{P}^1(\mathbb{C})$ *holomorphic outside of* 0, 1, *and* ∞ *has as vector space basis*

$$\{A_n, B_n \mid n \in \mathbb{Z}\}, \tag{120}$$

with the structure equations

$$\begin{aligned} A_n \cdot A_m &= A_{n+m}, \\ A_n \cdot B_m &= B_{n+m}, \\ B_n \cdot B_m &= A_{n+m} + 4A_{n+m+1}. \end{aligned} \tag{121}$$

Proof That the elements $A_{n,1}, A_{n,2}$ with $n \in \mathbb{Z}$ are a basis of $\mathcal{A}$ is a general fact by its very construction as Krichever–Novikov multi-point basis in [41] corresponding to our splitting of A. The transformation (119) is obviously a base change which even respects the homogeneous subspaces. By direct calculations the structure equations follow. □

Our splitting introduces a (strong) almost-grading for the algebra $\mathcal{A}$

$$\mathcal{A} = \bigoplus_{n\in\mathbb{Z}} \mathcal{A}_n, \qquad \mathcal{A}_n = \langle A_n, B_n\rangle_{\mathbb{C}}, \quad \dim \mathcal{A}_n = 2. \tag{122}$$

This is clear from the general construction. But it can be easily illustrated by (121) as

$$\mathcal{A}_n \cdot \mathcal{A}_m \quad \subseteq \quad \mathcal{A}_{n+m} \oplus \mathcal{A}_{n+m+1}. \tag{123}$$

Next we want to study central extensions of $\mathcal{A}$ (considered as Lie algebras) which are given by geometric cocycles as introduced in Sect. 6 and further discussed in Sect. 7. We showed that the 2 following cocycles $\gamma_0^{\mathcal{A}}$, $\gamma_\infty^{\mathcal{A}}$ defined by

$$\gamma_0^{\mathcal{A}}(f,g) = \mathrm{res}_0(fdg), \qquad \gamma_\infty^{\mathcal{A}}(f,g) = \mathrm{res}_\infty(fdg), \tag{124}$$

constitute a basis of the geometric cocycles. Recall that the set of geometric cocycles coincide with the $\mathcal{L}$-invariant respectively multiplicative cocycles. Note also that

$$\gamma_1^{\mathcal{A}}(f,g) = \mathrm{res}_1(fdg) = -\gamma_0^{\mathcal{A}}(f,g) - \gamma_\infty^{\mathcal{A}}(f,g) \tag{125}$$

gives the linearly dependent 3. cocycle. Recall also that $\mathrm{res}_a(fdg) = -\,\mathrm{res}_a(gdf)$, $a \in \Sigma_0$.

To calculate the cocycles it is enough to do the calculation for all type of pairs of basis elements A_n and B_m. See [53] for details of the calculation. We start with the result

Proposition 8.3

$$\begin{aligned}
\gamma_\infty^{\mathcal{A}}(A_n, A_m) &= 2n\,\delta_m^{-n},\\
\gamma_\infty^{\mathcal{A}}(A_n, B_m) &= 0,\\
\gamma_\infty^{\mathcal{A}}(B_n, B_m) &= 2n\delta_m^{-n} + 4(2n+1)\,\delta_m^{-n-1}\,.
\end{aligned}$$

Proposition 8.4

$$\begin{aligned}
\gamma_0^{\mathcal{A}}(A_n, A_m) &= -n\,\delta_m^{-n},\\
\gamma_0^{\mathcal{A}}(A_n, B_m) &= n\,\delta_m^{-n} + 2n\,\delta_m^{-n-1} + \sum_{k=2}^{\infty} n\,(-1)^{k-1}2^k\frac{(2k-3)!!}{k!}\delta_m^{-n-k}\,,\\
\gamma_0^{\mathcal{A}}(B_n, B_m) &= -n\delta_m^{-n} - 2(2n+1)\,\delta_m^{-n-1}\,.
\end{aligned}$$

Here $(2k-1)!! = 1\cdot 3\cdots(2k-1)$ is the double factorial.

Note that the second relation in the above proposition is expressed as a formal infinite sum. But for given n and m maximally one term will be non-zero. Hence, it has a well-defined value.

In accordance with the general results [47] about local cocycles only the cocycle $\gamma_\infty^{\mathcal{A}}$ is local. Here it will vanish for n, m outside of $-1 \le n+m \le 0$. Consequently, only the central extension, the Heisenberg algebra, defined via $\gamma_\infty^{\mathcal{A}}$ will admit an extension of the almost-grading to the central extension. Consequently, we obtain only in this case a triangular decomposition which is of importance for the representations appearing in field theory. This is not possible for the central extension defined by $\gamma_0^{\mathcal{A}}$ nor by $\gamma_1^{\mathcal{A}}$.

8.3 Current and Affine Algebra

Recall that for a finite-dimensional Lie algebra $\mathfrak{g}$ the current algebra of KN-type is defined by $\overline{\mathfrak{g}} = \mathfrak{g} \otimes \mathcal{A}$. In particular every choice of a basis in $\mathfrak{g}$ and a basis in $\mathcal{A}$ will yield a basis of $\overline{\mathfrak{g}}$. For $\mathcal{A}$ we choose the basis A_n, B_n, $n \in \mathbb{Z}$ introduced above. Automatically we get an almost-graded structure of $\overline{\mathfrak{g}}$ induced by the splitting $\{0, 1\} \cup \{\infty\}$. For its algebraic structure we obtain (via Proposition 8.2)

Proposition 8.5

$$
\begin{aligned}
[x \otimes A_n, y \otimes A_m] &= [x, y] \otimes A_{n+m}, \\
[x \otimes A_n, y \otimes B_m &= [x, y] \otimes B_{n+m}, \\
[x \otimes B_n, y \otimes B_m &= [x, y] \otimes (A_{n+m} + 4A_{n+m+1}).
\end{aligned}
$$

Its central extensions are given by geometric cocycles

$$
\gamma_\beta^{\overline{\mathfrak{g}}}(x \otimes f, y \otimes g) = \beta(x, y) \cdot \gamma^{\mathcal{A}}(f, g), \tag{126}
$$

with $\beta(.,.)$ a symmetric, invariant bilinear form for $\mathfrak{g}$ and $\gamma^{\mathcal{A}}$ a multiplicative 2-cocycle for the algebra $\mathcal{A}$. Recall that if $\mathfrak{g}$ is a simple Lie there exists a universal central extension $\widehat{\mathfrak{g}}$. In our case it has a two-dimensional center and will be given by

$$
\begin{aligned}
[x \otimes f, y \otimes g] = [x, y] \otimes (f \cdot g) &+ \alpha_0 \cdot \beta(x, y) \cdot \gamma_0^{\mathcal{A}}(f, g) \cdot t_0 \\
&+ \alpha_\infty \cdot \beta(x, y) \cdot \gamma_\infty^{\mathcal{A}}(f, g) \cdot t_\infty
\end{aligned} \tag{127}
$$

with $\alpha_0, \alpha_\infty \in \mathbb{C}$, t_0, t_∞ central in $\widehat{\mathfrak{g}}$, and β the Cartan-Killing form. The values of the cocycles for the introduced basis elements have been calculated above and will not be repeated here. In accordance with the general results [46] the centrally extended current algebra will be an almost-graded extension of the current algebra with respect to this basis if and only if $\alpha_0 = 0$. It is an easy task to write everything explicitely for special cases of the Lie algebra $\mathfrak{g}$.

8.4 Three-Point $\mathfrak{sl}(2, \mathbb{C})$-Current Algebra for Genus 0

As an example we will give the universal central extension of the 3-point $\mathfrak{sl}(2, \mathbb{C})$-current algebra. The general theory has been developed above. The 3-point $\overline{\mathfrak{sl}}(2, \mathbb{C})$ algebra is of relevance in quite a number of applications, we only name statistical mechanics [24, 26]. Hence, the explicite knowledge of the structure equations with respect to some basis might be of some interest. We take $\mathfrak{sl}(2, \mathbb{C})$ with the standard matrix generators

$$H := \begin{pmatrix} 1 & 0 \\ 0 & -1 \end{pmatrix}, \quad X := \begin{pmatrix} 0 & 1 \\ 0 & 0 \end{pmatrix}, \quad Y := \begin{pmatrix} 0 & 0 \\ 1 & 0 \end{pmatrix}, \tag{128}$$

and induced relations

$$[H, X] = 2X, \quad [H, Y] = -2Y, \quad [X, Y] = 2X\,. \tag{129}$$

As basis elements for the current algebra $\overline{\mathfrak{sl}}(2, \mathbb{C})$ with respect to the almost-grading introduced above we take the elements

$$Z^{(s)} = Z \otimes A_n, \quad Z^{(a)} = Z \otimes B_n, \qquad Z \in \{H, X, Y\}. \tag{130}$$

The structure of the current algebra comes via (121) from $\mathcal{A}$ (and of course from $\mathfrak{g}$). We need the Cartan-Killing form. Up to a normalization it is given by

$$\beta(A, B) = \operatorname{tr}(A \cdot B). \tag{131}$$

Hence,

$$\beta(X, Y) = \beta(Y, X) = 1, \quad \beta(H, H) = 2,$$
$$\beta(H, X) = \beta(H, Y) = \beta(X, H) = \beta(Y, H) = 0. \tag{132}$$

The universal central extension will have a two-dimensional center and will be given by

$$[Z\otimes f, W\otimes g] = [Z, W]\otimes f\cdot g + \alpha_\infty\cdot\beta(Z, W)\cdot\gamma_\infty^{\mathcal{A}}(f, g)\cdot t_\infty + \alpha_0\cdot\beta(Z, W)\cdot\gamma_0^{\mathcal{A}}(f, g)\cdot t_0,$$

with t_∞, t_0 central elements, $\alpha_\infty, \alpha_0 \in \mathbb{C}$. Recall that $\gamma_a^{\mathcal{A}}(f, g)$ can be calculated as $\operatorname{res}_a(f dg)$.

From the general theory we know that the central extension will be almost-graded with respect to the standard splitting if and only if $\alpha_0 = 0$.

All the data needed has been calculated already before. If we collect them we obtain the following results.

$$[H_n^{(s)}, H_m^{(s)}] = \alpha_\infty \cdot 4n \cdot \delta_m^{-n} \cdot t_\infty - \alpha_0 \cdot 2n \cdot \delta_m^{-n} \cdot t_0,$$

$$[H_n^{(s)}, H_m^{(a)}] = 2\alpha_0 \left(n\, \delta_m^{-n} + 2n\, \delta_m^{-n-1} + \sum_{k=2}^{\infty} n\, (-1)^{k-1} 2^k \frac{(2k-3)!!}{k!} \delta_m^{-n-k} \right) \cdot t_0.$$

$$[H_n^{(a)}, H_m^{(a)}] = \alpha_\infty (4n\delta_m^{-n} + 8(2n+1)\delta_m^{-n-1}) \cdot t_\infty - \alpha_0 (2n\delta_m^{-n} + 4(2n+1)\delta_m^{-n-1}) \cdot t_0$$

$$[H_n^{(s)}, X_m^{(s)}] = 2X_{n+m}^{(s)}, \quad [H_n^{(s)}, X_m^{(a)}] = 2X_{n+m}^{(a)}, \quad [H_n^{(a)}, X_m^{(a)}] = 2X_{n+m}^{(s)} + 8X_{n+m}^{(s)},$$

$$[H_n^{(s)}, Y_m^{(s)}] = -2Y_{n+m}^{(s)}, \quad [H_n^{(s)}, Y_m^{(a)}] = -2Y_{n+m}^{(a)}, \quad [H_n^{(a)}, Y_m^{(a)}] = -2Y_{n+m}^{(s)} - 8Y_{n+m}^{(s)},$$

$$[X_n^{(s)}, Y_m^{(s)}] = H_{n+m}^{(s)} + \alpha_\infty \cdot 2n \cdot \delta_m^{-n} \cdot t_\infty - \alpha_0 \cdot n \cdot \delta_m^{-n} \cdot t_0,$$

$$[X_n^{(s)}, Y_m^{(a)}] = H_{n+m}^{(a)} + \alpha_0 \left(n\,\delta_m^{-n} + 2n\,\delta_m^{-n-1} + \sum_{k=2}^{\infty} n\,(-1)^{k-1} 2^k \frac{(2k-3)!!}{k!} \delta_m^{-n-k} \right) \cdot t_0.$$

$$[X_n^{(a)}, Y_m^{(a)}] = H_{n+m}^{(s)} + 4H_{n+m+1}^{(s)} + \alpha_\infty (2n\delta_m^{-n} + 4(2n+1)\delta_m^{-n-1}) \cdot t_\infty$$
$$- \alpha_0 (n\delta_m^{-n} + 2(2n+1)\delta_m^{-n-1}) \cdot t_0 .$$

Of course, the elements t_∞ and t_0 are central and we have anti-symmetry. The local cocycle, i.e. the cocycle coming with t_∞ was given in [49] and reproduced in [51, Equ. 12.75]. Unfortunately, there the central terms related to $[H_n^{(.)}, H_m^{(.)}]$ were forgotten.

Remark 8.6 By Cox and Jurisich [16, Thm.2.4] a different form of a universal central extension for the $\mathfrak{sl}(2, \mathbb{C})$ current algebra was proposed. This form was taken up in [17]. An inspection of the structure equation shows that in the proposed form two independent cocycles coming with the central elements ω_0 and ω_1 show up. Both would be local. But this contradicts the uniqueness of the local cocycle class (up to rescaling) as obtained in [46], which was also recalled in the current article. A closer examination shows that if "$\omega_1 \neq 0$" the proposed structure constants do not define a Lie algebra.

The principal structure, as far as the central extension is concerned, in particular also that we have one unique local cocycle class (up to rescaling) does not depend in an essential manner on the simple Lie algebra. See also Bremner [9] for the example of the 4-point case. Here the universal central extension is 3-dimensional. One of the classes will be local with respect to the standard splitting, the other two not. The latter two are "coupled" with ultraspherical (Gegenbauer) polynomials. I like also to mention that in [5] also the 3-point case was considered in another basis exhibiting another symmetry useful in the context of statistical mechanics.

Remark 8.7 As an additional example we like to give the case of the current algebra of $\mathfrak{gl}(n, \mathbb{C})$ for the N-point case. Of course, as $\mathfrak{gl}(n, \mathbb{C})$ is not perfect it does not admit a universal central extension. But by the classification results we can give the maximal central extension for which the cocycles are multiplicative (or $\mathcal{L}$-invariant)

$$[x \otimes f, y \otimes g] = [x, y] \otimes f \cdot g + \sum_{i=1}^{N-1} \alpha_i \cdot \operatorname{tr}(x \cdot y) \operatorname{res}_{a_i}(f dg) \cdot t_i$$
$$+ \sum_{i=1}^{N-1} \beta_i \cdot \operatorname{tr}(x) \operatorname{tr}(y) \operatorname{res}_{a_i}(f dg) \cdot s_i . \tag{133}$$

Here $\alpha_i, \beta_i \in \mathbb{C}$ and t_i and s_i are central.

8.5 Vector Field Algebra

Recall that in the genus $g = 0$ case and $P_N = \infty$ the elements g for $\mathcal{F}^\lambda$ for $\lambda \in \frac{1}{2}\mathbb{Z}$ are given by

$$g(z) = a(z)dz^\lambda, \quad \text{with} \quad a(z) \in \mathcal{A}. \tag{134}$$

In particular, a basis of $\mathcal{A}$ induces a basis of $\mathcal{F}^\lambda$. We take as basis elements the elements

$$g_n^\lambda := A_{n-\lambda}dz^\lambda, \qquad h_n^\lambda := B_{n-\lambda}dz^\lambda, \quad n \in \mathbb{J}_\lambda. \tag{135}$$

The corresponding almost-grading reads as

$$\mathcal{F}^\lambda = \bigoplus_{n\in\mathbb{J}_\lambda} \mathcal{F}_n^\lambda, \qquad \mathcal{F}_n^\lambda = \langle g_n^\lambda, h_n^\lambda \rangle_{\mathbb{C}}. \tag{136}$$

For the vector field (i.e. forms of weight -1) we use

$$e_n := A_{n+1}\frac{d}{dz}, \qquad f_n := B_{n+1}\frac{d}{dz}, \quad n \in \mathbb{Z}. \tag{137}$$

The vector spaces $\mathcal{F}^\lambda$ are modules over $\mathcal{L}$. If $e = a\frac{d}{dz}$ and $g = bdz^\lambda$ then the module structure reads as

$$e \,.\, g = (a \cdot b' + \lambda\, b \cdot a')dz^\lambda. \tag{138}$$

With respect to our basis elements the structure equations are given by

Proposition 8.8

$$\begin{aligned}
e_n \,.\, g_m^\lambda &= (m + \lambda n)\, h_{m+n}^\lambda, \\
e_n \,.\, h_m^\lambda &= (m + \lambda n)\, g_{m+n}^\lambda + (4(m + \lambda n) + 2)\, g_{n+m+1}^\lambda, \\
f_n \,.\, g_m^\lambda &= (m + \lambda n)\, g_{m+n}^\lambda + (4(m + \lambda n) + 2\lambda)\, g_{n+m+1}^\lambda, \\
f_n \,.\, h_m^\lambda &= (m + \lambda n)\, h_{m+n}^\lambda + (4(m + \lambda n) + 2 + 2\lambda)\, h_{n+m+1}^\lambda.
\end{aligned}$$

For $\lambda = -1$ we get the vector field algebra structure.

Proposition 8.9

$$\begin{aligned}
[e_n, e_m] &= (m - n)\, f_{m+n}, \\
[e_n, f_m] &= (m - n)\, e_{m+n} + (4(m - n) + 2)\, e_{n+m+1}, \\
[f_n, f_m] &= (m - n)\, f_{m+n} + 4(m - n)\, f_{n+m+1}.
\end{aligned}$$

These expressions clearly exhibit the almost-graded structure. Observe that the algebra of global holomorphic vector fields is the subalgebra

$$\langle e_{-1}, f_{-1}, e_0 \rangle_{\mathbb{C}}, \tag{139}$$

which is isomorphic to $\mathfrak{sl}(2, \mathbb{C})$.

Next we calculate the universal central extension. We know that it has a two-dimensional center and can be given as

$$[\widehat{e}, \widehat{f}] = \widehat{[e, f]} + \alpha_0 \cdot \gamma_0^{\mathcal{L}}(e, f) \cdot t_o + \alpha_\infty \cdot \gamma_\infty^{\mathcal{L}}(e, f) \cdot t_\infty \tag{140}$$

with

$$\begin{aligned}
\gamma_0^{\mathcal{L}}(e, f) &= 1/2 \operatorname{res}_0(e \cdot f''' - f \cdot e''')dz = \operatorname{res}_0(e \cdot f''')dz = -\operatorname{res}_0(f \cdot e''')dz \\
\gamma_\infty^{\mathcal{L}}(e, f) &= 1/2 \operatorname{res}_\infty(e \cdot f''' - f \cdot e''') = \operatorname{res}_\infty(e \cdot f''')dz = -\operatorname{res}_\infty(f \cdot e''')dz\,.
\end{aligned} \tag{141}$$

First we consider the point ∞ and obtain in this way the local cocycle.

Proposition 8.10

$$\begin{aligned}
&\gamma_\infty^{\mathcal{L}}(e_n, e_m) = 2(n^3 - n)\,\delta_m^{-n} + 4n(n+1)(2n+1)\delta_m^{-n-1} \\
&\gamma_\infty^{\mathcal{L}}(e_n, f_m) = 0, \\
&\gamma_\infty^{\mathcal{L}}(f_n, f_m) = 2(n^3 - n)\,\delta_m^{-n} + 8n(n+1)(2n+1)\delta_m^{-n-1} + 8(n+1)(2n+1)(2n+3)\delta_m^{-n-2}
\end{aligned}$$

Proposition 8.11

$$\begin{aligned}
\gamma_0^{\mathcal{L}}(e_n, e_m) &= -(n^3 - n)\,\delta_n^{-m} - 2n(n+1)(2n+1)\delta_m^{-n-1} \\
\gamma_0^{\mathcal{L}}(e_n, f_m) &= (n^3 - n)\,\delta_m^{-n} + 6n^2(n+1)\delta_m^{-n-1} + 6n(n+1)^2\delta_m^{-n-2} \\
&\quad + \sum_{k \geq 3} n(n+1)(n+k-1)(-1)^k 2^k \cdot 3 \cdot \frac{(2k-5)!!}{k!}\delta_m^{-n-k} \\
\gamma_0^{\mathcal{L}}(f_n, f_m) &= -(n^3 - n)\,\delta_m^{-n} - 4n(n+1)(2n+1)\delta_m^{-n-1} - 4(n+1)(2n+1)(2n+3)\delta_m^{-n-2}\,.
\end{aligned}$$

See [27] for similar results.

Note also that we can express the term

$$\frac{(2k-5)!!}{k!}2^k \cdot 3 = \frac{12}{k(k-1)}\binom{2(k-2)}{k-2} \tag{142}$$

if this is more convenient. See Sect. 8.8 for some mathematical background on these calculations.

Again only the cocycle $\gamma_\infty^{\mathcal{L}}$ will be local with respect to the almost-grading introduced by our splitting, respectively by our basis. Hence, only for its corresponding central extension we have a triangular decomposition. Everything what was said for the function algebra case, remains true here.

8.6 *Differential Operator Algebra* $\mathcal{D}^1$

Recall that $\mathcal{D}^1$ is the (Lie algebra) semi-direct sum of $\mathcal{A}$ with $\mathcal{L}$ where $\mathcal{L}$ operates on $\mathcal{A}$ by taking the derivative. The homogeneous subspace of degree n is now

$$\langle A_n, B_n, e_n, f_n\rangle_{\mathbb{C}}. \tag{143}$$

The subalgebra $\mathcal{A}$ is abelian and Proposition 8.9 gives the structure equations for the vector fields. Proposition 8.8 specialized for $\lambda = 0$ yields the equations for the mixed terms.

Proposition 8.12

$$\begin{aligned}
[e_n, A_m] &= -[A_m, e_n] = m\, B_{m+n},\\
[e_n, B_m] &= -[B_m, e_n] = m\, A_{m+n} + (4m+2)\, A_{n+m+1},\\
[f_n, A_m] &= -[A_m, f_n] = m\, A_{m+n} + 4m\, A_{n+m+1},\\
[f_n, B_m] &= -[B_m, f_n] = m\, B_{m+n} + (4m+2)\, B_{m+n+1},
\end{aligned}$$

The geometric cocycles yield a 6-dimensional central extension. In addition to the 4 basis elements given by the pure cocycles given by the Propositions 8.3, 8.4, 8.10, 8.11 we have two additional basis cocycles corresponding to the mixing cocycle (70) ($e \in Ł$, $g \in \mathcal{A}$)

$$\gamma_0^{(m)}(e, g) = \operatorname{res}_0(eg''dz), \qquad \gamma_\infty^{(m)}(e, g) = \operatorname{res}_\infty(eg''dz). \tag{144}$$

Evaluated for the basis elements we obtain

Proposition 8.13

$$\begin{aligned}
\gamma_\infty^{(m)}(e_n, A_m) &= 0,\\
\gamma_\infty^{(m)}(e_n, B_m) &= -2n(n+1)\delta_m^{-n} - 4(n+1)(2n+1)\delta_m^{-n-1},\\
\gamma_\infty^{(m)}(f_n, A_m) &= -2n(n+1)\delta_m^{-n} - 4(n+1)(2n+3)\delta_m^{-n-1},\\
\gamma_\infty^{(m)}(f_n, B_m) &= 0\,.
\end{aligned}$$

Proposition 8.14

$$\gamma_0^{(m)}(e_n, A_m) = -n(n+1)\delta_m^{-n} - 2(n+1)^2\delta_m^{-n-1}$$
$$+\sum_{k\geq 2}(n+1)(n+k)(-1)^k 2^k \cdot \frac{(2k-3)!!}{k!}\delta_m^{-n-k}.$$

$$\gamma_0^{(m)}(e_n, B_m) = n(n+1)\delta_m^{-n} + 2(n+1)(2n+1)\delta_m^{-n-1},$$

$$\gamma_0^{(m)}(f_n, A_m) = n(n+1)\delta_m^{-n} + 2(n+1)(2n+3)\delta_m^{-n-1},$$

$$\gamma_0^{(m)}(f_n, B_m) = -n(n+1)\delta_m^{-n} - 6(n+1)^2\delta_m^{-n-1} - 6(n+1)(n+2)\delta_m^{-n-2}$$
$$+\sum_{k\geq 3}(n+1)(n+k)(-1)^{k-1}2^k \cdot 3 \cdot \frac{(2k-5)!!}{k!}\delta_m^{-n-k}.$$

By Theorem 7.12 the differential operator algebra admits a universal central extension and the introduced six geometric cocycles, each associated to a different central element will yield the universal central extension.

Proposition 8.15 *A cocycle class $[\gamma]$ for $\mathcal{D}^1$ will be local (with respect to the standard splitting) if and only if γ is cohomologous to a linear combination*

$$\gamma \quad \sim \quad \alpha_1 \cdot \gamma_\infty^{\mathcal{A}} + \alpha_2 \cdot \gamma_\infty^{\mathcal{L}} + \alpha_3 \cdot \gamma_\infty^{(m)}, \qquad \alpha_i \in \mathbb{C}. \tag{145}$$

In particular, the space of local cohomology classes is 3-dimensional.

This is a general result of [47, 51] which is for the 3-point illustrated by the above calculations. In the very general case (meaning arbitrary genus, arbitrary number of marked points, arbitrary splitting) the three cocycles in (145) are obtained by integrating over a separating cycle.

8.7 *The Lie Superalgebra*

In addition to the basis elements e_n and f_n of Ł we take

$$\varphi_n = A_{n+1/2}(dz)^{-1/2}, \quad \psi_n = B_{n+1/2}(dz)^{-1/2}, \qquad n \in \mathbb{Z} + 1/2. \tag{146}$$

Additionally to the structure constants of Proposition 8.9 we have

Proposition 8.16

$$
\begin{aligned}
[\varphi_n, \varphi_m] &= e_{n+m} \\
[\varphi_n, \psi_m] &= f_{n+m} \\
[\psi_n, \psi_m] &= e_{n+m} + 4\, e_{n+m+1} \\
[e_n, \varphi_m] &= (m - n/2)\, \psi_{n+m} \\
[e_n, \psi_m] &= (m - n/2)\, \varphi_{n+m} + (4m - 2n + 2)\, \varphi_{n+m+1} \\
[f_n, \varphi_m] &= (m - n/2)\, \varphi_{n+m} + (4m - 2n - 1)\, \varphi_{n+m+1} \\
[f_n, \psi_m] &= (m - n/2)\, \psi_{n+m} + (4m - 2n + 1)\, \psi_{n+m+1}\, .
\end{aligned}
$$

Similar expressions are given by Leidwanger and Morier-Genoud [35, Prop. 3.8].

Next we consider 2-cocycles. We have a two-dimensional space of geometric cocycles generated by γ_0^S and γ_∞^S obtained by taking the residue from (75) at 0 and ∞. For pairs of pure vector field type arguments we have the result of Proposition 8.10 and Proposition 8.11. For mixing of pure types it is zero. It remains to consider pairs of $-1/2$-forms. The calculations (see [53]) yields

Proposition 8.17

$$
\begin{aligned}
\gamma_\infty^S(\varphi_n, \varphi_m) &= 0 \\
\gamma_\infty^S(\varphi_n, \psi_m) &= -4(n - 1/2)(n + 1/2)\delta_m^{-n} - 8n(2n + 1)\delta_m^{-n-1}. \\
\gamma_\infty^S(\psi_n, \psi_m) &= 0
\end{aligned}
$$

Proposition 8.18

$$
\begin{aligned}
\gamma_0^S(\varphi_n, \varphi_m) = &-2(n + 1/2)(n - 1/2)\delta_m^{-n} + 4(n + 1/2)^2\delta_m^{-n-1} \\
&+ 2\sum_{k\geq 2}(n + 1/2)(n - 1/2 + k)(-1)^k 2^k \frac{(2k - 3)!!}{k!}\delta_m^{-n-k}
\end{aligned}
$$

$$
\gamma_0^S(\varphi_n, \psi_m) = 2(n - 1/2)(n + 1/2)\delta_m^{-n} + 4n(2n + 1)\delta_m^{-n-1}.
$$

$$
\begin{aligned}
\gamma_0^S(\psi_n, \psi_m) = &-2(n + 1/2)(n - 1/2)\delta_m^{-n} - 12(n + 1/2)^2\delta_m^{-n-1} \\
&- 12(n + 1/2)(n + 3/2)\delta_m^{-n-2} \\
&+ 2\sum_{k\geq 3}(n + 1/2)(n - 1/2 + k)(-1)^{k-1}2^k \cdot 3 \cdot \frac{(2k - 5)!!}{k!}\delta_m^{-n-k}\, .
\end{aligned}
$$

We illustrated again that only the cocycle γ_∞^S is local with respect to the standard splitting. The γ_∞^S is the one which was considered by Kreusch [32] (up to a different indexing of the basis elements).

Remark 8.19 Here we considered the central element to be even. We could have even dropped this assumption. In [50] we show that the corresponding cocycles (with odd central element) are cohomologically trivial.

8.8 Some Remarks on the Calculations of the Residues

Just to indicate the kind of arguments used above we will calculate the residues of $A_n dz$ and $B_n dz$ at the points where poles might be. In fact, everything can be finally reduced to calculating their values for the points 0, 1 and ∞. The results in the whole section are essentially based on such kind of residue calculations for products of of linear factors and their inverse (yielding Laurent series).

As starting point we need the Laurent series expansion of $(z-1)^m$ around zero. We collect the following well-known facts about binomial series.

The expansion

$$(z-1)^m = \sum_{k=0}^{\infty} \binom{m}{k} z^k (-1)^{m-k}, \qquad z \in \mathbb{C},\ |z| < 1 \tag{147}$$

is valid for all $m \in \mathbb{Z}$.

For negative exponents an equivalent expression is

$$(z-1)^{-n} = (-1)^n \sum_{k=0}^{\infty} \binom{n+k-1}{n-1} z^k, \qquad z \in \mathbb{C},\ |z| < 1, \tag{148}$$

where $n \in \mathbb{N}$. We have the easy relation

$$\binom{2k}{k} = \frac{(2k-1)!!}{k!} 2^k, \qquad k \in \mathbb{N}, \tag{149}$$

where $(2k-1)!! = 1 \cdot 3 \cdots (2k-1)$ is the double factorial.

Lemma 8.20 *For the residues at the point* 0 *we have*

$$\mathrm{res}_0(A_{-n}dz) = \begin{cases} 0, & n \le 0, \\ -1, & n = 1, \\ (-1)^n \dfrac{(2n-3)!!}{(n-1)!} 2^{n-1}, & n \ge -2. \end{cases}$$

Proof If $n < 0$ then there is no pole at $z = 0$. Hence let $n > 0$. We use (148) and calculate

$$A_{-n}(z) = (-1)^n \sum_{k=0}^{\infty} \binom{n+k-1}{n-1} z^{k-n}. \tag{150}$$

The residue at $z = 0$ is given by the coefficient paired with z^{-1}. Hence it is given by the coefficient for $k = n - 1$

$$\mathrm{res}_0(A_{-n}dz) = (-1)^n \binom{2(n-1)}{n-1}. \tag{151}$$

For $n = 1$ we obtain the value -1. For $n > 1$ we use (149) and obtain

$$\mathrm{res}_0(A_{-n}dz) = (-1)^n \frac{(2n-3)!!}{(n-1)!} 2^{n-1}. \tag{152}$$

This was the claim. □

From this lemma the non-locality of the cocycle defined via the residue at 0 follows.

Lemma 8.21

$$\mathrm{res}_0(B_m dz) = \begin{cases} 1, & m = -1, \\ 0, & \textit{otherwise}. \end{cases}$$

Proof We check that $B_m = (1/(m+1)A_{m+1})'$ if $m \neq -1$. Hence, for $m \neq -1$ the differential $B_m dz$ is an exact differential and hence does not have any residue. It remains $B_{-1}(z) = z^{-1}(z-1)^{-1}(2z-1)$ which obviously has as residue $+1$ at $z = 0$. □

Lemma 8.22

$$\mathrm{res}_1(A_m dz) = -\,\mathrm{res}_0(A_m dz)$$
$$\mathrm{res}_1(B_m dz) = \mathrm{res}_0(B_m dz)\ .$$

Proof We make a change of local coordinates $v = 1 - z$. and the point $z = 1$ corresponds to the point $v = 0$. □

Lemma 8.23

$$\mathrm{res}_\infty(A_m dz) = 0,$$
$$\mathrm{res}_\infty(B_m dz) = -2\,\mathrm{res}_0(B_m dz),$$
$$= \begin{cases} -2, & m = -1, \\ 0, & \textit{otherwise}. \end{cases}$$

Proof By the residue theorem [48] for a compact Riemann surface the sum over all residues of a meromorphic differential is zero. As our differentials have only poles at 0, 1, ∞ the claim follows from Lemmas 8.20, 8.21, and 8.22. □

References

1. A. Anzaldo-Meneses, Krichever-Novikov algebras on Riemann surfaces of genus zero and one with N punctures. J. Math. Phys. **33**(12), 4155–4163 (1992)
2. E. Arbarello, C. De Concini, V.G. Kac, C. Procesi, Moduli spaces of curves and representation theory. Commun. Math. Phys. **117**(1), 1–36 (1988). https://projecteuclid.org/euclid.cmp/1104161592
3. M.F. Atiyah, Riemann surfaces and spin structures. Ann. Sci. École Norm. Sup. (4) **4**, 47–62 (1971)
4. A.A. Belavin, A.M. Polyakov, A.B. Zamolodchikov, Infinite conformal symmetry in two-dimensional quantum field theory. Nucl. Phys. B **241**, 333–380 (1084)
5. G. Benkart, P. Terwilliger, The universal central extension of the three-point $\mathfrak{sl}_2$ loop algebra. Proc. Am. Math. Soc. **135**(6), 1659–1668 (2007)
6. L. Bonora, M. Martellini, M. Rinaldi, L. Russo, Neveu-Schwarz- and Ramond-type superalgebras on genus-g Riemann surfaces. Phys. Lett. B **206**(3), 444–450 (1988). https://doi.org/10.1016/0370-2693(88)91607-3
7. M.R. Bremner, Structure of the Lie algebra of polynomial vector fields on the Riemann sphere with three punctures. J. Math. Phys. **32**, 1607–1608 (1991)
8. M.R. Bremner, Generalized affine Kac-Moody Lie algebras over localizations of the polynomial ring in one variables. Can. Math. Bull. **37**(1), 21–28 (1994)
9. M.R. Bremner, Four-point affine Lie algebras. Proc. Am. Math. Soc. **123**, 1981–1989 (1995)
10. M.R. Bremner, On a Lie algebra of vector fields on a complex torus. J. Math. Phys. **31**, 2033–2034 (1990)
11. M.R. Bremner, Universal central extensions of elliptic affine Lie algebras. J. Math. Phys. **35**(12), 6685–6692 (1994)
12. P. Bryant, Graded Riemann surfaces and Krichever-Novikov algebras. Lett. Math. Phys. **19**(2), 97–108 (1990). https://doi.org/10.1007/BF01045879
13. A. Bueno, B. Cox, V. Futorny, Free field realizations of the elliptic affine Lie algebra $\mathfrak{sl}(2, \mathbf{R}) \oplus (\Omega_R/d\mathrm{R})$. J. Geom. Phys. **59**(9), 1258–1270 (2009)
14. B. Cox, Realizations of the four point affine Lie algebra $\mathfrak{sl}(2, R) \oplus (\Omega_R/dR)$. Pac. J. Math. **234**(2), 261–289 (2008)
15. B. Cox, X. Guo, R. Lu, K. Zhao, n-point Virasoro algebras and their modules of densities. Commun. Contemp. Math. **16**(3), 1350047 (2014)
16. B. Cox, E. Jurisich, Realizations of the three-point Lie algebra $\mathfrak{sl}(2, \mathcal{R}) \bigoplus (\Omega_{\mathcal{R}}/d\mathcal{R})$. Pac. J. Math. **270**(1), 27–48 (2014)

17. B. Cox, E. Jurisich, R. Martins, The 3-point Virasoro algebra and its action on Fock space (2015). arXiv: 1502.04102v1
18. R. Dick, Krichever-Novikov-like bases on punctured Riemann surfaces. Lett. Math. Phys. **18**, 255–265 (1989)
19. A. Fialowski, M. Schlichenmaier, Global deformations of the Witt algebra of Krichever-Novikov type. Commun. Contemp. Math. **5**(6), 921–945 (2003). https://doi.org/10.1142/S0219199703001208
20. A. Fialowski, M. Schlichenmaier, Global geometric deformations of current algebras as Krichever-Novikov type algebras. Commun. Math. Phys. **260**(3), 579–612 (2005). https://doi.org/10.1007/s00220-005-1423-5
21. A. Fialowski, M. Schlichenmaier, Global Geometric Deformations of the Virasoro algebra, current and affine algebras by Krichever-Novikov type algebras. Int. J. Theor. Phys. **46**(11), 2708–2724 (2007)
22. L. Guieu, C. Roger, *L'algèbre et le groupe de Virasoro* (Les Publications CRM, Montreal, 2007). Aspects géométriques et algébriques, généralisations. [Geometric and algebraic aspects, generalizations], With an appendix by Vlad Sergiescu
23. R.C. Gunning, *Lectures on Riemann Surfaces*. Princeton Mathematical Notes (Princeton University Press, Princeton, 1966)
24. B. Hartwig, P. Terwilliger, The tetrahedron algebra, the Onsager algebra, and the $\mathfrak{sl}_2$ loop algebra. J. Algebra **308**(2), 840–863 (2007)
25. N.S. Hawley, M. Schiffer, Half-order differentials on Riemann surfaces. Acta Math. **115**, 199–236 (1966). https://doi.org/10.1007/BF02392208
26. T. Ito, P. Terwilliger, Finite-dimensional irreducible modules for the three-point $\mathfrak{sl}_2$ loop algebra. Commun. Algebra **36**(12), 4557–4598 (2008)
27. E. Jurisich, R. Martins, Determination of the 2- cocycles for the three-point Witt algebra. arXiv:1410.5479.
28. V.G. Kac, Simple irreducible graded lie algebras of finite growth. (Russian). Izv. Akad. Nauk SSSR Ser. Mat. **32**, 1323–1367 (1968)
29. V.G. Kac, *Infinite-Dimensional Lie Algebras*, 3rd edn. (Cambridge University Press, Cambridge, 1990). https://doi.org/10.1017/CBO9780511626234
30. Ch. Kassel, Kähler differentials and coverings of complex simple Lie algebras extended over a commutative algebra, in *Proceedings of the Luminy Conference on Algebraic K-Theory (Luminy, 1983)*, vol. 34 (1984), pp. 265–275
31. C. Kassel, J.-L. Loday, Extensions centrales d'algèbres de Lie. Ann. Inst. Fourier (Grenoble). **32**(4), 119–142 (1982, 1983)
32. M. Kreusch, Extensions of superalgebras of Krichever-Novikov type. Lett. Math. Phys. **103**(11), 1171–1189 (2013). https://doi.org/10.1007/s11005-013-0628-3
33. I.M. Krichever, S.P. Novikov, Algebras of Virasoro type, Riemann surfaces and strings in Minkowski space. Funktsional. Anal. i Prilozhen. **21**(4), 47–61 (1987)
34. I.M. Krichever, S.P. Novikov, Algebras of Virasoro type, the energy-momentum tensor, and operator expansions on Riemann surfaces. Funktsional. Anal. i Prilozhen. **23**(1), 24–40 (1989). https://doi.org/10.1007/BF01078570
35. S. Leidwanger, S. Morier-Genoud, Superalgebras associated to Riemann surfaces: Jordan algebras of Krichever-Novikov type. Int. Math. Res. Not. IMRN **19**, 4449–4474 (2012). https://doi.org/10.1093/imrn/rnr196
36. R.V. Moody, Euclidean Lie algebras. Can. J. Math. **21**, 1432–1454 (1969). https://doi.org/10.4153/CJM-1969-158-2
37. V. Ovsienko, Lie antialgebras: prémices. J. Algebra **325**, 216–247 (2011). https://doi.org/10.1016/j.jalgebra.2010.10.003
38. A. Ruffing, Th. Deck, M. Schlichenmaier, String Branchings on complex tori and algebraic representations of generalized Krichever-Novikov algebras. Lett. Math. Phys. **26**, 23–32 (1992)
39. V.A. Sadov, Bases on multipunctured Riemann surfaces and interacting strings amplitudes. Commun. Math. Phys. **136**(3), 585–597 (1991). http://projecteuclid.org/euclid.cmp/1104202438

40. M. Schlichenmaier, Krichever-Novikov algebras for more than two points. Lett. Math. Phys. **19**(2), 151–165 (1990). https://doi.org/10.1007/BF01045886
41. M. Schlichenmaier, Krichever-Novikov algebras for more than two points: explicit generators. Lett. Math. Phys. **19**(4), 327–336 (1990). https://doi.org/10.1007/BF00429952
42. M. Schlichenmaier, Central extensions and semi-infinite wedge representations of Krichever-Novikov algebras for more than two points. Lett. Math. Phys. **20**(1), 33–46 (1990). https://doi.org/10.1007/BF00417227
43. M. Schlichenmaier, Verallgemeinerte Krichever–Novikov Algebren und deren Darstellungen. Ph.D. Thesis, University of Mannheim (1990)
44. M. Schlichenmaier, Degenerations of generalized Krichever-Novikov algebras on tori. J. Math. Phys. **34**(8), 3809–3824 (1993). https://doi.org/10.1063/1.530008
45. M. Schlichenmaier, Zwei Anwendungen algebraisch-geometrischer Methoden in der Theoretischen Physik: Berezin-Toeplitz-quantisierung und globale Algebren der zweidimensionalen Konformen Feldtheorie. Habilitation thesis, University of Mannheim (1996)
46. M. Schlichenmaier, Higher genus affine algebras of Krichever-Novikov type. Mosc. Math. J. **3**(4), 1395–1427 (2003)
47. M. Schlichenmaier, Local cocycles and central extensions for multipoint algebras of Krichever-Novikov type. J. Reine Angew. Math. **559**, 53–94 (2003). https://doi.org/10.1515/crll.2003.052
48. M. Schlichenmaier, in *An Introduction to Riemann Surfaces, Algebraic Curves and Moduli spaces*, 2nd edn. Theoretical and Mathematical Physics (Springer, Berlin, 2007). With an introduction by Ian McArthur
49. M. Schlichenmaier, Higher genus affine Lie algebras of Krichever-Novikov type, in *Proceedings of the International Conference on Difference Equations*. Special Functions, and Applications (World-Scientific, Munich, 2007), pp. 589–599
50. M. Schlichenmaier, Lie superalgebras of Krichever-Novikov type and their central extensions. Anal. Math. Phys. **3**(3), 235–261 (2013). https://doi.org/10.1007/s13324-013-0056-7
51. M. Schlichenmaier, in *Krichever-Novikov Type Algebras*. De Gruyter Studies in Mathematics, vol. 53 (De Gruyter, Berlin, 2014). https://doi.org/10.1515/9783110279641. Theory and applications
52. M. Schlichenmaier, Krichever-Novikov type algebras. An introduction, in *Lie Algebras, Lie superalgebras, Vertex Algebras and Related Topics*. Proc. Sympos. Pure Math., vol. 92 (Amer. Math. Soc., Providence, 2016), pp. 181–220
53. M. Schlichenmaier, N-point Virasoro algebras are multipoint Krichever-Novikov-type algebras. Commun. Algebra **45**(2), 776–821 (2017). https://doi.org/10.1080/00927872.2016.1175464
54. O.K. Sheinman, Elliptic affine Lie algebras. Funktsional. Anal. i Prilozhen. **24**(3), 51–61, 96 (1990). https://doi.org/10.1007/BF01077962
55. O.K. Sheinman, Highest weight modules of some quasigraded Lie algebras on elliptic curves. Funktsional. Anal. i Prilozhen. **26**(3), 65–71 (1992). https://doi.org/10.1007/BF01075634
56. O.K. Sheinman, Affine Lie algebras on Riemann surfaces. Funktsional. Anal. i Prilozhen. **27**(4), 54–62, 96 (1993). https://doi.org/10.1007/BF01078844
57. O.K. Sheinman, Highest-weight modules for affine Lie algebras on Riemann surfaces. Funktsional. Anal. i Prilozhen. **29**(1), 56–71, 96 (1995). https://doi.org/10.1007/BF01077040
58. O.K. Sheinman, in *Current Algebras on Riemann Surfaces*. De Gruyter Expositions in Mathematics, vol. 58. (Walter de Gruyter GmbH & Co. KG, Berlin, 2012). https://doi.org/10.1515/9783110264524. New results and applications
59. S. Skryabin, Degree one cohomology for the Lie algebra of derivations. Lobachevskii J. Math. **14**, 69–107 (2004)